Economic and Ecological Significance of Arthropods in Diversified Ecosystems

Akshay Kumar Chakravarthy
Shakunthala Sridhara
Editors

Economic and Ecological Significance of Arthropods in Diversified Ecosystems

Sustaining Regulatory Mechanisms

Springer

Editors
Akshay Kumar Chakravarthy
Division of Entomology and Nematology
Indian Institute of Horticultural Research (IIHR)
Bengaluru, Karnataka, India

Shakunthala Sridhara (retired)
Department of Entomology
University of Agricultural Sciences
Gandhi Krishi Vignana Kendra (GKVK)
Bengaluru, Karnataka, India

ISBN 978-981-10-1523-6 ISBN 978-981-10-1524-3 (eBook)
DOI 10.1007/978-981-10-1524-3

Library of Congress Control Number: 2016954124

Printed on acid-free paper

This Springer imprint is published by Springer Nature
The registered company is Springer Science+Business Media Singapore Pte Ltd.

Foreword

Arthropods are vital to the functioning of all ecosystems and to the survival of living beings on planet Earth. It becomes crucially important that arthropods are well studied, understood and conserved. Higher levels of specialization, predation, omnivory and diet shifts have allowed for higher species richness of arthropods in the tropics and subtropics, than in the temperate ecosystems. This, of course, may be with exceptions. However, arthropods in the tropics are comparatively less studied. Given the spectacular arthropod biodiversity and endemism in the tropics and subtropics, the species complex and their interactions seem to be scarcely documented. Their functions, ecological services and regulatory mechanisms are also not understood satisfactorily.

This book places emphasis on diversity and interactive relationships of arthropods with biotic and abiotic elements, vital for their conservation and management in wild and man-made habitats.

The book *Economic and Ecological Significance of Arthropods in Diversified Ecosystems: Sustaining Regulatory Mechanisms* explicitly deals with the diversity of arthropods in the different tropical and subtropical ecosystems so as to contribute towards their management and/or conservation. There are 20 chapters, most from India and a few from adjacent countries. The content of the book dwells on a wide range of topics embracing diversity, distribution, utility and interactions with other ecological elements in the ecosystem. This is a unique compilation which was hitherto missing from the international scenario. Obviously, its details and wide coverage make it a useful reference book for zoologists, entomologists, ecologists, scholars and scientists in conservation biology and arthropod science. Undoubtedly, it will generate interest and awareness in the scientific fraternity and the public for initiating monitoring systems and implementing conservation plans for arthropod populations.

Biologist, Researcher
Division of Apiculture- Institute of Animal Science
Hellenic Agriculture Org. "DEMETER"
Nea Moudania, Greece
December 2015

Fani Hatjina

Preface

Arthropods, no doubt, comprise the largest species group of animals on planet Earth. They are also the earliest animals on Earth. But until now they have received little research effort especially in the tropics and subtropics. As a result, to date, there have been not many up-to-date, comprehensive books on arthropods, despite their worldwide importance. Among arthropods, insects have received the maximum attention because of their presence as pests on crops, vectors of human and livestock diseases and as nuisance creatures. As a consequence biologists from time immemorial have been studying and laying emphasis only on insects of economic importance rather than on arthropods as a whole. This has resulted in undervaluing their roles and ecosystem services in different spheres, a vital void.

This compendium hopes to provide information on arthropods, their diversity, their interaction with other biotic and abiotic elements and their roles in the web of living organisms. Certainly, the book does not claim to present the vast body of information on arthropods in one volume. But it is a sincere attempt to focus on the importance and significance of arthropods in the tropics, aimed to urge their economic and ecological roles, and the vital links they forge with other biotic elements. It is imperative that humans consider arthropods as an important animal community and strive to sustain their activities and services in the ecosystems.

The book *Economic and Ecological Significance of Arthropods in Diversified Ecosystems: Sustaining Regulatory Mechanisms* has 20 chapters. Exhaustive data on arthropods associated with soil, forests and crops like rice, grain legumes, tea, coffee, mango, cotton, cashew and vegetables in the tropics are documented in this volume. The wealth of information provided on their population, roles and activities can be hopefully harnessed for human welfare and health of the ecosystems. There are also chapters on arthropod ecology, evolution, utility as food, medical and pharmaceutical value, diversity and distribution.

We hope this book will go a long way in fulfilling major gaps in our knowledge on arthropods and in improving environmental quality. This compilation is aimed to enthuse and convince the academicians as well as decision makers to utilize arthropods to not only improve yields but also sustain their ecosystem services.

Bengaluru, India

Akshay Kumar Chakravarthy
Shakunthala Sridhara

Prelude

Economic and Ecological Significance of Arthropods in Diversified Ecosystems: Sustaining Regulatory Mechanisms

Arthropods as a group represent three sub phyla (but only two exist today) and ten classes of incredibly diverse invertebrates that constitute more than 80 % of all living organisms on planet Earth. That they make up such unparalleled huge proportion of living animals on earth speaks of their resilience, adaptability and modifications in behaviour, feeding, reproduction and life history characteristics. It is true that many species, taxa and communities are studied exhaustively. But unfortunately most of the information is scattered in scientific publications, and only few publications are accessible. It was felt necessary to compile the data on different aspects of arthropod ecology, economic significance as well as sustaining regulatory mechanisms of this interesting group of animals. To connect pieces of widely scattered information on the huge group of arthropods and concisely present them would be, by any means, a Herculean task. It was quite intriguing how and why arthropods have not been often considered at a community or group level while dealing with specific, goal-oriented investigations such as species diversity, crop protection and crop productivity. This prompted us to edit a volume on ecological and economic significance of arthropods in diversified ecosystems. Inventories of arthropods in varied and diverse aquatic or terrestrial ecosystems are not available relative to their density. In most situations, studies on arthropods have revealed dominance of insects. Even on insects, several studies address the insects at individual and field scale level, across habitats. However, meta-analyses of arthropods would achieve significant positive results in agriculture like enhancement of natural enemies of pests, facilitation of pollination, reduction of crop damages and regulation of balanced populations of arthropods. It is in this direction that the book begins with a chapter on 'Ecology and Evolution of Arthropods'. Arthropods being dynamic have responded to changes in response to global warming, pesticides, introduced predators and parasitoids, changed land use pattern and other abiotic and biotic factors. Arthropod evolution is traced back to annelids, and changes in their ecology and adaptation makes a fascinating reading to both scientists and general readers. Such elucidations are ing expected to appreciate status of arthropods in the environment and their role in maintaining biodiversity of both flora and fauna. This prelude is expected to give the briefest overview of the contents of the book.

Bagyaraj and others contend that arthropods are ecosystem engineers and litter transformers. The authors in the chapter on soil biodiversity and arthropods have highlighted the role of arthropods on soil biodiversity and fertility. Although nematodes, earthworms, snails and slugs are not arthropods, their roles and mechanisms in processing and formation of soil particles are inseparable from arthropods. Arthropods carry out key functional roles in forests and contribute to forest regeneration. But their services are often ignored in conservation and forest development plans. George Mathew and others show the ways to sustain services of arthropods in forests. Seraj and Esfandiari compared arthropods in wild and cultivated ecosystems of Iran. From time immemorial, man is harnessing services of arthropods as pollinators, as medicine and as ornamental objects. Jayashankar and others have reviewed the published literature on utility of arthropods and have shown means to sustain them. An appraisal of select arthropod taxa is included to reflect arthropod diversity in Sri Lanka by Edirisinghe and co-workers. Arthropods serve as food for several groups of animals such as fishes, birds, small mammals and humans. In this compilation, values of arthropods as human food has been elaborately dealt with. Arthropods have interesting roles to perform in island ecosystems. Arthropods may serve as indicator species. Studies on butterfly communities of the Andaman and Nicobar Islands by Sivaperuman and Venkataraman are significant in this direction. For about 250 million years, arthropods have dominated terrestrial and aquatic habitats on earth. Thus, they had a vast timescale to adapt to varying landscapes, flora and fauna. In this context, arthropods thriving in various agro ecosystems such as rice, coffee, cashew, tea, cotton, jute and fibre crops, mango and vegetables have been documented. Several guilds of arthropods coexist in rice ecosystems, and it is crucially important to maintain their interrelationships. Cotton is one of the productive cultivated ecosystems, and cultivation of Old and New World cottons has given rise to two communities of arthropods that would be worthwhile to compare and decipher. Cotton ecosystem is endowed with unique biological attributes that influence arthropods in different ways. Reddy and Sreedevi have examined arthropod communities on mango and have indicated how sustaining rich biodiversity of arthropods can aid in harvesting better mango yields without using pesticides. Shivarama Bhat has called for not disrupting arthropod community on cashew as it serves as a perennial reservoir for beneficials like pollinators, predators and parasites. While several workers have examined arthropods of cultivated ecosystems, N. S. Aratchige and co-workers from Sri Lanka have looked specifically into coconut mites. That mites can be managed by eco-friendly methods is the issue dealt in the chapter. Similarly, Muraleedharan and Roy on arthropods of tea, Reddy on coffee, Binisha and co-workers on vegetables, Selvaraj and co-workers on jute and allied fibre crops and Prasannakumar and co-workers on nonleguminous vegetables have contributed chapters to enrich the existing knowledge on arthropods. These treatises are expected to benefit entomologists, agricultural and horticultural scientists, conservation biologists, policymakers, students and teachers in planning crop management, farming system plans, forestry management and biodiversity conservation and, finally, imparting teaching both introductory and advanced courses. At the end, attention is paid to the risk of health effects from pesticides for infants and children. Ranga Rao and others from ICRISAT, Hyderabad, India, reiterate to arrive at acceptable levels of pesticide residues in food and determine effective ways to increase food safety from pesticides.

Akshay Kumar Chakravarthy
Shakunthala Sridhara

Acknowledgements

The International Conference on Insect Science (ICIS-2013) was held in February 2013 at Bengaluru, South India. The conference provided a forum for crafting and expressing new ideas across a wide range of topics in entomology and related disciplines. Of the 536 abstracts received for presentation, more than 40 % were on biological control and conservation. So it was thought appropriate to bring out a book on arthropod biodiversity and conservation. Some of the participants who attended the conference have contributed chapters for the book. Besides, experts on select taxa of arthropods were also included to give a comprehensive and diversified coverage for the book. The editors are thankful to Dr. N. K. Krishna Kumar, DDG, Horticultural Science, Indian Council of Agricultural Research, New Delhi for all help and encouragement. The editors are also thankful to the Honourable Vice Chancellor of the University of Agricultural Sciences (UAS), GKVK, Bengaluru; officers and staff of the Department of Entomology, UAS, GKVK, Bengaluru; the Director of the Indian Institute of Horticultural Research (IIHR), Hesaraghatta, Bengaluru; and staff of the Division of Entomology and Nematology, IIHR, Bengaluru for extending all co-operation and help for the preparation of the book. Dr. C. T. Ashok Kumar, Dr. Abraham Verghese and Dr. N. E. Thyagaraj took personal interest in executing different works for the successful conduct of ICIS-2013. It would have not been possible to bring out a book of this magnitude and dimension without the help rendered by researchers and postgraduate students, viz. Vasudev Kammar, K. P. Kumar, Chandrashekaraiah, K. S. Nitin, A. T. Rani, Vijeth Arya, T. N. Madhu, Rajendra Prasad, Nethra, V. Sindhu and T. H. Savitha, and many others for their untiring efforts. The editors also immensely thank the contributors because information required for this unique book is highly scattered in literature and is often difficult to find. We acknowledge all the foreign and Indian delegates who sent manuscripts, their valuable contributions to this book, and those who not only participated but made ICIS-2013 happen so successfully.

We profusely thank the International Springer Group, New Delhi, for the deep interest and enthusiasm they have shown in publishing this book.

Akshay Kumar Chakravarthy
Shakunthala Sridhara

Contents

Contributors

Wasmia Al-Houty Department of Biological Sciences, Faculty of Science, Kuwait University, Safat, Kuwait

N.S. Aratchige Crop Protection Division, Coconut Research Institute of Sri Lanka, Lunuwila, Sri Lanka

Vijeth V. Arya Division of Entomology and Nematology, Indian Institute of Horticultural Research (IIHR), Hessaraghatta Lake Post, Bengaluru, Karnataka, India

D.J. Bagyaraj Department of Agricultural Microbiology, Gandhi Krishi Vignana Kendra (GKVK), University of Agricultural Sciences, Bengaluru, Karnataka, India

C.M.D. Bambaradeniya Ellicott City, MD, USA

Haseena Bhaskar Department of Agricultural Entomology, College of Horticulture, Kerala Agricultural University, Vellanikkara, Thrissur, Kerala, India

P.S. Bhat Division of Entomology and Nematology, Indian Institute of Horticultural Research (IIHR), Hessaraghatta Lake Post, Bengaluru, Karnataka, India

K.V. Binisha Plant Quarantine Station, Bengaluru, Karnataka, India

Akshay Kumar Chakravarthy Division of Entomology and Nematology, Indian Institute of Horticultural Research (IIHR), Hessaraghatta Lake Post, Bengaluru, Karnataka, India

M. Chandrashekaraiah Zonal Office, Central Silk Technological Research Institute, Bilaspur, Chhattisgarh, India

M. Charles St. Joseph's College (Autonomous), Bengaluru, Karnataka, India

B. Doddabasappa Department of Entomology, College of Horticulture, Kolar, Karnataka, India

J.P. Edirisinghe Department of Zoology, Faculty of Science, University of Peradeniya, Peradeniya, Sri Lanka

M. Esfandiari Department of Plant Protection, College of Agriculture, Shahid Chamran University of Ahvaz, Ahvaz, Iran

S.P. Gawande Ramie Research Station, Central Research Institute for Jute and Allied Fibre Crops (CRIJAF), Sorbhog, Assam, India

B.S. Gotyal Division of Crop Protection, Central Research Institute for Jute and Allied Fibre Crops (CRIJAF), Barrackpore, Kolkata, West Bengal, India

N.R. Gunawardene Curtin Institute for Biodiversity and Climate, Department of Environment and Agriculture, Curtin University, Perth, WA, Australia

Jayalaxmi Hegde Department of Entomology, University of Agricultural and Horticultural Sciences, Navule, Shivamogga, Karnataka, India

I.I. Hemachandra Department of Zoology, Faculty of Science, University of Peradeniya, Peradeniya, Sri Lanka

Sosamma Jacob Department of Agricultural Entomology, College of Horticulture, Kerala Agricultural University, Vellanikkara, Thrissur, Kerala, India

M. Jayashankar Division of Entomology and Nematology, Indian Institute of Horticultural Research (IIHR), Hessaraghatta Lake Post, Bengaluru, Karnataka, India

G.T. Jayasimha Department of Entomology, Agriculture College and Research Institute, Madurai, Tamil Nadu, India

Vasudev Kammar Department of Entomology, University of Agricultural Sciences, Gandhi Krishi Vignana Kendra (GKVK), Bengaluru, Karnataka, India

W.A.I.P. Karunaratne Department of Zoology, Faculty of Science, University of Peradeniya, Peradeniya, Sri Lanka

A.D.N.T. Kumara Division of Crop Protection, Coconut Research Institute of Sri Lanka, Lunuwila, Sri Lanka

K.P. Kumar Department of Agricultural Entomology, University of Agricultural Sciences (UAS), Gandhi Krishi Vignana Kendra (GKVK), Bengaluru, Karnataka, India

B. Ratna Kumari Acharya NG Ranga Agricultural University, Hyderabad, India

Nripendra Laskar Department of Agricultural Entomology, Uttar Banga Krishi Viswavidyalaya, Cooch Behar, West Bengal, India

Vijay Kumar Lingaraj Department of Entomology, College of Agriculture, University of Agricultural Sciences, Bangalore, VC Farm, Mandya, Karnataka, India

T.N. Madhu Department of Agricultural Entomology, Gandhi Krishi Vignana Kendra (GKVK), University of Agricultural Sciences (UAS), Bengaluru, Karnataka, India

G. Mathew Forest Health Division, Kerala Forest Research Institute, Peechi, Kerala, India

Narayanannair Muraleedharan Department of Entomology, Tocklai Tea Research Institute, Tea Research Association, Jorhat, Assam, India

Manja Naik Department of Agricultural Entomology, Gandhi Krishi Vignana Kendra (GKVK), University of Agricultural Sciences (UAS), Bengaluru, Karnataka, India

S. Onkara Naik Division of Entomology and Nematology, Indian Institute of Horticultural Research (IIHR), Hessaraghatta Lake Post, Bengaluru, Karnataka, India

C.J. Nethravathi Division of Entomology and Nematology, Indian Institute of Horticultural Research (IIHR), Hessaraghatta Lake Post, Bengaluru, Karnataka, India

K.S. Nitin Division of Entomology and Nematology, Indian Institute of Horticultural Research (IIHR), Hessaraghatta Lake Post, Bengaluru, Karnataka, India

N.R. Prasannakumar Division of Entomology and Nematology, Indian Institute of Horticultural Research (IIHR), Hessaraghatta Lake Post, Bengaluru, Karnataka, India

R.R. Rachana Division of Entomology, National Bureau of Agricultural Insect Resources (NBAIR), Bengaluru, Karnataka, India

B.S. Rajendra Prasad Division of Entomology and Nematology, Indian Institute of Horticultural Research (IIHR), Hessaraghatta Lake Post, Bengaluru, Karnataka, India

A.T. Rani Department of Agricultural Entomology, University of Agricultural Sciences (UAS), Gandhi Krishi Vignana Kendra (GKVK), Bengaluru, Karnataka, India

G.V. Ranga Rao International Crops Research Institute for the Semi-Arid Tropics, Hyderabad, India

T.N. Raviprasad Division of Entomology, ICAR-Directorate of Cashew Research, Puttur, Karnataka, India

Poluru Venkata Rami Reddy Division of Entomology and Nematology, Indian Institute of Horticultural Research (IIHR), Hessaraghatta Lake Post, Bengaluru, Karnataka, India

G.V. Manjunatha Reddy Department of Entomology, Central Coffee Research Institute, Coffee Research Station, Chikmagalur, Karnataka, India

G. Rohini Department of Zoology, Bangalore University, Bengaluru, Karnataka, India

Somnath Roy Department of Entomology, Tocklai Tea Research Institute, Tea Research Association, Jorhat, Assam, India

K.L. Sahrawat Acharya NG Ranga Agricultural University, Hyderabad, India

S.K. Sarkar Division of Crop Protection, Central Research Institute for Jute and Allied Fibre Crops (CRIJAF), Barrackpore, Kolkata, West Bengal, India

S. Satpathy Division of Crop Protection, Central Research Institute for Jute and Allied Fibre Crops (CRIJAF), Barrackpore, Kolkata, West Bengal, India

K. Selvaraj Division of Crop Protection, Central Research Institute for Jute and Allied Fibre Crops (CRIJAF), Barrackpore, Kolkata, West Bengal, India

A.A. Seraj Department of Plant Protection, College of Agriculture, Shahid Chamran University of Ahvaz, Ahvaz, Iran

P.R. Shashank Division of Entomology, Indian Agricultural Research Institute (IARI), New Delhi, India

C. Sivaperuman Zoological Survey of India, Andaman and Nicobar Regional Centre, Port Blair, Andaman and Nicobar Islands, India

Kolla Sreedevi Division of Entomology, Indian Agricultural Research Institute (IARI), New Delhi, India

V. Sridhar Division of Entomology and Nematology, Indian Institute of Horticultural Research (IIHR), Hessaraghatta Lake Post, Bengaluru, Karnataka, India

K.K. Srikumar United Planters Association of Southern India (UPASI), Tea Research Foundation, Tea Research Institute, Valparai, Coimbatore, Tamil Nadu, India

N.I. Suwandharathne Crop Protection Division, Coconut Research Institute of Sri Lanka, Lunuwila, Sri Lanka

N.E. Thyagaraj Department of Entomology, College of Agriculture, Hassan, Karnataka, India

K. Vanitha Division of Entomology, ICAR-Directorate of Cashew Research, Puttur, Karnataka, India

K. Venkataraman Zoological Survey of India, Prani Vigyan Bhawan, New Alipore, Kolkata, West Bengal, India

L.S. Vinesh Division of Entomology and Nematology, Indian Institute of Horticultural Research (IIHR), Hessaraghatta Lake Post, Bengaluru, Karnataka, India

S.P. Wani Acharya NG Ranga Agricultural University, Hyderabad, India

About the Editors

Dr. Akshay Kumar Chakravarthy is head and principal scientist at the Indian Institute of Horticultural Research (IIHR), Hesaraghatta, Bengaluru. With three decades of experience in teaching, research and extension, Dr. A. K. Chakravarthy has been the investigator for over 30 research projects and has guided more than 25 postgraduate students. He received Ph.D. from the Punjab Agricultural University, Ludhiana, and is a postdoc fellow of IARI, New Delhi. He is a fellow of the American Chemical Society, USA; is a member of several national and international scientific academia, an advisor, a panellist, a referee, a reviewer and an editor; and is associated with the publication of over 30 national and international journals worldwide. He has 400 publications in the form of books, chapters, monographs, bulletins, papers, short notes, commentaries, letters and meeting and project reports. A field-oriented, widely travelled biologist, he is actively working on novel approaches in integrated pest management, host-plant interaction, vertebrate pest management, biodiversity and environmental conservation issues. Currently, he has initiatives on nanotechnology too.

Shakunthala Sridhara after obtaining her Ph.D. in animal physiology from Bangalore University, joined a Ford Foundation project on vertebrate pest management in the University of Agricultural Sciences, Bengaluru, India, in 1973. Over the past 33 years, she has been researching on vertebrate pest management specially the control of rodents in the agricultural context. She has researched extensively on the ecology, population dynamics, food selection and feeding behaviour of rodents, toxicology of rodenticides and adoption of rodent pest management at village level. Her studies on behaviour relevant to management of avian and mammalian pests are pioneering in the Indian context and well acknowledged, culminating in adaptable technologies for their management. She has visited and interacted with specialists in the field across America and Europe several times. Keenly interested in animal behaviour studies and its application in pest and wildlife management, biodiversity conservation and animal produce, she is member of several national and international scientific bodies including the presidentship of Ethological Society of India, Indian representative in the International Council of Ethologists, IUCN species specialist group on rodents, etc. She has retired as professor and head of vertebrate biology (rodent control) in 2007 following a stint as Emeritus Scientist of Indian Council of Agricultural Research for two years in the University of Agricultural Sciences, Bengaluru.

Arthropods: Evolution and Ecology

1

A.K. Chakravarthy, Vasudev Kammar, and P.R. Shashank

Abstract

Arthropods constitute the dominant group in the animal kingdom and are a major part of global biodiversity. There are 1,302,809 species of arthropods described that include 45,769 fossil species. Arthropods are the most successful group found in almost all biogeographical regions and ecological zones and have a dominating influence on other elements of biodiversity. The Insecta have 1, 070,781 species and it alone accounts for over 80 % of all arthropods. Another major group is Arachnida having 114, 275 described species of which 55, 214 species are mites and ticks. Arthropods contribute to human food supply, pollinate crops, help maintain ecosystem sustainability by biologically suppressing destructive arthropods, but cause and transmit diseases to humans and livestock and incur crop losses. Invasive arthropods can negatively impact natural resources.

Keywords

Arthropods • Ecology • Evolution • Importance

A.K. Chakravarthy (✉)
Division of Entomology and Nematology, Indian Institute of Horticultural Research (IIHR), Hessaraghatta Lake Post, Bengaluru 560089, Karnataka, India
e-mail: chakravarthyakshay@gmail.com

V. Kammar
Department of Entomology, University of Agriculture Sciences, Gandhi Krishi Vignana Kendra (GKVK), Bengaluru 560065, Karnataka, India
e-mail: 4547vk@gmail.com

P.R. Shashank
Division of Entomology, Indian Agriculture Research Institute (IARI), New Delhi 110 001, India
e-mail: spathour@gmail.com

A.K. Chakravarthy, S. Sridhara (eds.), *Economic and Ecological Significance of Arthropods in Diversified Ecosystems*, DOI 10.1007/978-981-10-1524-3_1

1.1 Introduction

Today one single phylum of the animal kingdom is dominating on the planet Earth. On land, in the sea, even in the air itself, they are the true masters of the Earth. They are the arthropods. The name Arthropodsis is derived from the Greek *arthros* which means jointed and *poda* which means foot (Harper 2014) including the familiar arachnids, crustaceans, and insects. All arthropods have jointed appendages. This evolutionary innovation is probably the key to the stunning success of this diverse group. There are about ten billion arthropods alive at any one time.

Arthropods range in size from microscopic plankton to a few meters long. Their versatility and adaptability has rendered them to become the most species-rich members of all ecological guilds in most environments. There are over a million described species of arthropods, making up more than 80 % of all described living animal species. Out of these, insects alone form about three-fourths of the total organisms present on the Earth (Chapman 2009).

The evolutionary ancestry of arthropods dates back to the Cambrian period. The group is generally regarded as monophyletic, and many analyses support the placement of arthropods with cycloneuralians (or their constituent clades) in a superphylum. Overall, however, the basal relationships of Metazoa are not yet well resolved. Likewise, the relationships between and among phyla in Arthropoda still remain obscure.

The versatility of arthropods is such that they are dominant on land in both species richness and rank numerically the most prominent benthic, freshwater and marine ecosystems. Arthropods are immensely successful in every possible habitat on the land, be it equator or the poles or from high mountains to deep ocean trenches. Many arthropods cause economic loss to humans as most of them carry and spread diseased vectors and plant pests but also a boon as a veritable food resource (honey, edible food such as crabs, lobsters, and shrimps). They are also vital to the functioning of all ecosystems and a beneficial to humans in many ways. In addition to deriving nutrition from arthropods (e.g., directly or indirectly from bees, crabs, lobsters, and shrimps), humans probably could not survive ecologically without them (James 2003).

Arthropod is segmented which differs from annelids, but with an evolutionary tendency toward the fusion of several metameres into the body region (tagmata) with specialized crustaceans having two; however, myriapods (mostly millipedes and centipedes) lack tagmata. Arthropods have chitinous and proteinaceous exoskeletons which are frequently strengthened with calcium salts. Non-chitinized appendages of the exoskeleton project inward to aid muscular attachment. The continued somatic growth is rendered possibly by periodic ecdysis, a relatively strenuous and often dangerous process. In some arthropods, modified exoskeleton has aided in flying. The organ complexity of arthropods is very high among all invertebrates with the exception of molluscan cephalopods. The internal cavity is called hemocoel and accommodates all the vital organs. Blood called hemolymph circulates through an open circulatory system. However the coelom does not function as a hydrostatic organ as in annelids. Respiration is by diverse processes, viz., through the skin in small species, using gills in aquatic forms and through tracheae or book lungs in terrestrial arthropods. Reproduction is mostly dioecious although some forms exhibit parthenogenesis. Courtship and parental care, although less evident, are found in some members. With the exception of few aquatic forms which fertilize externally, internal fertilization is the norm. Most of the internal fertilization is by indirect transfer of the sperm via an appendage or by "suction" from the ground, rather than by direct injection. With the exception of scorpions which are viviparous, all arthropods lay eggs. The young ones are fully mature when born is some but in most arthropods eggs molt into immature forms without appendages, grow in cocoons, molt several times, and metamorphosize into adults. Similarly parental care ranges from non-existent to vivipary which further extends to the first molt of the young as in scorpions. The neural system in most arthropods is highly developed. In fact, the arthropod brain is one of the most complex of all living organisms.

The young of the Arthropods usually develop by cleaving of the cytoplasmic layer above a yolky sphere. Although larvae or discrete juveniles characterize terrestrial and aquatic forms, the aquatic larvae do not resemble the trochopore larvae of related phyla.

1.2 Evolutionary Relationships

Arthropods were earlier associated with the phylum Annelida because of their segmented body. But recent molecular data does not provide evidence for such a relationship, but places them closer to other phyla which shed their cuticle during ecdysis during growth such as the phyla Tardigrada, Onychophora, Nematoda, and Nematomorpha along with more distantly related Priapulida and Kinorhyncha (Resh and Carde 2009). The evolutionary lineages with other phyla with respect to similar characters are detailed below.

1.2.1 Segmentation

The embryos of all arthropods consist of a segmented body, built from a series of repeated modules. It is summarized that the last common ancestor of all living arthropods consisted of a series of undifferentiated segments, each segment with a pair of appendages that functioned as limbs. However, in both living and fossil arthropods, the segments are focused into tagmata in which segments and their appendages discharge specific functions (Ruppert et al. 2004), which is clearly evident in the three-part body of insect and the two-part bodies of spiders. In fact, the segmentation of body in mites is invisible. In addition arthropods have an acron in the front of the mouth and a telson at the rear end. Acron has eyes mounted on it (Ruppert et al. 2004) (Table 1.1).

Arthropod appendages originally bifunctional, those in the upper region functioning as gills while those on the lower side served as legs. In living arthropods, the appendages have been modified to discharge varied functions as gills, mouthparts, antennae (Gould 1990), or claws (Shubin et al. 2000). In many arthropods

Table 1.1 Diversity of arthropoda and two related phyla

Taxon	No. of estimated (~) or described species	Some biological features
Phylum Arthropoda	~2–6 million	Insects, arachnids, crustaceans, millipedes, and other invertebrates with segmented bodies and appendages on one or more segments; mostly with hard, chitinous exoskeleton that is periodically molted
Subphylum Trilobita	~15,000+	Extinct marine trilobites
Subphylum Chelicerata	~99,000–1 million	Originally marine but subsequent evolution has primarily been in terrestrial habitats
Order Xiphosura	5	
Class Arachnida	~98,000+	Marine horseshoe crabs
Class Eurypterida	300	Spiders, scorpions, and mites
Class Pycnogonida	1300	Extinct sea scorpions
		Sea spiders
Subphylum Myriapoda	~13,500	Terrestrial millipedes, centipedes, and others
Class Chilopoda	2800	
Subclass Epimorpha	1600	Predaceous centipedes mostly in tropical forest floor, 0.5–30 cm long
Subclass Anamorpha	1200	
Class Symphyla	160	Small (2–10 mm), mostly herbivorous, live in forest litter, sometimes called garden centipedes (pseudocentipedes), or just symphylans
Class Diplopoda	~10,000	
Subclass Penicillata	160	Millipedes
		Millipedes with a soft, noncalcified exoskeleton covered with tufts of setae or bristles (used as defense against ant and other predators)
Subclass Chilognatha	10,000	Millipedes with hard exoskeleton and chemical defenses against predators
Class Pauropoda	500	Minute (<1.5 mm) dwellers in soil and leaf litter
Subphylum Hexapoda	~5+ million	Insects, springtails, bristletails, etc.
Class Entognatha	11,000	Wingless springtails, two-pronged bristletails, and other wingless insects with internal mouthparts
class Elliplura	9600	Wingless entognathous (order Protura and Collembola or springtails)
Order Diplura	1000	
Class Insecta	~5+ million species	Blind, wingless inhabitants of forest litter, entognathous
Order Archaeognatha		
Subclass Dicondylia	500	Winged and wingless insects, all adults with six pairs of legs
	915,300+	Primitive, wingless insects, jumping bristletails
		Mostly winged insects (grasshoppers, true bugs, beetles, flies, butterflies, ants, etc.) and a few wingless species (silverfish)
Subphylum Crustacea	~52,000+	Shrimp, crabs, water fleas, barnacles, copepods, etc.

(continued)

Table 1.1 (continued)

Taxon	No. of estimated (~) or described species	Some biological features
Class Branchiopoda	500	Small crustaceans mostly confined to inland waters (freshwater through hypersaline), small, and common in lakes and ephemeral habitats without fish
Subclass Sarsostraca	200	
Subclass Phyllopoda	300	Brine and fairy shrimp
Class Remipedia	12	Water fleas (cladocerans), clam shrimp, shield shrimp
Class Cephalocarida	9	Remipedes; ancient, vermiform crustaceans found in marine caves; estimated diversity may be an order of magnitude higher
Class Maxillopoda	13,400	
Subclass Thecostraca	1320	Horseshoe shrimp; primitive, live in soft marine sediments
Subclass Tantulocarida	10	Crustaceans characterized by a reduced abdomen mostly lacking appendages
Subclass Branchiura	200	True barnacles and small groups of parasitic taxa
Subclass Pentastomida	100	Highly modified parasites of deep-sea crustaceans, estimated diversity over 1000 species
Subclass Mystacocarida	12	
Subclass Copepoda	11,690	Fish lice and carp lice; ectoparasites on fish and a few amphibians
Class Ostracoda	~8000+	Tongue worms, highly modified parasites of tetrapod vertebrates
Subclass Myodocopa	900	
Subclass Podocopa	7000	Interstitial species living in shallow or intertidal waters; reported diversity artificially low from similar external anatomy and habitat
Class Malacostraca	~29,000	Dominant crustaceans in zooplankton; a few parasites of marine fish and invertebrates
Subclass Phyllocarida	39	
Subclass Hoplocarida	400	Seed or mussel shrimps; enclosed in a bivalve chitinous carapace; total diversity is probably at least 20,000
Subclass Eumalacotraca	~29,000	Ostracodes with poorly calcified carapace; brood care within parent carapace
		Ostracodes with a hard carapace
		Crabs, water scuds, isopods, mantis shrimp, etc.
		Leptostracans; small filter feeders
		Mantis shrimp; they kill by smashing or spearing the prey
		Lobsters, king crabs, isopods (e.g., pill bugs), shrimp, amphipods; thoracic limbs jointed for walking or swimming
Phylum Onychopora	90	Velvet worms; mostly confined to tropical habitats
Phylum Tardigrada	1047	Water bears in aquatic and moist terrestrial habitats

Source: James (2003)

they have disappeared from some regions of the body, especially abdominal (Ruppert et al. 2004).

Tropical and subtropical systems have provided fascinating opportunities for studying evolutionary processes. For instance, the Indian subcontinent is an interesting entity given that it has been an island during much of its history following separation from Madagascar and currently is isolated from Eurasia by the Himalayas in the north and the Indian Ocean in the south. Recent molecular studies on a number of endemic taxa from India have reposted endemic radiations. These studies suggest that the uniqueness of Indian biota is not just due to diverse origin but due to evolution in isolation. The isolation of India has generated peculiarities typically seen on oceanic islands (Karanth 2015).

Among arthropods, insects have been extensively studied for evolutionary lineages. Fossil records have also facilitated understanding evolutionary relationships among arthropods and their relatives. There are two schools of thought for evolution of arthropods and insects separately. Arthropods identified as fossils are abundant in deposits dating back to the Cambrian period. The phylum Arthropoda is closely linked with two other phyla, the velvet worms (Onychophora) and water bears (Tardigrada). Both phyla have common characters of nonliving cuticle and they have appendages. However, the cuticle is not hard, and the appendages are not jointed.

The oldest insect fossil *Rhyniognatha hirsti* is estimated at 407–396 million years ago (Devonian period), which coincides with the age of fishes and growth of forests on dry land. The evolution and diversification of insects continued with new insect orders appearing through the Paleozoic, Mesozoic, and Cenozoic eras (Engel and Grimaldi 2004).

The change in global climate during the history of the Earth was invariably accompanied by changes in diversity of insects. The pterygotes were radiated in the Carboniferous, while the Endopterygota species underwent another major radiation in the Permian. The modern day insects are descendants of those that evolved from the survivors of the mass extinction at the Permian–Triassic boundary. Most modern insect families made their appearance during the Jurassic; they further evolved in the cretaceous. By the Tertiary, many more modern insect genera are believed to have been existed. It took about 100 million years for insects to evolve to forms similar to modern day forms (Resh and Carde 2009).

Appearance of all insect species, extant or extinct, are presented in Table 1.2 with the insect orders being presented at the fleas (Siphonaptera) appearing during Mesozoic is at the upper portion of the table, while the springtail fossils (order Collembola) are the oldest known fossils occurring at the Devonian period.

A reconstructed modular organism of the last known common ancestor of arthropods is a module covered with its own armored plate (sclerite) with a pair of biramous limbs. However there is some debate as to whether this organism was uniramous or biramous. It had a ventral mouth, preoral antennae, and dorsal positioned eyes. This arthropod was a generalist feeder of sediment (Bergström and Hou 2003).

Table 1.2 Select fossil records of class Insecta

Era	Period	Dates (MYA)*	Fossil record of insect order with comments
Cenozoic	Quaternary	3–0	No records
	Neogene	23–3	No records
	Paleogene	66–23	Mantophasmatodea (apterous carnivores, closest phylogenetic relationship with Grylloblattidae and Phasmatodea)
Mesozoic	Cretaceous	146–66	Siphonaptera (fleas, phylogenetically closer to Diptera and Mecoptera), Zoraptera (angel insects, Hemimetabola, these are link between orthopteroids and hemipteroids), Strepsiptera (endoparasites, they have twisted hind wings), Mantodea (praying mantis), and Isoptera (termites)
	Jurassic	200–145	Dermaptera (earwigs), Lepidoptera (moths and butterflies), Blattodea (cockroaches)
	Triassic	251–200	Hymenoptera (wasps), Trichoptera (caddish flies)
			Phasmatodea (leaf and stick insects), Odonata (dragonfly and damselfly)
Paleozoic	Permian	299–251	Raphidioptera (snake flies) Megaloptera (alderflies and dobsonflies), Thysanoptera (thrips), Diptera (flies), Neuroptera, Embiidina (web spinners), Mecoptera (scorpion flies), fossil hemipteroids
			Plecoptera (stone flies), Coleoptera (beetles)
			Grylloblattodea (rock crawlers, Hemimetabola, closely related to Orthoptera and Dermaptera)
	Carboniferous	359–299	Zygentoma (silverfish), Psocoptera (book lice)
			Hemiptera (bugs), Diplura, Miomoptera (extinct fossil order), Orthoptera (grasshoppers and crickets)
			Ephemeroptera (mayflies), Palaeodictyopteroidea (an extinct superorder of Paleozoic beaked insects and represents the important terrestrial herbivores)
			Protodonata
	Devonian	416–359	Archaeognatha (bristletails), Collembola (springtails)
	Silurian	444–416	No records
	Ordovician	488–444	No records
	Cambrian	542–488	No records

Sources: Gullan and Cranston
Engel and Grimaldi (2004)
**MYA* million years ago

Drosophila remains the best study model organism with widest availability of genetic tool. The bulk of genetic studies for behavioral isolation have been conducted on *Drosophila*. Species like *D. mauritiana*, *D. sechellia*, *and D. santomea* found in islands of Mauritius are known as "island endemic" species. *Drosophila simulans*, *D. mauritiana*, and *D. sechellia* diverged around quarter of a million years ago. Although relatively young, the species in this clade have diverged enough to show post-zygotic isolation. Hybridization between these species produces sterile males (according to Haldane's rule) but, conveniently for genetic analysis, produce fertile females.

1.2.2 Fossil Records

The earliest arthropods, the Ediacaran animals, namely, *Parvancorina* and *Spriggina*, lived 555 million years ago, while small arthropods with bivalve-like shells were found in fossil beds in China dating 541–539 million years ago (early Cambrian) (Lin et al. 2006; Mc Menamin 2003). The class was however quite diverse suggesting that they had been around for quite some time (Lieberman 1995). However, reexamination of the Burgess Shale, fossils belonging to 505 million years ago revealed many arthropods in the 1970s but they could not be placed in any of the existing arthropod group (Whittington 1979).

Earliest crustacean fossils are recorded at around 513 million years ago in the Cambrian (Budd et al. 2001) fossil shrimp date back to a little later period but occurred abundantly on the seabed (Callaway 2008) during Ordovician period (Zhang et al. 2007). Up till now Crustaceans have lived only in aquatic medium; probably they failed to develop excretory systems to conserve water.

With a jointed exoskeletons that could prevent desiccation, could support against gravity, and could aid locomotion outside the water, arthropods were predisposed to terrestrial life which they did about 450 million years ago as evidenced by terrestrial tracks left by them. The earliest identifiable terrestrial arthropods are from the Silurian period about 419 million years ago (Cowen 2000; Pisani et al. 2004) (Fig. 1.1).

Fossils of many spiders including many from the modern families date back to Jurassic and Cretaceous periods (Vollrath and Selden 2007). Aquatic forms with gills were found to live in Silurian and Devonian periods, while the earliest terrestrial forms with book lungs date back to early Carboniferous period (Jeram 1990). Around 420 million years ago, the oldest known arachnid (*Palaeotarbus jerami*) (Dunlop 1996) appeared. True spiders appeared in the late Carboniferous period about 299 million years ago. However, around 386 million years ago in the Devonian period, *Attercopus fimbriunguis*, bearing the earliest known silk-producing spigots appeared, but lacked spinnerets which characterize true spiders (Selden and Shear 1996).

The earliest insects seemed to have appeared in the Silurian period although the oldest insect fossil, that of *Rhyniognatha hirsti*, dated the Devonian era but it had mandibles similar to those of winged insects (Engel and Grimaldi 2004). By the late

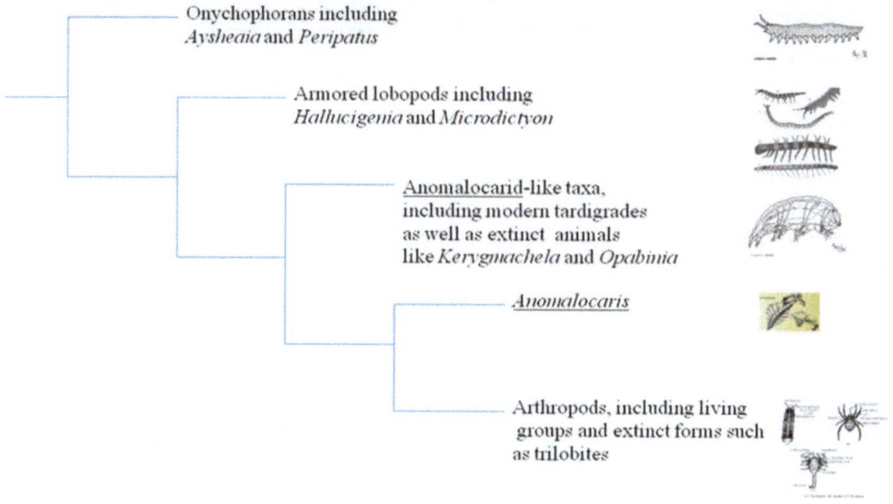

Fig. 1.1 Evolutionary family tree 1. Simplified and modified summary of Budd's "broad-scale" cladogram (Dunlop 1996)

Carboniferous, around 300 million years ago, there were already about 200 species of insects; some were enormously big and already exhibited different food habitat such as herbivory, detritivory, and insectivory. Among the social insects, ants and termites appeared in the early Cretaceous, and among the advanced social insects, bees evolved in late Cretaceous rocks, but they became abundant only in the Middle Cenozoic (Labandeira and Eble 2000).

The origin of arthropods presents an interesting evolutionary development of theories beginning 1952–1977 with the concept that they are polyphyletic and lack a common arthropod ancestor but evolved from three separate groups of "arthropods" originating from (a) wormlike ancestors, (b) chelicerates (spiders and scorpions), and (c) Uniramia (onychophorans, myriapods, and hexapods). This theory failed to include trilobites whose evolutionary relationships remained unexplained. The lack of common ancestry of the three groups was attributed to differential chemical means of hardening of the exoskeleton, differences in the anatomy of their compound eyes, and differential anatomy of the segments and appendages in the head; the supported argument emphasized the biramous limbs of crustaceans serving as gills and legs and the uniramous limbs of the other two groups serving as legs. Similarities between the three groups were attributed to convergent evolution (Gillott 1995).

However, research in the 1990s led to the acceptance of arthropods as monophyletic originating from a common arthropod-like ancestor. Budd after analyzing *Kerygmachela* (1993) and *Opabinia* (1996) argued that these two were similar to onychophorans and to "lobopods" of the early Cambrian. And in the "evolutionary tree," he presented these two as "aunts" and "cousins" of all the arthropods (Fig. 1.2). This interpretation made the meaning of arthropod "irrelevant." Claus Nielsen

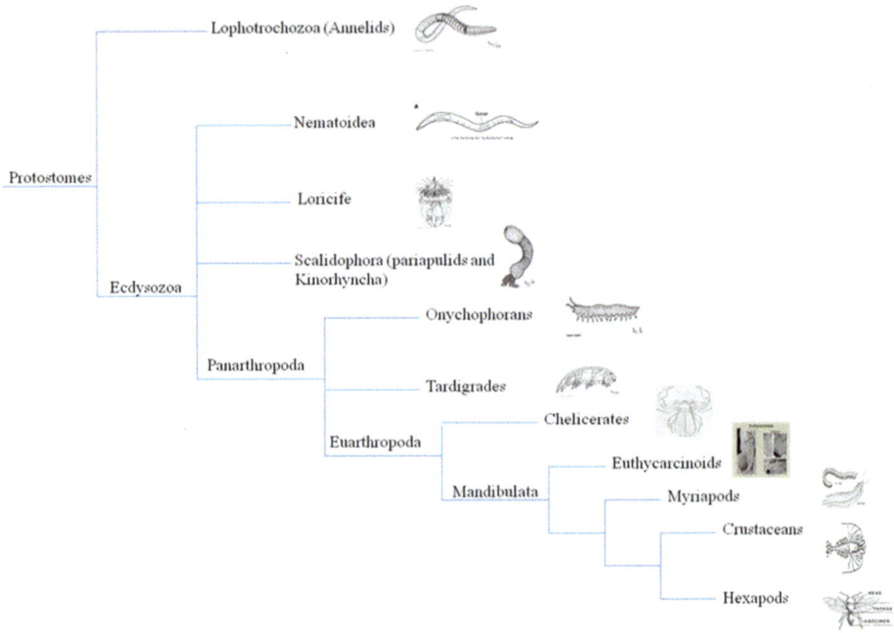

Fig. 1.2 Evolutionary family tree 2. Relationships of Ecdysozoa to each other and to annelids, including euthycarcinoids (Modified from Dunlop 1996)

proposed two nomenclatures: "Panarthropoda" (all arthropods) and "Euarthropoda" ("true arthropods"). The latter group was characterized by jointed limbs and hardened cuticle.

Even the argument that arthropods were a "sister group" to anomalocarids was unacceptable to Bergstrom and Hou (2003). The premitive arthropods were mud eaters, filtering food particles from mud using unspecialized number of appendages which acted as both gills and legs. In contrast Anomalocarids were big in size and had specialized mouth and grasping appendages; they had fixed number of body segments, some of which were modified into gills and tail fins. In contrast *Parapeytoia* was similar to earlier arthropods in having legs and a backward-pointing mouth and was considered a closer relative to arthropods than *Anomalocaris* (Bergstram et al. 2003). In a later publication, Hou et al. (2006) place arthropods closer to lobopods and tardigrades than anomalocarids.

For a long time, annelids were considered as the closest relatives of arthropods based solely on the common character of their segmented body. Together they were termed as Articulata. There were also views that arthropods are closely related to nematodes, priapulids, and tardigrades, but they lacked convincing relationships among them (Fig. 1.2) till the 1990s. But evidence from molecular phylogenetic analyses of DNA sequence revealed similarities between arthropods, nematodes, priapulids, and tardigrades except Annelids, all of which were placed under super-phylum labeled Ecdysozoa ("animals that molt"); further evidence for this grouping

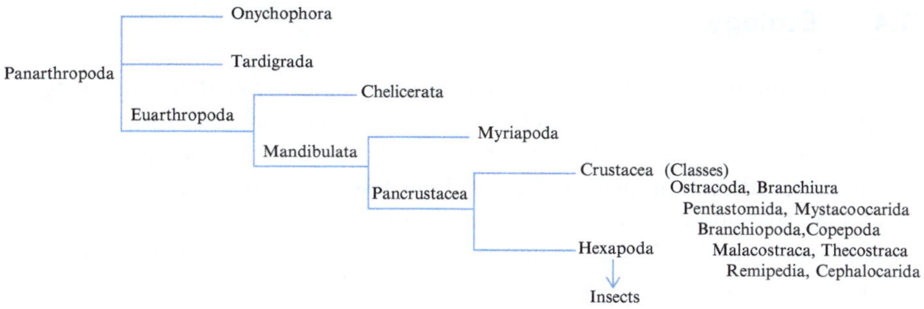

Fig. 1.3 Evolutionary family tree 3. Phylogenetic relationships of the major extant arthropod groups, traditional sub phyla Regier et al. (2010)

came from anatomy and development of member species. The annelids along with mollusks and brachiopods were placed in another superphylum, Lophotrochozoa (Schmidt-Rhaesa et al. 1999). If the Ecdysozoa theory is correct, then the common feature of arthropods and annelids, that is, having a segmented body, has evolved independently or inherited from an older ancestor. It is likely that it was lost during evolution in the non-arthropod members Ecdysozoa.

1.3 Classification (Fig. 1.3)

Arthropods are typically classified into five sub phyla ("Arthropoda" ITIS Report, 2006) which are shown below:

1. **Trilobites** are a group of formerly numerous marine animals that disappeared in the Permian–Triassic extinction event, though they were in decline prior to this killing blow, having been reduced to one order in the late Devonian extinction.
2. **Chelicerates** include spiders, mites, scorpions, and related organisms. They are characterized by the presence of chelicerae, appendages just above or in front of the mouth. Chelicerae appear in scorpions as tiny claws that they use in feeding, but those of spiders have developed as fangs that inject venom.
3. **Myriapods** comprise millipedes, centipedes, and their relatives and have many body segments, each bearing one or two pairs of legs. They are sometimes grouped with the hexapods.
4. **Crustaceans** are primarily aquatic (a notable exception being woodlice) and are characterized by having biramous appendages. They include lobsters, crabs, barnacles, crayfish, prawns, shrimp, and many others.
5. **Hexapods** comprise insects and three small orders of insect-like animals with six thoracic legs. They are sometimes grouped with the myriapods, in a group called Uniramia, though genetic evidence tends to support a closer relationship between hexapods and crustaceans.

There are a number of fossil forms from the early Cambrian which either lack affinity to any of the above sub phyla or show characters common to many of them and hence difficult to place, e.g., *Marrella* (Whittington 1971).

1.4 Ecology

Arthropods are biologically distinct group of invertebrates, found in a wide range of habitats. Having evolved long back they comprise about three-fourths of the animals living on Earth and include insects, spiders, crustaceans, centipedes, and millipedes. These diverse animals perform multiple functions and hence are ecologically important. Evolutionary biology and chemical-mediated ecology are today gaining ground mostly in developing countries, and most of the developing countries are situated in the tropics and subtropics. A deep understanding of the evolutionary processes is required to tackle the most pressing problems the human race faces on the planet Earth. Plants produce and emit a wide range of chemicals that provide information by a range of plant-associated arthropods. Plant body odors provide information on the ecology of plant–arthropod interactions that can be exploited for increasing crop productivity and sustaining biodiversity.

1.4.1 Urbanization and Arthropod Ecology

Traditional agricultural practices were in synchrony with nature and sustained natural balance; hence, their populations seemed to be a balance of existence between human beings, plants, and animals. But as the human race increased, the need for land also increased. Technological developments and advancements provided increased food production. Commercial farming was carried out over larger areas and land grew increasingly important as a source of food. This adversely affected the ecological balance of organisms resulting in the destruction of habitat of insects, arachnids, centipedes, and millipedes. Hedgerow vegetation at fringes and scrub patches were eliminated to expand the crop fields to extensive areas. This change in landscape forced terrestrial arthropods to migrate and seek food elsewhere, where it could be found abundantly. Before long they came back and discovered their old habitats but in a new form. The area had become a vast region of field crops, orchards, and vegetations providing rich natural functions. Arthropods started finding concentrate food supply in a single location and invaded these places in swarms. But their arrival was considered as plagues and infestations by humans. Moths, beetles, and locusts became problematic to farmers, since they were aggressive and attacked human beings who tried to drive them away. Several species of insects and other arthropods became pests and caused crop losses. This led to introduction of pesticides concocted from chemicals which could eradicate the pest's existence. Pesticides were successful as they slowly eliminated pest infestation and afforded protection to crops.

1.4.1.1 Ecological Importance of Terrestrial Arthropods

Very soon, pesticides were discovered to be harmful as they contaminated the soil, water, and air. Additionally, continuous agricultural studies revealed that crop yields were not as much as expected despite the absence of pest infestation. Scientists found winds to be insufficient to disperse the seeds. Since different crops had different systems of releasing their pollens and have different mechanisms of pollination and fertilization.

Several species of insects pollinate, and each species foraged for nectar and incidentally carried pollen. These species included bees, wasps, ants, butterflies, moths, flies, beetles, etc. Pollen grains became accidentally attached to their body, legs, and other body parts and were transferred to other agricultural crops. Soon it was discovered that most plants actually produced scents to send signals to insects that food in the form of nectar was available. Thus, plant volatiles and other semio-chemicals played a pivotal role in tritrophic interactions. Studies have found that only an estimated 10 % of pollen are used to produce flowers and in turn fruits. The rest are meant for the insects as enticements, to forage and open up their pollen sacs, chew on their seeds, or serve as agents for dispersal of pollens.

1.4.2 Pollination

Initially, the farmers went for large-scale production of maize because it did not require pollination, but this did not provide the ultimate solution to what was ailing the commercial agricultural industries. This happened in mostly temperate regions. In the tropics and subtropics, usually the new crop is cultivated in patches and later it occupies a large area.

Cattle, of the best livestock breed, are ideally fed with alfalfa, forage legumes, and silage and are the best sources of calcium and protein. Almost one-third of all cattle feeds are derived from agricultural products that relied on pollination in order to have good harvest. In developing countries in the tropics, cattle have alternate, diverse feeds, although they form poor substitutes.

As there were not enough insects foraging for food as inorganic fertilizers and pesticides affected their populations, most farms resorted to hand pollination. Even natural habitats of arthropods were no longer available because of rapid disturbance to natural settings. Forests became denuded as trees were being cut down, while most of the available land was put to alternate uses.

Soon farmers realized that manual pollination costed them as much as 25 % more in labor costs. An estimated annual cost of $11 million was being spent for hand pollination alone. The role of arthropods in pollination of crops, especially in arid and semiarid regions, is crucial. To remedy the situation, insects, mostly beetles, were imported and propagated in nest sites, which resulted in 20 % increase in crop yields and savings of about $115 million in dispensing with hand pollination. In tropics and subtropics, shrub and mild patches and roadside vegetation could help in sustaining pollinators and other beneficial arthropods.

Studies on management of plant pathogenic fungi and bacteria revealed the crucial role played by centipedes and millipedes in the process. The imbalance in the prey–predator relationship caused by inadequate number of centipedes and millipedes led to not only escalated plant diseases but also increased insect pests. The carnivorous arthropods which include cockroaches, spiders, mites, ticks, and all other insects which prey on smaller species maintain ecological balance. A good balance of these arthropods keeps the insect prey numbers under check.

In fact, ecological significance becomes economic importance as human beings find more use in the substances secreted or produced by different species of arthropods. Some of these are:

- Bees produce honey and their honeycombs contain beeswax.
- The pollen stored in honeycombs are rich mixture of vitamins, enzymes, and amino acids which have therapeutic benefits.
- Bee propolis, which is a dark resinous substance produced by bees as they feed on buds and barks of trees, was/is widely used for its antibiotic properties.
- Silk-producing arthropods, like those produced by caterpillars to protect their cocoons, were found to be strong enough to use and be woven into fabrics. Obviously, this discovery is the basis of ancient China's silk industry.
- In recent years, the spiders' web was discovered to have tensile strength; they became essential raw materials for Kevlar vests, fishing nets, surgical sutures, and adhesives as they contained natural antiseptics.
- The arthropod insect *Laccifer lacca* provides an organic resin known as shellac.

1.4.2.1 Ecological Role of Aquatic Arthropods

Aquatic arthropods comprise of members of sub phylum Crustacea such as shrimps, crabs, lobsters, prawns, water fleas, barnacles, and krill. The latter are small shrimp like aquatic creatures, and they represent the bottom of the aquatic arthropod's food chain. Arthropods in aquatic ecosystems play a vital role in the turnover of mineral elements and in regulating gaseous cycles. Aquatic arthropods that move about can easily migrate to places of food sources where they can find food to sustain their communities. Their appendages allow them to swim easily in the oceans' currents and movements. Krill, a basic food of larger aquatic arthropods, can be found abundantly almost anywhere, particularly where phytoplankton communities exist.

The main importance of these aquatic arthropod species to human beings is their function in the food chain and the indirect economic benefits mankind derives. For example, the blue crab provides livelihood to the coastal communities in the Gulf of Mexico, as major commercial capture fisheries provide jobs and other economic support. This has become the case in other tropical countries like India, Bangladesh, Sri Lanka, etc. Their annual catch is estimated at $40 million. However, due to the continuous spate of environmental problems, including the recent BP oil spill disaster in the Gulf, the presence of the blue crab in the communities' marsh edges and in sea grasses is said to be in critical conditions. Many aquatic arthropods are in peril due to pollution in major river systems.

References

Bergström J, Hou XG (2003) Arthropod origins. Bull Geosci 78(4):323–334
Budd GE (1993) A Cambrian gilled lobopod from Greenland. Nature 364(6439):709–711
Budd GE, Butterfield NJ, Jensen S (2001) Crustaceans and the Cambrian explosion. Science 294(5549):2047
Callaway E (2008) Fossilised shrimp show earliest group behaviour, new scientist, archived from the original on 15 October 2008, retrieved Oct 21, 2008
Chapman AD (2009) Numbers of living species in Australia and the world (PDF), 2nd edn. Australian Biological Resources Study, Canberra, pp 1–80
Cowen R (2000) History of life, 3rd edn. Blackwell Science, Malden, p 126
Dunlop JA (1996) "A trigonotarbid arachnid from the Upper Silurian of Shropshire" (PDF). Palaeontology 39(3):605–614
Engel MS, Grimaldi DA (2004) New light shed on the oldest insect. Nature 427(6975):627–630
Gillott C (1995) Entomology. Springer, London, pp 17–19. ISBN 0-306-44967-6
Gould SJ (1990) Wonderful life: the Burgess Shale and the Nature of History. Hutchinson Radius, London, pp 102–106
Harper D (2014) Arthropoda, Online Entomology Dictionary
Hou XG, Bergström J, Jie Y (2006) Distinguishing anomalocaridids from arthropods and priapulids. Geol J 41(3–4):259–269. doi:10.1002/gj.1050
James HT (2003) Arthropoda and related groups. In: Encyclopedia of insects. Academic Press, Amsterdam
Jeram AJ (1990) Book-lungs in a Lower Carboniferous scorpion. Nature 343(6256):360–361
Karanth KP (2015) An island called India: phylogenetic patterns across multiple taxonomic groups reveal endemic radiations. Curr Sci 108(10):1847–1851
Labandeira C, Eble GJ (2000) The fossil record of insect diversity and disparity. In: Anderson J, Thackeray F, van Wyk B, de Wit M (eds) Gondwana alive: biodiversity and the evolving biosphere. Witwatersrand University Press, Johannesburg
Lieberman RC (1995) Race, institutions, and the administration of social policy. Soc Sci Hist 19(4):514–515
Lin JP, Gon SM, Gehling JG, Babcock LE, Zhao YL, Zhang XL, Hu SX, Yuan JL, Yu MY, Peng J (2006) A Parvancorina-like arthropod from the Cambrian of South China. Hist Biol 18(1):33–45
Mc Menamin MAS (2003) *Spriggina* is a trilobitoid ecdysozoan (abstract), Abstracts with Programs. Geol Soc Am 35(6):105
Pisani D, Poling LL, Weiler LM, Hedges SB (2004) The colonization of land by animals: molecular phylogeny and divergence times among arthropods. BMC Biol 2:1
Regier, Shultz JW, Zwick A, Hussey A, Ball B, Wetzer R, Martin JW, Cunningham CW (2010) Arthropod relationships revealed by phylogenomic analysis of nuclear protein-coding sequences. Nature 463(7284):1079–1084
Resh VH, Carde RT (2009) Encyclopedia of insects, 2nd edn. Academic Press, Amsterdam. ISBN 0-12-374144-0
Ruppert EE, Fox RS, Barnes RD (2004) Invertebrate zoology, 7th edn. Thomson-Brooks, Belmont, pp 518–522
Schmidt-Rhaesa A, Bartolomaeus T, Lemburg C, Ehlers U, Garey JR (1999) The position of the Arthropoda in the phylogenetic system. J Morphol 238(3):263–285
Selden PA, Shear WA (1996) Fossil evidence for the origin of spider spinnerets. PNAS. doi:10.1073/pnas.0809174106, PMC 2634869, PMID 19104044
Shubin N, Tabin C, Carroll S (2000) Fossils, genes and the evolution of animal limbs. In: Gee H (ed) Shaking the tree: readings from nature in the history of life. University of Chicago Press, Chicago
Vollrath F, Selden PA (2007) The role of behavior in the evolution of spiders, silks, and webs (PDF). Annu Rev Ecol Evol Syst 38:819–846

Whittington HB (1971) Redescription of *Marrella splendens* (Trilobitoidea) from the Burgess Shale, Middle Cambrian, British Columbia. Geol Surv Can Bull 209:1–24

Whittington HB (1979) Early arthropods, their appendages and relationships. In: House MR (ed) The origin of major invertebrate groups, vol 12, The Systematics Association special. Academic, London, pp 253–268

Zhang XG, Siveter DJ, Waloszek D, Maas A (2007) An epipodite-bearing crown-group crustacean from the Lower Cambrian. Nature 449(7162):595–598

Soil Biodiversity and Arthropods: Role in Soil Fertility

2

D.J. Bagyaraj, C.J. Nethravathi, and K.S. Nitin

Abstract

Healthy productive soils are essential to meet the food requirement of humans and animals. Arthropods have important role in maintaining soil fertility. The major contribution of arthropods to soil is through decomposition and humification of all organic matter. In the soil, arthropods function as litter transformers, ecosystem engineers, and pulverizers. As much as 20 % of total animal litter input is processed by the activity of collembolans alone. Arthropods also stimulate mineralization of nutrients in soil. Soil practices in cultivated ecosystems significantly alter arthropod community which in turn has significant effect on soil productivity. Arthropods facilitate soil processes. Hence, understanding soil arthropod communities will prove useful in developing management plans for both wild and cultivated ecosystems.

Keywords

Decomposition • Soil arthropods • Soil biodiversity • Soil fertility

D.J. Bagyaraj
Department of Agricultural Microbiology, Gandhi Krishi Vignana Kendra (GKVK), University of Agricultural Sciences, Bengaluru 560065, Karnataka, India

C.J. Nethravathi (✉) • K.S. Nitin
Division of Entomology and Nematology, Indian Institute of Horticultural Research (IIHR), Hessaraghatta Lake Post, Bengaluru 560089, Karnataka, India
e-mail: nethraent9@gmail.com

A.K. Chakravarthy, S. Sridhara (eds.), *Economic and Ecological Significance of Arthropods in Diversified Ecosystems*, DOI 10.1007/978-981-10-1524-3_2

2.1 Introduction

The living organisms in soil range from microorganisms, small and large invertebrates to small mammals. Arthropods represent as much as 85 % of the soil fauna in species richness. They comprise a large proportion of the meso- and macrofauna of the soil. Macrofauna contribute to improve soil structure, aeration, and water infiltration. They predate on soil organisms and help to maintain biological equilibrium in soil. Mesofauna are important plant pathogens. Microfauna are important predators of bacteria and algae, thus regulating their population in soil. Arthropods function on two of the three broad levels of organization of the soil food web: they are plant litter transformers or ecosystem engineers. Litter transformers fragment, or comminute, and humidify ingested plant debris, which is deposited in feces for further decomposition by microorganisms, and foster the growth and dispersal of microbial populations. Large quantities of annual litter input may be processed, thus, for example, up to 60 % by termites. The comminuted plant matter in feces presents an increased surface area attack by microorganisms, which, through the process of mineralization, convert organic nutrients into simpler, inorganic compounds easily available to plants. Ecosystem engineers alter soil structure, mineral and organic matter composition, and hydrology. The burrowing by arthropods, particularly the subterranean network of tunnels and galleries that comprise termite and ant nests, improves soil porosity providing adequate aeration and water-holding capacity belowground, facilitates root penetration, and prevents surface crusting and erosion of topsoil. Also, the movement of particles from lower horizons to the surface by ants and termites aids in mixing the organic and mineral fractions of the soil. The feces of arthropods are the basis for the formation of soil aggregates and humus, which physically stabilize the soil and increase its capacity to store nutrients. Soil organisms are also responsible for many services like nutrient cycling, control of pests and diseases, formation of soil, degradation of wastes and harmful chemicals, and production of useful by-products like food, fuel, fiber, etc. In recent years industrial microbiologists have started looking into the soil for organisms capable of producing antibiotics, vitamins, hormones, and enzymes. The results obtained so far have revolutionized the industry in many countries. The objective of this paper is to review the state of knowledge on soil arthropods and its relevance to soil fertility and sustainable land use.

2.2 Pedogenesis and Composition

A handful of soil contains some minerals, organic matter, air, water, and living biota. Soil is one of the fundamental components for supporting life on earth. Soils originate and accumulate in a sequence of events that mark the stages of ecological succession, the development of biotic communities. Soil formation, or pedogenesis, involves a set of physical, chemical, and biotic processes. The properties of soils arise from the interactions of five basic factors: parent material, topography, climate, biota, and time (Jenny 1980).

The living organisms in soil range from microorganisms, small and large invertebrates, and small mammals. One teaspoon of garden soil contains thousands of species, millions of individuals, and a hundred meters of fungal networks. Bacterial biomass is particularly impressive and can amount to 1–2 t/ha which is roughly equivalent to the weight of one or two cows. Thus, for the soil biologists, soil is a living organism in an organic/mineral matrix. Soil biologists take two basic approaches for studying organisms: (i) taxonomic and (ii) process oriented (Bardgett et al. 2005; Gardi and Jeffrey 2009). Soil organisms have been classified on the basis of body width into microflora (1–100 μm, e.g., bacteria, fungi), microfauna (5–120 μm, e.g., protozoa, nematodes), mesofauna (80 μm^{-2} mm, e.g., Collembola, Acari), and macrofauna (500 μm–50 mm, e.g., earthworms, termites). The inventories of soil organisms have been considerably limited compared to above ground organisms, such as vascular plants. Another trend is that soil organisms of greater size (i.e., macrofauna) have received far greater attention than soil organisms of smaller size, despite greater diversity of the latter. The soil biota includes the macrofauna, mesofauna, microfauna, algae, fungi, prokaryotes, and rest of the microbiota, such as mycoplasmas, viruses, viroids, and prions (Turbe et al. 2010). Macrofauna include earthworms, mollusks, millipedes, earwigs, and insects. The major roles of macrofauna in soil are to accelerate organic matter breakdown and to mix organic matter and soil together. They contribute to improve soil structure, aeration, and water infiltration.

Biomass in soil means living organic material. While macrofauna may be big, they do not necessarily have the greater biomass in soil. The size and contribution to biomass are not necessarily related; neither are population density and importance to biological activity in soil. For example, two earthworms may have more effect on soil than a billion protozoa. Arthropods have long been recognized as important in the functioning of soil ecosystems, and a vast literature accordingly has accumulated, and principal roles played by arthropods in the processes that maintain soil fertility have been reviewed exhaustively (Culliney 2013).

2.3 Functional Grouping of Soil Biodiversity

Organisms found in soil can be categorized into three broad functional groups, viz., chemical engineers, biological regulators, and ecosystem engineers. Most of the species in soil are microorganisms such as bacteria, fungi, and protozoans, which are the chemical engineers of soil, responsible for the decomposition of plant organic matter into nutrients readily available for plants, animals, and humans (Gardi and Jeffrey 2009).

Soils also comprise a large variety of small invertebrates such as nematodes, pot worms, spring tails, and mites, which act as predators of plants, other invertebrates, or microorganisms, by regulating their dynamics in space and time. Most of these biological regulators are relatively unknown, contrary to the larger invertebrates, such as insects, earthworms, ants, termites, ground beetles, and small mammals, such as rats and mice, moles, and voles, which show fantastic adaptations to living in a dark below ground world. For instance, about 50,000 mite species are known to be soil living, but it has been estimated that up to 1 million species could be included in this group. Earthworms, ants, termites, and small mammals are also ecosystem engineers, since they modify or create habitats for smaller soil organisms. In this way, they also regulate the availability of resources for other soil organisms. Moles, for instance, are capable of extending their tunnel system by 30 cm per hour, and earthworms produce soil casts at rates of several hundreds of tonnes per ha each year (Barrios 2007).

Chemical engineers, biological regulators, and ecosystem engineers act mainly over distinct spatiotemporal scales, providing a clear framework for management options. This is because the size of organisms strongly determines their spatial aggregation patterns and dispersal distances. Thus, chemical engineers are typically influenced by local-scale factors, ranging from micrometers to meters and short-term processes, ranging from seconds to minutes (Brussaard et al. 2007).

2.4 Factors Influencing Soil Biodiversity

The activity and diversity of soil organisms are regulated by a hierarchy of abiotic and biotic factors. The main abiotic factors are climate, including temperature and moisture, soil texture and soil structure, and salinity and pH. Overall, climate influences the physiology of soil organisms, such that their activity and growth increase at higher temperature and soil moistures. As climatic conditions differ across the globe and at some places, between seasons, the climatic conditions to which soil organisms are exposed vary strongly. Soil organisms vary in their optimal temperature and moisture ranges, and this variation is life-stage specific, e.g., larvae may prefer other optima than adults. For instance, for spring tails, the optimum average temperature for survival is just above 20 °C, and the higher limit is around 50 °C, while some bacteria survive up to 100 °C in resistant forms. Soil texture and structure also strongly influence the activity of soil biota. For example, medium-textured loam and clay soils favor microbial and earthworm activity, whereas fine-textured sandy soils, with lower water retention potentials, are less favorable. Plants can strongly influence the activity and community composition of microorganisms in the vicinity of their roots, called the rhizosphere (Finlay 2006; Luster et al. 2009). In turn, plant growth may be limited or promoted by these soil microorganisms. Added to this, plants can influence the composition, abundance, and activity of regulators and ecosystem engineers, whereas these species in turn can influence vegetation composition and productivity. Finally, soil organisms induce plant defense responses to above ground pests and herbivores, and the above ground interactions can feed back in a variety of ways to the biodiversity, abundance, and activities of the soil organisms. In addition, within the soil food webs, each functional group can be controlled by bottom-up or top-down biotic interactions. Top-down effects are mainly driven by predation, grazing, and mutualist relationships. Bottom-up effects depend largely on competitive interactions for access to resources (Boellstorff 2008; Schils et al. 2008).

2.4.1 Organic Matter Decomposition

Organic compounds that reach the soil by way of animal and plant residues are made up of simple sugars, starch, cellulose, pectins, proteins, fats, waxes, lignin, phenols, tannins, alkaloids, pigments, and other products. The huge mass of the organic matter added to the soil is immediately acted upon by the soil biota (Bardgett et al. 2005):

- Macrofauna (millipedes, mollusks, wood lice, and fly larvae) attack the litter. They puncture the leaf epidermis in a process called fenestration and open the leaf to microorganisms. Larger the material's surface area, the more available it is for microbial attacks. If you exclude macrofauna from soil, organic matter decomposition slows. An interesting experiment that showed this effect involved incubating leaf tissue in mesh bags that excluded different organisms based on size and burying the bags in soil. About 95 % decomposition occurred in the 7.0 mm mesh that did not exclude compared to 35 % decomposition in the 0.5 mm mesh that excluded all but the smallest invertebrates and microbes.
- Mollusks, wood lice (isopods), millipedes, earwigs, fly larvae, and earthworms macerate the leaf litter and pulverize it.
- Earthworms, insects, and other burrowing creatures transport the litter into soil.
- Earthworms and pot worms further macerate the litter and mix it with soil.
- Microbial action turns the litter into an integrated component of the soil.

2.4.2 Symbiotic Nitrogen Fixation

Nitrogen is one of the major elements required for crop production. Nitrogen is present to the extent of 80 % in the atmosphere. Field trials in India have shown that rhizobial inoculation can increase yield of grain legumes by 20–50 % (Bagyaraj 2011).

2.4.3 Nonsymbiotic Nitrogen Fixation

A number of free-living organisms inhabiting soil are capable of fixing atmospheric nitrogen. Bacteria of genera *Azotobacter*, *Beijerinckia*, *Derxia*, and cyanobacteria (blue-green algae) are well known among these. Besides the ability to fix atmospheric nitrogen, *Azotobacter* is also known to synthesize biologically active substances such as B-vitamins, IAA, and gibberellins (Bengtsson et al. 2005).

2.4.4 Phosphorus Mobilization

Although a soil may have adequate quantities of phosphorus, it may be present in bound form unavailable to plants. Many fungi and bacteria (like *Aspergillus*,

Penicillium, and *Bacillus*) are potential solubilizers of bound phosphates as revealed by experiments in pure culture. These organisms produce organic acids like citric, succinic, lactic, and oxalic acids responsible for the solubilization of insoluble forms of phosphorus. A commercial preparation called "Phosphobacterin" containing *B. megaterium* has become popular. Another area attracting the attention of soil biologists is mycorrhiza. In nature the roots of most plants are invaded by fungi and are transformed into mycorrhiza which means "fungus root." The host and the fungus live in intimate symbiotic relationship. The fungi help in the phosphorus nutrition of plants through increased surface area of absorption, offer protection against moisture stress and some of the soil-borne plant pathogens, and enhance rooting and survival of cuttings through production of growth hormones (Bagyaraj 2014).

2.4.5 Biocontrol of Plant Pathogens

Eco-friendly organic farming technologies for plant protection have been gaining importance in recent years. Some of the plant diseases that can be controlled by antagonistic fungi and bacteria are as follows: rice seeds treated with *Pseudomonas aeruginosa* and *P. putida* reduced sheath blight infection (*Rhizoctonia solani*) in rice by 65–72 % in comparison to untreated check; *Pseudomonas fluorescens* was also found effective against banded leaf and sheath blight fungus (*R. solani* f. sp. *sasakii*). *Trichoderma harzianum* as fungal antagonist proved effective against *Macrophomina phaseolina* (charcoal rot) in several plant species (Powlson et al. 2012).

2.4.6 Pesticide Degradation

Degradation of pesticides in soil, by soil microorganisms, is another area of research in which scientists are actively engaged today. Some pesticides can be used as carbon and energy substrates by microorganisms and thus result in the disintegration of the pesticide (Powlson et al. 2012).

2.4.7 Extension to Other Fields

In recent years industrial microbiologists have started looking into the soil for organisms capable of producing substances including antibiotics, vitamins, hormones, and enzymes. The results obtained have revolutionized the industry in many countries. Several actinomycetes and some fungi and bacteria have been used in the industry for the production of antibiotics. *Torulopsis* spp. and *Ashbya gossypii* produce thiamin and *Streptomyces olivaceus* and *Streptomyces griseus* produce vitamin B_{12} in the fermentation broth. *Gibberella fujikuroi* has been utilized for the production of the plant hormone, gibberellic acid. Commercial proteolytic enzymes are produced using the soil fungus *Mortierella* spp. (Constanza et al. 1997).

2.5 Economic Value of Soil Biodiversity

In order to allow for performing cost-benefit analyses for measures to protect soil biodiversity, some economic estimates of the ecosystem services delivered by soil biodiversity need to be provided. Several approaches exist. The valuation can be based on the prices of the final products, such as food, fibers, or raw materials, or be based on the stated or revealed preference. The stated preference methods rely on survey approaches permitting people to express their willingness to pay for the services provided by biodiversity and its general contribution to the quality of life (e.g., aesthetical and cultural value). Alternatively, cost-based methods can be used; the value of a service provided by biodiversity is evaluated through a surrogate product. Thus, the "damage avoided" cost can be estimated, for instance, by the amount of money that should be spent to repair the adverse impacts arising in the absence of a functioning ecosystem (e.g., in the case of soil biodiversity, the cost of avoided floods). It has been estimated by economists that the world economic benefit of soil biodiversity is to the tune of several billion US dollars per year. For example, for waste recycling, it is 760×10^9 USD/year, soil formation 25×10^9 USD/year, nitrogen fixation 90×10^9 USD/year, degradation of chemicals 121×10^9 USD/year, pest control 160×10^9 USD/year, pollination 200×10^9 USD/year, etc. (Pimentel 1997; Barrios 2007; Brauman et al. 2007).

2.5.1 Soil Arthropods

Soils may harbor an enormous number of arthropod species. By one estimate, the soil fauna may represent as much as 23 % of all described organisms, with arthropods comprising 85 % of that number (Decaens et al. 2006). However, accurate figures have been difficult to come by, hampered, at least in part, by limitations in sampling methodology (Stork and Eggleton 1992). Arthropods comprise a large proportion of the meso- and macrofauna of the soil, animals with body lengths ranging from about 200 μm to 16 cm or more (Wallwork 1970). Of the hemiedaphon and euedaphon, those organisms live within the litter/humus boundary and lower in the soil profile (Eisenbeis and Wichard 1987).

2.5.1.1 Nematodes

One handful of soil from almost any area, among other forms of life, possesses elongate, threadlike, active animals which include the nematodes and other related forms. The word nematode is derived from the Greek root "NEMA" meaning thread. It is probably one of the oldest existing life forms on the basis of the fossil specimens discovered in Scotland (Maggenti 1981).

Based on their habitat, nematodes can be grouped into marine, soil, and freshwater forms. The marine nematodes form the largest group, about 50 % of the known nematode population. Soil nematodes, including the free-living, plant and animal parasitic, form the remaining 50 % of the total nematode species. The free-living forms have a variety of food habits. Some are predacious on soil microfauna including nematodes. Others use algae, fungi, and bacteria as their food source.

Nematode body is usually cylindrical to fusiform, tapering toward either end and somewhat wide in the middle. The females of some genera like *Heterodera* and *Meloidogyne*, however, may be spherical or oval. The soil and phyto nematodes may be 0.5–12 mm long and 20–30 μm wide. While the body is usually transparent, some may be, and rarely so, whitish or yellowish. Other colors may be due to the color of the ingested food. Externally the body is covered by a transparent, tough, cuticle invaginated at body openings such as oral, anal, excretory, and vaginal. In cross section, the body is more or less circular with its surface exhibiting pseudo segmentation on account of in folding of the cuticle. The nematode body is generally bilaterally symmetrical.

Organic matter is an essential component of all agricultural soils and influences the general microbial population of soil. Several hypotheses for the possible mechanisms involved in the management of nematodes have been put forth. One of these is that the products from decomposing organic matter are directly toxic. The second is that the microbivorous nematodes rapidly reproduce on the abundant bacteria and the decomposing products and help stimulate activity of natural decomposition involving changes in the soil. These changes promote host vigor by more efficient utilization of the soil nutrients (Sayre 1971).

There are countless microorganisms capable of decomposing plant and animal residues in soil. A succession of these microorganisms mediates a stepwise degradation of organic matter. The resulting products like fatty acids, enzymes, toxic gases (H_2S, NH_3), etc. have been found to be toxic, antibiotic, inhibitory, or attractive to nematodes.

Possibly the only but very interesting example of the influence of animal exudates is an odoriferous, volatile compound by the insect, *Scaptocoris talpa* Champ, protecting tomato seedlings from the attack of root-knot nematodes (Timonin 1961). Several plants (marigold, crucifers, asparagus, and citrus) have been demonstrated to have chemicals in their roots antagonistic to nematodes (Winoto Suatmadji 1969). The possibility of utilization of these agents in nematode management could materialize only when these principal chemicals in the exudates are characterized and their potential as nematicides explored.

The soil flora and fauna predacious on plant parasitic nematodes include fungi (*Arthrobotrys*, *Dactylaria*, *Dactylella*, *Catenaria*), other nematodes (*Diplogaster*, *Seinura*, *Mononchus*), Tardigrades (*Hypsibius myropus*) turbellarians (*Adenoplea* spp.), Collembola (*Isotoma* spp.), mites (*Onychiurus armatus*), enchytraeids, and protozoa (*Theratomyxa weberi*), whereas parasitic forms include viruses, some fungi (*Dactylella oviparasitica*), protozoa (*Duboscquia penetrans*), and bacteria (Barron 1977; Stirling and Mankau 1978).

Abiotic ecological factors influencing nematodes in soil environment include soil structure (particle size distribution, soil pores, and mechanical strength of soils), soil water (surface tension, suction, moisture characteristic, moisture profile, hysteresis, movement of water and osmotic potential), soil aeration, chemical properties of soil (soil reaction and soil chemicals), and soil temperature (Norton 1978).

2.5.1.2 Annelids

The terrestrial annelids are common earthworms, included under order Oligochaeta, the major soil fauna of the world. They have widespread distribution and show a high degree of adaptability to the prevailing soil conditions. Among the land inhabiting Oligochaeta, *Microdrili*, and other members belonging to terrestrial Enchytraeidae, are encountered in many habitats where true earthworms also occur.

2.5.1.3 Enchytraeids

The Enchytraeids differ from the lumbricids by their small size (5–15 mm) and whitish color. Overgaard Nielsen and Christensen (1959) have given the classification of Enchytraeids. They are mostly found in the top layer of somewhat moist, root-penetrated soil and in the litter of forests. They are absent in thick loamy soils having few pore space and in badly aerated wet soils. Their chief diet consists of dead plant residues. They can digest undecomposed plant residues but completely decomposed plant material is unsuitable. Occasionally Enchytraeids feed on pupae and eggs of insects and on small nematodes (Jegen 1920). Their distribution is mostly dependent on the type of soil.

2.5.1.4 Earthworms (Fig. 2.1)

Earthworms are considered as the most beneficial organisms to agriculture and are called "Nature's ploughman" (Darwin 1881). A large proportion of the energy of mature worms is used in cocoon production and the life span of the worms is directly related to the number of cocoons produced (Kale and Krishnamoorthy 1981). The size, shape, and color of the cocoons are species specific. The juveniles resemble the adults except in size. The time of attainment of reproductive phase varies in different species. Parthenogenesis in earthworms is recorded.

Earthworms are the major secondary decomposers in the soil (Fig. 2.1). The degradation of the leaf material commences from the time it detaches from the plant and drops to the ground to add to the litter. The worms feed upon litter materials partially degraded by microbes. Earthworms are found in association with other soil invertebrates like soil turbellarians, soil insects, and myriapods. Scarabaeid beetles, millipedes, centipedes, Turbellarians are considered as their natural enemies (Dindal 1970; Edwards and Lofty 1972).

2.5.1.5 Worm Cast Production and Physiochemical Properties of Soil

Worm casts are found in large amounts on soil surface during rainy season (Gates 1961; Dash and Patra 1979). Even in dry season, the dried-up crumbs of castings can easily be differentiated from the surrounding soil. The amount of casting produced on the soil surface is an index in assessing earthworm activity. The nature of worm cast is species specific. The earthworms that drill extensive burrows in the soil may use the casts with mucus to pack the walls of burrow. The castings of such burrow forming species are usually found on the burrow openings. *Pheretima posthuma* produces pellet-like casts and *Perionyx millardi's* castings are threadlike.

Earthworms (Annelida)

Pseudo scorpion (source: arthurevans.wordpress.com)

Snail (Mollusca)

Termites

Fig. 2.1 Macroarthropods

Hoplochaetella khandalaensis produces thick and long winding column which will be a hollow mound of 5 cm long and 2.5 cm wide. The biggest recorded casting is that of *Notoscolex birmanicus* from Burma. The dry weight of one casting of this worm weighed 1.6 kg after drying for 4 months (Tembe and Dubash 1961). *Perionyx excavatus* which does not form definite burrows in soils leaves the castings as minute noodle-like structures. The castings of *Lampito mauritii* are fine granular mounds of soil. *Pontoscolex corethrurus* and *Pheretima elongata* excrete the ingested soil as sticky, thick lumps on soil surface. The castings of *P. excavatus* are completely on the surface, whereas *P. elongata* and *P. corethrurus* use part of their castings to farm the burrow lining.

Physical-chemical properties of the castings differ from soil to soil. The large tower like castings of *Eutypheous woltoni* had coarser fractions than the surrounding soil. Small-sized castings were found in the same region and were superior to large castings and also the parent soil in their rate of percolation and dispersion coefficient. The aggregate size also differed in these two castings. The chemical composition of both the castings was found more favorable to plant growth than surrounding soil.

The pH of the castings was found higher than the surrounding soil. Total nitrogen, organic matter, nitrate nitrogen, phosphorous, potassium, sodium, and magnesium were at a higher level in the castings than in rest of the soil (Kale and Krishnamoorthy 1979; Dash and Patra 1979). Earthworms utilize the organic matter as food and release part of carbon as Co_2 in the process of respiration. The production of mucus and nitrogenous excrements enhances the level of nitrogen, and this in turn leads to the lowering of C:N (Senapati et al. 1980). The differences observed in the oxidizable carbon and other minerals in the castings are mostly due to selective feeding habit of the worms. They selectively feed on the decomposing particulate matter and defecate the partially digested material into the soil surface with coating of mucus (Kale and Krishnamoothy 1980). The castings of *Pontoscolex corethrurus* were found rich in soluble calcium and carbonates. Soluble carbonates contribute to the exchangeable base content of the castings (Kale and Krishnamoorthy 1980).

Earthworms are considered as biological converters of chemicals. Patel and Patel (1959) observed the high susceptibility of earthworms to soapnut (*Sapindus laurifolius*) extract. In fields, the application of soapnut extract was most effective against earthworms. The survivability, activity, and fecundity of the worms are highly influenced by residue concentrations. The concentration of carbaryl up to 100 ppm was favorable for worms (Kale and Krishnamoorthy 1979). In New Zealand, the effective control of chafer grubs was brought about by earthworms in transferring the pesticides into lower soil layers (Edwards 1973).

2.5.1.6 Snails and Slugs (Fig. 2.1)

Snails and slugs are more or less similar in structure except that the slugs lack the external hard spherical protective shell and their mantle is smoother. The color of slugs varies from gray to black. Both withdraw their body into shell when disturbed.

About 80,000 species of snails and slugs are described from the phylum Mollusca and are distributed throughout the world, inhabiting aquatic (including marine and freshwater) and terrestrial ecosystems. They are herbivorous, carnivorous, and also fungivorous. In terrestrial habitats, they live mostly in sandy, moist, damp places and feed on vegetable matter either live or dead. They are soft-bodied, unsegmented animals having an anterior head, a ventral muscular foot, and a dorsal visceral mass. Their body is surrounded by a specialized fleshy area known as mantle characteristic to this group, which is sheltered within an external limy shell.

The pulmonate slugs and snails feed voraciously on plant material, preferring the more succulent parts. Some slugs, like *Limax* spp., feed on bulbs, tubers, or roots, while *Arion subfuscus* feeds on fungi. In certain species powerful cellulase brings about complete digestion of the plant tissues, assuming that carbohydrates, lipase, and proteases are still being retained. Armed with this digestive tool, slugs and snails are able to cut and ingest comparatively large pieces of leaf and feed rapidly.

Rate of growth and reproduction may be high, especially in the tropics. The Giant African Snail, *Achatina*, is particularly formidable as a phytophagous pest. The snail *Partula*, occurring in Pacific islands, apparently subsists exclusively on the mycelia of fungi that grow on decaying plant materials.

2.5.1.7 Microarthropods

Soil arthropods measuring up to 10 mm in length can be considered as microarthropods. They are also considered as members of the mesofauna of the soil. These include Protura, Diplura, and Collembola of the class Insecta, Symphyla and Pauropoda of the class Myriapoda, and Tardigrada, Copepoda, and Isopoda of the class Arachnida.

Tardigrada

Tardigrada are peculiar arthropods, with worthy surface, possessing four pairs of stumpy, claw-bearing legs; the snout is conical. They normally live in moss cushions, but active tardigrades have been recorded from meadow soils and leaf litters. Their food consists partly of organic residues and partly of the contents of moss cells. They are also known to feed on nematode. *Macrobiotus* sp. feeding on *Rotylenchulus reniformis* and *Aphelenchus avenae* has been reported from India (Narayana Swamy and Nanje Gowda 1980). Females lay thick-shelled eggs in their skin carts. These anabiotic eggs are resistant to desiccation and may be blown up till the moisture is available. Species of the genera *Macrobiotus* and *Lydella* commonly occur in the soil.

Terrestrial Isopoda

The terrestrial isopods called woodlice (Oniscoidea, crustacean) play an important role in the breakdown of litter and wood residues (Kunhelt 1976). More than 3500 species in 518 genera have been described (Schmalfuss 2003). Despite their diversity, these animals are imperfectly adapted to a terrestrial existence. In particular, a set of structural (e.g., permeable cuticle) and physiological (gills) traits little modified from a marine ancestor means that the maintenance of water balance is of paramount importance to survival and is largely achieved through behavioral means. Isopods attain their greatest abundance in unmanaged temperate grasslands, numbers typically ranging from about 500–1000 m^{-2} (Curry 1994). A density of 7900 m^{-2} was estimated for *Trichoniscus pusillus* Brandt in scrub grassland in Britain, the maximum ever recorded for an isopod species (Sutton 1980).

They abound in decaying wood. The abdominal legs are adopted for aerial respiration. Their endopodites bear delicate branchiae traversed by minute tubes called "pseudotrachea." They feed on a variety of dead and decaying matter. They do not have an epicuticular wax layer, and, therefore, moisture is an important limiting factor in their distribution. In evergreen forests where abundant moisture is available, even during dry periods, they may play an important role in the breakdown of litter. Members of the genera *Porcellio*, *Oniscus*, and *Armadillidium* are common in the soil.

2.5.1.8 Arachnida

Pseudoscorpions

Pseudoscorpions (Figs. 2.1 and 2.2) are small, predaceous arachnids preying on small arthropods, psocids, collembolans, and mites. They live in habitats like soil cover, beneath barks, bird nests, etc. in small numbers. They easily escape notice by withdrawing their legs and palps playing possum.

Pseudoscorpions display elaborate courtship. Normally the female pseudoscorpions build nests during brooding. The nest is constructed with silk secreted by the silk glands present in the chelicerae which come out through the galea. Some species dwelling in the soil like *Compsaditha indica* do not build nests, and the brood sacs are firmly attached to the underneath of their abdomen.

The eggs undergo development within the brood sac and the time taken for the same varies from 3 days (*Compsaditha indica*) to 21 days (*Stenatemnus indicus*). During the last instar, the young ones receive copious secretion from the ovary, rich in protein yolks. They suck this secretion with the pumping organs which would have developed by now and get themselves bloated up as small balloons. Soon the brood sac gives way and protonymphs emerge out. The number of eggs carried by the female varies from 4 to 24.

Soil Acari

The Acari (Fig. 2.3) of the soil includes members that feed on dead plant materials and microflora (bacteria, fungi). In addition, species of Prostigmata and Mesostigmata prey upon micro- and mesofauna (e.g., nematodes, collembolans, enchytraeid worms). The oribatids are the numerically dominant group of Acari.

Populations of oribatids in the order of 105–106 individuals m^{-2} have been recorded in different forest types (Petersen and Luxton 1982), with densities higher in coniferous than in deciduous soils (Wallwork 1983). In desert ecosystems, their numbers comparatively are much reduced (Wallwork 1982), and roles in the turnover of organic matter insignificant (Whitford et al. 1983). A major abiotic factor constraining the distribution of oribatids is adequate moisture. With more than 9000 species in 172 families, most of which inhabit the soil/litter system (Norton and Behan-Pelletier 2009), the oribatids are considered the most successful of all soil arthropods. Species tend to exhibit low metabolic rates, slow development, lengthy life cycles, low numerical response capability, generally stable population densities, low fecundity, and iteroparity (Norton 1994). These traits suggest *K*-selection and are probably influenced in part by the low nutritional quality of the diet. Parthenogenesis may be common.

On the basis of vertical stratification, three "life forms" of Acari can be recognized, viz., soil forms, wandering forms, and arboreal forms. Soil forms in contrast to the other two forms live restrictedly in soil environment in a wide sense. Although soil also serves as a shelter for ticks during their life span, these are not considered as soil forms. On the basis of habitat, soil forms may be further grouped into two categories: edaphic, which live in the deeper soil layer (e.g., *Epilohmannia*), and hemiedaphic, which are inhabitants of litter and upper soil layers (e.g., *Scheloribates*).

Fig. 2.2 Soil arthropods

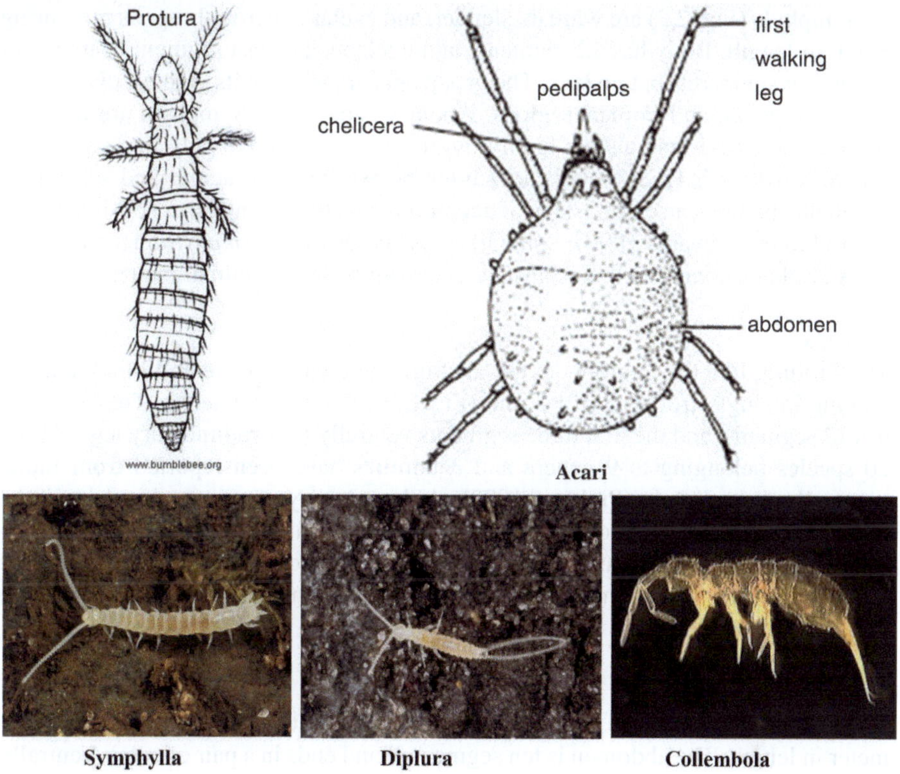

Fig. 2.3 Microarthropods

The feeding ecology of oribatids is diverse. Four main groups, based on modes of feeding, are commonly recognized: macrophytophages, which feed mainly on decaying higher plant material and rarely on fungi; microphytophages, those types feeding on fungi, bacteria, and other microflora; panphytophages, which have an expanded diet breadth, including plant matter as well as fungi; and coprophages, the diet of which includes fecal material. The majority of oribatids are obligate or facultative fungivores (Wallwork 1983). These animals have been contributing to the breakdown of plant litter since early in the Pennsylvanian Epoch of the Carboniferous period, perhaps 316 million years ago.

Symphyla

The Symphyla are small group of arthropods, with a reported 208 species in 13 genera and 2 families (Chapman 2009; Szucsich and Scheller 2011). Populations, however, may be large in some environments, in the order of 103–104 individuals / m^{-2}, and reach the highest densities in cultivated soils. Species also are common in grassland and forest soils. By one estimate, they may represent as much as 86 % of the total myriapod population in some soils, but are often overlooked because of their small size and wide dispersion through the soil profile.

Symphyla (Fig. 2.3) are whitish, slender, and eyeless microarthropods measuring 5 mm in length. Body has 12 segments and a telson. The last segment bears a pair of cerci or spinnerets but no legs. The group appears to reach its greatest diversity in warm temperate and tropical regions. About 12 species of Symphyla are reported from India. These animals are highly hygrotactic and survive only in a soil atmosphere of 100 % R.H. Symphyla are said to be extremely voracious and will attack vegetable matter at an earlier stage of decomposition than many other soil-inhabiting invertebrates (Edwards 1990). Some like *Scutigerella* and *Symphyllia* feed on fresh roots of plants. Scutigerellids consume decaying plant and animal matter.

Protura

The Protura, like the Pauropoda, are also minute creamy white and blind animals ranging in length from 0.5 to 1.5 mm (Fig. 2.3). They lack antenna. The abdomen has 12 segments and the first three segments ventrally bear rudimentary legs. About 20 species belonging to 9 genera and 3 families have been reported from India, especially from Kerala (Prabhoo 1972). Although they are more abundant in the forest soils, they are also common in grassland and plantation soils. They are apparently rare or absent in heavy soils. The Protura are known to feed on mycorrhizal fungi and for this reason they are likely to be more abundant in the feeder root zone of the plants.

Diplura

The Diplura (Fig. 2.3) are whitish, slender, blind insects reaching about one centimeter in length. The abdomen is ten segmented and ends in a pair of cerci. Ventrally on the abdomen segments 1(2)-7, coxal styli and bladders are present. Nine genera belonging to two families have been reported from India. *Japyx* is more widely distributed than *Anajapyx*, *Campodea* and *Heterojapyx* (Ghaisas and Ranade 1981). *Projapyx* is very rare. Both predatory and detritivorous forms are present under Diplura.

Collembola

The Collembola or the springtails are characterized by a six segmented abdomen bearing median appendages ventrally, i.e., ventral tube, the tenaculum, and the furcula (Fig. 2.3). The furcula and tenaculum may be reduced or absent in some families like Onychiuridae and Neanuridae. They formerly classified as primitively wingless insects (Boudreaux 1979), but now widely recognized as a lineage closely related to, but distinct from, the Insecta (Giribet and Edgecombe 2012). About 6500 species in 18 families have been described (Hopkin 1997). Like the oribatids, they also are extremely abundant in soil and leaf litter, with densities typically on the order of 104–105 individuals /m^{-2} and, again, higher in coniferous forests (Petersen and Luxton 1982), but are more numerous than oribatids in many soils. Agricultural soils may be rich in Collembola (Christiansen 1964). Edaphic species tend to be parthenogenetic (Hopkin 1997), life-history trait characteristic of animals living in stable environments. Average fecundity typically ranges between 50 and 100 eggs per female; depending on climate, there may be one to four generations annually.

Life spans of species living within the soil-litter system range between 2 and 12 months or more. Like soil-dwelling oribatids, Collembola require a soil atmosphere approaching saturation. The diet of Collembola is of considerable variation, including moss protonema, bacteria, fungal hyphae and spores, algae, protozoans, arthropod feces, pollen, decaying plant materials and humus, other Collembola (living or dead), and stored products. Species are divided between those that masticate their food and those that are fluid feeders. Majority of species are fungivorous (Hopkin 1997).

2.5.1.9 Millipedes and Centipedes

Diplopoda

Millipedes vary in body length from a few mm to 20 cm, with a minimum of 9 pairs of legs to a maximum of 200 pairs. Because of the wavelike motion of these legs, the animals appear to have thousand legs, hence their common name. About 12,000 species of millipede have been described and assigned to 2947 genera (Sierwald and Bond 2007). Millipedes are widely distributed; they are essentially the inhabitants of the forest floor (Blower 1951), with mull-type humus, and also are numerous in deciduous forests with a more humus formation. They tend to be more abundant and diverse in calcareous soils, in fairly moist habitats, and typically in the upper soil horizons. Densities of 1000–3000 /m^{-2} have been recorded. Individuals exhibit considerable longevity. Females of *Glomeris marginata* (Villers), for example, are known to live as long as 11 years. Fecundity as high as 2000 eggs per female has been recorded in some species (Hopkin and Read 1992).

Millipedes (Fig. 2.2) are detritivores, enrich soil system, acting as agents of decomposition by feeding on dead plant matter, such as leaf litter and wood, some also browsing on fungal mycelia and as accelerators in the nutrient release. Decomposition of the leaf litter by millipedes is by fragmentation and addition of microflora through fecal pellets (Kubiena 1955). The release of mineral nutrients into the soil is by feeding and defecation. Trace fossils attributable to burrowing by millipedes and including presumed fecal matter have been found in strata of late Ordovician (Richmondian) age, indicating that these animals have been active in soil processes for perhaps 445 million years (Retallack and Feakes 1987).

Among millipedes, the polydesmids digest woody plant residues. Millipedes belonging to the genus *Fontaria* are mull formers where as *Jonespeltis splendidus* is a humus-forming agent in the soil (Bano et al. 1976). Millipedes being largely responsible for mull production consume large quantities of litter and are instrumental in bringing down the carbon to nitrogen ratio from 30:1 to 10:1 (Blower 1956). With regard to ash mineralization, ash content increment is found in the excrements of the millipedes, *Glomeris marginata* and *Narceus annularis*. Increase in mineral content is also seen in the excrements of *Jonespeltis splendidus* (Bano and Krishnamoorthy 1977). Millipedes apart from being the litter feeders are also coprophagous. Coprophagy also leads to mineralization and improves carbon: nitrogen ratio in the soil. The feces of millipedes with high pH facilitate the growth and concentration of nitrogen-fixing bacteria.

Chilopoda

Centipedes are widely distributed in moist habitats throughout the tropical and temperate regions of the world. These are predominantly woodland species but are also common in grasslands and moorlands (Raw 1967). Centipedes are nocturnal and live in dark obscure place, under fallen logs, and in crevices in the soil. Since they cannot burrow into the soil, they utilize available crevices for shelter. Many species of centipedes are cave dwellers and a few of the geophilomorphs are marine.

Centipedes are generally predaceous (Fig. 2.2). The poisonous claws are used in securing their preys. Geophilids devour small soil-inhabiting arthropods and enchytraeid worms. *Lithobius variegatus* and *Lithobius* sp. feed on aphids, Collembola, mites, spiders, other centipedes, nematodes, and molluscus in addition to variable quantities of leaf litter. *Lithobius forficatus* in captivity readily accept flies, moths, and some noctuids. Scutigeromorphs are entirely insectivorous; scolopendromorps have a wide range of diet. *Scolopendra* are also known to feed on insect pests of various crops and slugs. Predatory associations of *Scutigera* sp. were found with termites, in the galleries of the fungus garden (Rajagopal and Veeresh 1981). So, to a certain extent, centipedes help in biological control of crop pests.

2.5.1.10 Isoptera (Termites)

Termites are more familiar insects in the tropics and subtropics (Fig. 2.1). They are eusocial, polymorphic insects and live in small to large colonies. The colony is made of various castes; workers and soldiers with perfect division of labor, each with a body wall suited for their specific function. Their role in the decomposition of wood and other cellulose material is very important. Termites digest cellulose by virtue of their possessing intestinal bacteria and protozoa. Termites also form a source of food for many animals including human beings.

Isoptera order has over 2600 described species in 281 genera (Kambhampati and Eggleton 2000). These social insects do*m*inate soil arthropod assemblages across much of the dry tropics and in dry temperate regions. Termites attain their highest diversity in the tropics (Bignell and Eggleton 2000). Termites are known to inhabit soil and wood. The soil-inhabiting termites are quite common as mound builders above the ground or as subterranean nest builders. Wood-inhabiting termites are mostly arboreal and construct nests either inside or on the trees. Soil-inhabiting termites form a dominant group of soil fauna which are known to play an important role in the rapid turnover of organic matter in the ecosystem. The transportation and transformation of mineral and organic components of soils and plants are related to their numbers.

Food is collected by the workers; food consists of plant materials, either living or entirely decomposed. This food is obtained from cellulosic materials like wood, grasses, herbs, leaf and plant litter, dung, humus, fungi, etc. Food is provided to the members of the colony in two ways: (1) nymphs and reproductives, incapable of feeding themselves, are fed by the workers either by the stomodeal or proctodeal food. Stomodeal food, a clear liquid or regurgitated food, is the sole nourishment for functional reproductives and soldiers. (2) Proctodeal food is a liquid excretion from rectal pouches and in the lower termites, it includes protozoan fauna. Young ones of lower termites are fed by proctodeal food as the protozoa is essential for digestion of cellulose. Workers collect food both for their consumption and also for the other individuals in the colony, which consists of any plant material.

Decomposition by Termites

Wood-eating termites influence decomposition of woody litter, like branches, logs, tree stumps, etc. Occasionally the decomposition of dung of grazing animals, a significant source of organic matter, is strongly influenced by the soil fauna, notably various Coleoptera, and, in some cases, termites (Ferrar and Watson 1970).

N. exitiosus, in dry sclerophyll forest, consumes 16–17 % of total annual fall of woody litter, and in Nigeria, the rate of decay of wood is largely dependent on the activities of termites. In a tropical rain forest, many fungus-growing termites carry much soil and microorganisms into wood. They are dominant litter consumers in the forest. Because of this they are excellent agents to combine organic matter with microorganisms. Veeresh et al. (1982) found significant reduction in the available carbon and potash in plots where termites (*O. wallonensis*) were allowed to feed on dung, leaf litter, and soft wood compared to plots protected from termite attack.

2.5.1.11 Ants

Ants occupy unique position among all insects on account of their dominance in both the number of species in a single family Formicidae (7600 species) and the number of individuals under each species. Numerically each colony has several thousands of individuals. In a colony of *Myrmicaria brunnea* (Myrmicinae), the number varies from 3800 to 13,500 individuals. In case of *Tetramorium caespitum*, the number of individuals varies from 1395 to 30,943.

Ants are found everywhere from the arctic regions to the tropics, from the timber line on the loftiest mountains to the shifting sands of the dunes and sea shores, and from the dampest forest to the driest deserts (Wheeler 1910). Ants do not depend on any one type of food. They build perfect nests like termites or bees. More dominant species have learnt to maintain more number of queens in a colony rather than depend on only one queen (*Solenopsis*, *Monomorium*, and others have more than one queen).

As a group, ants are beneficial and many species deserve our protection. They are useful in the following ways:

1. *Hasten the decomposition of organic substances*: The observations of Forel (1910) reveal that an ant colony brings 28 dead insects per minute which account for 100,000/day during its active period. The same will be more in tropical countries.
2. *Act as predators of other harmful insects*: Four Italian species, namely, *Formica lugubris*, *F. aquilaria*, *F. rufa*, and *F. polyctena*, have been used to protect the alpine coniferous forests against damage by insects (Pavan 1962).
3. *Help in moving soil while excavating nests*: According to Branner (1910), ants are so abundant that they replace earthworms as the chief earth movers in the

tropics. Studies made by Lyford (1963) have shown that they are nearly as important as earthworms in cold temperate forests as well. Many young naturalists have employed them for skeletonizing small vertebrates. In Europe, cocoon of Fallow ant has been used as bird food. Many years ago formic acid was distilled from worker ants. In Mexico the garments affected by caterpillars were freed by placing them on large hills of *Formica* and *Pogonomyrmex. Pogonomyrmex occidentalis* was useful in bringing fossil mammals to the surface. Honey ants (*Myrmecocystus melliger*) were used by the Indians for food and medicinal purposes. The huge heads of the soldiers of the South American leaf-cutting ant (*Atta cephalotes*) have been employed by the native surgeons in closing wound. *Oecophylla* was a source of food for many Indians.

2.5.1.12 Other Soil Insects

Other soil insects like, Crickets, Mole crickets, Grasshopper and grubs of scarabid beetles and caterpillar of moths and butterflies etc., also play on important role in maintaining soil fertility, structure and texture, and in rendering the soil porous. The porosity of soil helps in water percolation and this helps in the growth and development of roots of plants. Insects also attract microorganisms like soil bacteria, fungi which in turn variously affect soil biota. Some of these soil borne bacteria and fungi take part in decomposition which adds to the organic matter content of the soil. The insects also help in upturning surface soil to subsurface and also contribute to nutrient status of the soil. Thus, other insects impact soil conditions at microhabitat level.

2.6 Functional Role of Arthropods in Maintaining Soil Fertility

The term "soil fertility" denotes the degree to which a soil is able to satisfy plant demands for nutrients (including water) and a physical matrix adequate for proper root development. Arthropods function on two of the three broad levels of organization of the soil food web. They are "litter transformers" or "ecosystem engineers." Litter transformers, of which the micro arthropods comprise a large part, humidify ingested plant debris, improving its quality as a substrate for microbial decomposition and fostering the growth and dispersal of microbial populations. Ecosystem engineers are organisms that physically modify the habitat and regulate the availability of resources to other species (Jones et al. 1994). In the soil, this entails altering soil structure, mineral and organic matter composition, and hydrology. Ants and termites are the most important arthropod representatives of this guild, the latter group having received greater share of research attention (Lobry and Conacher 1990).

2.6.1 Influence of Arthropods on Nutrient Cycling

More than 90 % of net terrestrial primary production ultimately enters detritus food webs, where it is decomposed and recycled. Much of it originates in leaves and woody materials falling to the soil surface. Plant litter is a mixture of labile substrates (e.g., sugars, starch) easily digested by soil biota and other components (cellulose, lignins, tannins), more resistant to breakdown (Coleman et al. 2004). Decomposition of this material results from an interaction between physical and biological processes. Litter first must be physically weathered before it becomes suitable for further degradation by the soil microflora and fauna. Fungi are the important initial colonizers of plant litter (Harley 1971). With increasing disintegration and solubilization of the substrate, bacteria increase in importance. After this initial microbiological phase, the breakdown process slows and might come to a halt if not followed by animal activity. Saprophagous arthropods affect decomposition directly through feeding on litter and adhering microflora, thus converting the energy contained therein into production of biomass and respiration and, indirectly, through conversion of litter into feces and the reworking (reingestion) of fecal material, comminution of litter, mixing of litter with soil, and regulation of the microflora through feeding and the dissemination of microbial inoculum. Only a small proportion of net primary production is assimilated by soil arthropods (e.g., <10 % in oribatids, 4–20 % in millipedes and isopods). Thus, the indirect influence of these consumers on decomposition and soil fertility is of greater importance (Chew 1974). The influence of the soil fauna on decomposition processes is greatest in the humid tropics, where plant litter decomposition occurs most rapidly. This is due largely to the actions of the micro arthropods (Madge 1965).

2.6.1.1 Litter Feeding and Comminution

A major contribution of arthropods to the decomposition and humification processes is through the comminution of plant debris (Zimmer 2002). The physical fragmentation involved destroys the protective leaf cuticle, exposes cell contents, and increases water-holding capacity, aeration, and downward mobility of particulate and soluble substances. Comminution of plant litter is brought about largely by the feeding activity of saprophagous animals and, during passage through the digestive system, is accompanied by catabolic changes. The unassimilated residue from the commutative and catabolic processes is excreted as feces, typically smaller in size and of different chemical composition than the ingested food. The plant matter passing out in feces also presents an increased surface area to be attacked by microorganisms.

As much as 20 % of total annual litter input may be processed by the feeding activity of Collembola (Petersen 1994), a like proportion by that of oribatid mites, with about 3–10 % accounted for by Isopoda and Diplopoda and up to 60 % by termites (Collins 1981). Symphyla were estimated by Edwards (1973) to contribute about 2 % to annual litter turn over. Collins (1983) suggested that the biomass processed by termites commonly might be two or three times greater than the actual amount of litter consumed. Depending on the species involved, their densities, and the botanical composition, the mean litter comminution by millipedes in European forests ranged from 59 to 719 mg dry weight/ m^{-2}/ day^{-1}. Coprophagy is common in these arthropods and may be crucial to ensuring proper nutrition. In one study, the feces of millipedes showed an increase in pH, from 5.5 to 7.7, over that of the litter ingested and an eight fold increase in moisture content, providing a favorable substrate for increased microbial, particularly bacterial, activity. The bacterial biomass and digestion products improved the nutritional value of fecal pellets, which, on reingestion, provided a source of readily assimilable nutrients. Because of the large quantities of litter that they process, millipedes often are among the most important contributors of all soil invertebrates, to litter decomposition in moist, undisturbed habitats (Curry 1994). In habitats, in which earthworms are absent or rare, such as the acidic soils of coniferous forests, Collembola may assume a much greater role in physical breakdown of organic matter.

2.6.1.2 Mineralization of Nutrient Elements

Mineralization is the catabolic conversion of elements, primarily by decomposer organisms, from organic (i.e., bound in organic molecules) to inorganic form, such as the generation of CO_2 in the respiration of carbohydrates and breakdown of amino acids into ammonium (NH_4^+) and ultimately nitrate (NO_3). The direct or indirect actions of arthropods in processing plant litter increase available nutrient concentrations in the soil. Microorganisms efficiently convert the low-quality, recalcitrant resources of plant litter, such as the structural polymers comprising cell walls, into living tissue with much narrower carbon: nutrient ratios and of higher food value for animals, providing a rich source of nutrients at low metabolic cost to the consumer. A major proportion of the nutrients in the litter/soil system is concentrated and temporarily stored, or immobilized, in microbial biomass and subsequently in consumers, particularly the micro arthropods.

The microbial mineralization of nutrients may be stimulated by arthropod grazing. Several studies (Filser 2002) have demonstrated that grazing by Collembola has a strong stimulatory effect on fungal growth and respiration. Carbon mineralization by fungi and bacteria in comminuted litter was enhanced by an optimal level of grazing by micro- and macro arthropods; increased grazing pressure above the optimum inhibiting microbial respiration. Collembolan grazing on fungi can result in increased mobilization of available N and Ca, with implications for nutrient availability in particular environments, such as acidic forest soils, in which large nutrient pools tend to be immobilized in accumulated organic matter. Arthropod grazing on the microflora also acts to regulate the rate of decomposition, preventing sudden microbial blooms with the result that nutrients are mineralized and released from detritus, and made available for plant uptake, in a controlled and continuous fashion and their loss from the system minimized (Reichle 1977).

Arthropods influence the distribution of microbial populations in the soil by transporting microbes or their propagules on or in their bodies. For example, millipedes have large numbers of conidia of Streptomyces spp. adhering to their cuticle and dense populations of gut bacteria of the Actinomycetales, Bacillales, and Enterobacteriales, which were spread in feces. The microfloras have limited inherent capacities for movement and may remain for long periods in inactive resting stages if local conditions are unsuitable for growth and reproduction. Termites and their constructions function importantly in the replenishment of soil nutrients in natural ecosystems. The uniqueness of the involvement of termites in litter turnover and nutrient cycling was ably summed up by Lee and Wood (1961): "The combination of foraging for food over a wide radius from the nest and returning it to the nest…intense degradation of the plant tissue collected, and use of the excreted end products of digestion for mound-building resulting in their removal from participation in the plant/soil system for long periods, sets termites apart from other soil animals in their influence on soil organic matter." *Together with earthworms, termites are thought to contribute more to litter breakdown than all other soil-inhabiting invertebrates.*

Ant nests also may contain higher concentrations of nutrients than surrounding soil. Large amounts of organic matter from plant and animal (prey, carrion) sources accumulate in refuse dumps within nests. This material, combined with metabolic wastes and secretions from the ants themselves, may become incorporated into nest soil and undergoes decomposition and mineralization by the microflora, leading to an accumulation and local concentration of nutrients (Petal 1978). The organic matter content of soil worked by ants was reported to be 1.5 times greater than that of control soil. The increases in nutrient (e.g., cation) and organic matter content are likely factors influencing the generally observed shift in pH values toward neutral in ant-modified soil (Frouz and Jilková 2008). In infertile soils with low organic matter content and low rates of decomposition, predators, such as ants, speed the return to the soil of nutrients concentrated in the tissues of other animals (Petal 1978). This singular contribution of ants to organic matter turnover was recognized and emphasized more than a century ago by Wheeler (1910). A similar process results from the activities of those groups indirectly herbivorous in diet. The leaf-cutting attine ants (Myrmicinae: Attini), for example, are considered the chief agents for introducing organic matter, in the form of discarded fungus gardens, into the nutrient-poor soils of the New World tropics; this material then becoming available to other organisms for further decomposition. The activities of this tribe of ants clearly facilitate the decomposition of plant materials and promote nutrient cycling. Farji-Brener and Tadey (2009) concluded that the magnitude of the contribution of leaf-cutter ants to soil fertility was among the highest of any animal group.

2.6.2 Influence of Arthropods on Soil Structure

Biology plays a major role in the stabilization of soil structure (Oades 1993). Among the more biologically significant attributes of soil are the spatial organization of soil particles and of the pore spaces and voids among them, the combination of the particles into aggregates, and the stability of the aggregates in water. A favorable soil structure ensures adequate nutrient retention, aeration, and water-holding capacity below ground, facilitates root penetration, and prevents surface crusting and erosion of topsoil. Arthropods affect the structural properties of soils in various ways.

2.6.2.1 Soil Mixing and the Development of Pores and Voids

Biotic pedoturbation refers to the displacement or mixing of soil material through the actions of organisms (Wilkinson et al. 2009). In general, the mesofauna are not considered important in this process because they are too small to move most soil particles (although Collembola and oribatid mites are said to make active "micro-tunnels" in the soil matrix (Rusek 1985); these animals instead rely on existing cracks and crevices and the channels and spaces created by the larger fauna to aid their mobility within the soil (Oades 1993). The subterranean network of tunnels and galleries that comprise termite and ant nests plays an important role in enhancing aeration and water infiltration through the soil profile, increasing water storage, and retention of top soil. Termites have been reported to work the soil to depths of 50 m or more (Lepage et al. 1974). In experimental studies, Elkins et al. (1986) and Whitford (1991) found plots, from which subterranean termites had been eliminated, to have significantly reduced water infiltration and storage and increased run-off and sediment flow (bed load) compared to plots populated with termites. Mando et al. (1996), Mando and Miedema (1997), and Mando (1997) showed that active encouragement of termite activity through the application of surface mulches significantly improved the hydraulic properties of degraded soils. Under such conditions, soil, in which termites (*Macrotermes subhyalinus* [Rambur] and *Odontotermes* sp.) were active, had infiltration rates ranging from 2 to 6 to above 9 $cm^3 s^{-1}$, two to three times those in soil without termites (Léonard and Rajot 2001).

The infiltration pathways and sinks provided by ant nests limited post fire hill-slope erosion by reducing overland water flow rates following heavy rainfall events (Richards 2009). Experimental crop yield increased 36 % and infiltration rates threefold in plots supporting ant and termite populations over those in plots, from which the insects had been, excluded (Evans et al. 2011). The system of chambers and galleries comprising ant and termite nests, increases the porosity of soil, improving aeration and water infiltration, together with the organic matter (from feces, salivary, and other secretions). Further food remnants accumulating therein, enhances water-holding capacity and creates an environment favorable for the penetration of plant roots (Petal 1978) between closely packed soil particles and excavated burrows (Hopkin and Read 1992).

Jacot (1940) identified millipedes as one of the groups instrumental in mixing organic matter with the mineral soil. Oribatids also are thought to contribute to the deep mixing of organic material by their movements to, and deposition of feces in, lower layers of the soil (Wallwork 1970). Similarly, the tendency for symphyla to move rapidly up and down the soil profile serves to distribute their feces widely throughout the soil. Fecal materials are major constituents contributing to the formation of stable soil aggregates (see discussion below), which are important in maintaining adequate water infiltration and drainage (Tisdall and Oades 1982). The activities of ants and termites below ground may have a significant influence on the particle size distribution within the soil. In particular, the selective removal of the finer and smaller soil particles from lower in the profile may result in a greater contrast in texture between the topsoil and subsoil horizons, with implications for abiotic pedogenic processes through the soil profile, because they are more widespread than earthworms; ants, in the aggregate, had a much greater influence on soils. This view largely has been supported by a sizeable body of research over the ensuing years (Paton et al. 1995), which indicates global rates of ant-mediated pedoturbation of *ca.* 10 tonnes/ ha^{-1}/ $year^{-1}$. Lyford (1963) estimated that at least 60 g m^{-2} of soil in a New England (USA) forest was transferred annually by ants from the B horizon to the surface in the process of mound building and suggested that the entire A horizon (25–46 cm thick) of some of the virgin brown podzolic soils of the region was the product of such pedoturbation over a period of 3000–4000 years. An estimated 6 m^3 (7.4 tonnes dry weight) of soil ha^{-1} was moved by ants (*Formica cinerea* Mayr) from the B horizon to the surface to form the upper parts of extant mounds in a North American prairie. Bulk density (a function of porosity and organic matter content) of the deposited material was significantly lower than that of the surrounding soil (~0.29–0.53 vs. 1.04 g cm^{-3}). Nest building by *Atta cephalotes* (L.) in a tropical forest resulted in displacement to the surface of an estimated 460 tonnes of subsoil/ ha^{-1} over the average colony's life span. The authors concluded that leafcutter ants played a major role in soil genesis and development in New World tropics.

Termite mounds often are abundant in tropical regions, numbering in excess of 1000 /ha^{-1} and representing, in aggregate, many tonnes of soil (Lee and Wood 1961). In regions, in which mounds are numerous, an estimated 1 tonne or more / ha^{-1} of mineral soil from the lower horizons was deposited on the surface annually. The estimated amount of soil brought to the surface by subterranean termites in a temperate desert was somewhat lower, at about 750 kg /ha^{-1} /$year^{-1}$, and included a rich clay component (Nutting et al. 1987). The nest-building activity of termites was calculated to result in pedoturbation of a 20–37-cm-thick layer of soil in 1000 years, contributing significantly to soil turnover. Role of arthropods in structuring the soils is shown in Fig. 2.4.

2.6.2.2 Formation of Soil Aggregates

Soil aggregates, or peds, the basic units of soil structure, are formed by natural processes, commonly involving the activity of organisms (Hole 1981; Lynch and Bragg 1985). Fecal pellets, combining fine mineral particles with undigested organic matter, are the major contribution of invertebrates to the formation of soil aggregates (Rusek 1985; Pawluk 1987). Mucilaginous substances, by-products of microbial decomposition, bind the feces with other soil components into stable microstructures (Oades 1993; Harris et al. 1966). These organomineral complexes are substrates, on which inorganic nutrients may become adsorbed and so available to plants (Kunhelt 1976). The resulting humus, an amorphous colloidal material comprising partially decomposed organic matter that makes up topsoil and increases the soil's capacity to store nutrients (example, cations) and prevent their rapid leaching, thus is largely derived from animal feces (Ciarkowska and Niemyska-Łukaszuk 2002; Loranger et al. 2003). The humus of well-developed soils represents a significant pool of macronutrients, such as N, P, K, Ca, and Mg, which may be stored in amounts exceeding 1 tonne/ ha^{-1}. It also is involved in chelation reactions, which aid in the micronutrient nutrition of plants, buffers the soil against rapid changes in

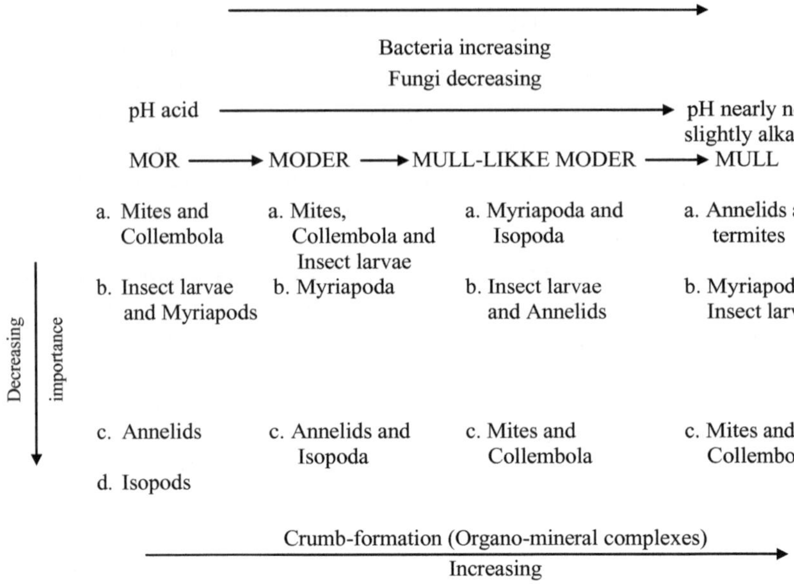

Fig. 2.4 Soil categories, pH, and biota responsible for soil structure (Wallwork 1970)

pH, and supports an abundance and diversity of microorganisms, promoting increased mineralization activity.

Organomineral microaggregate structures from the feces of soil-feeding termites also are thought to aid in increasing structural stability and porosity of tropical soils. In areas of high termite activity, these micropeds may comprise 20 % of the soil matrix (Kooyman and Onck 1987). Arthropod feces generally play a larger role in the formation of the moder and mor types of humus and in the formation of primitive soils. However, although earthworms generally are considered to dominate in mull-type soils (Bardgett et al. 2005), arthropods, such as millipedes, also may contribute substantially to these or to mull-like formations (Romell 1935). By contrast, a lack of millipedes, as well as earthworms, in the mor soils of coniferous forests is one of the main reason for the slow decomposition of pine needles (Hopkin and Read 1992). The volume of feces contributed may be considerable. For example, collembolan populations, at densities typical of forest soils, were estimated to produce around 175 cc of fecal pellets/ m^{-2} annually, equivalent to the formation of a soil layer of roughly 0.2 mm in depth. The production of fecal pellets by desert isopods (*Hemilepistus reaumuri*) ranged from 2 to 41 g/ m^{-2} /$year^{-1}$, depending on site conditions (primarily moisture regime), subsequently to be redistributed and mixed with soil during rainfall events (Yair and Rutin 1981). Striganova (1975) estimated annual millipede consumption of litter in woodland that results in a layer of fecal pellets 0.5–1 cm thick soil surface. On rocky sites, millipedes (Glomeridae) may facilitate the process of succession by consuming detritus accumulating in cracks and depositing excrement there, providing a substrate favorable for the

Table 2.1 List of minor soil insects and their role

Name and family	Feeding habit	Habitat preferred	Functional roles
Cricket, Gryllidae	On living/dry, dead organic matter	Dry hot niches, live in burrows/under logs	Pests of crops, upturn soil
Mole cricket, Gryllotalpidae	Carnivorous, crops	Soil dwelling, dig soil	Pests of crops, upturn soil
Grasshoppers, Acrididae	Phytophagous	Soil dwelling, dig soil	Pests of crops
Ground beetles, Carabidae	Carnivorous, crops	Under stones, rotting wood/under bark	Pests of crops, upturn soil
Tiger beetles, Cicindelidae	Carnivorous	Soil	Upturn soil
Rove beetles, Staphylinidae	Carnivorous, crop	Bark/decaying wood	Decompose decaying matter
Wasps, Vespidae	Carnivorous	Soil	Upturn soil

colonization of higher plants (Kunhelt 1976). The relative abundance of hemi- and edaphic arthropods in different soils is presented in Tables 2.1 and 2.2.

2.6.3 Current Threats to Soil Biodiversity

A majority of human activities result in soil degradation, impacting the services provided by soil biodiversity elements. Soil organic matter depletion and soil erosion are influenced by inappropriate agricultural practices, overgrazing, vegetation clearing, and forest fires. It has been observed, for example, that land without vegetation can be eroded more than 120 times faster than land covered by vegetation, thus lose less than 0.1 tonne of soil per ha/year. The activity and diversity of soil organisms are directly affected by the reduction of soil organic matter and indirectly by the reduction in plant diversity and productivity (Bengtsson et al. 2005). Land use management practices affect biodiversity. Within rural lands, soil biodiversity tends to decrease with the increasing intensification of farming practices (e.g., use of pesticides, fertilizers, heavy machinery) (Haygarth and Ritz 2009; Joris et al. 2013). Climate change may be the second most important factor affecting soil biodiversity. Climate change is likely to have significant impacts on services provided by soil biodiversity elements (Schils et al. 2008).

The pollution of soils is mostly a result of industrial activities and use of fertilizers and pesticides. Toxic pollutants can destabilize the population dynamics of soil organisms by affecting reproduction, growth, and survival (Gardi and Jeffrey 2009). Genetically modified crops (GMOs) may also be considered as a growing source of pollution for soil organisms. Most effects of GMOs are observed on chemical engineers, by altering the structure of bacterial communities, bacterial genetic transfer, and the efficiency of microbial-mediated processes. GMOs have also been shown to have effects on earthworm physiology, but to date little impacts on biological regulators are known (Daane et al. 1996).

Table 2.2 Approximate abundance (number m^{-2}) of hemi- and edaphic arthropods in different soils

Soil type							
Body-size group	Taxon	Tundra (arctic alpine)	Mor (boreal forest)	Mull (warm temperate forest)	Temperate grassland (prairie)	Tropical savanna	Tropical forest
Mesofauna	Microarthropoda	100,000	40,000	40,000	25,000	2000	15,000
	Symphyla	0	3000	600	1000	2000	800
Macrofauna	Diplopoda/Isopoda	0	500	1000	500	<1	400
	Formicidae	0	50	3000	1000	2000	800
	Isoptera	0	0	1000	1000	4000	5000

Due to globalization, invasives have become an important threat to soil biodiversity. Exotic species are called invasive when they become disproportionally abundant. Invasive species can have major direct and indirect impacts on soil services and native biodiversity. Invasive plants will alter nutrient dynamics and thus the abundance of microbial species in soil, especially of those exhibiting specific dependencies (e.g., mycorrhiza). Biological regulator populations tend to be reduced by invasive species, especially when they have species-specific relationships with plants. Soil biodiversity can serve as a reservoir of natural enemies against invasive plants. Setting up such biological control programs could save billions of rupees in prevention and management of invasive species (Seifert et al. 2009).

Soils are integral parts of ecosystems and are maintained in a fertile state largely through the actions of their constituent biota. Fertility is a function of a soil's capacity to provide plants not only with essential nutrients for growth and reproduction but also with a physical matrix that facilitates root growth and respiration and maintains its structural integrity against erosive forces. Arthropods influence soil fertility in two principal ways. First, they promote the decomposition of plant litter directly and indirectly, by transforming it physically and chemically into substrates amenable to further degradation. The higher assimilation efficiencies of termites allow these animals to convert a greater proportion of ingested litter directly into biomass that is possible for other soil arthropods, whereas the main contribution of the Collembola, Oribatida, Myriapoda, and Isopoda to nutrient cycling is via the indirect route, as secondary decomposers, conditioning litter, through comminution and passage through the gut, for further breakdown by the microflora. The nests of termites and ants, with their incorporated fecal materials, waste dumps, or fungal gardens, also provide rich substrates for the microbial degradation and mineralization of organic matter. The end result of these processes is the conversion of complex organic molecules into simpler, inorganic forms that can be used by plants. Arthropod grazing on microbial populations also may serve to regulate the availability of nutrients to plants, ensuring their release in a controlled and continuous manner and minimizing their loss from the root zone.

The second major way, in which arthropods contribute to the maintenance of soil fertility, is through their effects on the physical structure of the soil. Ants and termites are the pre eminent earth movers in many regions of the world and may surpass earthworms in this capacity in some cases. The pedoturbation resulting from their activities brings substantial amounts of subsoil to the surface, increasing the mineral content of the topsoil and providing sites for ion exchange in the root zone. The tunneling and burrowing of arthropods provide channels for air passage and water infiltration and also serve to mix organic matter into the upper soil layers. The feces of arthropods serve as nuclei for the accretion of soil aggregates and basic units of a soil's structure. This is important in maintaining integrity and are a significant factor in the formation of humus, which contributes to water and nutrient retention in the soil. A select list of endangered soil arthropods is provided (Table 2.3).

Table 2.3 A select list of endangered soil arthropods in South Asia

Class	Common name	Scientific name	Country	Reasons
Arachnida	Peacock tarantula	*Poecilotheria metallica*	India	Habitat quality is decreasing
Crustacea		*Perbrinckia punctata*	Sri Lanka	Habitat degradation
	Singapore freshwater crab	*Johora singaporensis*	–	Habitat development and degradation
Diplopoda	Major black millipede	*Doratogonus major*	South Africa	Habitat destruction
Insecta	Delta green ground beetle	*Elaphrus viridis*	California, USA	Invasive plants
	Cromwell chafer	*Prodontria lewisi*	New Zealand	Habitat degradation and predation by introduced species
	Dracula ant	*Adetomyrma venatrix*		Habitat loss
	Frigate Island giant tenebrionid beetle	*Polposipus herculeanus*	Frégate Island	Fungal diseases

Source: endangered species 2010: arthropods part 1 – Arachnids, Crustaceans and millipedes, part II insects by Thonoir

2.7 Conservation Priorities

Almost 98 % of earth's diversity is in terrestrial ecosystem, of which soil is the main component, so conservation of soil and its components like arthropods is critically important. Planting of trees on the easily eroded, easily compacted, and readily oxidized soils of the tropics is urgently required. Research in the temperate lands has shown that sowing of wildflower seed mixtures can be highly beneficial to insects (Samway 1994) and other arthropods. In the tropics regulating the human populations, plant refugia with indigenous species, restoration of wetlands/water bodies, judicious use of pesticides and chemicals, and preventing overexploitation of natural resources will go a long way in restoration of habitats. Restricting extensive landscape modifications and minimizing development and destruction of fertile soils is key to the arthropod conservation. Some of the endangered soil arthropods can be saved by captive breeding. Sustainable use of resources should form basis for soil restoration ecology. The best step would be to preserve as many landscapes as possible in natural state. Arthropods because of their small size, not appealing shapes, are generally not considered valuable for conservation. The knowledge of these creatures is lacking in the public. The landscapes which soil arthropods inhabit should be preserved or sustainably used, as they play a major role in the ecosystem.

References

Bagyaraj DJ (2011) Microbial biotechnology for sustainable agriculture, horticulture and forestry. NIPA Publishers, New Delhi

Bagyaraj DJ (2014) Interaction between arbuscular mycorrhizal fungi and the soil organisms and their role in sustainable agriculture. In: Singh DP, Singh HB (eds) Trends in soil microbial ecology. Studium Press LLC, Houston, pp 257–280

Bano K, Krishnamoorthy RV (1977) Changes in the composition of soils due to defecation by the millipede *Jonespeltis splendidus* (Verhoeff). Mysore J Agric Sci 11:561–566

Bano K, Bagyaraj DJ, Krishnamoorthy RV (1976) Feeding activity of millipede *Jonespeltis splendidus* Verhoeff and soil humification. Proc Indian Acad Sci 82:1–11

Bardgett RD, Usher MB, Hopkins DW (2005) Biological diversity and function in soils. Cambridge Univ Press, Cambridge, pp 172–188

Barrios E (2007) Soil biota ecosystem services and land productivity. Ecol Econ 64:269–285

Barron GL (1977) Topics in mycology No.1 The nematode destroying fungi. Canadian Biological Publications Ltd., Guelph, p 140

Bengtsson J, Ahnstrom J, Weibull AC (2005) The effects of organic agriculture on biodiversity and abundance: a meta-analysis. J Appl Ecol 42:261–269

Bignell DE, Eggleton P (2000) Termites in ecosystems. In: Abe T, Bignell DE, Higashi M (eds) Termites: evolution, sociality, symbioses, ecology. Kluwer Academic Publishers, Dordrecht, pp 363–387

Blower G (1951) A comparative study of the chilopod and diplopod cuticle. J Microsc Sci 92:141–161

Blower JG (1956) Some relations between millipedes and the soil. In Proceedings of 6th Congres International de la Science du Sol, Paris III, pp. 169–176

Boellstorff DL (2008) Estimated soil organic carbon change due to agricultural land management modifications in a semiarid cereal-growing region in Central Spain. J Arid Environ 73(3):389–392

Boudreaux HB (1979) Arthropod phylogeny with special reference to insects. Wiley, New York, p 320

Branner JC (1910) Geologic work of ants in tropical America. Bull Geol Soc Am 21:449–496

Brauman KA, Daily GC, Duarte TK, Mooney HA (2007) The nature and value of ecosystem services: an overview highlighting hydrologic services. Annu Rev Environ Resour 32:67–89

Brussaard L, de Ruiter PC, Brown GG (2007) Soil biodiversity for agricultural sustainability. Agric Ecosys Environ 121(3):233–244

Chapman AD (2009) Numbers of living species in Australia and the world, 2nd edn. Australian Government, Department of the Environment, Water, Heritage and the Arts, Canberra, p 80

Chew RM (1974) Consumers as regulators of ecosystems: an alternative to energetics. Ohio J Sci 74:359–370

Christiansen K (1964) Bionomics of collembola. Annu Rev Entomol 9:147–148

Ciarkowska K, Niemyska-Łukaszuk J (2002) Microstructure of humus horizons of gypsic soils from the Niecka Nidziańska area (South Poland). Geoderma 106:319–329

Coleman DC, Crossley DA Jr, Hendrix PF (2004) Fundamentals of soil ecology, 2nd edn. Elsevier Academic Press, Burlington, p 386

Collins NM (1981) The role of termites in the decomposition of wood and leaf litter in the Southern Guinea savanna of Nigeria. Oecologia 51:389–399

Collins NM (1983) Termite populations and their role in litter removal in Malaysian rain forests. In: Sutton SL, Whitmore TC, Chadwick AC (eds) Tropical rain forest: ecology and management. Blackwell Scientific Publications, Oxford, pp 311–325

Constanza R, d'Arge R, Groot R, Farbe S, Grasso M, Hannon B, Limburg K, Naeem S, O'Neil RV, Paruelo J, Raskin RG, Sutton P, van den Belt M (1997) The value of the world's ecosystem services and natural capital. Nature 387:253–260

Culliney TW (2013) Role of arthropods in maintaining soil fertility. Agriculture 3:629–659

Curry JP (1994) Grassland invertebrates: ecology, influence on soil fertility and effects on plant growth. Chapman & Hall, London, p 437
Daane LL, Molina JAE, Borry EG, Sadowsky MJ (1996) Influence of earthworm activity on gene transfer from *Pseudomonas fluorescens* to indigenous soil bacteria. Appl Environ Microbiol 62:515–521
Darwin C (1881) The formation of vegetable mould through the action of worms, with observations of their habits. Murry, London, p 326
Dash MC, Patra UC (1979) Worm cast production and nitrogen contribution to soil by a tropical earthworm population from a grass land site from Orissa India. Rev Ecol Biol Sol 16:79–83
Decaens T, Jimenez JJ, Gioia C, Measey GJ, Lavelle P (2006) The values of soil animals for conservation biology. Eur J Soil Biol 42:S23–S38
Dindal DL (1970) Feeding behavior of a terrestrial turbellarian *Bipalium adventitium*. Am Midl Nat 83:635–637
Edwards CA (1973) Biological aspects of the degradation and behavior of pesticides in soil. In Proceedings of 7th British insecticides and fungicide conference. pp. 811–823
Edwards CA (1990) Symphyla. In: Dindal DL (ed) Soil biology guide. Wiley, New York, pp 891–910
Edwards CA, Lofty JR (1972) Biology of earthworms. Chapman and Hall Ltd, London, p 283
Eisenbeis G, Wichard W (1987) Atlas on the biology of soil arthropods. Springer-Verlag, Berlin, p 437
Elkins NZ, Sabol GV, Ward TJ, Whitford WG (1986) The influence of subterranean termites on the hydrological characteristics of a Chihuahuan desert ecosystem. Oecologia 68:521–528
Evans TA, Dawes TZ, Ward PR, Lo N (2011) Ants and termites increase crop yield in a dry climate. Nat Commun 2, doi:10.1038/ncomms1257
Farji-Brener AG, Tadey M (2009) Contributions of leaf-cutting ants to soil fertility: causes and consequences. In: Lucero DP, Boggs JE (eds) Soil fertility. Nova Science Publishers, New York, pp 81–91
Ferrar P, Watson JAL (1970) Termites associated with dung in Australia. J Aust Entomol Soc 9:100–102
Filser J (2002) The role of collembola in carbon and nitrogen cycling in soil. Pedobiologia 46:234–245
Finlay R (2006) Identification of single species and communities. In: Luster J, Finlay R (eds) Handbook of methods used in Rhizosphere research. Swiss Federal Research Institute, Birmensdorf, pp 338–339
Forel A (1910) Glanures myrmecologianes. Ann Soc Entomol Belg 54:6–31
Frouz J, Jilková V (2008) The effect of ants on soil properties and processes (Hymenoptera: Formicidae). Myrmecol News 11:191–199
Gardi C, Jeffrey S (2009) Soil biodiversity. Joint Research Center, European Commission, Luxembourg, p 27
Gates GE (1961) Ecology of some earthworms with special reference to seasonal activity. Am Midl Nat 66:61–86
Ghaisas PS, Ranade DP (1981) Record of *Japyx solifugus* (Diplura) from Poona, UAS tech series – India, pp. 147–152
Giribet G, Edgecombe GD (2012) Reevaluating the arthropod tree of life. Annu Rev Entomol 57:167–186
Harley JL (1971) Fungi in ecosystems. J Appl Ecol 8(3):627–642
Harris RF, Chesters G, Allen ON (1966) Dynamics of soil aggregation. Adv Agron 18:107–169
Haygarth P, Ritz K (2009) The future of soils and land use in the UK: soil systems for the provision of land-based ecosystem services. Land Use Policy 26:S187–S197
Hole FD (1981) Effects of animals on soil. Geoderma 25:75–112
Hopkin SP (1997) Biology of the springtails (Insecta: collembola). Oxford University Press, Oxford, p 330
Hopkin SP, Read HJ (1992) The biology of millipedes. Oxford University Press, Oxford, p 233

Jacot AP (1940) The fauna of the soil. Q Rev Biol 15:28–58
Jegen G (1920) DieBedeutng der Encytraeiden fiir die Humus bildung. Landwirtsch Fb Schweiz 34:55–71
Jenny H (1980) The soil resource: origin and behavior. Springer-Verlag, New York, p 377
Jones CG, Lawton JH, Shachak M (1994) Organisms as ecosystem engineers. Oikos 69:373–386
Joris H, Caires E, Bini A, Scharr D, Haliski A (2013) Effects of soil acidity and water stress on corn and soybean performance under a no-till system. Plant and Soil 265:409–424
Kale RD, Krishnamoorthy RV (1979) Pesticidal effects of Sevin. R. (1-naphthyl-n-methyl carbamate) on the survivability and abundance of earthworm *Pontoscolex corethrurus*. Proc Indian Acad Sci 88B:391–396
Kale RD, Krishnamoorthy RV (1980) The calcium content of the body tissues and castings of the earthworm *Pontoscolex corethrurus* (Annelida: Oligochaeta). Pedobiologia 20:309–315
Kale RD, Krishnamoorthy RV (1981) Litter preference in the earthworm, *Lampito mauritii*. Proc Ind Acad Sci 90:123–128
Kambhampati S, Eggleton P (2000) Taxonomy and phylogeny of termites. In: Abe T, Bignell DE, Higashi M (eds) Termites: evolution, sociality, symbioses, ecology. Kluwer Academic Publishers, Dordrecht, pp 1–23
Kooyman C, Onck RFM (1987) The interactions between termite activity, agricultural practices and soil characteristics in Kisii District, Kenya. In Agricultural University Wageningen Papers 87-3, Agricultural University Wageningen: Wageningen, the Netherlands. pp. 1–120
Kubiena WL (1955) Animal activity in soil as a decisive factor in establishment of humus forms. In: McE Kevan DK (ed) Soil zoology, vol 2. Butterworths, London, pp 73–82
Kunhelt W (1976) Soil biology: with reference to the animal kingdom. Faber and Faber, London, p 483
Lee KE, Wood TG (1961) Termites and soils. Academic, New York, p 251
Léonard J, Rajot JL (2001) Influence of termites on runoff and infiltration: quantification and analysis. Geoderma 104:17–40
Lepage M, Morel G, Resplendino C (1974) Découverte de galeries de termites atteignant la nappe phréatique profonde dans le Nord du Sénégal. C R Acad Sci Sér D 278:1855–1858
Lobry de Bruyn LA, Conacher AJ (1990) The role of termites and ants in soil modification: a review. Aust J Soil Res 28:55–93
Loranger G, Ponge JF, Lavelle P (2003) Humus forms in two secondary semi-evergreen tropical forests. Eur J Soil Sci 54:17–24
Luster J, Göttlein A, Nowack B, Sarret G (2009) Sampling, defining, characterizing and modelling the rhizosphere – the soil science tool box. Plant and Soil 321:457–482
Lyford WH (1963) Importance of ants to brown podzolic soil genesis in New England. Harvard Forest Paper, Petersham
Lynch JM, Bragg E (1985) Microorganisms and soil aggregate stability. Adv Soil Sci 2:133–171
Madge DS (1965) Leaf fall and litter disappearance in a tropical forest. Pedobiologia 5:273–288
Maggenti AR (1981) Nematodes: development as plant parasites. Annu Rev Microbiol 35:135–154
Mando A (1997) The impact of termites and mulch on the water balance of crusted Sahelian soil. Soil Technol 11:121–138
Mando A, Miedema R (1997) Termite-induced change in soil structure after mulching degraded (crusted) soil in the Sahel. Appl Soil Ecol 6:241–249
Mando A, Stroosnijder L, Brussaard L (1996) Effects of termites on infiltration into crusted soil. Geoderma 74:107–113
Narayana Swamy BC, Nanjegowda D (1980) The Tardigrade *Macrobiotus* Sp. predaceous on plant parasitic Nematodes. Soil Biol Ecol News lett 1:9
Nielsen CO, Christensen B (1959) The enchytracidae, critical revision and taxonomy of European species. Natur Jutlandica 8–9:1–160
Norton DC (1978) Ecology of plant-parasitic nematode. Wiley, New York, p 268
Norton RA (1994) Evolutionary aspects of oribatid mite life histories and consequences for the origin of the astigmata. In: Houck MA (ed) Mites: ecological and evolutionary analyses of life-history patterns. Chapman & Hall, New York, pp 99–135

Norton RA, Behan-Pelletier VM (2009) Suborder oribatida. In: Krantz GW, Walter DE (eds) A manual of acarology. Texas Tech University Press, Lubbock, pp 430–564

Nutting WL, Haverty MI, LaFage (1987) Physical and chemical alteration of soil by two subterranean termite species in Sonoran desert grassland. J Arid Environ 12:233–239

Oades JM (1993) The role of biology in the formation, stabilization and degradation of soil structure. Geoderma 56:377–400

Overgaard Nielsen C, Christensen B (1959) The enchytraeidae – critical revision and taxonomy of European species. Naturhistorisk Museum, Aarhus

Patel HK, Patel RM (1959) Preliminary observations on the control of earthworms by soapnut (*Sapindus laurifolius* Vaw) extract. Indian J Entomol 21:251–255

Paton TR, Humphreys GS, Mitchell PB (1995) Soils: a new global view. Yale University Press, London, p 213

Pavan M (1962) La defence biologique des forest en italic avecless fourmis du groupe *Formica rufa* Symp. Genetica et biologica italic. Atti IV Congresso UIEIS Pavia 11:1–25

Pawluk S (1987) Faunal micromorphological features in moder humus of some western Canadian soils. Geoderma 40:3–16

Petal J (1978) The role of ants in ecosystems. In: Brian MV (ed) Production ecology of ants and termites. Cambridge University Press, Cambridge, UK, pp 293–325

Petersen H (1994) A review of collembolan ecology in ecosystem context. Acta Zool Fenn 195:111–118

Petersen H, Luxton M (1982) A comparative analysis of soil fauna populations and their role in decomposition processes. Oikos 39:288–388

Pimentel D (1997) Techniques for reducing pesticides: environmental and economic benefits. Wiley, Chichester

Powlson DS, Bhogal A, Chambers BJ, Coleman K, Macdonald AJ, Goulding KWT, Whitmore AP (2012) The potential to increase soil carbon stocks through reduced tillage or organic material additions in England and Wales: a case study. Agric Ecosyst Environ 146:23–33

Prabhoo NR (1972) South Indian protura. Rev Ecol Biol Sol 9:711–718

Rajagopal D, Veeresh GK (1981) Thermophiles and teritariophiles of *Odontotermes wallonensis* (Isoptera : Termitidae) in Karnataka, India. Colemania 1(2):129–130

Raw F (1967) Arthropods (except Acari and collembola). In: Burges A, Raw F (eds) Soil biology. Academic, London, pp 323–362

Reichle DE (1977) The role of soil invertebrates in nutrient cycling. In: Lohm U, Persson T (eds) Soil organisms as components of ecosystems. Swedish Natural Science Research Council, Stockholm, pp 145–156

Retallack GJ, Feakes CR (1987) Trace fossil evidence for late Ordovician animals on land. Science 235:61–63

Richards PJ (2009) *Aphaenogaster* ants as bioturbators: impacts on soil and slope processes. Earth Sci Rev 96:92–106

Romell LG (1935) An example of myriapods as mull formers. Ecology 16(1):67–71

Rusek J (1985) Die bodenbildende Funktion von Collembolen und Acarina. Pedobiologia 15:299–308

Samway MJ (1994) Insect conservation biology. Chapman and Hall, London, p 358

Sayre RM (1971) Biotic influence in soil environment. In: Zukerman BM, Mai WF, Rhode RA (eds) Plant parasitic nematodes, vol I. Academic, New York, pp 235–256

Schils R, Kuikman P, Liski J, van Oijen M, Smith P, Webb J, Alm J, Somogyi Z, van den Akker J, Billett M, Emmett B, Evans C, Lindner M, Palosuo T, Bellamy P, Alm J, Jandl R, Hiederer R (2008) Review of existing information on the interrelations between soil and climate change. Altera, Wageningen, p 205

Schmalfuss H (2003) World catalog of terrestrial isopods (Isopoda: Oniscidea). Stuttg Beitr Naturkunde Ser A (Biol) 654:1–341

Seifert EK, Bever JD, Maron JL (2009) Evidence for the evolution of reduced mycorrhizal dependence during plant invasion. Ecology 90(4):1055–1062

Senapati BK, Dash MC, Rana AK, Panda BK (1980) Observation on the effect of earthworm in the decomposition process in soil under laboratory conditions. Comput Physiol Ecol 5:140–142
Sierwald P, Bond JE (2007) Current status of the myriapod class Diplopoda (millipedes): taxonomic diversity and phylogeny. Annu Rev Entomol 52:401–420
Stirling GR, Mankau R (1978) *Dactylella oviparasitica*, a new fungal parasite of *Meloidogyne* eggs. Mycologia 70:771–783
Stork NE, Eggleton P (1992) Invertebrates as determinants and indicators of soil quality. Am J Altern Agric 7:38–47
Striganova BR (1975) Dispersion patterns of diplopods and their activity in the litter decomposition in the Carpathian foothills. In: Vaněk J (ed) Progress in soil zoology. Academia, Prague, pp 167–173
Sutton S (1980) Woodlice. Pergamon Press, Oxford, p 144
Szucsich N, Scheller U (2011) Symphyla. In: Minelli A (ed) Treatise on zoology—anatomy, taxonomy, biology: the myriapoda, vol 1. Koninklijke Brill NV, Leiden, pp 445–466
Tembe VB, Dubash PJ (1961) The earthworms: a review. J Bombay Nat Hist Soc 58:171–201
Timonin MI (1961) The interaction of plant pathogen and *Scaptocoris talpa* Champ. Can J Bot 39:695–703
Tisdall JM, Oades JM (1982) Organic matter and water-stable aggregates in soils. J Soil Sci 33:141–163
Turbé A, De Toni A, Benito P, Lavelle P, Ruiz N, Van der Putten WH, Labouze E, Mudgal S (2010) Soil biodiversity: functions, threats and tools for policy makers. Bio Intelligence Service, IRD and NIOO Report for European Commission (DG Environment), pp. 251
Veeresh GK, Rajgopal D, Mallik B (1982) Termites and fertility status of soil. VII International Colloquium of soil Zoology, 30th Aug–2nd Sept Louvain-La-Neuve, Belgium
Wallwork JA (1970) Ecology of soil animals. McGraw Hill, London, p 283
Wallwork JA (1982) Desert soil fauna. Prarger Publishers, New York, p 296
Wallwork JA (1983) Oribatids in forest ecosystems. Annu Rev Entomol 28:109–130
Wheeler WM (1910) Ants: their structure, development and behavior. Columbia University Press, New York, p 663
Whitford WG (1991) Subterranean termites and long-term productivity of desert rangelands. Sociobiology 19:235–243
Whitford WG, Freckman DW, Parker LW, Schaefer D, Santos PF, Steinberger Y (1983) The contributions of soil fauna to nutrient cycles in desert systems. In: Lebrun P, André HM, de Medts A, Grégoire-Wibo C, Wauthy G (eds) New trends in soil biology. Imprimerie J Dieu-Brichart: Ottignies-Louvain-la-Neuve, Belgium, pp 49–59
Wilkinson MT, Richards PJ, Humphrey GS (2009) Breaking ground, pedological geological and ecological implications of soil bioturbation. Earth Sci Rev 97:257–272
Winoto Suatmadji R (1969) Studies on effect of Tagetes species on plant-parasitic nematodes. Veenman and Zonen, NV, Wageningen
Yair A, Rutin J (1981) Some aspects of the regional variation in the amount of available sediment produced by isopods and porcupines, northern Negev, Israel. Earth Surf Proc Land 6:221–234
Zimmer M (2002) Nutrition in terrestrial isopods (Isopoda: Oniscidea): an evolutionary-ecological approach. Biol Rev 77:455–493

Butterfly Communities of Ritchie's Archipelago in Andaman and Nicobar Islands, India: Implications for Conservation of Arthropods and Their Habitats

3

C. Sivaperuman and K. Venkataraman

Abstract

The extensive distribution of butterflies in the Andaman and Nicobar islands is an important factor in the ecology of this island ecosystem. Butterflies are recognized as bioindicators and Lepidoptera are known to be susceptible to environmental changes. This work has been carried out on the islands of Ritchie's Archipelago. The Ritchie's Archipelago, a cluster of smaller islands, is situated 25–30 km east of Greater Andaman. This archipelago comprises mainly four larger islands, few smaller islands and several islets, which extends roughly as north-south chain parallel to the main great Andaman island group. The butterfly communities of Ritchie's Archipelago were investigated during 2008 and 2011, the line transect method by employing to assess the population of butterflies. A total of 84 species belonging to five families and 58 genera were recorded during the period of the study. Highest species was recorded in Nymphalidae family followed by families of Lycaenidae, Pieridae and Papilionidae. The Havelock and Neil islands recorded the highest shared species numbers (50 spp.). The diversity index (H′) ranged from 2.76 to 3.96, with the highest index of diversity observed in Havelock Island (3.96). The Ritchie's Archipelago with high representation of butterflies indicated its importance for in situ conservation.

Keywords

Butterfly • Distribution • Diversity • Ritchie's Archipelago

C. Sivaperuman (✉)
Zoological Survey of India, Andaman and Nicobar Regional Centre, Port Blair 744102, Andaman and Nicobar Islands, India
e-mail: c_sivaperuman@yahoo.co.in

K. Venkataraman
Zoological Survey of India, PraniVigyan Bhawan, M-Block, New Alipore, Kolkata 700053, West Bengal, India

A.K. Chakravarthy, S. Sridhara (eds.), *Economic and Ecological Significance of Arthropods in Diversified Ecosystems*, DOI 10.1007/978-981-10-1524-3_3

3.1 Introduction

Several studies have been conducted on butterflies in Andaman and Nicobar islands following the publication of Wood-Mason and de Niceville (1880, 1881a, b and 1882). Evans (1932) conducted studies on butterflies of Andaman and Nicobar islands. Later, many biologists have contributed on the butterfly biodiversity of Andaman and Nicobar islands (Talbot 1939, 1947; Ferrar 1948, 1952; Vane-Wright 1978, 1993; Arora and Nandi 1980, 1982; Chaturvedi 1982; Khatri 1989, 1991, 1992, 1993; Khatri and Singh 1988; Khatri and Mitra 1989a, b; Ripley and Beehler 1989; Veenakumari and Mohanraj 1991, 1996; Chaturvedi and Hussain 1991; Chandra and Khatri 1995; Davidar et al. 1995; Chandra and Rajan 1996; Mohanraj and Veenakumari 1996; Veenakumari et al. 1997; Devy et al. 1998; Sivaperuman et al. 2010, 2011). The present study was taken up to investigate the butterfly communities in Ritchie's Archipelago of Andaman and Nicobar islands.

3.1.1 Study Area

Ritchie's Archipelago is a conglomeration of small islands situated 25–30 km east of Great Andaman, the main centre island group of the Andaman Islands (Fig. 3.1). The archipelago consists of four larger islands, seven small islands and many islets. Across the Diligent Strait lies the Baratang Island and South Andaman Island lies on the western side.

3.1.1.1 Climate

The climate is wet tropical. It is warm and humid most of the year. Seasons can be divided into dry and rainy seasons. The extreme winter and summer are practically unknown, but there is a general nip in the air during December, January and February. During March, April, May and October, the weather can be uncomfortable due to high humidity although the temperature is not high. The average annual temperature ranges from 26.85 to 33.5 °C. The humidity varies from 65 to 91 %. The highest humidity is experienced from May to November during the southwest monsoon. The rainfall ranges from 2,020 to 3,774 mm per year. The southwest monsoon which brings most of the precipitation normally begins in May and ends in October. The northeast monsoon starts during November and ends in December.

3.1.2 Study Area

The following islands were surveyed to assess the community structure of butterflies, namely, Havelock, John Lawrence, Henry Lawrence, South Button, Middle Button, North Button, Inglis, Outram and Neil islands.

Coordinates of transects in different islands: The coordinates of the study area are given in Table 3.1.

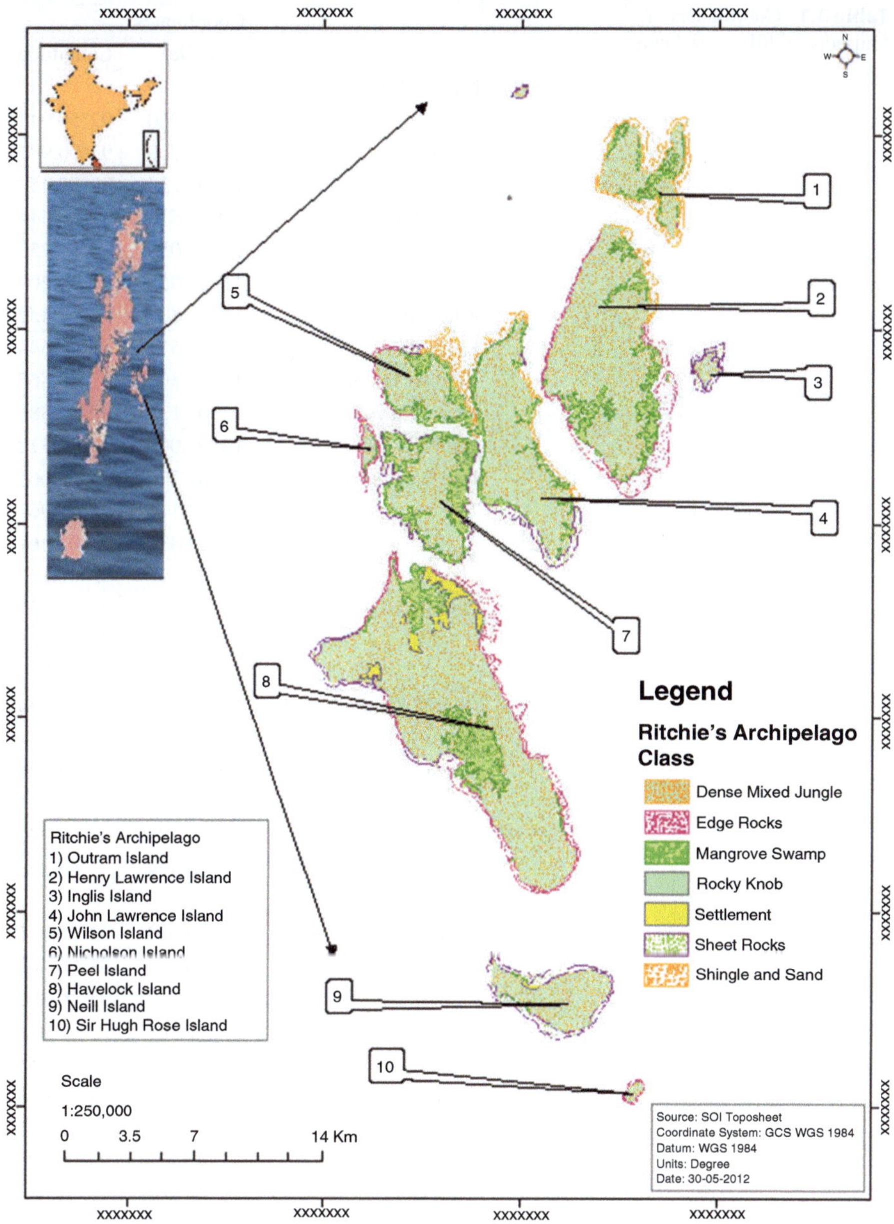

Fig. 3.1 Ritchie's Archipelago, Andaman and Nicobar islands

Table 3.1 Coordinates of transects in different islands

Location	Coordinates	
	Latitude	Longitude
Havelock Island		
Havelock	12° 01.960′	92°50.940′
Kalapathar	11° 58.769′	93°00.980′
Kalapathar	12° 00.235′	93°00.452′
Kalapathar	11° 58.511′	93°00.344′
Radha Nagar	11° 59.050′	92°57.253′
Radha Nagar	11° 59.059′	92°57.209′
Radha Nagar	11° 59.837′	92°57.452′
Kalapathar	11° 58.511′	93°00.344′
Krishna Nagar	11° 59.305′	92°58.865′
Henry Lawrence Island	12° 05.137′	92°04.386′
	12° 05.000′	93°06.312′
John Lawrence Island	12° 04.276′	93°03.063′
	12° 03.116′	93°02.967′
	12° 02.830′	93°02.461′
	12° 03.221′	93°02.146′
Outram Island	12° 13.761′	93°06.055′
	12° 13.537′	93°04.415′
	12° 00.574′	92°56.808′
Middle Button Island	12° 16.473′	93°01.334′
South Button Island	12° 13.467′	93°01.244′
North Button Island	12° 18.974′	93°03.826′
Inglis Island	12° 08.586′	93°06.651′
	12° 08.683′	93°07.252′
	12° 08.454′	93°06.556′
Neil Island		
Neil	11° 59.305′	92°58.865′
Neil	11° 50.571′	93°00.868′
Neil	11° 50.527′	93°00.899′
Sitapur	11° 49.168′	93°03.382′
Sitapur	11° 48.897′	93°03.058′
Sitapur	11° 49.411′	93°03.688′
Sitapur	11° 49.347′	93°02.735′
Rampur	11° 49.229′	93°02.296′
Ramnagar	11° 49.202′	93°02.901′
Lakshmanpur	11° 50.057′	93°01.407′

3.2 Materials and Methods

This study was conducted between November 2008 and December 2011. The estimation of butterfly species is made by 600 m transect, which can be traversed in an hour (Pollard 1977; Pollard and Yates 1993). The enumeration of transects was made between 0600 and 1100 h and data was not collected during heavy rain or strong winds. Identification of butterflies was made using the characteristics of physical features keeping field guides and reference books as identification tools (Evans 1932; Ferrar 1948; Kunte 2000; Kehimkar 2008). Butterflies observed in the particular island were recorded individually. Collections of unfamiliar species for identification were made. A separate note of species observed outside the transects and forest edges was recorded. Butterflies observed in the transects only were considered for further statistical analyses.

Species richness: The total number of species collected in different islands was considered as species richness in the present study.

Diversity indices: Shannon index, Simpson index and Hill's diversity numbers N1 and N2 were calculated for select islands deploying the formula of Ludwig and Reynolds (1988). Computations were done using the programme SPDIVERS.BAS developed by Ludwig and Reynolds (1988).

Shared species: The shared indices were calculated using the computer programme "SPADE" developed by Chao and Shen (2010).

Species rank abundances: The species rank abundances was calculated deploying the biodiversity software (Lambshead et al. 1997).

3.3 Results

3.3.1 Occurrence of Species

A total number of 84 butterfly taxa were recorded in the study period, belonging to five families and 58 genera (Fig. 3.3). Of these, Nymphalidae was the most common family with 29 species, followed by Lycaenidae (17 species), Pieridae (15 species), Papilionidae (13 species) and Hesperiidae (10 species) (Table 3.2). The dominant family Nymphalidae accounted for 52.62 % of individuals, followed by Pieridae (18.67 %) and Papilionidae (16.59 %).

3.3.2 Species Richness and Abundance of Butterflies

Variation of species richness and abundance of butterflies is observed in different islands. The highest record of species richness and abundance was observed in Havelock followed by Neil and Outram (Table 3.3). Highest individuals have been recorded from the family Nymphalidae in Havelock and Neil islands.

3.3.3 Overall Diversity Indices

The overall diversity parameters are presented in Table 3.4. Diversity indices (H′) were 3.97 and (λ) 0.02. Species richness indices (R1) were 9.95 and (R2) 2.03. Likewise, higher values of indices (N1) were 53.16 and Hill's numbers (N2) were 44.25. Evenness indices (E1) were 0.93 and (E2) 0.174.

Table 3.2 Family-wise composition, percentage of species and individuals observed

Family	Number of species	Percentage	Number of individuals	Percentage
Hesperiidae	10	11.90	32	2.47
Papilionidae	13	15.48	215	16.59
Pieridae	15	17.86	242	18.67
Lycaenidae	17	20.24	125	9.65
Nymphalidae	29	34.52	682	52.62
	84		1296	

Table 3.3 Difference observed in line transect method in the study sites

Variables	Havelock	John Lawrence	Henry Lawrence	Inglis	South Button	North Button	Middle Button	Outram	Neil
Butterfly individuals observed	466	110	102	97	69	61	78	110	224
No. of species observed	79	39	37	35	21	25	27	41	56
Sampling effort (km walked)	32	12	12	18	10	10	10	15	30
Percentage of species identified	94.04	46.43	44.05	41.67	25	29.76	32.14	48.80	66.66

Table 3.4 Overall diversity indices in Ritchie's Archipelago

Richness indices		Diversity indices		Hill's number		Evenness indices	
R1	R2	λ	H′	N1	N2	E1	E2
9.95	2.03	0.02	3.97	53.16	44.25	0.93	0.74

3.3.4 Diversity Indices of Butterflies in Different Islands

The proportional abundance of species based on indices is the best approach in measuring the diversity. Most commonly used diversity indices like Shannon index, Simpson's index, Hill's numbers N1 and N2 and Evenness E1 and E2 have been determined. The diversity index (H′) ranged from 2.76 to 3.96, with the highest in Havelock (3.96) (Table 3.5).

3.3.5 Family-Wise Diversity Indices of Butterflies in Different Islands

Nymphalidae showed high values of diversity (H′) in Havelock (3.18), followed by John Lawrence (2.55), Henry Lawrence (2.63), Inglis (2.48), Outram (2.41) and

Table 3.5 Diversity indices of butterflies in different islands of Ritchie's Archipelago

	Richness indices		Diversity indices		Hill's number		Evenness indices	
Islands	R1	R2	λ	H′	N1	N2	E1	E2
Havelock	11.51	3.40	0.02	3.96	52.22	46.74	0.93	0.74
John Lawrence	6.18	2.87	0.04	3.18	23.95	24.24	0.93	0.80
Henry Lawrence	6.54	3.13	0.03	3.27	26.26	30.01	0.95	0.85
Inglis	6.82	3.30	0.03	3.31	27.38	31.55	0.95	0.86
South Button	4.49	2.41	0.06	2.76	15.82	16.88	0.92	0.79
North Button	4.90	2.73	0.06	2.76	15.73	16.11	0.91	0.75
Middle Button	5.10	2.66	0.05	2.91	18.39	19.53	0.93	0.80
Outram	7.23	3.34	0.03	3.35	28.53	30.86	0.94	0.82
Neil	9.36	3.53	0.03	3.61	37.05	33.45	0.92	0.73

Neil (2.74) islands. The South and North Button islands showed high values in Pieridae, followed by Lycaenidae and Papilionidae (Table 3.6). Similarly, the other indices also showed high values in Nymphalidae.

3.3.6 Similarity Indices Between Islands

Using the qualitative data, similarity indices for different islands were computed (Table 3.7). The most abundant species which influences the similarity index is depicted by the high similarity of the community between John Lawrence and Henry Lawrence and the least similarity between John Lawrence and South Button Island.

3.3.7 Shared Species Statistics and Similarity Coefficients

The shared species statistics across pairs of the nine islands are provided in Table 3.8. The species numbers observed in each island and the species encountered in both islands were compared. Havelock and Neil islands showed the highest number of shared species (50 species), as these two islands support large extent of agricultural fields and home gardens.

3.3.8 Species Rank Abundance

To find abundance pattern of the butterfly species, rank abundance curve was plotted using grand mean of the all species. The common species is being displayed on the left and the rare species on the right. While ranking the overall abundance, 20 species were identified more common which is followed by few abundant species, and greater proportion of rare species was observed (Fig. 3.2).

Table 3.6 Family-wise diversity indices of butterflies in different islands of Ritchie's Archipelago

		Richness indices		Diversity indices		Hill's number		Evenness indices	
Islands	Family	R1	R2	λ	H′	N1	N2	E1	E2
Havelock	Hesperiidae	2.06	1.89	0.08	1.55	4.71	12.00	0.96	0.94
	Papilionidae	2.60	1.44	0.18	2.00	7.40	5.63	0.81	0.62
	Pieridae	3.34	1.85	0.06	2.60	13.49	15.55	0.96	0.90
	Lycaenidae	3.15	2.09	0.09	2.28	9.76	10.67	0.92	0.81
	Nymphalidae	4.67	1.67	0.04	3.18	23.97	24.05	0.96	0.89
John Lawrence	Hesperiidae	0.91	1.15	0.25	0.64	1.89	4.00	0.92	0.94
	Papilionidae	1.80	1.32	0.22	1.60	4.94	4.45	0.82	0.71
	Pieridae	1.61	1.44	0.13	1.59	4.90	7.94	0.99	0.98
	Lycaenidae	0.91	1.15	0.25	0.64	1.89	4.00	0.92	0.94
	Nymphalidae	3.11	1.74	0.07	2.55	12.75	14.27	0.96	0.91
Henry Lawrence	Papilionidae	2.26	1.71	0.12	1.93	6.92	8.63	0.93	0.86
	Pieridae	2.40	2.12	0.06	1.73	5.66	15.75	0.97	0.94
	Lycaenidae	1.54	1.51	0.13	1.35	3.86	8.00	0.98	0.97
	Nymphalidae	3.37	1.88	0.06	2.63	13.83	16.00	0.97	0.92
Inglis	Hesperiidae	1.44	1.50	0.13	1.04	2.83	7.50	0.95	0.94
	Papilionidae	2.47	1.94	0.11	1.92	6.83	9.00	0.92	0.85
	Pieridae	2.27	1.87	0.09	1.87	6.50	10.83	0.96	0.93
	Lycaenidae	2.09	1.81	0.08	1.77	5.86	12.00	0.99	0.98
	Nymphalidae	2.99	1.75	0.07	2.48	11.91	13.62	0.97	0.92
South Button	Papilionidae	2.06	1.89	0.08	1.55	4.71	12.00	0.96	0.94
	Pieridae	2.06	1.46	0.14	1.89	6.59	7.13	0.91	0.82
	Lycaenidae	2.50	2.11	0.07	1.89	6.64	15.00	0.97	0.95
	Nymphalidae	1.47	1.10	0.17	1.67	5.29	5.99	0.93	0.88
North Button	Papilionidae	2.28	2.00	0.10	1.68	5.35	10.00	0.94	0.89
	Pieridae	2.15	1.57	0.16	1.81	6.12	6.37	0.87	0.76
	Lycaenidae	1.44	1.50	0.13	1.04	2.83	7.50	0.95	0.94
	Nymphalidae	1.00	0.89	0.25	1.28	3.60	3.99	0.92	0.90
Middle Button	Papilionidae	2.28	2.00	0.15	1.58	4.86	6.67	0.88	0.81
	Pieridae	1.94	1.49	0.12	1.86	6.43	8.05	0.96	0.92
	Lycaenidae	2.01	1.73	0.13	1.68	5.35	7.94	0.94	0.89
	Nymphalidae	1.41	1.01	0.16	1.71	5.54	6.12	0.96	0.92
Outram	Hesperiidae	2.06	1.89	0.08	1.55	4.71	12.00	0.96	0.94
	Papilionidae	2.77	2.12	0.09	2.06	7.86	11.54	0.94	0.87
	Pieridae	2.15	1.57	0.14	1.92	6.79	7.34	0.92	0.85
	Lycaenidae	1.82	1.67	0.10	1.58	4.86	10.00	0.98	0.97
	Nymphalidae	2.97	1.72	0.08	2.41	11.13	12.30	0.94	0.86
Neil	Hesperiidae	1.03	1.13	0.21	1.08	2.94	4.80	0.98	0.98
	Papilionidae	2.56	1.56	0.18	1.94	6.95	5.68	0.81	0.63
	Pieridae	2.79	1.83	0.14	2.08	8.01	7.04	0.87	0.73
	Lycaenidae	2.72	2.06	0.09	2.07	7.92	11.25	0.94	0.88
	Nymphalidae	3.68	1.78	0.06	2.74	15.50	16.00	0.95	0.86

Table 3.7 Similarity matrix for select islands in Ritchie's Archipelago

Islands	Havelock	John Lawrence	Henry Lawrence	Inglis	South Button	North Button	Middle Button	Outram	Neil
Havelock	0.00	0.73	0.83	0.84	0.47	0.42	0.55	0.66	0.87
John Lawrence		0.00	0.99	0.55	0.18	0.35	0.28	0.46	0.57
Henry Lawrence			0.00	0.65	0.36	0.40	0.39	0.58	0.57
Inglis				0.00	0.69	0.66	0.69	0.69	0.79
South Button					0.00	0.77	0.86	0.52	0.41
North Button						0.00	0.77	0.61	0.37
Middle Button							0.00	0.73	0.40
Outram								0.00	0.71
Neil									0.00

Table 3.8 Shared species statistics and similarity coefficients between pairs of nine islands

First sample	Second sample	Shared species	Jaccard classic	Sorensen classic	Morisita-Horn
Havelock	John Lawrence	29	0.40	0.57	0.62
Havelock	Henry Lawrence	31	0.43	0.61	0.68
Havelock	Inglis	32	0.45	0.62	0.68
Havelock	South Button	19	0.26	0.41	0.39
Havelock	North Button	21	0.29	0.46	0.35
Havelock	Middle Button	23	0.32	0.48	0.45
Havelock	Outram	35	0.49	0.66	0.55
Havelock	Neil	50	0.69	0.82	0.77
John Lawrence	Henry Lawrence	23	0.60	0.75	0.79
John Lawrence	Inglis	17	0.38	0.54	0.44
John Lawrence	South Button	9	0.22	0.36	0.15
John Lawrence	North Button	12	0.31	0.47	0.29
John Lawrence	Middle Button	11	0.26	0.42	0.23
John Lawrence	Outram	16	0.33	0.49	0.37
John Lawrence	Neil	20	0.32	0.49	0.47
Henry Lawrence	Inglis	19	0.43	0.60	0.50
Henry Lawrence	South Button	11	0.28	0.43	0.29
Henry Lawrence	North Button	13	0.33	0.50	0.32
Henry Lawrence	Middle Button	12	0.29	0.44	0.31
Henry Lawrence	Outram	19	0.40	0.58	0.45
Henry Lawrence	Neil	20	0.32	0.49	0.46
Inglis	South Button	16	0.44	0.62	0.55
Inglis	North Button	15	0.39	0.56	0.51

(continued)

Table 3.8 (continued)

First sample	Second sample	Shared species	Jaccard classic	Sorensen classic	Morisita-Horn
Inglis	Middle Button	17	0.45	0.62	0.54
Inglis	Outram	17	0.34	0.51	0.54
Inglis	Neil	23	0.38	0.55	0.64
South Button	North Button	13	0.46	0.63	0.55
South Button	Middle Button	14	0.48	0.64	0.69
South Button	Outram	11	0.25	0.40	0.41
South Button	Neil	16	0.29	0.45	0.34
North Button	Middle Button	16	0.57	0.72	0.61
North Button	Outram	16	0.40	0.76	0.48
North Button	Neil	16	0.28	0.44	0.30
Middle Button	Outram	20	0.52	0.68	0.58
Middle Button	Neil	17	0.29	0.45	0.33
Outram	Neil	28	0.48	0.65	0.58

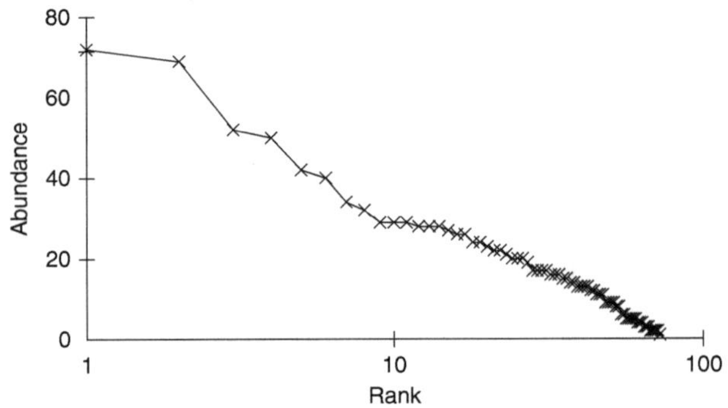

Fig. 3.2 Species rank abundance of butterfly species in Ritchie's Archipelago

3.4 Discussion

A total of 84 species of butterflies belonging to five families of the Order Lepidoptera were recorded. Of these, Nymphalidae with 29 species was the most common family, followed by Lycaenidae (17 species), Pieridae (15 species), Papilionidae (13 species) and Hesperiidae (10 species). Nymphalidae was the dominant family group accounting for 52.62 % individuals. A similar distribution pattern was also reported in Western Ghats (Kunte 1997; Kunte et al. 1999; Eswaran and Pramod 2005; Krishna Kumar et al. 2008; Devy and Davidar 2001; Padhye et al. 2006). The Shannon index showed high value in Havelock and Neil islands. This reflects on the variability in the efficiency of different butterfly species to efficiently utilize the available habitat.

The structural complexity of habitat and diversity of vegetation forms are correlated with species diversity of insects and other animals (Gardner et al. 1995). Southwood (1975) suggested that the quality of food influences herbivores to a considerable extent. Host plant utilization corresponds to the availability of sufficient adult resources only (Grossmueller and Lederhouse 1987). For successful butterfly habitat, it is dire essential to have sufficient larval as well as adult food resources. In the current study, the maximum number of butterfly species and individuals were observed in Havelock and Neil islands, where diverse species of plants were available. The changes in the diversity of different islands in the present study area are also evident from the given data. This is due to the variations in the microhabitat, floristic structure and other habitat parameters. Habitat preference of butterfly species can be directly related to the availability of food plants (Thomas 1995). Some of the islands in Ritchie's Archipelago are occupied by thick mangrove swamps and sandy beaches. In the latter, the littoral or beach forest consists of flowering bushes and this habitat supports more number of species. It was also observed that nymphalids and pierids regularly visited the seashore and settled on damp patches for a few seconds, while others like sailers, lacewings and blues confined themselves to the forested area. The skippers remain within the forest area. From the conservation point of view, several endangered and endemic butterfly species have been recorded from the study area. Many species are specific in their habitat requirements and, hence, are good indicators of habitat diversity of a locality (Kocher and Williams 2000). Localities with greater diversity of habitats generally support a greater number of butterfly species (Kocher and Williams 2000). The Ritchie's Archipelago in Andaman and Nicobar islands support most local species. Scientific management and maintenance of these habitats are very important to maintain and enhance the habitat quality for butterflies.

Andaman and Nicobar islands are unique in endemism (Fig. 3.3); the following species were recorded during the investigation: Giant Red Eye (*Gangara thyrsis*), tailed jay (*Graphium agamemnon*), Andaman Mormon (*Papilio mayo*), Andaman clubtail (*Atrophaneura rhodifer*), Andaman crow (*Euploea andamanensis*), Hewitson Andaman Viscount (*Tanaecia cibaritis*), Andaman birdwing (*Troides helena*), Andaman swordtail (*Graphium epaminondas*), Andaman map (*Cyrestis thyodamas*) and Andaman chestnut palmfly (*Elymnias scottonis*). The habitat degradation is the major threat to biodiversity which includes butterflies. The protected areas' life plays a vital role in the preservation of diversity of butterflies, where populations of some species are contracted largely or entirely to protected areas (Thomas 1995). The high representation of species in Ritchie's Archipelago reinforces the importance of in situ conservation. Long-term monitoring studies are essential where emphasis on host plants as well as the factors influencing the distribution, abundance and diversity of butterflies is taken into consideration.

3.4.1 Butterflies: Conservation Concerns

The Andaman and Nicobar islands prehistorically were endowed with rich tropical humid forests. The islands were in pristine condition until 1788. More than 1,500 flowering plant species of the islands of varying diversity offered suitable habitat for butterflies of which about 301 species are endemic plants. This rich biodiversity of the region is similar to that of Western Ghats and Sri Lanka and hence should also

Fig. 3.3 Butterfly communities of Ritchie's Archipelago in Andaman and Nicobar islands, India: (**a**) Common rose, *Atrophaneura aristolochiae* (Fabricius); (**b**) Andaman Viscount, *Tanaecia cibaritis* (Hewitson); (**c**) Clipper, *Parthenos sylvia* (Cramer); (**d**) lime butterfly, *Papilio demoleus* (Linnaeus); (**e**) tailed jay, *Graphium agamemnon*; (**f**) peacock pansy, *Junonia almana*; (**g**) Andaman Mormon, *Papilio mayo*; (**h**) lemon pansy, *Junonia lemonias*; (**i**) white-banded awl, *Hasora taminatus*; (**j**) striped tiger, *Danaus genutia* (Cramer); (**k**) common albatross, *Appias albino* (*Linnaeus*); (**l**) Andaman birdwing, *Troides helena*

Fig. 3.3 (continued)

be included as one of the major biodiversity "hotspots" (Myers 1990). This closed ecosystem was perturbed by its first major human impact in the year 1858 because of the penal settlement, followed by a gradual expanding of populations over the decades. The impacts gradually increased in their magnitude in the beginning of 1941 when convicts and repatriates from Sri Lanka, Myanmar and India settled in the islands. In the post-independence scenario, ex-service men, more and more refugees from South India, West Bengal and Sri Lanka inflicted further degeneration of habitats. Following the allotment of land to the settlers for agriculture and housing, as well as the illegal encroachment of surrounding lands by these settlers, coupled with clearing the woods for generating income, resulted in severe environmental degradation. Implementation of development schemes increased on several islands accelerating environmental deterioration further.

Table 3.9 Rare (R), very rare (VR) and straggler (S) species of the Andaman and Nicobar islands

		Andaman			Nicobar		
Family	Subfamily	R	VR	S	R	VR	S
Papilionidae		3	–	2	2	1	–
Pieridae		2	–	4	3	–	4
Lycaenidae		11	8	–	15	1	–
Nymphalidae	Nymphalidae	11	2	3	4	3	3
	Danainae	2	1	4	3	–	2
	Amathusinae	2	–	–	–	–	–
	Satyrinae	2	–	–	2	1	–
Hesperiidae		19	2	–	6	1	–
Total		52	13	13	35	7	9

Table 3.10 Endemic subspecies of the Andaman and Nicobar islands

Family	Subfamily	Andaman	Nicobar	Both
Papilionidae	Papilioninae	12	9	–
Pieridae	Pierinae	10	11	1
Lycaenidae		58	34	–
Riodinidae		1	–	–
Nymphalidae	Nymphalinae	34	19	2
	Danainae	8	17	2
	Amathusinae	2	–	–
	Satyrinae	7	6	–
Hesperiidae		38	13	8
Total		170	109	13

The butterfly fauna of the Andaman and Nicobar islands is insular with its origins in the faunas of the Indo-Myanmar and Indo-Malayan regions. The Andaman elements have closer affinities to Myanmar and mainland elements but the Nicobar elements appear most closely related to Malayan elements. The undisturbed ecology with optimal conditions led to evolution of several local as well as endemic taxa because of long isolation of the islands from the mainland and other countries of Asian continent. Evans (1932) described about 260 forms of butterflies, followed by Ferrar (1951) who described 268 forms from Andaman and Nicobar islands. At present there are 214 species, 236 subspecies which is included in 116 genera belonging to five families and three subfamilies. These islands are endemic to more than 50 % of the taxa. Many endemic taxa in each family are rare, very rare or stragglers (Tables 3.9, 3.10 and 3.11). The Indian Wildlife Protection Act (1972) has not included these taxa in Schedule I, II or III so they cannot be stated as "threatened" or "endangered". Lycaenid recorded the maximum number of endemic subspecies compared to other butterfly families (Table 3.10). Most of the lycaenid subspecies may be specialists requiring specialized niche. Hesperids were next in the order of abundance of endemic subspecies. Papilionidae recorded the maximum (Table 3.11)

Table 3.11 New records of butterflies during the last decade from the Andaman and Nicobar islands

Family	Subfamily	Andaman	Nicobar
Papilionidae	Papilioninae	7	4
Pieridae	Pierinae	4	3
Lycaenidae		2	2
Nymphalidae	Nymphalinae	2	–
	Danainae	1	1
	Amathusinae	1	–
	Satyrinae	2	–
Hesperiidae		1	–
Total		20	10

number of new species for the Andaman and Nicobar islands (Table 3.11). This is possibly because papilionids are comparatively larger sized butterflies and it is easier to detect them compared to butterflies belonging to other families. Modernization of the islands resulted in habitat destruction because of deforestation, with correlative decreases in population viability of the butterfly fauna. Because of host specificity of several butterflies, their inability to adapt to ecological changes results in habitat loss and in turn their survival. Many number of species here may already be extinct due to continuing destruction of habitat during the past 50 years. The Government of India has initiated several programmes which include forest-based industries, agroforestry, agriculture and tourism. Such programmes pose immediate threats not only to the irreplaceable endemic butterflies but to most other components of the endemic biota. There is a dire need to preserve and conserve the butterfly biodiversity for which immediate implementation of monitoring schemes to both evaluate present status and dynamics over time. Thus far the government action has included declaration of a biosphere reserve, designation of national parks and sanctuaries and implementation of the Wildlife Protection Act 1972. These actions do not appear effective in checking the degradation of most butterfly species and the natural resources upon which they are dependent. Additional measures are essential to counter the impacts of the human activities and stress on conservation of the critical habitats to prevent mass extinction of many endemic species and subspecies from the biogeographically rich Andaman and Nicobar islands.

References

Arora GS, Nandi DN (1980) On the butterfly fauna of Andaman and Nicobar Islands (India) I. Papilionidae. Rec Zool Sur India 77:141–151

Arora GS, Nandi DN (1982) On the butterfly fauna of Andaman and Nicobar Islands India II. Pieridae. Ibid 80: 1–15

Chandra K, Khatri TC (1995) Butterflies of great Nicobar Islands. Indian J For 18(4):276–273

Chandra K, Rajan PT (1996) Butterflies of North, Middle and South Button Island National Park, Andaman. Insect Environ 2(3):110–111

Chao A, Shen TJ (2010) Program SPADE (species prediction and diversity estimation). Program and user's guide, http://chao.stat.nthu.edu.tw

Chaturvedi N (1982) Butterflies of Andaman Islands with some new records. J Bombay Nat Hist Soc 79(3):702–704

Chaturvedi N, Hussain SA (1991) Some butterflies of Narcondam Island (Andaman). J Bombay Nat Hist Soc 88(3):463

Davidar P, Devy S, Yoganand TRK, Ganesh T (1995) Reserve size and implications for the conservation of biodiversity in the Andaman Islands. In: Boyle TJB, Boontawee B (eds) Measuring and monitoring biodiversity in tropical and temperate forests. CIFOR, Jarkarta, pp 287–302

Devy MS, Davidar P (2001) Response of wet forest butterflies to selective logging in Kalakad-Mundanthurai Tiger Reserve: implications for conservation. Curr Sci 80(3):400–405

Devy MS, Ganesh T, Davidar P (1998) Patterns of butterfly distribution in the Andaman Islands: implications for conservation. Acta Oecologia 19(6):527–534

Eswaran R, Pramod P (2005) Structure of butterfly community of Anaikatty Hills, Western Ghats. Zoos' Print J 20(8):1939–1942

Evans WH (1932) The identification of Indian butterflies. Bombay Natural History Society, Bombay, p 454

Ferrar ML (1948) Butterflies of Andaman and Nicobar Islands. J Bombay Nat Hist Soc 47(3):470–491

Ferrar ML (1951) On the butterflies of Andaman and Nicobar Islands. J Bombay Nat Hist Soc 47:470–491

Ferrar ML (1952) On the butterflies of the Andaman and Nicobar Islands. J Bombay Nat Hist Soc 47:470–491

Gardner SM, Cabido MR, Valladares GR, Diaz S (1995) The influence of habitat structure on arthropod diversity in Argentine semiarid Chaco forest. J Veg Sci 6:349–356

Grossmueller DW, Lederhouse RC (1987) The role of nectar source distribution in habitat use and oviposition by the tiger swallowtail butterfly. J Lepidop Soc 41(3):159–165

Kehimkar I (2008) The book of Indian butterflies. Bombay Natural History Society, Mumbai, p 497

Khatri TC (1989) A revised list of butterflies from Bay Islands. J Andaman Sci Assoc 5:57–61

Khatri TC (1991) On some Nymphalidae (Rhopalocera: Lepidoptera) from the Andaman and Nicobar Islands. Islands on Mar 3: 82–94

Khatri TC (1992) On some Lycaenids (Rhopalcocera: Lepidoptera) from Andaman and Nicbar Islands. Islands on Mar 6: 8–16

Khatri TC (1993) Butterflies of the Andaman and Nicobar Islands: conservation concerns. J Res Lepidop 32:170–184

Khatri TC, Mitra B (1989a) On some Danaidae (Lepidoptera: Rhopalocera) from the Andaman and Nicobar Islands. Hexapoda 1:109–116

Khatri TC, Mitra B (1989b) Some Pieridae (Lepidoptera: Rhopalocera) from Andaman and Nicobar Islands. Hexapoda 1:127–137

Khatri TC, Singh RK (1988) Some Papilionidae (Lepidoptera: Rhopalocera) from Andaman and Nicobar Islands. J Andaman Sci Assoc 4(1):39–46

Kocher SD, Williams EH (2000) The diversity and abundance of North American butterflies vary with habitat disturbance and geography. J Biogeogr 27:785–794

Krishnakumar N, Kumaraguru A, Thiyagesan K, Asokan S (2008) Diversity of papilionid butterflies in the Indira Gandhi wildlife sanctuary, Western Ghats, southern India. Tigerpaper 35:1–8

Kunte K (1997) Seasonal patterns in butterfly abundance and species diversity in four tropical habitats in northern Western Ghats. J Biosci 22(5):593–603

Kunte K (2000) India – a lifescape butterflies of Peninsular India. Indian Academy of Science, Bangalore, Universities Press, Bangalore, p 270

Kunte K, Joglekar A, Ghate U, Padmanabhan P (1999) Patterns of butterfly, bird and tree diversity in the Western Ghats. Curr Sci 77:577–586

Lambshead PJD, Paterson GLJ, Gage JD (1997) BioDiversity professional beta. The Natural History Museum and the Scottish Association for Marine Science Bombay. Mar Ecol Syst 1(2):60–66

Ludwig JA, Reynolds JF (1988) Statistical ecology, a premier on methods and computing. A Wiley-Intersceince publication, New York, p 337

Mohanraj P, Veenakumari K (1996) Host plants, phenologies and status of swallowtails (Papilionidae: Lepidoptera) in the Andaman and Nicobar Islands, Bay of Bengal. Indian Ocean Biol Con 78:215–221

Myers (1990) The biodiversity challenge: expanded hot spots analysis. Environmentalist 10:243–256

Padhye AD, Dahanukar N, Paingankar M, Deshpande M, Deshpande D (2006) Season and landscape wise distribution of butterflies in Tamhini, Northern, Western Ghats, India. Zoo's Print J 21(3):2175–2181

Pollard E (1977) A method for assessing changes in the abundance of butterflies. Biol Con 12:115–134

Pollard E, Yates TJ (1993) Monitoring butterflies for ecology and conservation. Chapman and Hall, London, p 292

Ripley SD, Beehler BM (1989) Ornithogeographic affinities of the Andaman and Nicobar Islands. J Biogeogr 16:232–232

Sivaperuman C, Shah SK, Raghunathan C (2010) Diversity of butterflies in Great Nicobar Biosphere Reserve (GNBR), A and N Islands. In: Proceedings of 3rd Asian Lepidoptera conservation symposium and training programme, Zoo Outreach Organisation Coimbatore, Tamil Nadu, India pp 21

Sivaperuman C, Shah SK, Raghunathan C, Ramakrishna (2011) Structure and species composition of butterflies in Great Nicobar Biosphere Reserve, Andaman and Nicobar Islands. In: Tyagi BK, Veer V (eds) Entomology: ecology and biodiversity. Scientific Publisher, Jodhpur, pp 168–179

Southwood TRE (1975) The dynamics of insect populations. In: Insects science and society. Academic Press, New York, pp 151–199

Talbot G (1939) The fauna of British India including Ceylon and Burma (butterflies). Taylor and Francies, London, p 74

Talbot G (1947) Fauna of British India, vol II, Butterflies. Today & Tomorrow Printers and Publishers, New Delhi, p 273

Thomas JA (1995) The ecology and conservation of Maculineaarion and other European species of large blue butterfly. In: Pullin AS (ed) Ecology and conservation of butterflies. Chapman & Hall, London, pp 180–197

Vane-Wright RI (1978) Ecological and behavioursal origins of diversity in butterflies. In: Diversity of insect faunas. Mound LA, Waloff N (eds) Symp R Entomol Soc London 9: 56–70

Vane-Wright RI (1993) Milkweed butterflies (Lepidoptera: Danainae) and conservation priorities in the Andaman and Nicobar Islands, India. Butterflies 4:21–36

Veenakumari K, Mohanraj P (1991) *Erionota thrax* L. (Lepidoptera: Hesperiidae), a new record to Andaman Islands. J Andaman Sci Assoc 7(2):91–92

Veenakumari K, Mohanraj P (1996) Why Ferrar failed to find second specimen of Polyura Schreiber tisamenus Fruhstorfer (Lepidoptera: Nymphalidae) in the Andaman Islands, Bay of Bengal, Indian Ocean. Entomologist 115(3):159–160

Veenakumari K, Mohanraj P, Sreekumar PV (1997) Host plant utilization by butterfly larvae in the Andaman and Nicobar Islands (Indian Ocean). J Insect Conserv 1:235–246

Wood-Mason J, de Niceville L (1880) List of diurnal Lepidoptera from Port Blair. Andaman Islands. J Asiat Soc Bengal 49(2):223–243
Wood-Mason J, de Niceville L (1881a) List of diurnal Lepidoptera from Port Blair. Andaman Islands. J Asiat Soc Bengal 49(2):223–243
Wood-Mason J, de Niceville L (1881b) Second list of Rhopalocerous Lepidoptera from Port Blair. Andaman Islands. J Asiat Soc Bengal 50(4):243–262
Wood-Mason J, de Niceville L (1882) Second list of Rhopalocerous Lepidoptera from Port Blair. Andaman Islands. J Asiat Soc Bengal 11:14–20

Documenting Arthropods in Select Wild and Cultivated Ecosystems in Iran and Kuwait

4

A.A. Seraj, M. Esfandiari, and Wasmia Al-Houty

Abstract

Studies were conducted on insect fauna in Karkheh Wildlife Refuge (KWLR), southwest Iran. Insects were sampled from May to October. Species of insects were classified into morph species and allotted to the species as far as possible. A total of 2207 insects were collected deploying a combination sampling methods, e.g. netting, beating, hand-picking and trapping. Insects sampled belonged to 100 species, 47 families and 13 orders of Insecta. Based on the information gathered, Coleoptera, with 32 species, had the highest diversity index of $H' = 0.318$ and were the most diverse and abundant among sampled insects. The staphylinid beetle *Achenium debile* (Erichson) was the major species, with relative abundance of 23.9 %. The bug *Pyrrhocoris apterus* Linnaeus was categorized as the second dominant species, with abundance of 8.4 %. The Shannon-Wiener and Simpson's indices were 3.286 and 0.91, respectively. Evenness of species was 0.7, using Pielou's index. Results of studies on insects from other parts of Iran and Kuwait are also discussed in this chapter.

Keywords

Dominant species • Insect biodiversity • Southwest Iran • Shannon-Wiener

A.A. Seraj (✉) • M. Esfandiari
Department of Plant Protection, College of Agriculture, Shahid Chamran University of Ahvaz, Ahvaz, Iran
e-mail: seraja@gmail.com

W. Al-Houty
Department of Biological Sciences, Faculty of Science, Kuwait University, P.O. Box-5969, Safat 13060, Kuwait
e-mail: w.alhouty@ku.edu.kw

A.K. Chakravarthy, S. Sridhara (eds.), *Economic and Ecological Significance of Arthropods in Diversified Ecosystems*, DOI 10.1007/978-981-10-1524-3_4

4.1 Introduction

Insects constitute more than half of the world's known animal species. Insects have a crucial and often central role in ecosystem functioning and define the complex nature of biodiversity, and their living quality is an index of ecosystem health (Jana et al. 2006). There are few studies on insect biodiversity in southwest Iran, e.g. Soleimanne Zhadian (2009) showed that planting strips of alfalfa around sugarcane fields could increase biodiversity indices and resulted in reduction of the damages caused by *Sesamia* spp. (Lepidoptera: Noctuidae) in Khuzestan. Other research studies on Thysanoptera in wheat fields of Khuzestan revealed that due to high diversity of this order and interspecific competition, thrips species cannot outbreak in the mentioned wheat fields (Ramazani 2010).

Karkheh Wildlife Refuge (KWLR) is located in the west of Khuzestan Province (southwest Iran) along Karkheh River and encompasses over 13,000 ha. This area is excessively hot and dry in the summer and the annual average rainfall does not exceed 300 mm. KWLR possesses green and thick plantations which contain *Populus* spp., *Tamarix* sp. and *Lycium* sp. Such dense forests have suitable habitat conditions by which one can conserve the diversity of wildlife. KWLR has crucial part in the ecology and biodiversity of the region. It is the habitat of Persian fallow deer, *Dama mesopotamica* (Brooke), which is nearly extinct today (Shalbaf 2011). At present, 30 Persian fallow deers are kept in a 70 ha protected site. The aim of this survey was to estimate diversity, evenness and species richness of insect populations at the habitat of Persian fallow deer. There was no data on insects of KWLR earlier to the current study.

4.2 Materials and Methods

4.2.1 Study Area

This investigation was initiated at the protected site of Persian fallow deer (70 ha), located at Karkheh Wildlife Refuge (KWLR), at approximate coordinates of 31° 55′ N 48° 15′ E. The main vegetations of the area are *Populus* sp., *Tamarix* sp., *Lycium* sp., dewberries and grasses of different types.

4.2.2 Sampling Plan

Sampling was conducted during April to September 2010, twice a month: a 2 h day sampling and an hour night sampling. Sampling methods were non-specific according to Southwood (1978) and included netting, beating, hand-picking and trapping (pitfall and light traps). Netting was carried out with a 30 cm net swept over the topmost 20–30 cm of vegetation (grasses, low shrubs, etc.).

4.2.3 Material Identification

Specimens were sorted and identified to the lowest taxonomic level possible, at least to family level based on available resources and keys. Samples were sorted to recognizable taxonomic units (RTUs), according to Oliver and Beattie (1993). A total of 100 morphospecies belonging to 13 orders and 47 families were separated.

4.2.4 Data Analysis

All the indices were assessed following the software Species Diversity and Richness (SDR) version 3.0 (Pisces Conservation) (Henderson and Seaby 2002). The following diversity indices were used to calculate insect diversity at the study sites as per Magurran (2004): one is the Shannon-Wiener index, $H = \sum (pi - \text{In}(pi)$, where S is the number of species recorded in a segment of study area and pi is the proportion of individuals of ith species, and the other is Simpson's $1 - D = 1 - \sum [ni(ni-1)/N(N-1)]$, where ni is the number of the individuals of the species and N is the total number of individuals per unit study area. Evenness was calculated by Pielou's index, $E = H^{'} / \ln S$, where H' is the Shannon-Wiener function and S is the total number of species recorded.

Dominance index:

$$C = \sum (\frac{ni}{N})^2$$

Insects were also sampled from other regions of Iran and some parts of Kuwait by entomologists and are included in this paper.

4.3 Results

The research studies were conducted from April to September in 2010. A total of 2207 specimens were captured, belonging to species numbering 100, 47 families and 13 insect orders (Table 4.1). It is worth mentioning that the absolute numbers provided are an underestimate of the total diversity as all microhabitats in the study area were not sampled. Insects such as borers in plants, detritivores living in nests and soil-dwelling orders like Collembola were not sampled. Results (Table 4.1) indicated that 76 % of the recognizable taxonomic units (RTUs) belonged to just five orders: Coleoptera, Lepidoptera, Hemiptera, Homoptera and Hymenoptera.

Depending on the data collected, Coleoptera, with 32 species and diversity index of H′ = 0.318, were the most diverse and abundant order among collected insects. Hemiptera and Lepidoptera were ranked as second and third after Coleoptera, respectively (Table 4.1). Staphylinidae and Scarabaeidae were the most abundant as well as diverse families among Coleoptera families based on H′ value. The staphylinid beetle *Achenium debile* (Erichson) was the predominant species, with 23.9 %

Table 4.1 Number of families, RTUs and H′ values for each insect order collected at KWLR (2010)

Orders	Number of families	Number of RTU	H′ value
Ephemeroptera	1	1	0.009
Odonata	2	2	0.006
Orthoptera	3	6	0.129
Dictyoptera	2	2	0.012
Dermaptera	1	1	0.012
Hemiptera	8	14	0.310
Homoptera	6	6	0.050
Neuroptera	4	8	0.078
Coleoptera	8	32	0.318
Diptera	1	2	0.003
Trichoptera	1	2	0.115
Lepidoptera	7	18	0.202
Hymenoptera	3	6	0.150

Table 4.2 Diversity indices of Shannon-Wiener and Simpson, evenness and abundance of insects in KWLR (2010)

Sampling dates	Abundance	Species diversity (H′)	Species diversity (1–D)	Evenness (E)
May	405	2.34	0.64	0.66
June	999	1.97	0.72	0.56
July	422	2.75	0.91	0.78
August	148	2.13	0.83	0.75
September	108	2.05	0.81	0.76
October	125	2.7	0.91	0.86
Total	2207	3.28	0.91	0.71

of relative abundance. Diurnal firebug *Pyrrhocoris apterus* Linnaeus was categorized as the next dominant species, with relative abundance of 8.4 %. Dermaptera and Ephemeroptera, each of them with one species, had the least species number.

Shannon-Wiener and Simpson's biodiversity index values were 3.286 and 0.91, respectively, in this study. Evenness was 0.7, deploying Pielou's index. Overall biodiversity indices revealed higher biodiversity in the region, but existence of dominant species (e.g. *Achenium debile* and *Pyrrhocoris apterus*) decreased biodiversity and evenness indices in some monthly samples (Table 4.2).

4.4 Discussion

It is difficult to compare our results to regional biodiversity data since most of the previous studies deal with a single taxon and data on similar habitats are unavailable. However, the fauna of Noctuidae (Lepidoptera) of sugarcane fields (Esfandiari

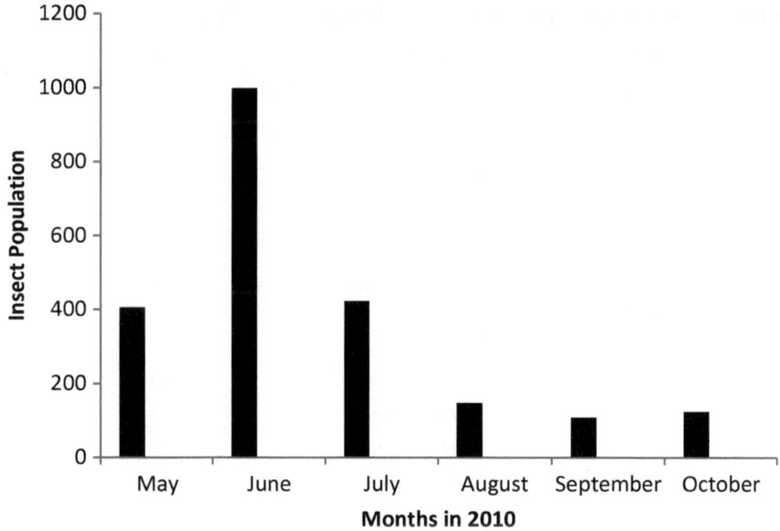

Fig. 4.1 Monthly changes in the population of insects at KWLR in 2010 (6 months)

et al. 2010), 30 km to KWLR, shows similarity to the collected noctuids at KWLR: e.g. *Clytie* spp., which feed on *Tamarix* spp., was common in night samples. Consequently, the biodiversity (diversity index, species richness and evenness) of insects in KWLR was significant because of the vegetation of study sites. Vegetation has a crucial role for the presence of insect community by providing food and breeding sites for insects.

Based on this example, it seems that excessive warm weather in midsummer decreased diversity and abundance of insects due to its effect on development and reproduction of insects. Suitable ecological conditions and climatic factors such as optimum temperature should increase the insects' abundance gradually after summer (Fig. 4.1).

This work was initiated to document features of biodiversity of insect populations of KWLR. Further study is required in this context, and further collections are necessary for having a detailed periodic assessment of biodiversity of insects and establishment of standard monitoring practices for estimating the environmental quality in this area.

4.5 Biodiversity of Aquatic Insects in Isfahan Province, Iran

Insects in water bodies are the dominant group of arthropods that have a crucial part in spread of human and animal diseases. In 2011, aquatic insects were sampled using aquatic net in Zayandeh-Rood River and its branches, Isfahan Province, Iran (Shayeghi et al. 2014). A total of 741 aquatic insects were sampled and identified. They comprised seven families and 12 genera representing two orders: Diptera

Table 4.3 The prevalence of aquatic insects in the study area, 2011

Order	Family	Genus	No.	Percent (%)
Diptera	Culicidae	*Culex* (*Culex theileri*)	384	51.82
	Syrphidae	*Eristalis*	4	0.54
	Chironomidae	*Chironomus*	296	39.95
Coleoptera	Gyrinidae	*Gyrinus*	6	0.81
	Dytiscidae	*Agabus*	5	0.67
		Dytiscus	13	1.75
		Hydroporus	6	0.81
	Haliplidae	*Haliplus*	4	0.54
		Peltodytes	3	0.40
	Hydrophilidae	*Laccobius*	10	1.35
		Enochrus	4	0.54
		Hydrobius	6	0.81
Total	7 families	12 genera	741	100

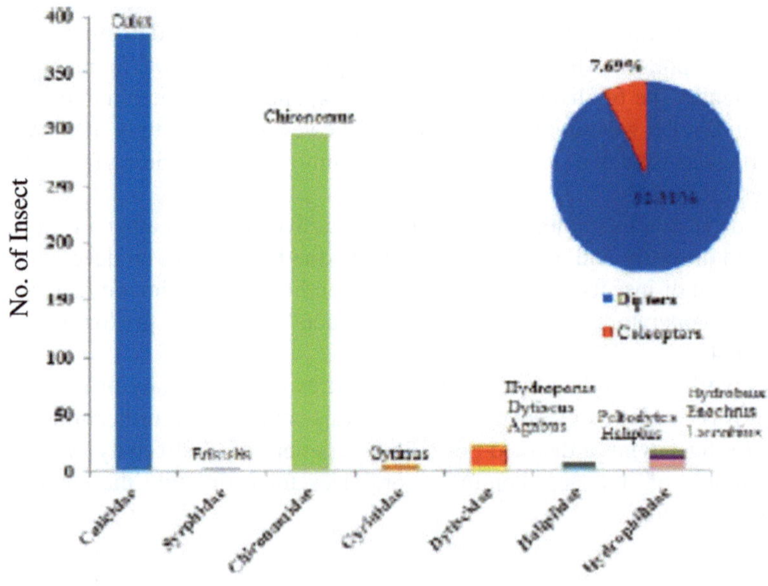

Fig. 4.2 Aquatic insect genus and family composition in study area

(92.31 %) and Coleoptera (7.69 %). The families Culicidae, Syrphidae and Chironomidae from Diptera and Gyrinidae, Dytiscidae, Haliplidae and Hydrophilidae from Coleoptera were classified (Table 4.3; Figs. 4.2 and 4.3).

Fig. 4.3 Families of collected aquatic insects: (**a**) Hydrophilidae, (**b**) Haliplidae, (**c**) Gyrinidae, (**d**) Dytiscidae, (**e**) Culicidae, (**f**) Chironomidae and (**g**) Syrphidae

4.5.1 Diversity of the Genus *Dolichopus* Latreille in Three Select Habitats of East Azerbaijan Province, with New Records for Iran

Diversity of *Dolichopus* Latreille in three different habitats of East Azerbaijan Province, with new records for Iran, was surveyed by a standard entomological net in forest, grassland and wetland areas in northwest Iran during 2013. Based on the data gathered, the forest with the highest diversity indices ($H' = 2.53$, 14 species, and $H' = 2.19$, ten species, in Chichakli and Keleybar regions, respectively) had the most number of species and dominant abundant species, followed by grassland and

Table 4.4 Comparison of relative abundance of *Dolichopus* species in East Azerbaijan Province, Iran

Study area				
Species	Chichakli	Kandovan	Keleybar	Qurigol
Dolichopus austriacus	5.25	0	0	0
Dolichopus malekii	17.35	0	9.09	0
Dolichopus campestris	5.25	10	9.09	0
Dolichopus clavipes	5.25	6.67	9.09	0
Dolichopus griseipennis	0	0	0	4.76
Dolichopus immaculatus	2.35	0	0	0
Dolichopus kiritshenkoi	0	20	0	0
Dolichopus longitarsis	12.58	26.67	20	0
Dolichopus nubilus	5.25	0	0	0
Dolichopus perversus	5.25	0	1.82	71.43
Dolichopus plumipes	10.45	0	10.91	0
Dolichopus salictorum	0	0	0	4.76
Dolichopus siculus	7.17	0	12.72	0
Dolichopus signifier	6.24	13.33	7.27	0
Dolichopus simplex	7.17	23.33	14.55	19.05
Dolichopus subpennatus	7.17	0	5.46	0
Dolichopus ungulatus	3.26	0	0	0

wetland. The dominant species in the study area were *Dolichopus longitarsis* and *D. simplex*. In addition, three species (*D. siculus*, *D. kiritshenkoi* and *D. plumipes*) were recorded from Iran as a first report (Table 4.4).

4.5.2 Species Diversity in East Azerbaijan Province

Diversity indices for several sites are depicted in Table 4.2. The species number was maximum in Chichakli and subsequently Keleybar and Kandovan, while Qurigol had the minimum number of species. The diversity and evenness indices revealed a significant difference among the study sites at 5 % level ($P < 0.05$). They were maximum in Chichakli followed by Keleybar and Kandovan, while Qurigol had the minimum diversity index (Table 4.5).

4.6 Insect Fauna in Kuwait

With the Iraqi invasion in Kuwait and the subsequent Gulf War, there have been changes in the climate and habitats altering the fauna and flora of Kuwait. A survey of insects collected from January 1992 to December 2009 repealed occurence of 174 species representing 146 genera. When added to the previous checklist published in 1997, the total species documented from Kuwait is now 648 species

Table 4.5 Shannon-Wiener and Simpson's diversity indices and Pielou's *J* evenness index for *Dolichopus* in East Azerbaijan Province, Iran

Study area				
Diversity indices	Chichakli	Kandovan	Keleybar	Qurigol
Species number	14	6	10	4
Shannon-Wiener (H')	2.53	1.69	2.19	0.89
Simpson (1–D)	12.48	5.85	9.70	1.93
Evenness (J)	0.90	0.60	0.77	0.31

Table 4.6 Number of insect species recorded from 1980 to 1990 and from 1991 to 2009

	1980–1990		1991–2009	
Order	Genus	Species	Genus	Species
Thysanura	2	2	2	2
Ephemeroptera	–	–	1	1
Odonata	7	11	7	11
Orthoptera	17	24	5	5
Dictyoptera	7	8	3	4
Embioptera	–	–	1	1
Isoptera	2	2	2	2
Dermaptera	2	2	–	–
Mallophaga	4	4	–	–
Anoplura	2	3	–	–
Thysanoptera	1	1	–	–
Hemiptera	20	23	37	45
Homoptera	12	12	12	12
Neuroptera	7	9	1	1
Coleoptera	87	148	56	73
Strepsiptera	1	1	–	–
Trichoptera	1	1	–	–
Lepidoptera	57	75	15	15
Diptera	62	82	14	16
Siphonaptera	2	4	2	4
Hymenoptera	45	62	13	14
Total		474		206

belonging to 489 genera in 127 families and 21 orders. Two species found in each genera of one family and one order are wingless (Apterygota), and the remaining 646 species are having wings (Pterygota). Of the total, 145 species, 123 genera in 42 families and 12 orders are Exopterygota, the remaining 501 species belonged to 364 genera and 48 families and 8 orders of Endopterygota. Documentation of arthropods in Iran and Kuwait will certainly help other workers to carry out studies in the two countries. As there is no baseline data on arthropods, it is essential that conservation of arthropods be prioritized (Table 4.6).

Acknowledgements We are grateful to Iranian environmental organization authorities at Khuzestan Province who supplied the collecting permits. Supports provided by Shahid Chamran University of Ahvaz and Islamic Azad University of Ahvaz are greatly acknowledged. We also thank Dr. Volker Assing, Hannover, Germany, for his help in identification of the dominant species (*Achenium debile*).

References

Esfandiari M, Mossadegh MS, Shishehbor P (2010) The Noctuidae *s.l.* (Lepidoptera) in sugarcane fields of Iran. In: IXth European congress of entomology, book of abstracts. Hungarian Natural History Museum Budapest, Hungary pp 144

Henderson PA, Seaby RMH (2002) Species diversity and richness-Projecto Mamiraua, Version 3.0. Pisces Conservation Ltd

Jana G, Misra KK, Bhattacharya T (2006) Diversity of some insect fauna in industrial and non-industrial areas of West Bengal India. J Insect Conserv 10:249–260

Magurran AE (2004) Measuring biological diversity. Blackwell Publishing, Oxford, p 256

Oliver I, Beattie AJ (1993) A possible method for the rapid assessment of biodiversity. Conserv Biol 7:562–568

Ramazani L (2010) Thysanoptera of Khuzestan: their biodiversity and population dynamics of dominant species in wheat crops. Ph.D dissertation, Shahid Chamran University of Ahvaz, Iran pp 237

Shalbaf SH (2011) Survey on insect biodiversity and the determination of dominant species in the habitate of *Dama mesopotamica* in Karkheh Wild Life Refuge. MSc. thesis of Islamic Azad University of Ahvaz-Science and Research Branch, Iran pp 95

Shayeghi M, Vatandoost H, Gorouhi A, Sanei-Dehkordi AR, Salim-Abadi Y, Karami M, Jalil-Navaz MR, Akhavan AA, Shiekh Z, Vatandoost S, Arandian MH (2014) Biodiversity of aquatic insects of Zayandeh-rood river and its branches, Isfahan Province, Iran. Arthropod-Borne Dis 8(2):197–203

Soleimanne zhadian E (2009) Alfalfa planting beside sugarcane and its effect on biodiversity and sugarcane stemborers damage. Plant Prot (Sci J Agric) 32(1):1–13

Southwood TRE (1978) Ecological methods with particular reference to the study of insect population. Chapman and Hall, London, p 525

An Appraisal of Select Insect Taxa in Sri Lanka

5

J.P. Edirisinghe, W.A.I.P. Karunaratne, I.I. Hemachandra, N.R. Gunawardene, and C.M.D. Bambaradeniya

Abstract

The chapter provides information on diversity of selected insect taxa, namely, Isoptera, Aphididae, Thysanoptera, Formicidae, and Apidae, and of the rice fields in Sri Lanka. Isoptera of the island comprises 76 species in 29 genera and 4 families, with 33 species restricted to the island. Fauna is rich in wood-feeding foragers and fungus-growing wood feeders and poor in humus and soil feeders. Seventy-four species of aphids in 40 genera and 8 subfamilies have been documented with the exception of subfamily Anoeciinae. Host plant specificity is not so pronounced among aphids except for a few restricted to endemic plants and specific weeds. Thysanoptera are represented by 113 species in 63 genera. Among them are many cosmopolitan pests and several potential viral vector species with vegetables, ornamentals, and cut flowers harboring a large majority of thrips. Ants comprise 181 species in 61 genera and 12 subfamilies. Of special interest is the endemic SF Aneuretinae, solely represented by the relict ant, *Aneuretus simony*, now known to be widely distributed and abundant. Bees comprise 144 species of pollen bees and 4 species of honeybees in 38 genera and 4 families. Pollen bees are best known for their nesting habits as ground, hollow stem (leaf-cutter bees), and wood (carpenter bees) nesters. Among them are several

J.P. Edirisinghe (✉) • W.A.I.P. Karunaratne • I.I. Hemachandra
Department of Zoology, Faculty of Science, University of Peradeniya, Peradeniya, Sri Lanka
e-mail: jpediri48@gmail.com; inokap@pdn.ac.lk; indujinah@yahoo.com

N.R. Gunawardene
Curtin Institute for Biodiversity and Climate, Department of Environment and Agriculture, Curtin University, GPO Box U1987, Perth, WA 6845, Australia
e-mail: n.gunawardene@curtin.edu.au

C.M.D. Bambaradeniya
No. 7620, Oldfield Lane, Ellicott City, MD 21043, USA
e-mail: cbambaradeniya@gmail.com

A.K. Chakravarthy, S. Sridhara (eds.), *Economic and Ecological Significance of Arthropods in Diversified Ecosystems*, DOI 10.1007/978-981-10-1524-3_5

specialist bees. Rice fields are rich in insects, due to their habitat heterogeneity, harbored 317 species, belonging to 19 orders and 104 families, during the two cultivation cycles, *Yala* and *Maha*.

Keywords
Endemism • Natural history • Plant relationships • Species composition

5.1 Introduction

5.1.1 Geological History

Sri Lanka (Ceylon) was once part of the Indo Malayan and Afro-Madagascan complex. About 135 million years ago, Sri Lanka in common with India, Australia, and Africa formed a single land mass, the Gondwanaland after the breakup of the single land mass, the super continent Pangea. When Gondwanaland broke up forming a series of plates about 100 million years ago, the Indian plate formed consisted of Peninsular India and Sri Lanka. In the late Miocene, about 12 million years ago, Sri Lanka got separated from South Indian Peninsula by the Palk Strait. Sri Lanka's indigenous flora and fauna are shared largely with those in the Indian Peninsula. The Western Ghats in India (located in the South Indian states of Karnataka and Kerala) and the Sinharaja forest (in Southwestern Sri Lanka) together forms a global biodiversity hot spot due to the high biological diversity and severe threat to its fauna and flora. In the whole of Asia, Sri Lanka has the highest diversity of fauna per unit area of land mass.

5.1.2 Location and Climate

Sri Lanka lies in the Indian Ocean between longitudes E 79° 39′ and 81° 53′ and latitudes N 5° 54′ and 9° 52′. It is 435km long and 240km at its widest, encompassing an area of 65,525 km^2, of which 64,742 km^2 are land and the rest is inland waters. The island lies on three peneplains at 0–122 m (low elevation), 305–762 (mid elevation), and 914–2438 m (high elevation). Since Sri Lanka lies in the equatorial belt, only slight variation in temperature, relative humidity, and day length are experienced. The dominating climatic factors are the South West monsoons (May–September) and the North East trade winds (November–February).

5.1.3 Insect Diversity and Documentation

Sri Lanka's geological history together with its climate and vegetation accounts for its unique fauna and flora with high endemicity seen especially in the Central Highlands (Montane region) and the South western region of the island. Sri Lanka has a rich and unique fauna within a small land mass of nearly 65,500 km^2 comprising of relict, endemic, indigenous, and exotic fauna with many threatened and rare species. The flora of Sri Lanka is well documented and has been compiled into ten volumes in the Trimen Series. About 28 % of the island's flora is endemic. The most comprehensive documentation of the fauna of Sri Lanka appears in the Series Fauna of British India including Sri Lanka. Of the fauna, the vertebrates are well known and 16 % of the land vertebrates are endemic. The invertebrates are not so well documented. The present knowledge of Sri Lanka's arthropod fauna, especially of the Insecta, is a result of taxonomic studies by specialists from Europe and North America. A review by Wijesekara and Wijesinghe (2003) gives the history of insect collections and insect diversity in Sri Lanka. Recent studies by several local scientists have added many previously unrecorded taxa to the Sri Lankan checklists.

5.2 Isoptera

5.2.1 Taxonomy

Termites are currently classified into nine extant families under the orders Blattaria and Isoptera (Krishna et al. 2013). Classically, termites are grouped into lower termites (all families except the Termitidae) and higher termites (family Termitidae). Termitidae constitute majority (84 %) of the world's termites. The world fauna comprises 2933 living species under 282 genera and 9 families (Krishna et al. 2013).

Systematic and descriptive work on termites of Sri Lanka dates back to the early and mid-twentieth century. Among others, Wasmann (1902), Desneux (1904), Holmgren (1910a, b, 1911, 1912), Kemner (1934), and Roonwal (1970) have contributed mostly to the current knowledge of termites in Sri Lanka. Studies thereafter, focused on termites of economic importance in plantation crops and timber. Information on termites of Sri Lanka spanning a period of 100 years (1893–1988) has been compiled by Hemachandra et al. (2012) into a checklist of 64 species in 27 genera and 4 families. The most recent publication on termites of the world by Krishna et al. (2013) includes 61 species of termites in 27 genera and 4 families for Sri Lanka, comprising Hodotermitidae (1 sp.) Kalotermitidae (14 spp.), Rhinotermitidae (9 spp.), and Termitidae (37 spp.). The slight disparity in the number of species in the two lists arises due to the exclusion of four species of Termitidae and inclusion of one species of Kalotermitidae in the publication by Krishna et al. (2013). Among the Isoptera of Sri Lanka (Table 5.1) are several species that are cosmopolitan pests, and among them, *Cryptotermes dudleyi* is a well-known house pest. Also, the endemic, *Cryptotermes perforans*, is a pest of wooden furniture.

Following a recent study on forest termites of Central Sri Lanka, several more species (15 spp.) and two more genera, *Bulbitermes* (Fig. 5.2) and *Grallatotermes* (Fig. 5.2), not documented previously have been reported (Hemachandra 2014). Accordingly, the present taxonomic composition of Isoptera of Sri Lanka comprises 76 species in 29 genera and 4 families. Isoptera of Sri Lanka is not taxonomically or

Table 5.1 Current taxonomic list of Isoptera of Sri Lanka

Family	Subfamily	Genera	No. of species	Feeding habit
Hodotermitidae		*Anacanthotermes*	*1*	*G*
Kalotermitidae		*Postelectrotermes*	*1*	*WN*
		Neotermes	*3*	*WN*
		Kalotermes	*1*	*WN*
		Glyptotermes	*3*	*WN*
		Bifiditermes	*1*	*WN*
		Cryptotermes	*6*	*WN*
Rhinotermitidae	Coptotermitinae	*Coptotermes*	*5*	*W*
	Heterotermitinae	*Heterotermes*	*2*	*W*
	Rhinotermitinae	*Prorhinotermes*	*1*	*W*
	Termitogetoninae	*Termitogeton*	*1*	*W*
Termitidae Latreille	Macrotermitinae	*Macrotermes*	*1*	*F*
		Odontotermes	*14*	*F*
		Hypotermes	*3*	*F*
		Microtermes	*2*	*F*
	Apicotermitinae	*Eurytermes*	*2*	*H*
		Speculitermes	*1*	*H*
	Nasutitermitinae	*Nasutitermes*	*6*	*W*
		Bulbitermes	*2*	*W*
		Ceylonitermes	*2*	*W*
		Hospitalitermes	*1*	*E*
		Grallatotermes	*1*	*E*
		Trinervitermes	*2*	*G*
		Ceylonitermellus	*2*	*S*
	Termitinae Latreille	*Synhamitermes*	*2*	*W*
		Microcerotermes	*5*	*W*
		Angulitermes	*1*	*H*
		Dicuspiditermes	*3*	*H*
		Pericapritermes	*1*	*H*
Total number 29			76	

G grass feeders, *WN* wood-nesting wood feeders, *W* wood feeders, *F* fungus feeders, *H* humus feeders, *E* microepiphyte feeders, *S* soil feeders

Fig. 5.1 Aphids on a Gramineae

Fig. 5.2 *Ceylonitermellus hantanae*, a soil-dwelling and soil-feeding endemic genus (**a**) Nest (**b**) Soldier

biologically diverse unlike those of India, most likely due to Sri Lanka's small land area. Within Isoptera, *Odontotermes* (Termitidae) is the most diverse being represented by 14 species that have a wide distribution.

5.2.2 Endemism

According to literature, 33 species of Isoptera are restricted to Sri Lanka. They are as follows:

Kalotermitidae: Postelectrotermes militaris, Neotermes greeni, N. kemneri, Kalotermes jepsoni, Glyptotermes ceylonicus, G. dilatatus, G. minutes, Cryptotermes ceylonicus, C. perforans

Rhinotermitidae: Heterotermes colonics, Termitogeton umbilicatus

Termitidae: Odontotermes koenigi, O. *preliminaries, O. taprobanes, Hypotermes winifredi, Microtermes macronotus, Eurytermes ceylonicus, Nasutitermes ceylonicus, N. horni, N. lacustris, N. oculatus, Ceylonitermes escherichi, Hospitalitermes monoceros, Ceylonitermellus hantanae,* C. *kotuae, Synhamitermes ceylonicus, S. colombensis, Microcerotermes bugnioni, M. cylindriceps, M. greeni, Angulitermes ceylonicus, Dicuspiditermes hutsoni, Pericapritermes ceylonicus*

However, much caution has to be taken when expressing the number of endemic termite species, as more field studies get under way in the Indian subcontinent, many previously unrecorded species are likely to be encountered. Of the genera endemic to Sri Lanka, the genus *Ceylonitermellus* with two species, *C. hantanae* and *C. kotuae* warrants special mention. The former species, since its original description, was recently collected from its type locality Hantana as well as from forests in Gannoruwa and lower montane region in the Knuckles, in Central

Sri Lanka (Hemachandra, 2014) (Fig. 5.2). The latter species, *C. kotuae*, is known only from its type species whose type locality is Galle-Kotuwa (Galle Fort in Southern Sri Lanka) (Chhotani 1997; Krishna et al. 2013).

5.2.3 Habitats and Microhabitats

Termites are not usually seen in the open as their soft bodies are subjected to desiccation. They lead a concealed life in wood, carton nests, soil, or earthen mounds. The non-foraging species nest and feed on the same medium that they live in. The foraging species move in search of food in protected earthen galleries they construct or unprotected in the open, under shade. Based on where termites live, they may be grouped into the following categories:

1. Wood dwellers

   Family: Kalotermitidae; Genera *Postelectrotermes*, *Neotermes*, *Kalotermes*, *Glyptotermes*, *Bifiditermes*, and *Cryptotermes*

   Family: Rhinotermitidae; Genera *Coptotermes*, *Termitogeton*, and *Prorhinotermes*
2. Ground dwellers

   Family: Hodotermitidae; Genus *Anacanthotermes*

   Family: Rhinotermitidae; Genera *Coptotermes*, and *Heterotermes*

   Family: Termitidae; Genera *Macrotermes Odontotermes*, *Hypotermes*, *Microtermes*, *Eurytermes*, *Speculitermes*, *Nasutitermes*, *Ceylonitermes*, *Trinervitermes*, *Ceylonitermellus*, *Synhamitermes*, *Microcerotermes*, *Dicuspiditermes*, and *Pericapritermes*
3. Arboreal carton nesters

   Family: Termitidae; Genera *Nasutitermes*, *Hospitalitermes*, *Bulbitermes*(?), and *Grallatotermes* (?)
4. Earthen mound dwellers

Family: Termitidae; Genera *Macrotermes*, *Odontotermes*, and *Hypotermes*

Ground-dwelling termites are known to occupy several microhabitats, reflecting their ecological/niche diversity. Availability of microhabitats therefore, may influence their distribution in natural habitats such as forests. The forest floor in particular harbors numerous microhabitats. A study of microhabitats of termites on the forest floor in the Knuckles and Kandy regions revealed that majority of the species (42 of 68 spp.) inhabit the soil. Thirty-six species occurred in logs and stumps, 34 in fallen branches and sticks, and 27 species in leaf litter and humus (Hemachandra 2014). Termites inhabiting these microhabitats are largely ground-dwelling or earthen mound-dwelling foragers that wander away from their nest sites. Of the different forest vegetation types examined in the Knuckles region, lowland semievergreen forests harbored the most number of termite species (27 of 28 spp.) that inhabit the forest floor litter, reflecting the role of microhabitats in termite distribution (Hemachandra 2014).

Table 5.2 Feeding habits among Isoptera of Sri Lanka

Feeding habit	No. of species
1. Grass feeders	03
2. Wood feeders and wood nesters	15
3. Wood feeders	26
4. Fungus growers and feeders and wood nesters (Macrotermitinae)	20
5. Microepiphyte feeders	02
6. Humus feeders	08
7. Soil feeders	02
Total	76

5.2.4 Feeding Habits

Feeding habits are generally deduced from the generic identity of termites. Based on these criteria, Isoptera of Sri Lanka has rich wood-feeding foragers, followed by fungus-growing wood-feeding species and is poor in humus and soil-feeding species (Table 5.2). The three species of live-wood-feeding termites, *Postelectrotermes militaris, Neotermes greeni, and Glyptotermes dilatatus* in the family Kalotermitidae, are of special interest as tea pests at all elevations: in upcountry, mid country, and low country plantations. Yet another species, *Kalotermes jepsoni* attacks deadwood and live wood and is also a tea pest. The endemic *Glyptotermes ceylonicus* confined to 500 and 650 m elevation, is also considered a plantation pest as it has been found on freshly dead branches of rubber, cocoa, *Acacia* and at the foot of live tea bushes. Several other species, *Glyptotermes dilatatus*, *Cryptotermes perforans*, and *Coptotermes ceylonicus*, are well-known timber termites of Sri Lanka. Of the ground-living termites, those living in the soil-wood interface, which are the humus feeders account for an insignificant number of species (eight). They come under five genera that includes *Dicuspiditermes,* according to the recent study of forest termites. The soil-living true soil feeders account for only two species, *Ceylonitermellus hantanae* and *C. kotuae*, both being endemic to Sri Lanka. According to these records, the Isoptera of Sri Lanka is poor in species that contribute to soil fertility. As more field studies get underway in Sri Lanka, more light will be shed on the important and beneficial soil termites. The tree-living lichen feeders (microepiphyte feeders) comprise of two species, endemic *Hospitalitermes monoceros* (Fig. 5.3) and *Grallatotermes* sp. The genera that represent the grass feeders are poorly known, and comprise of *Anacanthotermes viarum*, *Trinervitermes biformis*, and *T. rabidus* that include both primitive species and higher termites. These species have not been recorded since their description (Fig. 5.4).

Fig. 5.3 *Hospitalitermes monoceros*, an endemic microepiphyte feeder (**a**) Nest (**b**) Trail (**c**) Soldier

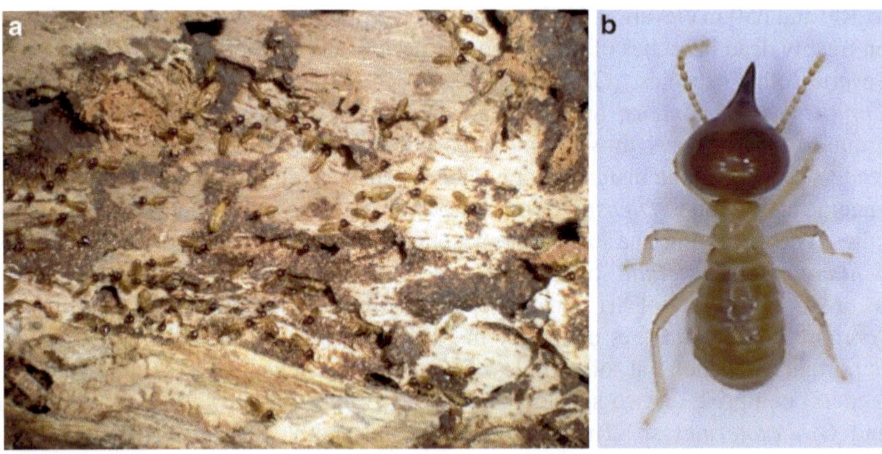

Fig. 5.4 *Grallatotermes*, new genus of microepiphyte feeders (**a**) Trail (**b**) Soldier

5.2.5 Distribution of Forest Termites

The recent field survey (Hemachandra 2014) shed light on the distribution of forest termites of Central Sri Lanka with respect to altitude and vegetation type. At altitudes >1300 m in the upper montane regions of the Knuckles range, paucity of subterranean, ground-living, dead-wood-dwelling, and foraging species were evident. *Postelectrotermes militaris* is the only live-wood-dwelling species recorded for this region. However, presence of more live-wood-dwelling species cannot be ruled out as collection methods destructive to live plants were not used in this survey. The field survey in Central Sri Lanka covering 30 forest locations in 14 vegetation types inferred that a large number of species are restricted to a particular vegetation type, while a few species such as *Odontotermes bellahunisensis*, *O. guptai*, and *O. horni* have a wide distribution in several vegetation types (Hemachandra 2014).

5.3 Aphididae

5.3.1 Taxonomy

Aphids are a group of small, soft-bodied insects that infest plants and feed on the plant sap. They are classified under the order Homoptera, superfamily Aphidoidea, and family Aphididae. Worldwide, more than 4000 aphid species are known and they are distributed among 10 subfamilies. Of the aphid subfamilies, Aphididae includes majority of the species. Aphids being exclusive plant sap feeders, about 455 of the species are plant pests with 250 species being serious pests of crops world over. Aphids have a worldwide distribution but there are more species in temperate region than in the tropics.

5.3.2 Aphids

Studies on aphids of Sri Lanka date back to 1890, with the recording of *Greenidea (=Siphonophora) artocarpi,* the very first species to be identified from Sri Lanka (Westwood 1890). This work was followed by many others including Judenko and Eastop (1963) who added 24 new species to the list of Sri Lankan aphids. Work thereafter, on aphids focused on pest species and the damage they cause. A compilation of previously described aphids (up to late 1990s) amounted to 72 species in 38 genera distributed among 8 subfamilies (except Anoeciinae and Drepanosiphinae) (Wijerathna and Edirisinghe 1999) (Table 5.3). The subfamily Aphidinae comprised majority of the species (53) and genera (25) (Table 5.3.1). Among the previously described aphids are seven species endemic to Sri Lanka.

Although aphids on economic plants are well documented in Sri Lanka, general surveys of aphids covering the natural vegetation and weeds have received little attention. This gap in information was addressed through a survey of nearly 1000 plants covering six agro ecological regions of the country (Wijerathne 1997). The survey yielded a total of 47 species of aphids in 28 genera distributed among 5 subfamilies. Aphids of the subfamilies Lachinae, Chaitophorinae, and Calaphidinae were not recorded during this survey, but a single species (*Tinocallis kahawaluokalani*) in the subfamily Drepanosiphinae was documented. Of the aphids collected during this general survey, 12 species in 10 genera are new records for Sri Lanka.

Table 5.3 Aphid taxa known from Sri Lanka

		Past[a]		**Recent		Overall	
	Subfamily	Genera	Species	Genera	Species	Genera	Species
1	Aphidinae	25	53	21	40	25	53
2	Drepanosiphinae	–	–	01	01	01	01
3	Greenideninae	03	03	03	04	03	04
4	Hormaphidinae	03	04	02	02	03	05
5	Lachininae	04	05	–	–	04	05
6	Pemphiginae	01	02	01	01	01	02
7	Chaitophorinae	02	03	–	–	02	03
8	Calaphidinae	01	01	–	–	01	01
	Total	38	72	27	48	40	74

**Recently recorded by Wijerathna (1997)
Source: [a]Wijerathna (1997)

Table 5.3.1 Aphids endemic to Sri Lanka

Subfamily	Species
Aphidinae	*Macrosiphum minutum*
	Matsumaraja capitophoroides
	Micromyzus dispersum
	Micromyzus eastopi
	Micromyzus judenkoi
	Micromyzus nigrum
Greenideninae	*Greenideoidea ceyloniae*
Hormaphidinae	*Astegopteryx insularis*
Lachininae	*Lachnus greeni*
Pemphiginae	*Ceratopemphigus zehntneri*

Furthermore, three rare species of aphids and three other species not found in India were also recorded. Of the seven species, originally known to be endemic to Sri Lanka, only two were recorded; others have not been documented since their first description. Currently, the aphid fauna documented from Sri Lanka amounts to 74 species in 40 genera and 8 subfamilies (Table 5.3) (Wijerathna 1997), Anoeciinae being the only subfamily not represented in Sri Lanka.

5.3.3 Host Plant Relationships

Host associations and host specificity are important aspects of aphid-host plant relationships. This aspect in Sri Lanka was addressed through a general survey of aphids (Wijerathna 1997). Of over 1000 species of plants examined during the survey, only one third of the plant species (in 71 families) had aphids living on them. Vegetables in particular harbored well-known pest species (Wijerathna and Edirisinghe 1999). Host specificity was not so pronounced among the aphids, with only 15 species being confined to specific plants, among which are bamboos (*Bambusa vulgaris*), medicinal plants (*Artemisia vulgaris*), economic plants (coffee, carrot, cardamom,

Table 5.3.2 Natural enemies of aphids recorded from Sri Lanka

Order	Family	Genera	Species
Coleoptera	Coccinellidae	09	24
Coleoptera	Chrysomelidae	05	06
Diptera	Syrphidae	06	06
Neuroptera	Chrysopidae	02	02
Hemiptera	Plataspidae	01	01
Hymenoptera	Aphidiidae	01	01
Hymenoptera	Braconidae	03	03
Hymenoptera	Encyrtidae	02	02
	Total	29	45

roses), endemic plants (*Mesua ferrea*), and common weeds (*Vernonia cinerea* and *Wikstroemia indica*). The aphid,*Sitobion wikstroemiae* on the latter weed species is a rare aphid previously recorded only from South Africa and Mauritius. Its host plant, *Wikstroemia indica* had been introduced to Sri Lanka from East Africa and has since become a weed in and around Kandy.

5.3.4 Natural Enemies

Natural enemies of aphids play a key role in their control. As aphids generally feed in exposed locations such as young shoots and leaves, they are susceptible to a variety of natural enemies, both predators and parasitoids (Table 5.3.2). However, studies on aphid natural enemies in the Indian subcontinent are few. General predators include lacewings, lady beetles, hover flies, midges, several bugs, and beetles. Parasitoids that attack aphids include ichneumon and braconid wasps. During the survey of natural and cultivated plants for aphids, a rich natural enemy community of 25 species was recorded comprising predatory Coccinellidae (12 spp.), Chrysomellidae (4 spp.), Neuroptera (2 spp.), Hemiptera (1 sp.), and parasitoids: Braconidae (3 spp.), Aphidiidae (1 spp.), and Syrphidae (2 spp.) (Wijerathna and Edirisinghe 1999). Previous studies too have confirmed the presence of a rich fauna of aphid predators numbering 37 species. Thus, at present, the total number of natural enemy complex associated with aphids is 45 species in 29 genera (Table 5.3.2). This aspect remains to be harnessed in Sri Lanka for the biological control of aphids, replacing use of insecticides against pest species.

5.4 Thysanoptera

5.4.1 Taxonomy and Classification

Thysanoptera comprises of minute phytophagous insects of considerable economical importance. According to the widely accepted traditional classification, Thysanoptera includes two suborders – Tubulifera and Terebrantia. Tubulifera comprises thrips that lay eggs on the external surfaces of host plants and are placed in Phlaeothripidae with two subfamilies, Idolothripinae and Phlaeothripinae. Suborder Terebrantia includes thrips that lay eggs in the internal tissues of plants and are distributed across eight families. The world fauna of Thysanoptera consist of about 6000 species in 768 genera with 3500 species (in 450 genera) of Tubulifera and 2412 species (in 320 genera) of Terebrantia, distributed across 6 subfamilies and 9 families. Majority of the world species and genera are represented by the subfamilies Phlaeothripinae (2800 spp. in 370 genera) and Thripinae (1700 spp. in 225 genera) (Mound 2007).

Systematic and taxonomic work on Thysanoptera of Sri Lanka dates back to Schmutz (1913) who described 43 species from collections made in Sri Lanka by Heinrich Uzel, the European author who published the very first extensive account of the order Thysanoptera. In later years, Ananthakrishnan (1963) is known to have dealt with Terebrantia of the Indo-Ceylonese region. Studies by several authors thereafter were largely on thrips of economic importance, particularly on ornamental plants (Pitkin 1976). Tillekaratne et al. (2007) compiled all the published past literature on thrips of Sri Lanka spanning 1859–2007 into a checklist of 78 species, in 46 genera in 3 families. Thrips, have been considered principally in relation to their agricultural importance, and no serious attempts had been made to survey the thrips fauna of the island in a variety of habitats, until the recent study by Tillekaratne (2010). This study resulted in adding several previously unrecorded species (24 species in 19 genera) to the Sri Lankan checklist and recorded the host plants of thrips. Presently, Thysanoptera of Sri Lanka comprises 113 species in 63 genera, of which 52 species (26 genera) belong to suborder Tubulifera and 61 species (37 genera) to Terebrantia. Suborder Terebrantia in Sri Lanka is represented by the family Thripidae (with three subfamilies). Thus, Thysanoptera of Sri Lanka comes in three families: Phlaeothripidae (52 species in 26 genera), Aeolothripidae (one species), and Thripidae (60 species in 36 genera) (Table 5.4) (Tillekaratne et al. 2007).

5.4.2 Special Features

Thrips recorded from Sri Lanka have zoogeographic affinities with India and adjacent countries in the region, and no species is known to be endemic to the island. Thrips fauna is dominated by the subfamilies Phlaeothripinae (35 spp. in 18 genera) and Thripinae (40 spp. in 20 genera) with the genus *Thrips* having the highest number of species (16). The subfamily Thripinae is known to contain many species injurious to crop plants. The Sri Lankan fauna comprises of many cosmopolitan pests (42 spp.) and several potential viral vector (5 spp.) species, with cultivated vegetables, ornamental plants, and cut flowers harboring a large majority of the species. During a recent survey (Tillekaratne 2010), 13 species of thrips were identified from cut flowers, *Thrips flavus* being the most common. These species cause damage to flowers as well as to foliage indicating the risk to export trade. With the expansion of the cut flower and foliage export industry in recent times in Sri Lanka, thrips have become a major concern and a potential threat. The major pest thrips of

Table 5.4 Species and genera of Thysanoptera in Sri Lanka

	No. of species from previous records (No. of genera)[a]	No. of species from recent study (No. of genera)[b]	Overall species (genera) based on previous and recent records
Suborder Tubulifera			
F. Phlaeothripidae	17 (8)	4 (2)	17 (08)
SF. Idolothripinae			
SF. Phlaeothripinae	22 (10)	16 (11)	35 (18)
Suborder Terebrantia			
F. Aeolothripidae	1 (1)	None	1 (1)
F. Thripidae	14 (10)	12 (11)	18 (14)
SF. Panchaetothripinae			
SF. Dendrothripinae	2 (2)	2(2)	02 (02)
SF. Thripinae	23 (12)	31 (18)	40 (20)
Total	79 (43)	65 (44)	113 (63)

[a]Tillekaratne et al. (2007)
[b]Tillekaratne (2010)

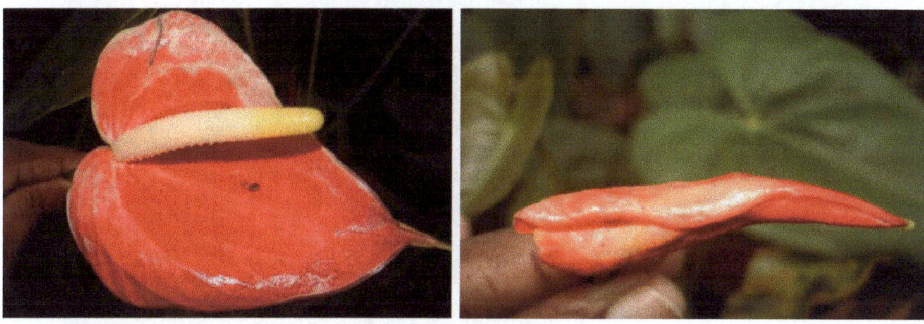

Fig. 5.4.1 Flower and bud of *Anthurium andraeanum* damaged by *Chaetanaphothrips orchidee*

rose, asters, and chrysanthemum worldwide is *Frankliniella occidentalis*, the western flower thrips. Although this species had been recorded from Sri Lanka, it had been found on plants other than cut flowers (Tillekaratne et al. 2007). Thrip damage to anthuriums and roses by specific species (Figs. 5.4.1 and 5.4.2) is a concern to growers. In countries where gerberas are grown, heavy infestations of thrips have been reported, while in Sri Lanka thrips damage to gerberas is minimal. Among thrips infestations in natural vegetation, leaf damage in *Ficus* due to *Gynaikothrips* species is characteristic (Fig. 5.4.3).

While majority of thrips show a wide distribution within the country, being present wherever their host plants are found, a few species, however, tend to confine themselves to certain elevations and specific crops and flora found therein. *Thrips flavus* in particular was the most widely distributed and abundant species in the upcountry (>900 m elevation) wet zone (>2500 mm rainfall) being found on almost all the flowering plants, while *Thrips simplex* was found confined to this region, occurring on ornamentals grown there.

Fig. 5.4.2 Flowers of *Rosa indica* damaged by *Thrips hawaiiensis*

Fig. 5.4.3 Leaf fold damage in *Ficus benjamina* due to infestation by *Gynaikothrips ficorum* (*Inset*)

While majority of thrips tend to be generalists, specific relationships between thrips and host plants were evident from field studies. Species such as *Liothrips karnyi on Piper nigrum*, *Sciothrips cardamomi* on *Elettaria cardamomum*, and *Praepodothrips sp*. on *Lycopersicon esculentum* have been identified as host specific in Sri Lanka (Tillekaratne et al. 2011).

Varying types of feeding damages are seen in thrip-infested plants, such as scarring, browning, and discoloration of flowers, streaking of leaves, stunted terminal shoots, deformed fruits, leaf galls, bronze coloration in leaves, and leaf curl, where much of the latter type of damage could be due to viral infection following feeding by vector species. The genus *Gynaikothrips* includes well-known causatives of leaf gall (Fig. 5.4.4). Among the 180 species of thrips in the world (Mortiz et al. 2001), 14 species are known to be pests that damage vegetables even in Sri Lanka.

Fig. 5.4.4 Simple leaf fold galls caused by *Gynaikothrips* sp.

5.5 Formicidae

5.5.1 Classification and Taxonomy

Bolton (2003) documented 21 subfamilies and 283 genera of ants worldwide. However, species and genus descriptions are still continuing. According to estimates of many taxonomists, there are more than 15,000 species of ants worldwide (Table 5.5), with some estimates going as high as 20,000. However, especially in the tropical Asia, many new ant species await description.

Ants of Sri Lanka have been poorly documented, with the degree of endemicity within the family unknown (Dias 2002). The first comprehensive survey of ants in Sri Lanka was published by Bingham (1903), as part of the fauna of British India. According to historical data and current museum collections, 181 species of ants in 61 genera and 12 subfamilies have been recorded for Sri Lanka (Dias 2006) and later updated to 64 genera and 202 species (Dias et al. 2012). According to them, subfamilies (and species number) recorded for Sri Lanka currently are Aenictinae (5 spp.), Amblyoponinae (4), Aneuretinae (1), Cerapachyinae (7), Dolichoderinae (12), Dorylinae (3), Ectatomminae (1), Formicinae (58), Leptanillinae (4), Myrmicinae (80), Ponerinae (37), and Pseudomyrmicinae (4). Myrmicinae, by far, is the most species-rich subfamily followed by Formicinae and Ponerinae. The formicines tend to be dominated by members of *Camponotus* and Ponerinae by the genus *Leptogenys*.

5.5.2 Endemism

While many ants from Sri Lanka have specific names related to their locality (e.g., *Paratopula ceylonica* Emery, *Dolichoderus taprobanae* Smith, *Diacamma ceylonense* Emery), most species are shared with the Indo Malayan region. However, Sri Lanka is of special interest to myrm ecologists as it is the home to the relict ant, *Aneuretus simoni* Emery (Fig. 5.5.1). It is the only extant representative of the subfamily Aneuretinae, of which the other genera are extinct, known only from fossil remains found in the Baltics, Russia, and the USA (Bolton 2003). This widely

Table 5.5 Common ant species in Sri Lanka based on the Global Invasive Species Database

Subfamily	Genus	Species	Notes
Dolichoderinae	*Tapinoma*	*melanocephalum*	A household pest with native range not known. It is considered a tramp ant throughout the world
Dolichoderinae	*Technomyrmex*	*albipes*	This species is most likely a native species to Indo Malaya. It has large colonies and often inhabit undisturbed as well as disturbed areas
Dolichoderinae	*Technomyrmex*	*bicolor*	This species is most likely a native species to Indo Malaya. It has large colonies and often inhabit undisturbed as well as disturbed areas
Formicinae	*Anoplolepis*	*gracilipes*	Most likely a native to Indo Malaya, it can form supercolonies and has caused grave impacts on island habitats in other parts of the world but they appear to
Formicinae	*Oecophylla*	*smaragdina*	This is a native to all of Asia and Australia; it is a successful colonizer of forest edges and plantation forests; however, it is not considered invasive as it is found
Formicinae	*Paratrechina*	*longicornis*	The brown crazy ant, very common tramp ant from Africa found in association with human-disturbed habitat throughout the world
Myrmicinae	*Leptomyrmex*	*quadri spinosis*	A common inhabitant of disturbed forest, it is a native to Asia and Australia
Myrmicinae	*Meranoplus*	*bicolor*	Found throughout tropical Asia, it is generally arboreal
Myrmicinae	*Monomorium*	*pharaonis*	The pharaoh ant is a well-known global invasive, originally from Africa. Found in most tropical areas of the world; it is of some human concern as has been shown
Myrmicinae	*Monomorium*	*floricola*	The diminutive flower ants is a native of the Indo Malayan region but is a well-known worldwide tramp species
Myrmicinae	*Myrmicaria*	*brunnea*	This ground-dwelling ant is commonly found inhabiting pathways through disturbed forest. It is widespread species throughout Asia
Myrmicinae	*Solenopsis*	*geminata*	Common name – tropical fire ant. Native to Central and South America, it is on the global invasive list and is a species easily spread by humans

(continued)

Table 5.5 (continued)

Subfamily	Genus	Species	Notes
Myrmicinae	*Tetramorium*	*bicarinatum*	Native to Asia, it has become another tramp ant and is found worldwide
Ponerinae	*Diacamma*	*rugosum*	Widespread in Asia, this "queenless" ant species inhabits disturbed habitats such as cocoa plantations
Ponerinae	*Harpegnathos*	*saltator*	Found throughout India and Sri Lanka, this charismatic ant is often seen walking along trails and paths, and its characteristic jumping behavior is easily observed
Ponerinae	*Leptogenys*	*processionalis*	The *Leptogenys* have "army ant"-type behavior with foraging columns that are easily seen near semi-near semi-forested habitats in both Sri Lanka and India
Ponerinae	*Odontomachus*	*simillimus*	Also known as *O. haematodes*, it is widespread in Asia and appears to be able to colonize disturbed and undisturbed forest habitats

Source: http://www.issg.org/database/welcome/) or on Antweb.org

Fig. 5.5.1 The relict ant, *Aneuretus simoni* (Source: Antweb.org. Photograph by A. Nobile)

distributed subfamily is now confined to the island of Sri Lanka, and the extant species was thought to be very rare within the island. However, more recent work has found it to be quite abundant in disturbed forests in the country's southwest (Jayasuriya and Traniello 1985) and central hills (Karunarathna and Karunaratne 2013). It is still on the IUCN Red Data List (IUCN 2001). Phylogenetically, it has been placed between the primitive and the modern lineages of ants, and it has generated interest as to its exact placement within the ant subfamily tree.

Stereomyrmex horni Emery is another species known only from Sri Lanka, first collected and described by Emery in 1901 and subsequently collected by Bingham

Fig. 5.5.2 The newly described Sri Lankan endemic ant, *Tyrannomyrmex legatus* Alpert (Source: Antweb. org. Photograph by M. Branstetter)

(1903) and most recently by Dias et al. (2011). Interestingly, unlike *A. simoni*, *S. horni* has been consistently collected from only one locality within Sri Lanka indicating a far more restricted range than the relict ant. Not much is known about its biology, but rarity in collections indicates that it is likely to be a subterranean dweller with small colonies.

The most recent endemic species to be described for Sri Lanka is *Tyrannomyrmex legatus* Alpert (Fig. 5.5.2). Alpert (2013) described the third species for this relatively new genus from single specimen collected from leaf litter in lowland wet forests of the south west.

5.5.3 Habitats and Distribution

Ants, being ubiquitous, inhabit urban landscapes to undisturbed forests. A variety of habitats in Sri Lanka have been examined for ants by several workers: effects of logging and elevation in relation to tree species distribution in lowland wet forest in the south west of the island (Gunawardene et al. 2008, 2010, 2012), the dry zone habitats in the north of the country (Dias and Kosgamage 2012), the "home garden" in the central province of Sri Lanka (Amarasinghe 2010), the rice agro ecosystem (Bambaradeniya et al. 2004), and mixed agricultural system (Harindra et al. 2007).

Ants are increasingly considered as indicator species of habitat disturbances. Many of the ant species found in the agro ecosystem were also found inhabiting the edges of a protected forest reserve (Gunawardene et al. 2008, 2010 and 2012). These species are widespread throughout the Asian tropics and can be considered as either disturbance specialists or very generalized in their habitat preferences. Such species would not be considered invasive as their provenance is often Asian and hence likely to be native species that have colonized human-disturbed ecosystems well. Table 5.1 lists the most common ant species.

Fig. 5.5.3 *Leptogenys* sp. crowding together around a temporary nest; their pupae can be seen loosely sitting on top of the leaf litter

5.5.4 Ant Specialists

Most ants are generalists and opportunistic in their food preference. However, there are many that are specialists in environments they inhabit (e.g., hot and dry) or with respect to food preference. The blind Asian army ant *Aenictus* spp. is well known for long winding columns through vegetation in search of nests of other ant species which they raid. They prey on eggs and larvae of other ant species and have large mobile colonies similar to the well-known *Eciton* army ants of South America. In Sri Lanka, eight such species have been recorded so far. Another group of "true" army ants found in Sri Lanka are the Dorylinae. They are most speciated in Africa where they form large raiding columns, causing immense destruction to invertebrate fauna. The Asian members of this family are less studied as they are subterranean, living and hunting exclusively inside the soil.

The genus *Leptogenys* (Fig. 5.5.3) is another ant group with an army ant-like life history, with *Leptogenys processionalis* (Jerdon) being a widespread species in Sri Lanka. This ant is often seen in a short column passing through paths and across gardens, both in India and Sri Lanka. They form temporary nests in leaf litter and are predators of termites and other litter-dwelling invertebrates.

5.5.5 Ants in Agriculture

As many ants are generalists, they do not always produce benefits to agriculture. Some species are specialist "farmers" of plant-sucking hemipterans such as coccids or mealybugs. They actively protect and translocate their food source (honeydew-producing mealybugs) and transmit lethal plant viruses associated with the mealybugs from plant to plant (Delabie 2001). The cacao swollen shoot virus (Peiris 1953) in Sri Lanka is transmitted by mealybugs (by arboreal ants such as *Crematogaster* spp.), and the infection is incurable once a plant is infected.

Although *Oecophylla smaragdina* F. has proven to be an important component of natural pest management in fruit trees throughout Asia, studies are lacking in Sri Lanka. Rickson and Rickson (1998) concluded that in cashew plantations in Sri Lanka, those that had reduced pesticide spraying had the most diverse ant assemblage and the lowest pest load. They supported the idea that ants and cashew trees benefited from each other's presence to the point of eliminating the need for costly pesticide spraying. Way et al. (1989) found that the small myrmicine ants *Monomorium floricola* (Jerdon) and *Crematogaster* spp. protected coconut plantations in Sri Lanka from outbreaks of the coconut caterpillar *Opisina arenosella* Walker (Lepidoptera: Xyloryctidae).

Ant species in Sri Lanka are still being discovered. Ant diversity in Sri Lanka is still underestimated and many areas of the island are not sampled. Arboreal and subterranean ants require further research as currently utilized methods focus on ground-dwelling species. There is a greater chance of many more rare and unusual ants to be discovered. However, the rate of landscape change in Sri Lanka is high with many natural areas under threat from development. Sri Lanka is a red hot biodiversity hot spot as a result of the combination of high levels of endemism and high levels of landscape change.

5.5.6 Apoidea

5.5.6.1 Taxonomy and Classification

Bees are grouped under the superfamily Apoidea with sphecoid wasps. There are more than 20,000 identified bee species in the world (Michener 2000), and they are classified into 7 families and 443 genera. Bees are broadly divided in to two groups, honeybees and pollen bees. Honeybees (Family Apidae) of the world comprise 7 genera and 53 species. They form large colonies and construct impressive nests/hives in which they live, lay eggs and rear their young, and also store honey and pollen. A bee colony consists of a large number of individuals belonging to different casts: a single queen, several males, and many sterile females that are workers. Pollen bees constitute majority of the world's bees and they do not form colonies nor live in hives. Pollen bees have no eusocial organization or caste system and comprise only of males and females that nest in ground and hollows of stems and wood.

5.5.7 Bees of Sri Lanka

Information on bees of Sri Lanka dates back to the British colonial period where Dalla Torre (1896) listed 17 species of bees and Bingham (1897) recorded and described 42 species of bees. During the post-colonial period, Sakagami and Ebmer (1987), Schwarz (1990), Sakagami (1978, 1991), Sakagami et al. (1996 and 1998), Snelling (1980), and Baker (1996) worked on bees of Sri Lanka. The Smithsonian – Sri Lanka Insect Survey conducted from 1969 through 1975 up to 1987 resulted in

Table 5.6 Bee taxa recorded from Sri Lanka

Family	Genera	Species
Apidae	09	58
Halictidae	19	53
Megachilidae	09	35
Colletidae	01	02
Total	38	148

the identification of several more bee species and led to a series of publications. The more recent publication on "Bees of the world" (Michener 2000) includes 29 bee species from Sri Lanka. The very first field survey of bees by a local scientist (Karunaratne 2004) conducted during 2004–2013 added 5 new genera and 20 species of bees to the Sri Lankan bee fauna. This survey identified a new bee species, *Lipotriches edirisinghe* Pauly 2005. Following these studies, an updated checklist of bees of Sri Lanka was published by Karunaratne et al. (2005) that included 137 species in 35 genera. Presently, 148 species of bees are known from Sri Lanka, and they come under 38 genera and are classified into 4 families (Table 5.6). Apidae includes both honeybees and anthophorid bees. Halictidae includes the largest number of bee genera.

5.5.8 Diversity of Bees

Although the bee fauna of Sri Lanka is much smaller than that of neighboring India, it comprises species having diverse nesting habits, foraging behavior, and floral relationships.

5.5.9 Honeybees

There are only four species of honeybees in two genera in Sri Lanka, *Apis* (three spp.) and *Trigona* (one identified species and one unidentified species). Of the *Apis* bees, *A. cerana* is the most common and the domesticated honey bee. *Apis dorsata*, the giant honeybee, is the most aggressive. The genus *Trigona*, the stingless honeybees, is presently represented by *T. iridipennis*.

5.5.10 Pollen Bees

The large majority of bees in Sri Lanka are pollen bees represented by143 species in 36 genera. Pollen bees are best known for their nesting habits, broadly categorized as ground nesters, stem nesters (leaf-cutter bees), and wood nesters (carpenter bees).

Fig. 5.6.1 *Amegilla* sp. with a pollen load on the hindleg

5.5.11 Ground-Nesting Bees

There are over 70 species of ground-nesting bees described from Sri Lanka, and they come under several families and genera. Family Halictidae includes many ground-nesting genera: *Halictus*, *Homalictus*, *Lasioglossum*, *Patellapis*, *Austronomia*, *Curvinomia*, *Hoplonomia*, *Leuconomia*, *Gnathonomia*, *Pachynomia*, *Lipotriches*, *Maynenomia*, *Nomia*, *Steganomus*, *Ceylalictus*, and *Systropha*. Family Apidae includes the two ground-nesting genera: *Amegilla* (Fig. 5.6.1) and *Tetralonia*. Females of ground-nesting species make intricate system of galleries in the soil that end as cells and are lined with cellophane-like material to prevent the walls from collapsing. Eggs are laid on pollen balls stored in these cells. Larvae that hatch from eggs feed on the pollen and develop inside the cells until they pupate and become adults. The emerging young bees use the same nest that their mother built to nest or build nests close to the natal nest. This brings about the aggregation of several nests in the same location resulting in a "bee village." Such large bee villages are found in Ussangoda, a seaside plateau in the southwestern coast of the island. Here, thousands of nests of *Pseudapis oxybeloides* and *Ceylalictus* sp. are found, where female bees are seen actively collecting pollen from the common creeping herb, *Evolvulus alsinoides* (F: Convolvulaceae).

5.5.12 Carpenter Bees

There are 20 species of carpenter bees (F. Apidae) in two genera that nest in wood and wooden structures. *Xylocopa*, the giant carpenter bees, are represented by 13 species in Sri Lanka. They use their sharp hard mandibles to drill tunnels in wood, mostly in construction timber, to make nests. Species in forested areas nest in dead trees or in hollows of dry bamboo stems. The largest among bees of Sri Lanka is the carpenter bee *X. tenuiscapa* that nests in open rafters of old buildings. *Ceratina* that includes the dwarf carpenter bees comprise six species that nest in dry stems. Of

Fig. 5.6.2 Leaf-lined nest of *Megachile* sp. on the hind leg in a bamboo stem

them, females of *Ceratina binghami* are the most colorful being bright metallic green, while the males are bright metallic blue. This species is common in agricultural tracts as well as weedy habitats and prefers to nest in dry stems of *Gliricidia*.

5.5.13 Leaf-Cutter Bees

Family Megachilidae includes the leaf-cutter bees in the genera *Megachile* (Fig. 5.6.2), *Lithurgus*, and *Heriades*. Nests are constructed in preexisting holes in wood or inside hollow stems including bamboo. The nests are lined with pieces of leaves cut from plants in the vicinity, using their strong mandibles with sharp cutting edges. Leaves of rose plants and leaflets of Leguminosae plants are popularly used. The female bee partitions the nest gallery constructed in wood with the cut leaves forming few chambers/cells. Each cell is provisioned with pollen gathered by the bee, where she lays a single egg in each chamber. Thereafter, the entrance hole to the stem nest is closed with a plug of leaves until development is completed. About 21 days later, the newly developed adults emerge from the nest, cutting through the leaf plug with their mandibles.

5.5.14 Cuckoo Bees

Cuckoos or parasitic bees, do not construct their own nests but instead use provisioned (food laden) nest of another bee (referred to as the host bee) to lay eggs. Cuckoo bee larvae that hatch from these eggs feed on the pollen provisioned by the host bee and develop inside the host nest. As such, cuckoo bees do not need to collect pollen to provision their nests. As a result, cuckoo bees have lost the pollen-carrying hairs on the body, giving the bee a wasp-like appearance. In Sri Lanka there are about 25 species of cuckoo bees, each having its own specific host bee species. The most common cuckoo bee genus in Sri Lanka is *Thyreus* (Fig. 5.6.3) that lays eggs in the nests of *Amegilla*. Both species of bees belong to the same

Fig. 5.6.3 Cuckoo bee, *Thyreus insignis*

Fig. 5.6.4 *Lasioglossum alphenum* buzzing on curled anthers of *Osbeckia octandra* (Melastomataceae)

family. Another genus of cuckoo bees is *Coelioxys* that lays eggs in nests of *Megachile*, and both species belong to the family Megachilidae. *Sphecodes* lays eggs in the ground nests of *Lasioglossum* (F. Halictidae). The recent survey in Sri Lanka recorded five genera of parasitic bees, namely, *Coelioxys*, *Thyreus*, *Euaspis*, *Nomada*, and *Sphecodes*.

5.5.15 Specialist Bees

In certain flowering plants, mostly in the families Solanaceae, Melastomataceae, and Leguminosae, anthers do not split open along their length to release pollen. Their flowers have long tubular anthers with terminal pores through which pollen grains are released by sonicating (buzzing) bees, upon vibration. Certain bee species in the families Halictidae and Anthophoridae are capable of buzz pollination. *Amegilla* spp. *Curvinomia iridescens*, *Lasioglossum alphenum* (Fig. 5.6.4), *Patellapis kaluterae*, and *Xylocopa collaris* pollinate flowers through buzz pollination. Buzz-pollinating bees range in size from the largest carpenter bee, *Xylocopa tenuiscapa*, to one of the smallest bees, *Patellapis* species.

Lithurgus atratus (family: Megachilidae) is yet another specialist bee that pollinates flowers producing larger pollen grains. Its pollen-carrying structure (scope) consists of sparsely set long hairs that can accommodate large pollen grains. It thus collects pollen from plants such as *Abelmoschus esculentus* (okra) and *Argyreia populifolia. Lithurgus atratus* is considered as an important pollinator of okra.

5.5.16 Endemicity

According to published information on Sri Lankan bees and Indian bees by Gupta (2003), 20 species of bees have not been recorded from elsewhere and hencecan be tentatively considered as endemic to Sri Lanka. They are *Anthidiellum butarsis*, **A. krombeini*, **Systropha tropicalis*, *Tetralonia conmixtana*, *T. taprobanicola*, *T. fumida*, **Lasioglossum alphenum*, **L. kandiense*, **L. bidentatum*, **L. aulacophorum*, *Hylaeus krombeini*, **Amegilla puttalama*, **Amegilla subinsularis*, **Patellapis kaluterae*, *P. sigiriellus*, **Austronomia krombeini*, *A. austella*, *A. ustula*, **Nomada wickwari* **N. antennata*, and **Lipotriches edirisinghei*. Of those considered to be endemic, 13 species (marked with an asterisk) were collected during the recent survey (Karunaratne 2004).

Majority of the endemic species have a restricted distribution in Sri Lanka. The four species of *Lasioglossum*, L. (*Sudila*) *alphenum*, *L.* (*Sudila*) *aulacophorum*, *L.* (*Sudila*) *bidentatum*, and *L.* (*Evylaeus*) *carnifrons*, are confined to >1500 m altitude except *L.* (*Sudila*) *kandiense* (Sakagami, et al. 1998). The latter species had been collected between 100 and 200 m in Kandy (Sakagami et al. 1998) and also from >400 m at Sinharaja Forest Reserve (Karunaratne 2004). Thus, *L. kandiense* appears to be rare and needs to be conserved. Yet another special feature of the endemic bees is the relationship between the oligolectic endemic bees, *Tetralonia* sp. 1 specialized on pollen of the endemic plant *Argyreia populifolia* (Convolvulaceae). Similarly, the endemic plant *Osbeckia octandra* takes advantage of the buzz-pollinating endemic bee, *Lasioglossum alphenum*.

5.5.17 Threats and Conservation

In Sri Lanka, natural habitats are fast declining due to urbanization. Cleared habitats are encroached by invasive, wind-pollinated tall grasses such as *Panicum maximum* that does not produce nectar. A single genus of bees, *Lipotriches*, has become specialized to collect pollen from grasses as soon as the anthers of flowers dehisce early in the morning. A semi-agricultural habitat (11 ha) with a diverse vegetation cover in the Peradeniya University Park harbored more than 50 species of bees during 2000–2003 signifying the importance of partially disturbed habitats and vegetation in the conservation of bees. Complete clearance of this land in later years resulted in the growth of tall grasses and reduced the number of bee species to less than 20. Surveys conducted across the country found that weedy habitats with a diverse plant cover provide a year-round supply of nectar and pollen that attract many wild bee species in large numbers. Agricultural habitats with weedy bunds and adjacent mixed vegetation cover attract bees including ground nesters that find such sites with loose soil suitable for nesting. Slashing rather than clean weeding of agricultural habitats allow new growth of weeds ensuring flowering and would also keep their ground nests undisturbed. Abandoned land having a mixed weed cover attracts many wild bees, and such land should be set aside to conserve bees. Pesticides used in agriculture and horticulture are harmful to bees. Bees that forage in agricultural areas will get directly exposed to sprays and thereafter to spray residues left on plants. When pollen from flowers contaminated with insecticides are collected by bees and deposited in their nests, the developing young would be affected. Hence, application of pesticides, if carried out in the late afternoons and early mornings when the bees are less active and flowers are not yet opened or have closed, would protect pollinator bees.

5.6 Insects of the Rice Fields

5.6.1 Taxa Recorded

Rice fields harbor a rich and varied fauna dominated by arthropods, mainly insects and spiders. A study conducted by Bambaradeniya (2000a, b) highlighted the taxonomic composition, structure, and colonization of insects in an irrigated rice field. Insect fauna recorded consisted of 317 species, belonging to 19 orders and 104 families. Majority belonged to Hymenoptera (81 species in 26 families), dominated by bees and ants (Fig. 5.7.1). The second largest insect order recorded was Lepidoptera consisting of 58 species, in 7 families, dominated by Nymphalidae (24 spp.).

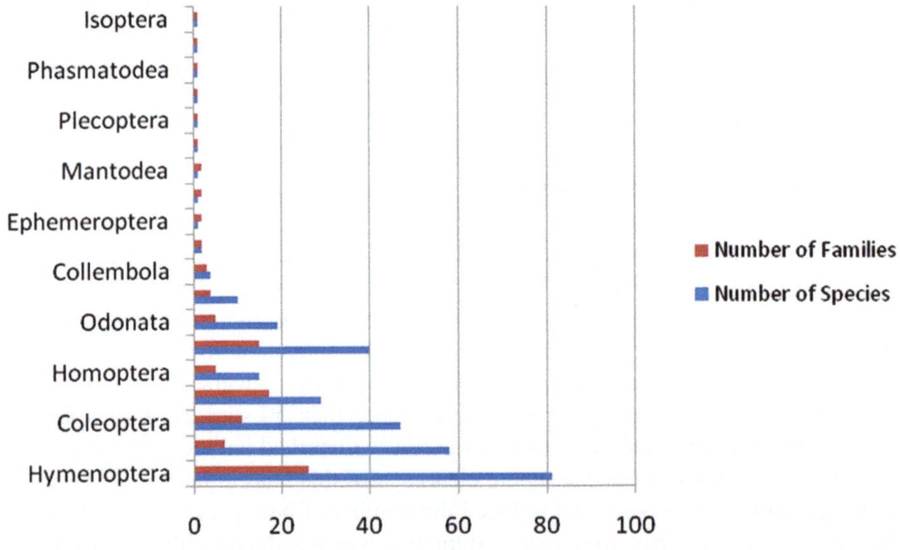

Fig. 5.7.1 Species composition and taxonomy of insects in a rice field

Coleoptera was the third largest insect order, with 47 species in 11 families. Carabids (18 spp.) were dominant among the Coleoptera. The orders Homoptera and Heteroptera together included 44 species in 22 families. The homopterans were dominated by Cicadellidae (ten spp.), while the heteropterans were dominated by Pentatomidae (seven spp.). In Diptera, of the 40 species in 15 families, Culicidae (14 spp.) was the dominant. Odonata included 19 species in 5 families, dominated by Libellulidae (nine spp.). Orthoptera included ten species in four families, where the Acrididae (six spp.) was the dominant. The Collembola included four species, in three families. The orders Strepsiptera, Thysanoptera, Ephemeroptera, and Mantodea included two species each. The remaining six insect orders (Dermaptera, Plecoptera, Phasmatodea, Blattoidea, Neuroptera, and Isoptera) included one species each.

5.6.2 Rice Field Microhabitats and Insect Fauna

The flooded rice fields go through an aquatic, semiaquatic, and dry phases during a single cultivation cycle with the growth of the rice plants from seedling through booting, flowering, and grain maturing stages. The field bunds that are an integral part of the rice ecosystem contribute to habitat heterogeneity and insect diversity. Sixteen species were confined to the rice habitat during the aquatic phase, while 74 were exclusively confined to the weedy bunds. Adult aquatic insects of the flooded rice habitat belonged to the orders Heteroptera (nine species, eight families) and Coleoptera (seven species, three families). Aquatic larvae comprised 41 species in four orders (Diptera, 19 species, 4 families; Odonata, 19 species, 4 families; Ephemeroptera, 2 species, 1 family; Plecoptera, 1 species, 1 family). Bund vegetation harbored mostly Lepidoptera; Papilionidae, Nymphalidae, Lycaenidae and Pieridae, and Hymenoptera (Apidae).

5.7 Insect Guilds

Based on food habits of insects, five arthropod guilds were identified (Table 5.7). Majority were predators (89 spp.) dominated by Coleoptera and Hymenoptera, each with 25 species, followed by Odonata with 19 species. Of the 130 species of phytophagous insects, majority (76 spp.) were visitors associated with rice field weeds. Phytophagous guild was dominated by Lepidoptera (50 spp.) followed by Hymenoptera with 15 bees species. Other phytophagous insects comprised 55 species of rice pests represented by sap feeders, leaf feeders, stem feeders, and root feeders (Table 5.7). Homopterans (14 spp.) were the dominant phytophagous pest group, closely followed by heteropteran pests (10 spp.). The parasitoid guild comprising 46 species of insects was dominated by hymenopterans (40 spp.). The scavenger/decomposer guild contained the fewest number of species (16 spp.), dominated by Diptera (10 spp.), followed by Collembola (4 spp.). The overall species composition reflects a high richness of insect natural enemies (predators and parasitoids) in relation to pests, where the natural enemy to pest ratio was 2.5:1. A majority of the parasitoids attacked pests of rice.

Table 5.7 Insect families in different feeding guilds in rice fields, Bathalagoda

Guild	Phytophages (rice pests)				Phytophages (non-rice pests)	Predators	Parasitoids	Scavengers/decomposers
Order	SF	DFM	SB	RF				
Blattoidea								1 (1F)
Coleoptera		6 (1F)		3 (2F)	5 (2F)	25 (4F)		1 (1F)
Collembola								4 (3F)
Dermaptera						1 (1F)		
Diptera		5 (4F)				2 (2F)	4(2F)	10 (4)
Heteroptera	10 (3F)				4 (4F)	9 (7F)		
Homoptera	14 (4F)				1 (1F)			
Isoptera				1 (1F)				
Hymenoptera		1 (1F)			15 (4F)	25 (5F)	40 (17)	
Lepidoptera		4 (3F)	4(1F)		50 (5F)			
Mantodea						2 (1F)		
Neuroptera						1(1F)		
Odonata						19 (5F)		
Orthoptera		6 (1F)				4 (3F)		
Phasmatodea						1 (1F)		
Strepsiptera							2 (2F)	
Thysanoptera	1 (1F)				1 (1F)			
Total	25 (8F)	22 (10F)	4 (1F)	4 (3F)	76 (17F)	89 (30F)	46 (20F)	16 (9F)

Pests: *SF* sap feeders, *DFM* defoliators/miners, *SB* stem borers, *RF* root feeders, *F* families

The data on insects collected from rice fields and bund habitats over two consecutive rice cultivation cycles (*Yala* and *Maha*) was used to document the relative abundance of insect taxa under different feeding guilds. Among the phytophagous pests on rice, Homoptera were generally dominated by Cicadellidae (Table 5.7.1). The abundance of delphacids showed a considerable increase during the *Maha* cycle. Among the cicadellids, the green leafhoppers (*Nephotettix* spp.) were the predominant species, while the delphacids were dominated by the white-backed plant hopper (*Sogatella furcifera*). The abundance of the mirid bug *Cyrtorhinus lividipennis*, known to prey on eggs and nymphs of homopteran pests (especially the delphacids), showed a considerable increase during the *Maha* cycle (Table 5.7.1). The parasitoids were dominated by the hymenopteran, Mymaridae. The trichogrammatids showed a considerable increase during the *Maha* cycle.

Among the insects inhabiting the bund habitats sampled with a sweep net, *Leptocorisa oratorius* was the most abundant (Table 5.7.2). The homopterans were dominated by Cicadellidae. Among the predatory insect groups in the field bunds, Odonata were dominated by the damselflies (Coenagrionidae), while the Coleoptera were dominated by Coccinellidae. Hymenopteran parasitoids in the bunds were dominated by braconids, followed by Chalcididae and Ichneumonidae (Table 5.7.2).

5.7.1 Colonization and Succession

The colonization and succession of major arthropod taxa in the rice fields followed a uniform pattern in relation to the growth stages of the rice crop as well as the phases of the field (Table 5.7.3). The colonization and buildup of insect communities observed in the rice field was characterized by early occurrence of terrestrial insect predators (e.g., Staphylinidae, Formicidae, Vespidae) and dipteran scavengers for a short period during the field preparatory stage, followed by a rapid colonization and multiplication of aquatic predatory insects (e.g., Mesoveliidae, Veliidae, Gerridae, Hydrometridae, Dytiscidae, Noteridae), aquatic heteropterans that feed on algae (e.g., Corixidae), coleopteran scavengers (e.g., Hydrophilidae), ephemeropteran scavengers (e.g., Baetidae), and collembolan scavengers (e.g., Isotomidae, Sminthuridae) in the flooded fields ready for crop establishment. The flooded fields were also visited by odonates (e.g., Libellulidae, Coenagrionidae) and dipteran flies (e.g., Culicidae) for breeding. The sap-sucking and defoliating pest phytophages (e.g., Cicadellidae, Delphacidae, Acrididae) rapidly colonized the fields and multiplied during crop establishment and tillering stages. This in turn facilitated an increase in the numbers of predatory insects (e.g., Coccinellidae, Miridae) and parasitoids (e.g., Mymaridae, Scelionidae, Braconidae, Trichogrammatidae) during the late tillering and booting stages of the crop. The grain sap feeding phytophages (e.g., Alydidae, Pentatomidae) colonized the fields during the flowering and milk grain stages. Non-pest phytophages consisting of many lepidopterans (e.g., Lycaenidae, Papilionidae, Hesperiidae, Pieridae) visited the bunds covered with weeds during the reproductive stage of the crop. In general,

Table 5.7.1 Relative proportions (%) of major arthropod species in different guilds during *Yala* and *Maha cycles* (based on blower-vac sampling)

Guild/taxa	*Yala* (*n* = 90)	*Maha* (*n* = 100)
Phytophages (pests)		
Homoptera		
Cicadellidae	81.3	53.5
Nephotettix virescens	35.4	41.0
N. nigropictus	38.7	26.4
Recilia dorsalis	10.8	19.9
Cofana spectra	14.2	12.6
Delphacidae	18.7	46.5
Nilaparvata lugens	23.4	22.1
Sogatella furcifera	76.6	77.9
Heteroptera		
Leptocorisa oratorius	79.0	89.7
Pentatomidae	21.0	10.3
Diptera		
Chironomidae	81.7	28.3
Orseolia oryzae	9.7	66.2
Muscidae	7.3	4.0
Orthoptera		
Acrida exaltata	68.0	71.4
Predators		
Heteroptera		
Cyrtorhinus lividipennis	33.3	76.9
Coleoptera		
Paederus alternans	38.5	36.3
Carabidae	34.6	36.3
Coccinellidae	27.0	27.8
Parasitoids		
Mymaridae	39.3	39.0
Scelionidae	23.3	18.2
Trichogrammatidae	19.3	32.8
Braconidae	8.0	2.5
Pteromalidae	4.0	2.4
Diapriidae	3.0	2.5
Ichneumonidae	2.0	1.5
Eulophidae	1.1	1.1

n number of samples

Table 5.7.2 Relative proportions (%) of major insect taxa in the non-rice bund habitat of the rice field during *Yala* and *Maha cycles* (data from sweep net samples)

Guild/taxa	*Yala* ($n = 45$)	*Maha* ($n = 50$)
Phytophages (pests)		
Homoptera		
Cicadellidae	91.5	87.0
Delphacidae	8.5	13.0
Heteroptera		
Leptocorisa oratorius	90.0	94.0
Pentatomidae	8.0	2.8
Orthoptera		
Acrida exaltata	43.5	42.5
Oxya japonica	32.5	34.3
Predators		
Heteroptera		
Reduviidae	58.9	90.0
Miridae	41.1	10.0
Coleoptera		
Coccinellidae	88.5	94.1
Micraspis discolor	65.0	71.9
Carabidae	11.5	5.9
Odonata		
Libellulidae	25.5	21.7
Diplacodes trivialis	37.5	38.5
Coenagrionidae	70.5	69.5
Ceriagrion spp.	46.1	65.2
Hymenoptera		
Formicidae	96.0	95.5
Solenopsis spp.	65.2	45.8
Orthoptera		
Conocephalus longipennis	70.0	57.1
Gryllidae	28.0	41.0
Parasitoids		
Hymenoptera		
Braconidae	26.5	40.6
Chalcididae	38.2	32.2
Ichneumonidae	26.5	25.0
Others	8.8	2.2

n = sweep net replicates (with 20 sweeps/replicate)

Table 5.7.3 Major insect taxa that colonized the different stages/phases of the rice field

	Major insect taxa colonized			
Cultivation stage/phase	Rice pests	Predators	Parasitoids	Scavengers
Field preparation (semiaquatic)		*Solenopsis* spp., *Camponotus* spp., *Odontomachus* spp., *Paederus alternans*, *Euborellia* spp., *Liris* spp., *Delta campaniformes*, *Tridactylus* spp.		*Blatella germanica*, Diptera (Tabanidae)
Nursery (aquatic)	*Nephotettix* spp., *Recilia dorsalis*, *Baliothrips biformis*	*Solenopsis* spp., *Camponotus* spp., *Paederus alternans*		
Flooded fields prior to be transplanted (aquatic)		*Microvelia* spp., *Mesovelia* spp., *Hydrometra greeni*, *Gerris adelaides*, Dytiscidae, Odonata (Libellulidae, Lestidae, Coenagrionidae)		Hydrophilidae, Baetidae, Diptera, Collembola
Active tillering (aquatic)	*Orseolia oryzae*, *Cnaphalocrocis medinalis*, *Sogatella furcifera*, *Nilaparvata lugens*, *Cofana spectra*, Orthoptera (Acrididae), Diptera (Chironomidae)	*Opheonia* spp., Orthoptera (Gryllidae).	Hymenoptera (Mymaridae, Scelionidae)	Collembola
Booting (aquatic)	*Scirpophaga incertulas*	*Cyrtorhinus lividipennis*, *Conocephalus longipennis*, Coleoptera (Coccinellidae)	Hymenoptera (Trichogrammatidae, Braconidae)	
Flowering (aquatic)	*Leptocorisa oratorius*	Heteroptera (Reduviidae)		
Milk Grain (aquatic)	*L. oratorius*, Heteroptera (Pentatomidae)			
Grain ripening (semiaquatic)		Coleoptera (Carabidae)		Collembola
Mature crop (dry)	*Coptotermes* spp.	Hymenoptera (Formicidae)		
Fallow period (dry)	*Cofana spectra*, *Nephotettix spp.*, *Coptotermes* spp.	*Paederus alternans*, *Micraspis discolor*, Formicidae		

the pest phytophages increased in numbers faster than predators and parasitoids, and predators arrived faster than parasitoids.

In conclusion, the study highlights the fact that the composition and structure of the insect communities in a rice ecosystem are characterized by (1) a high turnover of species, (2) rapid waves of colonization and uniform pattern of succession during consecutive rice cultivation cycles, (3) presence of species well adapted to specific niches and feeding guilds, (4) presence of species tolerant to short-lived but drastic physical changes in the rice field and (5) species that are specific to a particular growth stage of the rice plant or particular phase of the rice field, and (6) a higher species richness of predatory and parasitoid biocontrol insects compared to rice pest insects. Among the insects recorded, *Brachystegus decoratus* (Turner) (Hymenoptera: Sphecidae) is a new record to Sri Lanka. This species has been previously recorded from India (Krombein 1998).

References

Alpert GD (2013) A new species of *Tyrannomyrmex* (Hymenoptera: Formicidae) from Sri Lanka. Zootaxa 3721:286–290

Amarasinghe HE (2010) Species composition and nesting habits of ants in a hill-country home garden in Sri Lanka. Asian Myrmecol 3:9–20

Ananthakrishnan NR (1963) Biology and control of Scirtothrips bispinosus (Bagnall) on tea.-Rep. UPASI Sci. Dept. (Tea Sect.) Coonoor 37–53

Baker DB (1996) Notes on some palearctic and oriental Systropha, with descriptions of new species and a key to the species (Hymenoptera: Apoidea: Halictidae). J Nat Hist 30:1527–1547

Bambaradeniya CNB (2000a) Ecology and biodiversity in an irrigated rice field ecosystem in Sri Lanka. Ph.D. Thesis, University of Peradeniya, Peradeniya, Sri Lanka

Bambaradeniya CNB (2000b) (ed) Fauna of Sri Lanka: status of taxonomy, research and conservation. The World Conservation Union and Government of Sri Lanka, Colombo, pp 43–52

Bambaradeniya CNB, Edirisinghe JP, De Silva DN, Gunatilleke CVS, Ranawana KB, Wijekoon S (2004) Biodiversity associated with an irrigated rice agro-ecosystem in Sri Lanka. Biodivers Conserv 13:1715–1753

Bingham CT (1897) The Fauna of British India including Ceylon and Burma, Hymenoptera vol. 1, wasps and bees. Taylor and Francis, London, pp 414–567

Bingham CT (1903) The Fauna of British India, including Ceylon and Burma. Hymenoptera, vol. II. Ants and Cuckoo-wasps. Taylor and Francis, London

Bolton B (2003) Synopsis and classification of Formicidae. Memoirs Am Entomol Inst 71:1–370

Chhotani OB (1997) Fauna of India-Isoptora (termites). Published by Director, Zoological Survey of India, Calcutta

Dalla Torre CG (1896) Catalogus Hymenopterorum hucusque descriptorum systematicus et synonymicus. X:Apidae (Anthophila) – Leipzig pp 634

Delabie JHC (2001) Trophobiosis between Formicidae and Hemiptera (Sternorrhyncha and Auchenorrhyncha): an overview. Neotropical Entomol 30:502–516

Desneux J (1904) A propos de la phylogénie des Termitides. Ann Soc Entomologique Belg 48:8278–28

Dias RKS (2002) Ants of Sri Lanka with a brief description of field and laboratory methods. Sarvodaya Vishvalekha Press, Ratmalana

Dias RKS (2006) Current taxonomic status of ants (Hymenoptera: Formicidae) in Sri lanka and their host plants. MAB Series No.21. NSF Publication ISBN 1391-5010, 21: 2–4

Dias RKS, Kosgamage KRKA (2012) Occurrence and species diversity of ground-dwelling worker ants (Family: Formicidae) in selected lands in the dry zone of Sri Lanka. J Sci Univ Kelaniya 7:55–72

Dias RKS, Peiris HAWS, Ruchirani HPGRC (2011) Discovery of *Aneuretus simony* Emery in a disturbed forest in Kalutara, and *Stereomyrmex horni* Emery in Anuradhapura Sanctuary, Sri Lanka. Asian Myrmecol 4:99–102

Dias RKS, Kosgamage KRKA, Peiris HAWS (2012) The taxonomy and conservation status of ants (Order: Hymenoptera, Family: Formicidae) in Sri Lanka. In: Weerakoon DK, Wijesundara S (eds) The National Red List 2012 of Sri Lanka; conservation status of the fauna and flora. Ministry of Environment, Colombo, pp 11–19

Gunawardene NR, Majer JD, Edirisinghe JP (2008) Diversity and richness of ant species in a lowland wet forest reserve in Sri Lanka. Asian Myrmecol 2:71–83

Gunawardene NR, Majer JD, Edirisinghe JP (2010) Investigating residual effects of selective logging on ant species assemblages in Sinharaja Forest Reserve, Sri Lanka. For Ecol Manag 259:555–562

Gunawardene NR, Majer JD, Edirisinghe JP (2012) Correlates of ant (Hymenoptera: Formicidae) and tree species diversity in Sri Lanka. Myrmecological News 17:81–90

Gupta RK (2003) An annotated catalogue of bees of the Indian Region, online http://geocities.com/BeesInd2/braunsapis.htm

Harindra AK, Amarasinghe E, Edirisinghe JP (2007) Sri Lankan ant assemblages on multiple use land. Biodiversity 8:7–14

Hemachandra II (2014) Diversity, distribution and ecology of termites among selected vegetation types in Kandy and Knuckles regions. Ph. D thesis, University of Peradeniya

Hemachandra II, Edirisinghe JP, Karunaratne WAIP, Gunatilleke CVS, Fernando RHSS (2012) Diversity and distribution of termite assemblages in montane forests in the Knuckles Region, Sri Lanka. Int J Trop Insect Sci 34 (No.1): 41–52

Holmgren N (1910a) Das System der Termiten. Zool Anz 35(9):10284–10286

Holmgren N (1910b) Isoptera [The Percy Sladen Trust Expedition to the Indian Ocean in 1905 under the leadership of Mr. J. Stanley Gardiner]. Trans Linn Soc Lond Zool 14(2):8135–8148

Holmgren N (1911) Termitenstudien: 2. Systematik der Termiten. Die Familien Mastotermitidae, Protermitidae und Mesotermitidae. Kungliga Svenska Vetenskapsakademiens Handlingar 46:61–86

Holmgren N (1912) Termitenstudien: 3. Systematik der Termiten. Die Familie Metatermitidae. Kungliga Svenska Vetenskapsakademiens Handlingar 48:141–166

IUCN (2001) IUCN Red list categories: version 3.1. Prepared by the IUCN Species Survival Commission. IUCN, Gland, Switzerland and Cambridge, UK

Jayasuriya AK, Traniello JFA (1985) The biology of the primitive ant *Aneuretus simoni* (Emery) (Formicidae :Aneuretinae) I. Distribution, abundance, colony structure, and foraging ecology. Insect Soc 32:363–374

Judenko, Eastop (1963) A list of Sinhalese aphids. Entomologists Monthly Mag 3:62–63

Karunaratne WAIP (2004) Taxonomy and natural history of bees in selected areas of Sri Lanka. PhD thesis (unpublished). PGIS, University of Peradeniya, Sri Lanka

Karunaratne WAIP, Edirisinghe JP, Pauly A (2005) An updated checklist of bees of Sri Lanka with new records. Publication No. 23. MAB handbook and occasional paper series, National Science Foundation, Sri Lanka

Karunarathna DAGNB, Karunarathe WAIP (2013) Two new localities of Sri Lankan Relict Ant *Aneuretus simoni* Emery, 1893 (Formicidae: Aneuretinae) with the very first record in the intermediate zone. J Threatened Taxa 5:4604–4607

Kemner NA (1934) Systematische und biologische Studien über die Termiten Javas und Celebes. Kungliga Svenska Vetenskapsakademiens Handlingar 13:41–241

Krishna K, Grimaldi DA, Krishna V, Engel MS (2013) Treatise on the Isoptera of the World. Bulletin of the American Museum of Natural History

Krombein KV (1998) *Brachystegus decoratus* (Turner) (Hymenoptera: Sphecidae). An Indian Wasp New to Sri Lanka. Proc Entomol Soc Wash 100(4):828

Michener CD (2000) The bees of the world. Johns Hopkins University Press, Baltimore, p 913

Moritz G, Morris DC, Mound LA (2001) Thrips ID – pest thrips of the world. CD-Rom Published for ACIAR by CSIRO Publishing, Melbourne

Mound LA (2007) Thysanoptera (Thrips) of the world – a checklist. Available from: http://www.ento.csiro.au/thysanoptera/worldthrips.html

Peiris JWL (1953) A virus disease of Cacao in Ceylon. Trop Agric 109:135–138

Pitkin BP (1976) The hosts and distribution of British thrips. Ecol Entomol 1(1):41–47

Rickson FR, Rickson MM (1998) The cashew nut, *Anacardium occidentale* (Anacardiaceae), and its perennial association with ants: extrafloral nectary location and the potential for ant defense. J Bot 85:835–849

Roonwal ML (1970) Biology of termites. J Zool Soc India 21(1):57–66

Sakagami SF (1978) Tetragonula stingless bees of the Continental Asia and Sri Lanka (Hymenoptera, Apidae). J Fac Sci Hokkaido Univ Ser VI Zool 21(2):165–247

Sakagami SF (1991) The halictid bees of Sri Lanka and the vicinity. II. *Nesohalictus* (Hymenoptera: Halictidae). Zool Sci 8:169–178, Hokkaido University. Japan

Sakagami SF, Ebmer PA (1987) Taxonomic notes on Oriental halictine bees of the genus Halictus (Subgen. Seladonia) (Hymenoptera Apoidea). Linzer Biol Beitr 19(2):301–357

Sakagami SF, Ebmer AW, Tadauchi O (1996) The Halictine bees of Sri Lanka and the vicinity. III. *Sudila* (Hymenoptera: Halictidae) Part 1. Esakia 36:143–189

Sakagami SF, Ebmer AW, Tadauchi O (1998) The Halictine bees of Sri Lanka and the vicinity. III. *Sudila* (Hymenoptera: Halictidae) Part 2. Esakia 38:55–83

Schmutz K (1913) Zur Kentniss der Thysanopteran fauna von Ceylon. Sitzungsber Akad Wiss Wien 122:991–1089

Schwarz HF (1990) Beitrag zur Kenntnis orientalischer Nomada Arten. Entomofauna Suppl, cited in Michener, C. D. (2000) Bees of the world. Baltimore, Maryland USA

Snelling RR (1980) New bees of the genus Hylaeus from Sri Lanka and India (Hymenoptera: Colletidae). Contributions in Science, Natural History Museum of Los Angeles County, No. 328: 1–18

Tillekaratne K (2010) Taxonomy, distribution and host plant relationships in selected sites in Sri Lanka. M.Phil. Thesis, University of Peradeniya, Sri Lanka pp 239

Tillekaratne K, Mound LA, Strassen RZ, Edirisinghe JP (2007) List of thrips (Thysanoptera) recorded form Sri Lanka. J Natl Sci Found Sri Lanka 35(3):197–205

Tillekaratne K, Edirisinghe JP, Gunatilleke CVS, Karunaratne WAIP (2011) Survey of Thrips in Sri Lanka: a checklist of thrips species, their distribution and host plants. Ceylon J Sci (Biol Sci) 40:89–108

Wasmann E (1902) Termiten, Termitophilen und Myrmekophilen, gesammelt auf Ceylon von Dr. W. Horn 1899, mit anderm ostindischen Material bearbeitet. Zoologischer Jahrbücher, Abteilung für Systematik, Ökologie und Geographie der Tiere 17:99–164

Way MJ, Cammell ME, Bolton B, Kanagaratnam P (1989) Ants (Hymenoptera: Formicidae) as egg predators of coconut pests especially in relation to biological control of the coconut caterpillar, *Opisina arenosella* Walker (Lepidoptera: Xyloryctidae), in Sri Lanka. Bull Entomol Res 79:219–233

Westwood (1890) On a species of aphideous insects infesting the Bred fruit trees in Ceylon. Trans R Entomol Soc Lond 4:647–651

Wijerathna MAP (1997) A survey of aphids and their natural enemies on economic and other plants. M. Phil. Thesis, University of Peradeniya, Sri Lanka

Wijerathna MAP, Edirisinghe JP (1999) A checklist of aphids and their host plants from Sri Lanka. MAB Checklist and Handbook Series. Publication No. 21. National Science Foundation, Sri Lanka

Wijesekara A, Wijesinghe DP (2003) History of insect collection and a review of insect diversity in Sri Lanka. Ceylon J Sci (Bio Sci) 31:43–59

Utility of Arthropods by Indigenous Communities: Sustaining Natural Resources

6

M. Jayashankar, M. Charles, Vijeth V. Arya, and Jayalaxmi Hegde

Abstract

The use of natural resources for therapeutic purposes is as old as the humankind and continues around the world to this day. Collection of plant and animal products still prevails in different indigenous communities over the world. Ethnobiological knowledge has been passed on from generation to generation. Food habits and diet composition are adaptations to particular environmental and social conditions. Preferences given to arthropods utilized as food by humans depend on the palatability, availability, and nutritional cum medicinal values as well as on local traditions and customs. A concise compilation of arthropod use among indigenous communities in the oriental region is documented. Knowledge-based management where the benefits of biodiversity and ecosystems are acknowledged will be needed to prevent environmental degradation and ensure survival of arthropods and human communities as well.

Keywords

Arthropods • Entomophagy • Ethnoentomology • Traditional knowledge

M. Jayashankar (✉) • V.V. Arya
Division of Entomology and Nematology, Indian Institute of Horticultural Research (IIHR), Hessaraghatta Lake Post, Bengaluru 560089, Karnataka, India
e-mail: jay81zoology@gmail.com

M. Charles
St. Joseph's College (Autonomous), Bengaluru 560027, Karnataka, India
e-mail: mcharlesmsc@gmail.com

J. Hegde
Department of Entomology, University of Agricultural and Horticultural Sciences, Navule, Shivamogga 577216, Karnataka, India

A.K. Chakravarthy, S. Sridhara (eds.), *Economic and Ecological Significance of Arthropods in Diversified Ecosystems*, DOI 10.1007/978-981-10-1524-3_6

6.1 Introduction

Food gathering like hunting and fishing depends upon edible wild plants and animals and is the oldest human food-getting technology. Today, however, only a small number of societies practice it and tend to inhabit marginal environments. It is widely believed that population growth and seasonal variation in rainfall could have caused the transition from food collection to food production (Ember and Ember 1993). Food habits and diet composition are adaptations to particular environmental and social conditions.

6.2 Traditional Knowledge

Scheduled tribes who mostly inhabit forests constitute 8.61 % of the total population in India, numbering 104.28 million (2011 Census), and cover about 15 % of the country's area (http://tribesindia.com/). Ethnographers and researchers have studied the utilization of plants and animals as medicine and food in India by indigenous communities, which has led to reviews elaborating utilities of arthropods in India. The fact that tribal people need special attention is reflected from their low social, economic, and participatory indicators. Their traditional knowledge (TK) on medicinal significance can be used for improving their economic status as it is still prevalent among tribal communities playing a major role in health care. Traditional knowledge (TK) is variously referred as traditional ecological knowledge, local knowledge, or folk knowledge which is knowledge developed by local and indigenous communities over time in response to the needs of their specific local environment (Chouhan 2012).

Protections of the TK of the local and indigenous communities is one of the most contentious and complicated issues. TK is receiving a lot of attention nowadays due to its utility all over the world. It has become a focus in international forums. The protection under intellectual property rights (IPRs) of Traditional and Indigenous Knowledge (TIK) has received growing attention since the adoption of the Convention on Biological Diversity (CBD) in 1992 (Hirwade and Hirwade 2012). Most indigenous people have traditional songs, stories, legends, dramas, methods, and practices as means of transmitting specific human elements of traditional knowledge. Sometimes it is preserved in artifacts handed from father to son or mother to daughter. TK can be found in multitude fields such as nutrition, agriculture, fisheries, human health, veterinary care, handicrafts, performing arts, folk songs, religion, astrology, and many other day-to-day customs and practices. It is unique to every culture or society.

Arthropods are beneficial to humans by producing honey, silk, and wax and important as pollinators of crops, natural enemies of pests, scavengers, and food for other creatures. Arthropods and the substances extracted from them have been used as food, medicine, and ornament by human cultures over the world. Besides medicine, these organisms have also played mystical and magical roles in the treatment of several illnesses in many cultures. Ethnoentomology (study of the relationship between insects and people) focuses on how insects have been or are being used in human societies around the world. This includes insects used for food, rituals, and medicine. The consumption of insects in case of food scarcity and for their taste has been age-old practices. In recent years, in few places, consuming insects has been declining because insect-eating practice is considered as old-fashioned, dirty, and unhealthy. But edible insects and their traditional practices are still carried on by various groups for protein, fat, minerals, and vitamins (Patrick et al. 2008). A concise compilation of arthropod use among indigenous communities in India is documented in the present article. Many reviews (Kumar et al. 2008; Mahawar and Jaroli 2008; Pushpangadan et al. 2014; Akalesh et al. 2014) are available about the topic. However, the present review focuses on arthropods in ethnomedicine, entomophagy, and cultural aspects. A total of 56 insect species were utilized by 30 ethnic groups in edible, medicinal, cultural, aesthetic, and ornamental categories (Lokeshwari et al. 2011).

6.3 Culinary Utilities

Arthropods represent a traditional food category in many cultures of the world and have played an important part in the history of human nutrition in Africa, Asia, and Latin America. Almost 2000 insect species are consumed globally (Van 2013), of which many are regarded as delicacy (Gondo et al. 2010) and could be eaten in preference to fresh meat (Durst et al. 2010). Insects used in food include caterpillars, silkworms, grubs, beetle, moth's larvae, crickets, grasshoppers, locusts, arachnids, spiders, and scorpions. They can be eaten on their own or mixed with other ingredients. The term entomophagy refers to the use of insects as food. Insects, as the most species-rich taxon of all animals, exhibit an enormous biodiversity and represent a colossal biomass in nature. Mitsuhashi (2008) arrived at a figure of at least 1900 species of edible insects worldwide. Preference given to insect species utilized as food by humans depends on the insect's palatability, availability, and nutritional value as well as on local traditions and customs. Most of the edible insects, some of which are crop pests, but at the same time possess high nutritional qualities, constitute an important part of the local daily diet. Insects are very important source of nutrition for livestock as well. Ants, bees, termites, caterpillars, water bugs, beetle larvae, flies, crickets, katydids, cicadas, and dragonfly nymphs are among a long list of edible insects that provide nutrition for the people of Asia, Australia, Africa, South America, the Middle East, and the Far East (Srivastava et al. 2009). In Asia, around 349 edible insect species have been recorded which constitute 20 % of total insects consumed (Dennis 2008). Wasp consumption is an age-old practice in China; deep frying and frying the wasps with chicken egg and then offering to guests are seen even today. In Thailand there are 81 types of insects eaten which include larvae of different beetles when compared to other countries of Asia. Indonesia, Malaysia, Myanmar, the Philippines, and Vietnam have around 150–200 edible insect species (Dennis 2008). Nepal, Pakistan, India, and Sri Lanka have 57 edible insects. To meet the needs of local people, edible insect farming is in full swing in countries like northeast Thailand (Dennis 2008).

The Muria, one of the primitive tribes in the Bastar district in southeastern Madhya Pradesh, is very fond of an insect larva known as "chind kira." These yellowish-white larvae, each weighing about 50.0 g, are collected from young date palms (known as "chind"). Eggs of ants are collected from the leafy nests and considered as a delicacy. "Gurmurikira," which is collected from its nests by holding a lighted torch, is also eaten by many of the tribal people (Roy and Rao 1957). Insects are collected mainly from forests, and collections are done sustainably since it is a livelihood practice. Indigenous people depending on forest edible insect collections get good income and livelihood opportunities (Patrick et al. 2008). Edible insects, viz.,*Dorylus orientalis*, *Acheta domestica*, *Lethocerus indicus*, *Odontotermes obesus*, *Apis indica*, *Vespa orientalis*, *Hydrochera ricksseckeri*, *Hieroglyphus banian*, *Neoconocephalus palustris*, *Philosamia ricini*, *Antheraea assama*, and *Bombyx mori* (Borkakti 2005). Commonly consumed insects of Assam are *D. orientalis*, *Gryllus sp.*, *Lethocerus grandis*, *O. obesus*, *A. indica*, *Vespa* sp., *Agabetes acuductus or H. ricksseckeri*, *H. banian*, *N. palustris*, *P. ricini*, *A. assama*, and *B. mori* (Nath et al. 2005). Insect larvae (chind kira or gurmurikira) are mostly fried. Most of the indigenous communities are also habituated with the consumption of giant water bug (*Lethocerus indicus*), cricket, locusts, honeybee brood, and especially late-instar larvae and pupae (Hazarika 2008).

Around 29 species of edible insects are consumed by indigenous communities in greater Chabua of Dibrugarh of Assam (Das et al. 2011). At least 81 species of local insects, belonging to 26 families and five orders of insects, namely, *Coleoptera* (24 species), *Orthoptera* (17 species), *Hemiptera* (16 species), *Hymenoptera* (15 species), and *Odonata* (9 species), are being used as food among members of two tribal societies (i.e., the Nyishi of East Kameng and the Galo of West Siang) of Arunachal Pradesh (Chakravorhty et al. 2011). So the practices of entomophagy and entomotherapy by members of the Nyishi and Galo tribes, two ethnic groups of Arunachal Pradesh (Northeast India), are quiet ancient. The selection of the food insects among the Nyishi and Galo is passed by traditional tribal beliefs as well as the taste and availability of the insects. Depending on the species, only particular or all developmental stages are consumed. Some food insects may be included in the local diet throughout the year, others only when seasonally available. Commonly, specimens are being prepared for consumption by roasting, frying, or boiling. Members of the Galo use a greater number of insect species for remedial purposes than the Nyishi. In Assam, Mizoram, Manipur, and Tripura, the cinnamon bug, *Ochrophora* (*Udonga*) *montana* (Distant) (*Heteroptera*: *Pentatomidae*), is fried in oil and consumed (Fig. 6.1; Thakur and Firake 2012). With the degradation of natural resources, rapid population growth, and increasing influence of "westernization," the traditional wisdom of entomophagy and entomotherapy is at risk of being lost (Chakravorty et al. 2011).

Fig. 6.1 Entomophagy in Assam and the central region, India (http://thecsrjournal.in/a-bug-licious-solution-for-food-shortage/)

6.3.1 Termites

Consuming winged termites is also known to have nutrients and is in high demand during its seasons. Kannikaran and Palliyan tribes of South India have been using termites as food to enhance lactation in women, due to its high content of iron and zinc, and it is also the cheapest source of animal protein in Manipur (Gope and Prasad 1983; Rajan 1987). Winged adult termites like *Macrotermes subhyalinus* are high in magnesium and copper. In some South Indian tribes, every boy 12–14 years old is said to be given a termite queen to eat, after which he runs a distance of 2 or more miles; having done this he will be able to endure fatigue and run well (Gope and Prasad 1983). Rajan (1987) reports that after the first showers of the rainy season, the winged adult sexual forms of termites fly into houses in swarms, attracted by the lights – oil lamps, electric lamps, or tube lights in South India. Termites fallen are swept up, cleaned, fried, and eaten. In the North Arcot district of Tamil Nadu, the winged termites are known as *eesal* in Tamil. At the first hint of rain in the district, a forest tribe known as Irumbars sets three or four lighted lamps around each termite mound. The next morning the dead termites are swept up, cleaned, brought to the nearest big town, and sold to merchants who fry part of the catch and put it up for sale. Part of it is mixed with fried groundnut. Bengal gram, puffed rice, salt, and spices are fried and put up for sale. The fried pulses, spices, and salt enhance the taste (Rajan 1987).

In Karnataka, the winged termite known as "eechalu hula" is collected using lights. Rajan (1987) notes that in some villages, the queen termite is collected and fed raw to weak children. Among the early ancestors, consuming termites was widely seen (John 2007) and along with termites, ants, bees, and caterpillar that were also consumed based on the seasonal availability. In Meghalaya, termites are served as a source of proteins and carbohydrates (Paul and Dey 2011). The sexual forms of termites are popular probably because of the higher protein (87 %), carbohydrate (2.7 %), and amino acid (6.7 %) contents which are greater than those in worker forms (82 % proteins, 1.3 % carbohydrates, and 4.7 % amino acids) (Paul and Dey 2011). Further, the mineral content of sexual forms of termites is greater than conventional vegetarian food, salmon fish, and broiler chicken (Paul and Dey 2011). In Odisha, termites are eaten alone or together with rice (Srivastava et al. 2009). Live termite queens are given to the old to strengthen their backs.

6.3.2 Grasshopper Species

Acrididae (short-horned grasshoppers), *Acrida gigantea*, *Acridium malanocorne*, *A. peregrinum*, *Locustam ahrattarum*, *Mecopoda elongata*, and *Schistocerca gregaria* are consumed. Locusts are appreciated in many parts of India, and it is said that dried locusts form an ingredient of curries even in Calcutta (Maxwell-Lefroy 1971). Locusts are salted and eaten in India. Indian sepoys made a famous curry with locusts as ingredients (Bargagli 1877). Das (1945) analyzed *S. gregaria* for the use both as food and as fertilizer in India. The locusts were high in crude protein (61.75 %, air dried) and fat (16.95 %). The dried locusts could be tinned in quantity to ensure keeping them indefinitely. It is observed that dried locusts might be utilized for insectivorous cage and game birds. As fertilizer, locusts have fairly high nitrogen, phosphate, and potash. They are a menace to crops and contribute as fertilizer after death (Das 1945).

6.3.3 Honeybees and Ants

Sema Nagas, who occupy part of the watershed that divides Assam from Burma, collect both honey and larvae of wild rock bees. The nests are considered the private property of the finder; the bee brood is commonly eaten in the comb in India (Meyer-Rochow and Changkija 1997). Murries of Bastar use the red ant, *Oecophylla smaragdina*, as a regular item of food. The nests are collected throughout the year, but especially during the dry season, and torn open and the contents shaken into a cloth. The insects, both mature and immature, are beaten into a pulpy mass and enclosed in a pocket made of sal leaves. They are mixed with salt, turmeric, and chillies, ground between stones, and then eaten raw with boiled rice. They are also sometimes cooked with rice flour, salt, chillies, etc. into a thick paste, which is said to give great powers of resistance against fatigue and the sun's heat. Ants captured from the nests in the trees are covered and tied up in cups made of leaves and roasted. After the ants have been roasted, these are squeezed into a paste and baked with salt and chillies to make "chutney." Larvae of bees are also eaten. Sometimes these are killed and dried in the sun. Sun-dried ants are powdered and stored for future use. The powder is sour to taste and is used for the preparation of vegetable and meat curry. It is very clearly understood that insects like ants, bees, termites, caterpillar, honeybee larva, and bee products were highly used as food sources in China compared to other Asian countries. The Nyishi and Galo tribes use ants to cure scabies, malaria, toothaches, stomach disorders, blood pressure anomalies and other ailments in humans, and foot and mouth disease as well as worm infections in cattle (Chakravorty et al. 2011).

6.3.4 Beetles

It is reported that 42 species of insects like *Cerambycidae*, *Batocera rubra*, *Coelosterma scabrata*, *Coelosterma* sp., *Neocerambyx paris*, *Xystrocera globosa*, and *Xystrocera* sp. are used as food by the Ao Nagas in northeastern India; also there are two species of edible spiders for sale in a local market in Kohima (Meyer-Rochow and Changkija 1997). Both larvae and adults of the dytiscid, *E. sticticus*, are consumed. It breeds in brackish ponds. Larvae are gathered as they leave the water to pupate in the soil and also the newly emerged adults as they attempt to return to water (Meyer-Rochow and Changkija 1997). *Scarabaeidae* (scarab beetles), *Oryctes rhinoceros* larva, and *Xylotrupes gideon* are also processed as food. It is reported that *Aspongopus nepalensis*, found under stones in the dry riverbeds of Assam, is much sought after for the use as food, pounded up, and mixed with rice (Distant 1902). In the East Kameng district, Nyishi people consume *Batocera* spp. Some scarabaeid species, belonging to the genera *Lepidiota*, *Anomala*, and *Propomacrus*, are consumed by the Nyishi, but not the Galo people. *Catharsius* sp. is one of the favorite insect food items of the Galo, but not the Nyishi people (Chakravorty et al. 2011).

6.3.5 Bugs

The tribes Miris, Mishmas, Abors, and some Nagas consume *Aspongopus nepalensis*, *A. chinensis*, and *Cyclopelta subhimalayensis*. Also, it is said that these bugs are paralytic if eaten without first removing the red bilobed stink gland lying between the abdomen and metathorax (Distant 1902). The Nagas in Assam also consume the edible *Aspongopus nepalensis*, *Coridius chinensis*, and *C. subhimalayensis* (Hoffmann 1947). The large bugs of the genus *Aspongopus* are eaten with rice. The painted bug, *Bagrada picta*, is eaten by the Ao Naga, and its vernacular name is tsüngi (Meyer-Rochow and Changkija 1997).

6.3.6 Lepidopterans

Prepupae are removed after the cocoons have been completely formed and are considered a nutritious delicacy. Insects are also used as animal feed or fodder. *Bombyx mori* pupae are used in poultry extensively. De-oiled silkworm pupa meal from Mysore showed the highest level of crude protein (76.0 %, with free amino acids removed), lysine (5.36 %), histidine (1.94 %), and arginine (4.13 %). These pupae were low in calcium and phosphorus compared to fish meals and some other products (Chopra et al. 1970). In chick feeding trials (white leghorn) of 8 weeks of age, Ichhponani and Malik (1971) found that half of the fish meal (5 %) and half of the groundnut cake (10 %) in the ration can be replaced by de-oiled silkworm pupae and corn-steep fluid, respectively, with no reduction in final weight or feed/gain ratio. The supplies of fish meal are irregular and costly in Assam, while the silkworm pupae of muga silkworm, *Antheraea assamensis* is economical. *Dytiscidae* (predaceous diving beetles), *Curculionidae* (weevils, snout beetles), *Scarabaeidae* larva (scarab beetles), *Belostomatidae* (giant water bugs), *Apidae* (honeybees), *Apis dorsata* larva and pupa, winged termites, termite queens, Arctiidae (tiger moths), *Antheraea paphia* pupa, and *Antheraea proylei* pupa are consumed by various tribes in Manipur. Food for poultry farming in tribal areas in India is different, and tribals use insects, maggots, worms, and tender leaves along with cereals (http://www.poulvet.com/poultry/articles/feeding_strategies.php). In the northeast region, the Ahom community consumes silkworm pupae in the mature stage, whereas other tribes (Galo, Naga, Bodo, Missing, Rabiha, Kachari) prefer these insects in prepupal form (Sarmah 2011). The most favorable insect life stages utilized by indigenous communities are the caterpillars and pupae of the mulberry silkworm, *Bombyx mori*, and non-mulberry silkworms (*Lepidoptera*: *Saturniidae*), viz., *A. pernyi*, *A. assamensis*, *Attacus ricini*, and *Samia ricini* Donovan (Paul and Dey 2011).

6.3.7 Spiders

Using spiders for pest control is an age-old practice of a specific tribal community – *Nooka Dora* – of Andhra Pradesh and Orissa, border villages in the North Eastern Ghats, India. *Konda Dora* tribal farmers control black ants with tiny domestic red ants in Jowar fields (Shankar 2002). The Nochmaniin in the Nicobar Islands consumes mostly beetles and worms but also favor some kinds of spiders, centipedes, and locusts (Williamson 2005).

6.3.8 Crustaceans

The Katkari (Maharashtrian tribe) women draw crabs from their holes during the summer months by rubbing two stones to imitate the sound of thunder showers. Crabs think it is about to rain and leave their holes only to be grabbed by the tribal women (http://en.wikipedia.org/wiki/Katkari_people). Tribes in Nagaland are known to consume crabs in combinations with pork and bamboo shoots (http://idiva.com/opinion-ifood/a-treat-of-the-seven-sisters/25539). Crabs are a delicacy among the Andamanese tribes as well (http://www.indianmirror.com/tribes/andamanesetribe).

In the future insects eating may be accepted favorably by processing and mixing insects with other food items. It is well understood that, when there will be shortage of food, people will be going for edible insects. Edible insects may be used as a space-travel food in distant future (Jun 2008). The future of insect consuming by indigenous people in various parts of the world will be to its maximum. In Japan, Tama Zoo Insectarium, Tokyo insect-eating events are organized (Jun 2008). When insects are consumed without one's knowledge along with regular food, it causes no harm or allergy. This has led people to accept and adapt insect-consuming practices. So the future of insect as human food is likely to be more.

6.4 Medicinal Utilities

The most ancient and complete record of the use of insects for medicinal preparations is available in China (Read 1935). In many parts of the world, different sections of the society are using medico-entomological drugs to this day. Animal-based medicines have always played a significant role in the healing practices, magic rituals, and religions of indigenous tribes. A number of studies have in recent years drawn attention to the therapeutic value of insects and their products (Chakravorty et al. 2011). Traditional ethnobiological knowledge and the habit of accepting insects as food and as an integral part of local therapies are nowadays confined to the traditionally living, largely indigenous societies of regions that until now have experienced only a limited amount of "westernization." The therapeutic uses of insects are often a closely guarded secret and only passed on to certain individuals from one generation to another by word of mouth. Transfer of knowledge in this way is an age-old practice and a well-accepted sociocultural attribute among the ethnic societies of Northeast India. These communities with their local biological resources have a considerable understanding of nature and thus possess deep ethnobiological knowledge. The tribes are totally dependent for livelihood on the forests and resources and collect certain plants and animals for food and folk medicinal purposes which is an age-old practice for them.

About 11 species of insects were used by Kaniyar and Paliyar tribes in Tirunelveli district, Tamil Nadu, South India (Singh and Padmalatha 2004). Traditional therapeutics use animals among tribal population of Tamil Nadu for the treatment of over 17 kinds of ailments, including asthma, arthritis, epilepsy, paralysis, hydrocele, and leprosy (Solavan et al. 2004). Jamir and Lal (2005) describe the traditional method of treating many kinds of ailments using twenty-six animal species and their products by different Naga tribes. Singh and Padmalatha (2004) documented the ethno-entomological practices in Tirunelveli district, Tamil Nadu, where 11 species of insects were used to prepare traditional medicine. Twelve species of insects were used in South India and used in traditional method (Dixit et al. 2010), while ten species of insects were used in therapeutic purposes by the Karbis and Rengma Nagas – two ethnic tribes of Karbi Anglong district of Assam (Ronghang and Ahmed 2010). Ethnobiological knowledge has been passed on from generation to generation.

Arthropods represent a rich and largely unexplored source of new medicinal compounds (Dossey 2010). The medicinal uses of insects were often defined by the doctrine of signatures, which stated that an organism bearing parts that resemble human body parts, animals, or other objects was thought to have useful relevance to those parts, animals, or objects. This doctrine is common throughout traditional and alternative medicine (En.wikipedia.org/wiki). Though insects were widely used throughout history for medical treatment on nearly every continent, relatively little medical entomological research has been conducted since the revolutionary advent of antibiotics. Heavy reliance on antibiotics, coupled with discomfort with insects in Western culture, limited the field of insect pharmacology until the rise of antibiotic-resistant infections sparked pharmaceutical research to explore new resources. Below is the summary of such knowledge.

6.4.1 Termites

Termites are said to cure a variety of diseases, both specific and vague. Typically the mound or a portion of the mound is dug up, and the termites and the architectural components of the mound are together ground into a paste which is then applied topically to the affected areas or, more rarely, mixed with water and consumed. This treatment was said to cure ulcers, rheumatic diseases, and anemia; it was also suggested to be a general pain reliever and health improver (Chakravorty et al. 2011). Termites as medicine have played a major role in curing asthma. As a remedial measure, some South Indian tribes roast the winged termites in an earthen pot and consume as such in the evenings for 3 days. The antimicrobial properties of the termite *Odontotermes formosanus* against various bacterial strains confirm the antimicrobial properties of the termite and thus support tribal remedial measures.

6.4.2 *Lepidoptera*

The Jatropha leaf miner, a lepidopteran which feeds preferentially on Jatropha, is an example of a major insect agricultural pest which is also a medicinal remedy. The larvae are harvested, boiled, and mashed into a paste. This is administered topically and said to induce lactation, reduce fever, and sooth gastrointestinal tracts (Srivastava et al. 2009).

6.4.3 Honeybees

Honey is therapeutically used since time immemorial due to its antibacterial, anti-inflammatory, and wound healing properties (Chakraborty and Debnath 2003). Honeybee products are used medicinally across Asia. Honey can be applied to the skin to treat excessive scar tissue, rashes, and burns and applied as a poultice to eyes to treat infection (Chakravorty et al. 2011). It is also consumed for digestive problems and can be heated and consumed to treat head colds, cough, throat infections, laryngitis, tuberculosis, and lung diseases. Honeybee venom is used by direct stings to relieve arthritis, rheumatism, polyneuritis, and asthma (Ramos-Elorduy de Concini and Pino Moreno 1988). Propolis, a resinous, waxy mixture collected by honeybees and used as a hive insulator and sealant, is often consumed by menopausal women because of its high hormone content, and it is said to have antibiotic, anesthetic, and anti-inflammatory properties. Royal jelly is used to treat anemia, gastrointestinal ulcers, arteriosclerosis, hypo- and hypertension, and inhibition of sexual libido (Ramos-Elorduy de Concini and Pino Moreno 1988). Bee bread, or bee pollen, is eaten as a general health restorative. Apitherapy is the medical use of honeybee products such as honey, pollen, bee bread, propolis, royal jelly, and bee venom. One of the major peptides in bee venom, called melittin, has the potential to treat inflammation in sufferers of rheumatoid arthritis and multiple sclerosis. It is administered by direct insect sting or as intramuscular injections. Bee products demonstrate a wide array of antimicrobial factors, and laboratory studies have shown to kill antibiotic-resistant bacteria, pancreatic cancer cells, and many other infectious microbes. Beeswax is used in several dermatologic disorders, and royal jelly has been used to treat postmenopausal symptoms. Tribes like Irular, Toda, Kannikaran, Kurimbas, and Palliyan in Tamil Nadu use honey and winged termites for curing asthma (Solavan et al. 2004).

6.4.4 Bugs

The Biate tribe race is the prominent inhabitant of Meghalaya (http://en.Wikipedia.org/wiki/Biatepeople). Using animals and animal parts in traditional therapy in Dima Hasao district is common for instance, Cockroach, *Periplaneta americana* is crushed and consumed three times every week for 1 month to cure tuberculosis. For treating malaria *Cimex lectularius* is swallowed thrice daily for 1 week. Oil extracted from *Ochrophora* (*=Udonga*) *montana* (*Hemiptera*: *Pentatomidae*) is believed to cure many health problems in spite of its bad smell. To avoid undesirable reactions caused by the consumption of certain species of insects, sometimes highly specific preparation methods exist. Frequently appendages that cause some allergic reactions and, in the case of some bugs, parts of the abdomen that may contain hallucinogens or neurotoxins are removed by the Galo people before eating.

6.4.5 Beetles

The Nyishi and Galo use blister beetle to help against skin allergies (Chakravorty et al. 2011).

6.4.6 *Arachnida*

Arachnids have also been used for thousands of years in traditional medical practices. It is believed that the early Indians used the tail spines of the *Tachypleus gigas* crab as spear tips. After grinding the body, it was used as fertilizer for their fields and ponds. Some of the tribes inhabiting the northeast coast of Orissa still use the tail piece for relief from different types of pain. It has been reported that the tail tips are used for healing arthritis or other joint pains and they are sold by faith healers in West Bengal. It is believed that Indians in the early days used to eat the appendages of the horseshoe crab. Additionally, the dead carapace of the crab is boiled with mustard oil and used for treating rheumatic pain (Chatterji and Vijayakumar 1987; Chatterji 1994; Huma 2007).

The reluctance of the traditional healers to reveal their secrets is due to the belief that revealing the properties and secrets renders the medicine ineffective. Most of the people prefer this traditional cure to the modern pharmaceuticals, as it is less expensive and claimed to be more effective. Thus, zoo therapy is an important and integral part of the traditional health-care system of a tribal community. However, overexploitation and lack of regulation and monitoring to safeguard for sustainable utilization is a point to consider seriously from the conservation point of view. But in India this traditional knowledge is fast eroding due to modernization. Thus, there is an urgent need for inventories and recording all ethnobiological information among the different ethnic communities before the traditional cultures are lost.

6.5 Ornamental Utilities

Cultural entomology treats the influence of insects upon the "essence of humanity as expressed in the arts and humanities" (Hogue 1980). Weaving and silkworm rearing is another part of Bodo culture. Many families rear their own silkworms. The cocoons of the silkworms are spun into silk. Women weave and wear their own Dokhnas, which is the traditional dress of the Bodo women.

India has been a rich source of beetle wing embellished textiles and ornaments for many centuries. Most likely, there is a long history of beetle wings used as decoration by indigenous peoples in many parts of India. Well-known objects such as "desert jewelry," dolls and playthings, decorative fans, Jain sacred book covers, and torans (hung over niches or doors) decorated with beetle wings have been made by Rajasthanis. In India some of the oldest documented uses of beetle wings are found in early Basholi school miniature paintings from the Pahari Hills, an area which now lies in Himachal Pradesh, Jammu, Punjab Hills, and Uttar Pradesh. Insects were used in ornaments in Indus Valley Civilization, and beetles' wings were treated as jewel among Nagas in India. Molted blocks with beetles and other hard insects are used to decorate key chains.

6.6 Utility and Conservation

Indigenous communities still use animal products and by-products for cure of diseases, viz., honey is used as expectorant, and cattle urine has been used as a therapeutic medicine. Information on medicinal and nutritional value of arthropods would be imperative in the future to solve the hunger- and health-related problems. Studies conducted on tribals living in different parts of India have reported them to be socially and economically disadvantaged groups and their diets nutritionally deficient (Singh and Rajyalakshmi 1993; Mishra et al. 2002; Murugesan and Ananthalakshmi 1991). The availability of all types of modern food stuffs and the degradation of resources make ethnic people worldwide inclined to abandon their traditions and discard their rich indigenous knowledge. From a nutritional aspect, the traditional food is often not only healthier; it is also the product of generations of harmonious coexistence between tribe and environmental resource. Due to unprecedented population increases, the resources of the forest, including food insects, can become over-exploited resulting in the diminishing of biotic resources (Chakravorty et al. 2011). Consequently, preserving traditional food practices enriches and conserves the traditional habits of tribal community. Dependence of a community on the ecosystem should be protected from anthropogenic modification. In the past, fear-based traditions sufficed for sustainable environmental management, but as communities develop, knowledge-based adaptive management where the benefits of biodiversity and ecosystems are acknowledged will be needed to prevent environmental degradation and ensure the survival of arthropods.

References

Akalesh KV, Surya BP, Thengtom R, Jashodeb A (2014) Traditional healing with animals (zootherapy) by the major ethnic group of Karbi Anglong district of Assam, India. Int J Pharm Pharm Sci 6(8):593–600

Bargagli P (1877) Insetti Comestibili Letturanell. Soc Entomol Ital Revista Europaea-Riv Internaz Fasc 16(6):11

Borkakti R (2005) Uttar Purbanchalar Manuhar Priya Kadhya – Patang Prantik, 16th July, pp 33–35

Chakraborty N, Debnath PK (2003) Bees make medicine for mankind. Indian J Tradit Knowl 2(1):22–26

Chakravorty J, Sampat G, Benno V, Meyer-Rochow (2011) Practices of entomophagy and entomotherapy by members of the Nyishi and Galo tribes, two ethnic groups of the state of Arunachal Pradesh (North-East India). J Ethnobiol Ethnomed 7:5

Chatterji A (1994) The horseshoe crab-a living fossil, a project. Swarajya Publication, Orissa, p 157

Chatterji A, Vijayakumar R (1987) The amazing horseshoe crab. Science Today. September pp 62–64
Chopra AK, Malik NS, Makker GS, Ichhponani JS (1970) Evaluation of poultry feeds available in India. I. Proximate analysis, energy values and basic amino-acid contents of feed ingredients. J Res Ludhania 8:232–236
Chouhan VK (2012) Protection of traditional knowledge in India by patent: legal IOSR aspect. J Humanit Soc Sci 3(1):35–42
Das S (1945) Locust as food and manure. Indian Farming 6:412
Das S, Chetia Mondeep, Shah Rajesh Kumar (2011) A study on Edible insects used by the tea tribes in Greater Chabua area, Dibrugarh. Abstract of International Seminar on Bioresources and Human Sustenance. Zoology Department, Cotton College, Guwahati, Assam, p 154
Dennis VJ (2008) The contribution of edible forest insects to human nutrition and to forest management: current status and future potential. Forest insects as food: humans bite back, Proceedings of a workshop on Asia-Pacific resources and their potential for development. 19–21 February 2008, pp 5–22
Distant WL (1902) The Fauna of British India: Rhynchota. Vol. I. Heteroptera. Taylor and Francis, London, p 283
Dixit AK, Kadavul K, Rajalakshmi S, Shekhawat MS (2010) Ethno-medicobiological studies of South India. Indian J Tradit Knowl 9(1):116–118
Dossey AT (2010) Insects and their chemical weaponry: new potential for drug discovery. Natural Products Reports, Issue 12
Durst PB, Johnson DV, Leslie RN, Shono K (2010) Forest insects as food: humans bite back. FAO Regional Office for Asia and the Pacific, Bangkok; http://www.fao.org/docrep/012/i1380e/i1380e00.pdf website
Ember RC, Ember M (1993) Anthropology, 6th edn. Prentice/Hall, New Delhi, p 554
Gondo T, Frost P, Kozanayi W, Stack J, Mushongahande M (2010) Linking knowledge and practice: assessing options for sustainable use of mopane worms in southern Zimbabwe. J Sustain Dev Afr 12:127–145
Gope B, Prasad B (1983) Preliminary observation on the nutritional value of edible insect of Manipur. J Adv Zool 4:55–67
Hazarika R (2008) Peculiar palate, horizon. The assam tribune. Saturday, May, p 24
Hirwade M, Hirwade A (2012) Traditional knowledge protection: an Indian prospective DESIDOC. J Libr Inform Technol 32(3):240–248
Hoffman WE (1947) Insects as human food (India: Pentatomidae). Proc Entomol Soc Wash 49:233–237
Hogue CL (1980) Commentaries in cultural entomology-definition of cultural entomology. Entomol News 91(2):33–36
Huma A (2007) The Indian horseshoe crab, *Tachypleusgigas* (Muller) and its biomedical applications Thesis Submitted to Goa University, p 173
Ichhponani JS, Malik NS (1971) Evaluation of de-oiled silkworm pupae meal as protein source in chicken rations. Br Poult Sci 12:231–234
Jamir NS, Lal P (2005) Ethno scientific practices among Naga tribes. Indian J Tradit Knowl 4(1):100–104
John RM (2007) General entomology: N C State University. Last update 18 August, 2014
Jun M (2008) The future use of insects as human food. Forest insects as food: humans bite back, Proceedings of a workshop on Asia-Pacific resources and their potential for development. 19–21 February 2008, pp 115–122
Kumar S, Barthakur ND, Rao ML (2008) Bioprospecting with reference to medical insects and tribes in India; an overview. Indian Forester 134(12):1575–1591
Lokeshwari RK, Shantibala T, Singh TK (2011) Insect biresources in Manipur: ethno-practices and exploring new horizons. Abstract of International Seminar on Bioresources and Human Sustenance, Cotton College and Zoological Society of Assam, 20th–22nd Oct, 2011, p 152

Mahawar MM, Jaroli DP (2008) Traditional zootherapeutic studies in India: a review. J Ethnobiol Ethnomed 4:17
Maxwell-Lefroy (1971) Insect as food. Chapter 22, Springer, South central Asia
Meyer-Rochow VB, Changkija S (1997) Uses of insects as human food in Papua New Guinea, Australia, and north-east India: Cross-cultural considerations and cautious conclusions. Ecol Food Nutr 36(2–4):159–185
Mishra CP, Singh N, Chakravarty A (2002) Dietary pattern of a tribal community of Naugarh block. Tribal Health Bull 8(1):6–11
Mitsuhashi J (2008) Sekaikonchushokotaizen. Yasaka Shobo, Tokyo
Murugesan TP, Ananthalakshmi A (1991) Dietary practices of the Paliyar tribal group and the nutrient content of unconventional foods. Indian J Nutr Diet 28:297
Nath DHK, Gogoi R, Gogoi G (2005) Insects as human food with special reference to Assam and North East India. Asian Agri-Hist 9(2):119–127
Patrick B, Durst, Kenichi S (2008) Edible forest insects: exploring new horizons and traditional practices. Forest insects as food: humans bite back, Proceedings of a workshop on Asia-Pacific resources and their potential for development. 19–21 February 2008, pp 1–4
Paul D, Dey S (2011) Assessment of the nutritive value of some wild edible insects of Meghalaya, North East India. J Entomol Res 35:353–358
Pushpangadan P, George V, Sreedevi P, Ijinu TP, Ninawe A (2014) Ethnozoological knowledge of Indian scheduled tribes, scheduled caste and rural communities. Indian J Tradit Knowl 13(4):735–741
Rajan TKC (1987) Tiny wild fauna and human food. My For 23:177–180
Ramos-Elorduy de Concini JJM, Moreno P (1988) The utilization of insects in the empirical medicine of ancient Mexicans. J Ethnobiol 8(2):195–202
Read BE (1935) Insect drugs. Peking Nat Hist Bull 94:8–85
Ronghang R, Ahmed R (2010) Edible insects and their conservation strategy in Karbi Anglong District of Assam, North East India. The Bioscan 2:515–521
Roy JK, Rao RK (1957) Investigation on the diet of the Muria of Bastar District. Govt. of India, Calcutta. Bull Dept Anthropol 6(1):33–45
Sarmah MC (2011) Eri pupa: a delectable dish of North East India. Curr Sci 100(3):279
Shankar KJNG (2002) Experiments with spiders, ants and other indigenous practices. LEISA Mag 4(3):36
Singh RAJA, Padmalatha (2004) Ethno-entomological practices in Tirunelveli District, Tamil Nadu. Indian J Tradit Knowl 3(4):442–446
Singh AK, Rajyalakshmi C (1993) Status of tribal women in India. Soc Chang 23(4):1–18
Solavan A, Paulmurugan R, Wilson V, Ranjith AJA (2004) Traditional Therapeutic uses of animals among tribal population of Tamil Nadu. Indian J Tradit Knowl 3(2):182–186
Srivastava SK, Babu, Pandey (2009) Traditional insect bioprospecting-as human food and medicine. Indian J Tradit Knowl 8:85–494
Thakur ANS, Firake DM (2012) *Ochrophora montana* (Distant): a precious dietary supplement during famine in northeastern Himalaya. Curr Sci 102(6):845–846
Van HA (2013) Potential of insects as food and feed. Ann Rev Entmol 58:563–583
Williamson PJ (2005) Displaced by Tsunami, Ancient Tribe has unusual eating habits for National Geographic News January 20
http://en.wikipedia.org/wik.
http://en.Wikipedia.org/wiki/Biatepeople.
http://en.wikipedia.org/wiki/Katkari_people.
http://idiva.com/opinion-ifood/a-treat-of-the-seven-sisters/25539.
http://thecsrjournal.in/a-bug-licious-solution-for-food-shortage.
http://www.poulvet.com/poultry/articles/feeding_strategies.php.

Insects as Human Food

7

A.K. Chakravarthy, G.T. Jayasimha, R.R. Rachana, and G. Rohini

Abstract

Entomophagy, the consumption of insects by humans, is practiced in many countries around the world but predominantly in Asia, Africa, and Latin America. The practice has many environmental, health, social, and livelihood benefits. Insects are good source of protein, fat, minerals, and vitamins. Worldwide, nearly 1700 insect species are edible; four insect orders in rank sequence which predominate human consumption, viz., Coleoptera, Hymenoptera, Orthoptera, and Lepidoptera, account for 80 % of the species eaten. In Africa, ants, termites, beetle grubs, caterpillars, and grasshoppers are eaten. Eri silkworm, wasps, bamboo caterpillars, crickets, and locusts predominate insect diet of Asians. Important Latin American insect food comprises leaf-cutter ants, palm weevil larvae, bee, and wasp brood. Practicing entomophagy does carry certain risks like allergic reactions to the people consuming them. It is an age-old practice that continues to this day in many parts of the world. Science increasingly provides data corroborating the nutritional and health benefits of entomophagy, according to a

A.K. Chakravarthy (✉)
Division of Entomology and Nematology, Indian Institute of Horticultural Research (IIHR), Hessaraghatta Lake Post, Bengaluru 560089, Karnataka, India
e-mail: chakravarthyakshay@gmail.com

G.T. Jayasimha
Department of Entomology, Agriculture College and Research Institute, Madurai, Tamil Nadu, India
e-mail: jayaavavaa@gmail.com

R.R. Rachana
Division of Entomology, National Bureau of Agricultural Insect Resources (NBAIR), Bengaluru 560024, Karnataka, India
e-mail: vavarachana@gmail.com

G. Rohini
Department of Zoology, Bangalore University, Jnana Bharathi Campus, Bengaluru 560056, Karnataka, India

A.K. Chakravarthy, S. Sridhara (eds.), *Economic and Ecological Significance of Arthropods in Diversified Ecosystems*, DOI 10.1007/978-981-10-1524-3_7

broader acceptance of this practice, while giving due consideration to certain risk factors. Rewards in terms of long-term food security, income potential, pesticide reduction, and insect conservation are conceivable, and thus entomophagy has the potential of becoming an important factor in sustainable development.

Keywords
Asia • Africa • Entomophagy • Latin America

7.1 Introduction

Entomophagy is the consumption of insects by humans. It is heavily influenced by culture and religion of the people and is practiced in many countries around the world, predominantly in parts of Asia, Africa, and Latin America. Insects supplement the diets of approximately two billion people and have always been a part of human diets (FAO 2013). However, it is only recently that entomophagy has captured the attention of the media, research institutions, chefs and other members of the food industry, legislators, and agencies dealing with food. The Edible Insects Programme at FAO also examines the potential of arachnids (spiders and scorpions) for food and feed, although by definition these are not insects. Entomophagy offers advantages which are environmental, health, economic, livelihood, and social in nature and are predicated to play a vital role in future food security. The most commonly eaten insect groups are beetles, caterpillars, bees, wasps, ants, grasshoppers, locusts, crickets, cicadas, leaf and plant hoppers, scale insects and true bugs, termites, dragonflies, and flies.

7.2 History

The earliest citing of entomophagy is found in biblical literature. However, eating insects was, and still is, taboo in many Westernized societies (FAO 2013). The practice of entomophagy has, at one time or other, been established on every continent except Antarctica, although evidence from Europe is sparse and generally restricted to its more southern and eastern regions (Tommaseo-Ponzetta 2005). All tropical continents and even North America to this day have certain epicenters of entomophagy, such as the American Southwest and neighboring Mexico, the Amazon basin in South America, Central and Southern Africa, Southeast Asia, and aboriginal Australia. As biodiversity, activity, and the density of many insect groups tend to increase toward the equator, so do opportunities for entomophagy, and the menu generally is richest in the humid regions (Paoletti 2005).

About three decades ago, in response to surging human populations and concomitant threats to biodiversity and food security, numerous rural development initiatives started in developing countries, including science-driven efforts to explore

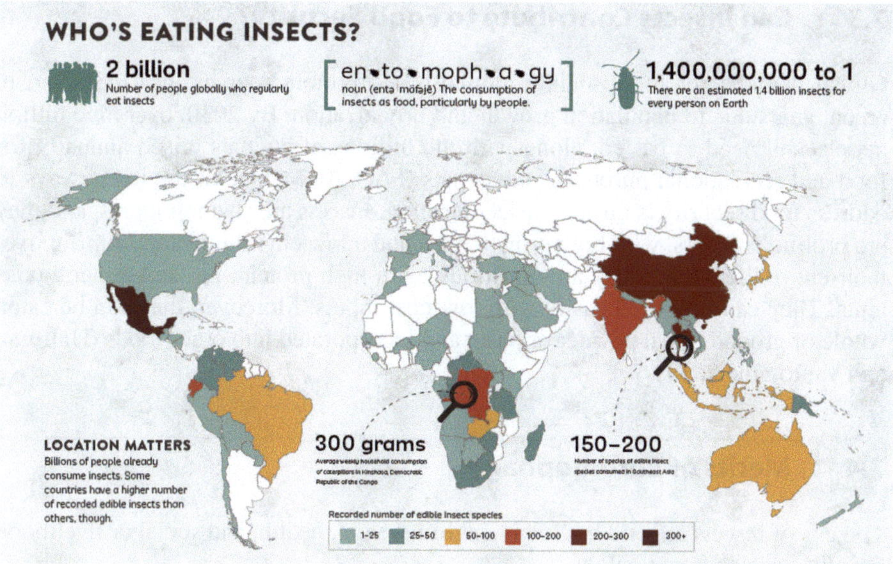

Fig. 7.1 Human populace and entomophagy (http://thecsrjournal.in/a-bug-licious-solution-for-food-shortage/)

insects as feed for livestock (Fig. 7.1). By serendipity, the nutritional merits of entomophagy also came under close scrutiny, and supportive data led to attempts to promote this practice as well. In 1988, Gene DeFoliart started publishing *The Food Insects Newsletter*, which eventually reached readers at least in 82 countries. This partially humorous but sufficiently serious publication became a forum for documentation of contemporary as well as historic information on entomophagy from remote corners of the world. In the process, earlier suggestions for using edible insects as minilivestock (Osmaston 1951) were reviewed (Paoletti 2005).

The resulting flurry of publicity spawned serious research efforts and resulted in numerous synopses on entomophagy for specific people, regions, countries, and continents. Much of this growing body of knowledge is reflected in the references listed by contributors in Paoletti (2005). Entomophagy was incorporated in certain university curricula, featured in movies, and insect feasts (bug banquets) were staged in conjunction with entomological conferences, nature centers, state fairs, zoo and museum exhibits, school events, parties, and military or wilderness survival training exercises. In Australia, "bush tucker" supplies became commercially available, a restaurant chain and an airline adopted insects as signature food, several relevant books were published, and there were TV shows (Yen 2005). There were even two conferences on "Insects as a Food Resource."

7.3 Can Insects Contribute to Food Security?

Global demand for food, mainly animal-based protein sources, has increased in recent years due to population growth and urbanization. By 2030, over nine billion people will need to be fed, along with the billions of animals raised annually for food and recreational purposes and as pets (FAO 2013). One of the many ways to address food security is through insect farming. Insects are cosmopolitans, and they are prolific breeders, with high growth rates and a low environmental footprint over their entire life cycle. They are nutritious, with high protein, fat, and mineral contents. They can be reared easily with low-cost inputs. Moreover, they can be eaten whole or ground into a powder or paste and incorporated into other foods (Halloran and Vantomme 2013).

7.4 Merits of Entomophagy

The use of insects as food has many environmental, health, and social or livelihood benefits, as elaborated below.

7.4.1 Environmental Benefits

Insects have high feed conversion efficiency because they are cold-blooded. On average, insects can convert 2 kg of feed into 1 kg of insect mass, whereas cattle require 8 kg of feed to produce 1 kg of body weight gain. The production of greenhouse gases and ammonia by most insects is lower than that of conventional livestock. For example, pigs produce 10–100 times more greenhouse gases per kg of weight than mealworms. They use significantly less water than conventional livestock. Insect farming is less land dependent than conventional livestock farming (Halloran and Vantomme 2013).

In order to feed people, in the future, farmers will be compelled to expand their agricultural lands which unfortunately will be done by clearing the existing forest area leading to adverse ecological effects. In fact according to the entomophagy advocate David Gracer (2008), insect farming in little spaces could be an ideal form of urban agriculture, in view of increased urbanization. In addition, insects can be reared on organic sidestreams including human and animal waste. This can help in biodegradation. Some edible insects are also pests. Collecting them from fields could solve the hunger problem to certain extent as well as protect crops and reduce and mitigate the need for pesticides (FAO 2013).

7.4.2 Health Benefits

Insects provide high-quality protein and nutrients comparable with meat and fish. Insects are particularly important as a food supplement for undernourished children because most insect species are high in fatty acids, comparable with fish. The UN Report (FAO 2013) adds that the unsaturated omega-3 and omega-6 fatty acid compositions in meal worms are comparable with that in fish (and higher than in cattle and pigs). They are also rich in fiber and micronutrients such as copper, iron, magnesium, manganese, phosphorous, selenium, and zinc as well as riboflavin, pantothenic acid, biotin, and, in some cases, folic acid. For example, fried or dried termites contain 32–38 % proteins (Santos Oliveira et al. 1976, Nkouka 1987). In the Bolivarian Republic of Venezuela, soldiers of *Syntermes* species (e.g., *Syntermes aculeosus*) are renowned for their high nutritional value whose protein content is a remarkable 64 %; the genus is also rich in essential amino acids such as tryptophan, iron, calcium, and other micronutrients. They pose a low risk of transmitting zoonotic diseases (Halloran and Vantomme 2013). Other than nutrition, insects like melon bug produce oil which is used in medicine, to cure skin lesions (Mariod et al. 2004). These insects also possess antibacterial properties. Due to its high antibacterial activity, melon bug oil can be used as a preservative in meat and meat products to control gram-positive bacteria (http://www.fao.org/docrep/018/i3253e/i3253e02.pdf).

7.4.3 Livelihood and Social Benefits

Insect gathering and rearing offer important livelihood strategies. They can be directly and easily collected in the wild. Minimal technical or capital expenditure is required for basic harvesting and rearing equipment. They can be gathered in the wild, cultivated, processed, and sold by the poorer members of the society, such as women and landless people. These activities can directly improve diets and provide cash income through the selling of excess production as street food. For example, in Zimbabwe, stink bugs are a valuable source of income for the Norumedzo community and are essential for buying household items and covering school fees (http://www.fao.org/docrep/018/i3253e/i3253e02.pdf). Besides serving as sources of food, insects provide humans with a variety of other valuable products like honey and silk.

7.4.4 Feed for Livestock

Ingredients for both animal and fish feed usually include fish meal, fish oil, soybeans, and several other grains which can be substituted by insects. The conventional feed (both plant and fish based) represent 60–70 % of production costs (FAO 2013). Chicken can be seen picking up tiny worms and insects in fields in the natural environment. Likewise, fish too hunt insects. Evidently, insects can play their role as feed for animals. Grasshoppers, crickets, cockroaches, termites, lice, stink bugs, cicadas, aphids, scale insects, psyllids, beetles, caterpillars, flies, fleas, bees, wasps, and ants have all been used as complementary food sources for poultry (Ravindran and Blair 1993). In India where the poultry industry is a fast-growing agrobusiness, the use of expensive maize as a feed ingredient is expensive for the farmers. Feeding poultry with sericulture waste, which until now has only been used for biogas production and composting, showed better conversion rates than those obtained through the use of conventional feed stock (Krishnan et al. 2011).

7.5 Status of Insects as Human Food

Ramos-Elorduy (1997) mentioned 1391 insect species eaten worldwide, of which 524 are eaten in 34 countries of Africa representing 38 % of all species consumed. He also reported in Mexico about 348 species being consumed, which is the highest number recorded for one country in the world. Of the 1391 species listed by Ramos-Elorduy (1997), most belong to the Coleoptera (24 %), followed by the Hymenoptera (22 %), Orthoptera (17 %), Lepidoptera (16 %), Heteroptera (7 %), Homoptera (5 %), Isoptera (3 %), Diptera (2 %), and others (4 %). Huis (2003) reported that in sub-Saharan Africa, edible insects are from Lepidoptera (30 %), Orthoptera (29 %), Coleoptera (19 %), Homoptera (7 %), Isoptera (6 %), Hymenoptera (5 %), Heteroptera (3 %), Diptera, and Odonata (1 %). Malaisse (1997), after intensive studies in the region inhabited by the Bemba, listed only 38 different species of caterpillars as edible.

Another study lists more than 1900 edible insect species being consumed around the world. However, this number continues to grow as more research is conducted. From the data that are available, the most commonly consumed insects (Fig. 7.2) are beetles (Coleoptera) (31 %), caterpillars (Lepidoptera) (18 %), and bees, wasps, and ants (Hymenoptera) (14 %). These are followed by grasshoppers, locusts, and

Fig. 7.2 Commonly consumed insects, order wise (http://thecsrjournal.in/a-bug-licious-solution-for-food-shortage/)

crickets (Orthoptera) (13 %); cicadas, leaf and plant hoppers, scale insects, and true bugs (Hemiptera) (10 %); termites (Isoptera) (3 %); dragonflies (Odonata) (3 %); flies (Diptera) (2 %); and other orders (5 %) (Halloran and Vantomme 2013).

7.5.1 Africa

In parts of Africa, ants, termites, beetle grubs, caterpillars, and grasshoppers are eaten. Some insects such as termites are eaten raw soon after hatching, while others are baked or fried before eating. Commonly eaten species in Angola are the termite, *Macrotermes subhyalinus*; the palm weevil larva, *Rhynchophorus phoenicis*; and the saturniid caterpillar, *Ustater psichore*. Termites, caterpillars of many species, grasshoppers, and other orthopterans are popular and sold in their markets. The saturniid caterpillar, *C. forda*, is the most widely marketed edible insect in Nigeria. Other widely marketed insect foods are palm grubs (*R. phoenicis*), termites, and *Anaphes* larvae. More than 60 species of insects in at least 15 families and 6 orders have been reported as food in Zambia. Among them the most important ones are honeybees, a saturniid caterpillar known locally as mumpa (DeFoliart 1999).

According to Chavunduka (1975), winged insects and giant crickets (*Brachytrupes membranaceus*) are frequently consumed in Zimbabwe. These insects are processed by grilling or frying without additional fat or they can be eaten raw. Sago grubs are popular for cooks in Papua New Guinea, most often boiled or roasted over an open fire. Other edible insects eaten in this country include larvae of moths, wasps, butterflies, dragonflies, beetles, adult grasshoppers, cicadas, stick insects, moths, and crickets. In the Bikita District of Zimbabwe, *Encosternum delegorguei* which is commonly known as "harurwa" is much sought after and can be bartered for grain. Other species that are consumed in Zimbabwe include *Pentascelis remipes* (local name "magodo") as well as *P. wahlbergii* (local name "nharara"). Caterpillars, termites, locusts, honeybees, and ants are among the favorites and largely consumed insects. In Southern Africa, the most widely consumed insects are mopane worms (*Imbrasia belina*), locusts, bugs, termites, honeybees, and crickets (Halloran and Vantomme 2013).

7.5.2 Asia and Oceania

The giant water bug roasted and eaten whole is a favorite food in Asia. It is easily collected around lights at night near bodies of water. For the tribal peoples in Northeastern India, the pupa of the eri silkworm, *Samia ricini*, is highly popular as food. Insects represent the cheapest source of animal protein in Manipur. More than 80 species of insects in at least 35 families have been reported as food in Thailand. Some insects, such as wasps, bamboo caterpillars, crickets, and locusts, are sold as delicacies in the finest restaurants and food shops. Fried locusts and locust fritters appear widely in city markets of Thailand. In both historical and modern Japan, the most popular and widely eaten insects have been the rice-field grasshoppers (mainly

Oxya yezoensis japonica). They are fried and slightly seasoned with soy sauce to prepare a luxury dish called "inago."

The second most widely eaten insect food in modern Japan is "hachinoko," bee or wasp larvae, which may be eaten raw, boiled down in soy sauce, or served over boiled rice. Bee and wasp brood are among the many canned insects available in the market. Another widely available product is "zazamushi," the name for aquatic insects inhabiting gravel beds in rivers and usually consisting mainly of larval Trichoptera. In South Korea the widely eaten insects have been the rice-field grasshopper (*O. velox*) and canned silkworm (*Bombyx mori*) pupae. Popular insects eaten in the Philippines are June beetles, grasshoppers, ants, mole crickets, water beetles, katydids, locusts, and dragonfly larvae. They can be fried, broiled, or sautéed with vegetables. A great variety of insects is eaten in Papua New Guinea, mainly the famous sago grub, *Rhynchophorus ferrugineus papuanus*. Another important insect exploited here as food is cerambycid grub, *Hoplocerambyx severus*. In Australia, the most widely used are witchetty grubs (Cossidae), the bogong moth (Noctuidae), the bardee larva (Cerambycidae), honeypot ants, honey and brood of the stingless bees, and the sweet manna of various lerp insects (Homoptera) (DeFoliart 1999).

7.5.3 Latin America

Important Latin American insect food comprises leaf-cutter ants (*Atta* spp.), palm weevil larvae (*Rhynchophorus* spp.), and bee and wasp brood (Apidae and Vespidae). Roasting is the usual method of cooking. Other edible species are stingless bees of the genera *Melipona*, *Scaptotrigona*, and *Trigona*. Both the honey and the brood are utilized. Wasp broods are sold in the market while still in the combs. In Mexico immature stages of the ants, *Liometopum apiculatum* and *L. occidentale* var. *luctuosum* popularly known as "escamoles," have great demand as food. They are best served when fried with onions and garlic. Three lepidopterans, viz., giant skipper butterfly (*Aegiale hesperiaris*), the red agave worm (*Comadia redtenbachi*), and *Eucheira socialis*, are widely exploited as food in Mexico. More than 20 species of grasshoppers and locusts are used as food, of which species of the genus *Sphenarium* are particularly important. They are mixed with onion, garlic, and chili powder and then boiled and dried in the sun or fried. Columbian citizens enjoy eating a variety of insects such as termites, palm grubs, and leaf-cutter ants (*Atta* spp.). Ants are ground up and used as a spread on breads. In Brazil lea-cutter ants (*Atta* spp.) make a valuable contribution to the diet of indigenous population (Defoliart 1999).

7.6 Nutritional Composition of Insects

Entomophagy lies in the recognition that insects are extremely nutritious. They are an excellent source of protein, vitamins, and minerals and tend to be low in carbohydrates. Insect diet has outstanding advantages (Fig. 7.3). A recent research

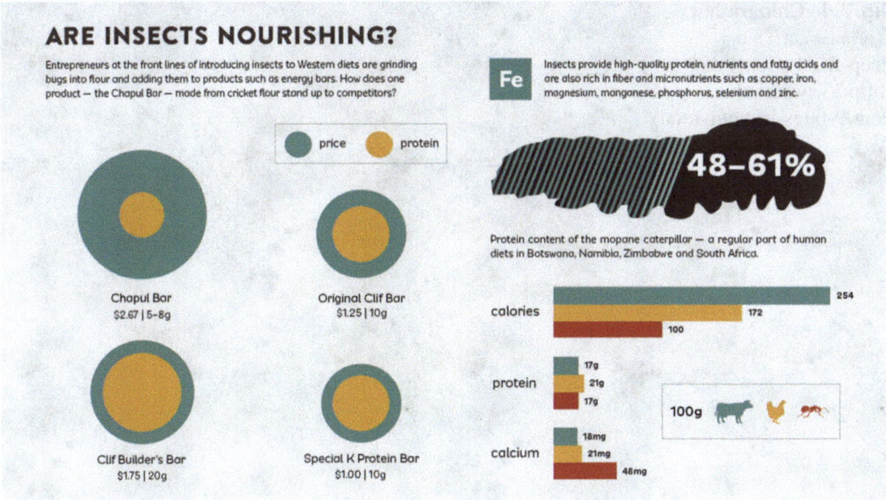

Fig. 7.3 Nutrients in insects (http://thecsrjournal.in/a-bug-licious-solution-for-food-shortage/)

concludes that many edible insect species provide satisfactory caloric, protein, and amino acid content for human diets while being high in monounsaturated fats and/or polyunsaturated fats and rich in micronutrients such as "copper, iron, magnesium, manganese, phosphorous, selenium, and zinc as well as riboflavin, pantothenic acid, biotin, and in some cases folic acid" (http://onlinelibrary.wiley.com/doi/10.1002/mnfr.201200735/abstract). Different insects used as human food are provided in Figs. 7.4, 7.5, 7.6, 7.7, 7.8, and 7.9.

Insects are important source of protein (DeFoliart 1997). They have been shown to contain high concentrations of good quality proteins and high digestibility (DeFoliart 1989). Caterpillars contain 50–60, palm weevil larvae 23–36, Orthoptera 41–91, ants 7–25, and termites 35–65 in g protein/100 g dry weight. The amino acid composition of dried mopane worms is relatively complete, with high proportions of lysine and tryptophan (limiting in maize protein) and of methionine (limiting in legume seed proteins). The nutritional content of the mopane worm has been found to comprise 60.70 % crude protein, 16.70 % crude fat, and 10.72 % minerals, on a dry matter basis, and it is therefore a highly nutritious supplement to the diet of people consuming them (Dreyer and Wehmeyer 1982).

Insects in general are rich in fat, in particular caterpillars, palm weevil larvae, and termites. They are a good source of iron, the A and B vitamins. Being an animal food, they contain even more bio-efficacious micronutrients than vegetables (Bukkens 1997). One hundred grams of caterpillars provides 76 % of an individual's daily protein requirement and more than 100 % of the daily requirements for many of the vitamins and minerals (Santos Oliveira et al. 1976).

Ramos-Elorduy et al. (1997) analyzed nutrient composition of 78 species of edible insects representing 23 families from Mexico. They include orders Anoplura, Diptera, Orthoptera, Hemiptera, Homoptera, Lepidoptera, Coleoptera, and

Fig. 7.4 Chingritthot (จิ้งหรีดทอด), a Thai deep-fried cricket dish (http://www.eatthis.com/9-bugs-in-your-food)

Fig. 7.5 The start-up six foods' chips, made from beans, rice, and cricket flour (http://www.vox.com/2014/4/30/5664782/insects)

Fig. 7.6 Taco de chapulines—a grasshopper taco—is a traditional Mexican delicacy (http://www.vox.com/2014/4/30/5664782/insects)

Fig. 7.7 Separating meal worms from chaff, the Netherlands (FAO 2013)

Fig. 7.8 Irulas—a south Indian tribe harvesting termites (Photo: Rom Whitaker (http://www.thehindu.com/features/metroplus/a-moveable-feast/article2903369.ece)

Fig. 7.9 Lizards, scorpions, and bugs displayed at a food stall in Beijing (http://www.thehindu.com/todays-paper/tp-national/biting-into-a-bug/article4786557.ece)

Hymenoptera. The dry basis protein content ranged from 15 % to 81 %. The highest was found in a wasp of the genus *Polybia*. Fat content ranged from 4.2 % (several grasshopper species) to 77.2 % in the larvae of a butterfly *Phassus triangularis*. The insect richest in carbohydrates was found to be the ant, *Myrmecocystus melliger*, with 77.7 %. Protein digestibility varied between 76 % and 98 %. The calorie contribution varied from 293 to 762 kcal/100 g, the highest value was for the butterfly larvae of *Phassus triangularis which* constituted a significant component of the diet of rural communities in Oaxaca.

7.7 The Risks

Practicing entomophagy does carry certain risks. Most of the insects eaten around the world are wild harvested, which means that no one can be sure of what the insects themselves have been exposed to. Where and how insects live and feed is quite important: even cicadas, which are sedentary for most of their lives, may have fed from roots of trees that absorbed drenched chemicals. It is probably best to avoid eating insects, until you figure out your tolerance levels. One should also never eat raw insects unless they've been bred and raised by a trusted source, because it is impossible to detect if a raw insect is tainted with pesticides. In addition, it is also difficult to know if a raw insect is carrying germs, and lastly some insects store certain compounds that may cause sicknesses, just as some insects are poisonous (www.insectsarefood.com).

There are no known cases of transmission of diseases or parasitoids to humans from the consumption of insects (on the condition that the insects were handled under the same sanitary conditions as any other food). Compared with mammals and birds, insects pose less risk of transmitting zoonotic infections to humans, livestock, and wildlife (Halloran and Vantomme 2013). Science increasingly provides data corroborating the nutritional and health benefits of entomophagy, suggesting broader acceptance of this practice while giving due consideration to certain risk factors.

Insects, whose thought itself is sufficient to null someone's appetite, are omnipresent in the food supplies. Some examples include raisins (golden) which can have 10 or more whole or equivalent insects and 35 fruit fly eggs per 8 oz and mushrooms (canned), average of over 20 or more maggots of any size per 100 g which are considered harmless. FDA allows up to 30 fly eggs per every 100 g of tomato paste; hops can have an average of 2500 aphids per 10 g (http://www.mainstreet.com/slideshow/17-foods-scary-surprises/page/4 and http://blog.chron.com/sciguy/2011/05/top-10-grossest-food-defects-the-fda-deems-safe-for-humans/). All this suggests that humans are consuming insects indirectly. To make it more consumable and appealing, processed insect products in the form of pastes and extracts can be marketed.

To enhance food security and potentially generate extra income, edible insects can be managed at various levels of intensity, from mini game in situ to more intensive management of semi- or fully domesticated mini livestock ex situ. Currently, the tropical Americas still seem to rely on edible insects as mini game to a considerable extent, while semi domestication is progressing in Africa and in Asia and full domestication is most advanced. Where natural or near natural forests still exist or can be restored, certain insects can be treated like other game animals. This applies especially to those with limited potential for domestication, such as univoltine insects, species with low fecundity, long developmental periods, and only random or periodic abundance. To guide extraction levels and other criteria, their population status and seasonal fluctuations in populations must be documented. Rules and regulations can be finalized for extraction and policy matters must be given due consideration. Real aspects of specimen collection and processing must also be considered. Rewards in terms of long-term food security, income potential, and social aspects must also be given consideration.

References

Bukkens SGF (1997) The nutritional value of edible insects. Ecol Food Nutr 36:287–319

Chavunduka DM (1975) Insects as a source of food to the African. Rhod Sci News 9:217–220

Gracer D (2008) The Colbert Report in 2008, Stephen Colbert's character drew joking parallels between entomophagy and cannibalism. Available at http://io9.gizmodo.com/should-you-really-start-eating-insects-509900937

DeFoliart GR (1989) The human use of insects as food and as animal feed. Bull Entomol Soc Am 35:22–35

DeFoliart GR (1997) An overview of the role of edible insects in preserving biodiversity. Ecol Food Nutr 36:109–132

Defoliart GR (1999) Insects as food: why the western attitude is important. Ann Rev Entomol 44:21–50

Dreyer JJ, Wehmeyer AS (1982) On the nutritive value of mopanie worms. Sali J Sci 78:33–35

FAO (2013) Edible insects: future prospects for food and feed security. FAO, Rome

Gene DeFoliart (1988) Food Insect Newsl 1(2):10–12
Halloran A, Vantomme, P (2013) The contribution of insects to food security, livelihoods and the environment. Available from http://www.fao.org/forestry/edibleinsects/en/
Huis AV (2003) Insects as food in sub-Saharan Africa. Insect Sci Appl 23(3):163–185
Krishnan R, Sherin L, Muthuswami M, Balagopal R, Jayanthi C (2011) Seri waste as feed substitute for broiler production. Sericologia 51(3):369–377
Malaisse F (1997) Se nourrirenforêtclaireafricaine: Approcheécologiqueetnutritionelle. Les PressesAgronomiques de Gembloux, Gembloux, p 384
Mariod AA, Matthaus B, Eichner K (2004) Fatty acid, tocopherol and sterol composition as well as oxidative stability of three unusual Sudanese oils. J Food Lipids 11:179–189
Nkouka E (1987) Les insectes comestibles dans lessociétés d'Afrique Centrale. Revue Scientifique et Culturelle du CICIBA. Muntu 6:171–178
Osmaston HA (1951) The termite and its uses for food. Uganda J 15:80–82
Paoletti MG (2005) Ecological implications of mini livestock. Science Pub, Enfield
Ramos-Elorduy J (1997) Insects: a sustainable source of food. Ecol Food Nutr 36:247–276
Ramos Elorduy J, Moreno JMP, Prado EE, Perez MA, Otero JL, De Guevara OL (1997) Nutritional value of edible insects from the state of Oaxaca, Mexico. J Food Compos Anal 10:142–157
Ravindran V, Blair R (1993) Feed resources for poultry production in Asia and the pacific world's poultry. Sci J 49:219–235
Santos Oliveira JFS, Passos de Carvalho J, Bruno de Sousa RFX, MadalenaSimão M (1976) The nutritional value of four species of insects consumed in Angola. Ecol Food Nutr 5:91–97
Tommaseo-Ponzetta M (2005) Insects: food for human evolution. In: Paoletti MG (ed) Ecological implications of mini livestock. Science Pub, Enfield, pp 141–161
Yen A (2005) Insect and other invertebrate foods of the Australian aborigines. In: Paoletti MG (ed) Ecological implications of minilivestock. Science Pub, Enfield, pp 367–387

Websites

http://blog.chron.com/sciguy/2011/05/top-10-grossest-food-defects-the-fda-deems-safe-for-humans/
http://onlinelibrary.wiley.com/doi/10.1002/mnfr.201200735/abstract
http://thecsrjournal.in/a-bug-licious-solution-for-food-shortage/
http://www.fao.org/docrep/018/i3253e/i3253e.pdf
http://www.fao.org/docrep/018/i3253e/i3253e02.pdf
http://www.mainstreet.com/slideshow/17-foods-scary-surprises/page/4
http://www.thehindu.com/features/metroplus/a-moveable-feast/article2903369.ece
www.insectsarefood.com

Arthropod Community on Rice: A Blend of Aquatic and Terrestrial Species

8

Vijay Kumar Lingaraj, K.S. Nitin, and B.S. Rajendra Prasad

Abstract

Ricefields hold high arthropod diversity, which at present unfortunately is threatened. Traditional systems of ricefields along with other rural elements like canals and ponds formed a composite unit in a rural, countryside setup. These systems played a pivotal role in the productivity of a nation. Massive landscape changes and development of aquatic lands with anthropogenic factors have incurred damage to biodiversity in paddy fields. This has resulted in unstable and uniform arthropod communities. The impinging factors on the ecology of rice-cultivated ecosystems are abundant in several countries but but more so in tropical Asia. Fertilizers and pesticides considerably reduced beneficial arthropods in rice fields so also the modern rice cultivation practices. Deploying biological indicators in rice fields is suitable for arthropod preservation. Multispecies mixed planting, less use of pesticides and balanced application of fertilizers, public awareness and the use of bioagents will help in restoring arthropod diversity in rice fields and assist in mitigating emissions of greenhouse gases. The most successful example of rice integrated pest management (IPM) is in Indonesia where trained farmers use non-chemical methods and sustain arthropod communities.

Keywords

Arthropods • Diversity • Natural enemies • Rice fields

V.K. Lingaraj (✉)
Department of Entomology, College of Agriculture, University of Agriculture Sciences, Bangalore, VC Farm, Mandya 571405, Karnataka, India
e-mail: vkumaruasb@gmail.com; vijigall@gmail.com

K.S. Nitin • B.S. Rajendra Prasad
Division of Entomology and Nematology, Indian Institute of Horticultural Research (IIHR), Hessaraghatta Lake Post, Bangalore 560089, Karnataka, India

A.K. Chakravarthy, S. Sridhara (eds.), *Economic and Ecological Significance of Arthropods in Diversified Ecosystems*, DOI 10.1007/978-981-10-1524-3_8

8.1 Introduction

Rice (*Oryza sativa* L.) fields represent one of the common aquatic habitat in cultivated ecosystems. Rice paddies also constitute the most traditional farming systems since human civilization began with man domesticating landscapes near water resources/aquatic habitats. It started during 3300–1300 BC in the Indo-Gangetic plains of Indus Valley Civilization in the Indian subcontinent and in Yangtze River Delta during Neolithic period in China. Ricefields harbour one of the highest arthropod biodiversity (Yufeng et al. 2014), but, however, now this diversity of species is seriously threatened and endangered today. Conversion of land for agriculture and anthropogenic works has impacted increasingly heavy loss to rice fauna and flora, leading to uneven and uniform populations in biotic entities (Yufeng Luo et al. 2014). In countries like India, Pakistan and Sri Lanka, large tracts of paddy-growing areas are being converted to commercial crops like banana, ginger, turmeric and nurseries of fruit and vegetable crops, the contributors being acute shortage of water and labour.

Traditional rice fields serve as artificial wetlands providing food for humans, food and shelter for wildlife and feeding and breeding grounds for a number of plant and animal taxa. Rice fields play crucial role in the ecology of the aquatic ecosystems. However, massive landscape changes and modern agricultural practices have tremendously increased the homogenization of the rice ecosystems. Extensive and intensive agricultural practices, particularly pesticide and fertilizers and excessive landscape transformations, have not only poisoned the soil, water and environment but have also collapsed the biological communities and niche created by rice fields. It is proven that paddy biodiversity supplies the resources to sustain stable rice fields and surrounding ecosystems and contributing substantially to the ecosystem services that rice fields supply, creating economic and ecological value to society in the process (Wang 2002).

The pressures on the ecology of paddy fields are documented for several countries, with Japan providing a typical instance (Washitani 2007; Koganezawa 2009). The conditions are worse in China (Yufeng Luo et al. 2014) and Vietnam too (Pham and Giang 2006). The current review on arthropod community of paddy fields in tropics and subtropics is aimed to contribute towards restoration measures that can help to maintain biodiversity in paddy fields vis-à-vis environmental quality.

8.2 Rice Ecosystem Biodiversity

Rice biodiversity refers to all biotic elements embracing plant, animal, microbial, genetic diversity and related processes (Koganezawa 2009; Giang et al. 2009). Wild plants, weeds, crops, pests and natural enemies form vital constituents of rice ecosystem and all of which maintain stability of rice ecosystems. Rice ecosystems along with associated elements like ridges, refugia, ponds, canals, ditches and irrigation systems from rural landscape in rural environments form a composite unit. These traditional ecosystems played crucial role in the ecology and productivity of

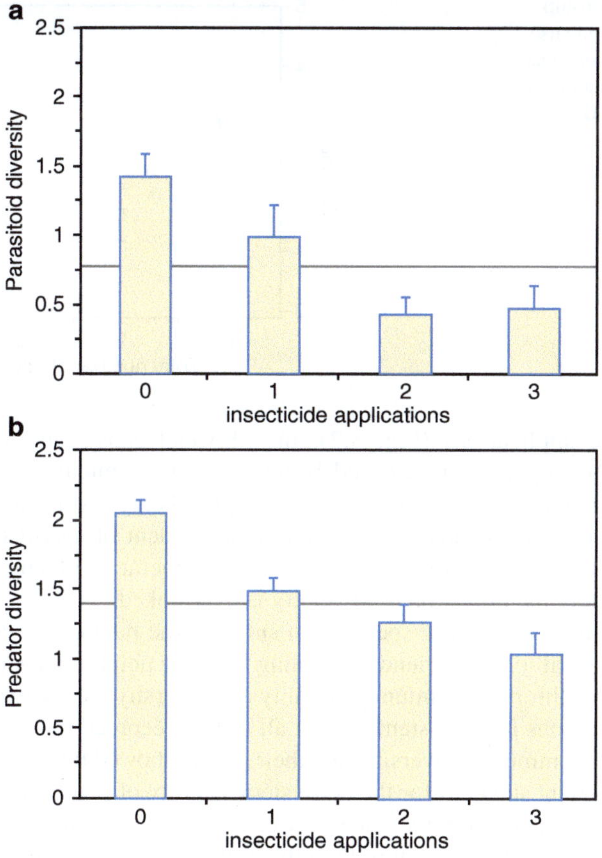

Fig. 8.1 Relationship between insecticide use (total number of applications during the growing season) and diversity of natural enemies measured with Shannon's index (above, parasitoids; below, predators (Ueno 2013))

a county. For instance, in Asia, rice ecosystems provide habitats for a number of diverse taxa. This means rice-cultivated patches support, at least in part, biodiversity of the region or locality. Bioresources in agroecosystems have been given emphasis because they play key part in crop productivity and system functioning (Dudley et al. 2005; Feng 2002). For instance, parasites and predators serve agroecosystem function regulating pest populations (Barbosa 1998; Hajek 2004; Jarvis et al. 2007). In cultivated fields, crop production depends generally on biological suppression given by natural enemies and other beneficials. Injudicious use of pesticides results in decline of biodiversity of such beneficials and results in pest resurgence (Pimentel 1997; Dent 2000). Ueno (2013) developed a link between insecticide use and diversity of beneficials. As the number of insecticide applications increased, the diversity of parasitoid and predators decreased (Fig. 8.1). He further found that density of the bioindicator, *Itoplectis naranyae*, also depended

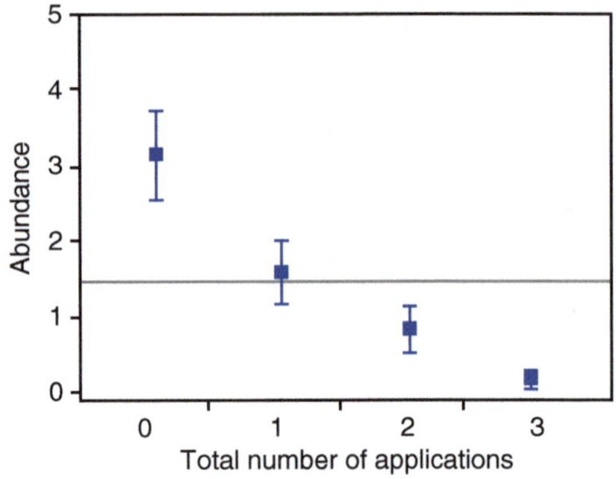

Fig. 8.2 Relationship between insecticide use and density of *Itoplectis naranyae* in farmers' rice paddies (Ueno 2013)

on insecticide applications (Fig. 8.2). In cultivated ecosystem, yield of crops depends on bio-suppression-afforded beneficials and inimical factors that affect pests and pathogens. Across tropics and subtropics, being eco-friendly is in strong demand in agribusiness today. Sustainable management of agri-resources is also crucial. But often eco-friendly term is desired for generating of safe food for man, but unfortunately its impact on biodiversity is overlooked.

In fact the use of indicator species can simplify the public concern to consider what is 'ecological' or 'eco-friendly' farming from the point of view of biodiversity conservation. Arthropod community stability and diversity are the two indicators of ecological functions in ecosystems. Jin et al. (1990) recorded the spatial strata and fluctuations of community diversity, and their results showed that community diversity is an important standard for the ecosystem stability of rice because as diversity index declines to less than three, the ecosystem stability becomes poor. In contrast, when the diversity index was more than three, the ecosystem community was productive and stable. The succession of one set of arthropod taxa by another set, of rice field arthropod community, was due to the growth of rice and seasonal variations in community structure, and relative population is a dynamic characteristic. Hu et al. (1997) divided the field arthropod community with the growth and development of rice plant into three phases: expansion, fluctuation and decline, leading to loss.

8.3 Arthropod Diversity

In Tamil Nadu, South India, Kandibane et al. (2007) determined diversity and community structure of aquatic arthropods between weeded and partially weeded rice ecosystems during 2000–2001. Observations revealed a total of 12, 2, 6 and 3 species of Odonata, Ephemeroptera, Hemiptera and Coleoptera, respectively. *Agriocnemis femina* Brauer, *Pantala flavescens* Fab., *Crocothemis servilia* Drury and *Diplocodes trivialis* Rambur dragonflies were the major species at both sites. A sum of 18 weeds was documented in partially weeded sites. Guild of aquatic arthropods in rice ecosystem revealed that predators, scavengers and phytophages dominated the community. Coleoptera, Odonata, Hemiptera and invertebrates were the predatory group. Ephemeroptera, Coleoptera and Diptera constituted the scavengers. The weeded rice ecosystem in Tamil Nadu had comparatively lower density of predatory arthropods. Kumar and Khanna (1983) too found dragonfly naiads feeding on primary consumers, viz., tadpoles of frog, toads and mosquito larvae. Vinson and Hawkins (1998) stated that dipteran flies preferred permanent and intermittent stream in a stable paddy-cultivated area for development and survival. Photographs of some of the common natural enemies of major pest insects found in paddy ecosystems of Southern India are provided (Figs. 8.12a, 8.12b and 8.12c).

The arthropod community, structure, functions and diversity in rice ecosystems in Guangdong Province, China, were studied in 2009. The workers collected 114 species of arthropods, comprising of 58 species of spiders, 16 species of predatory insects, 25 species of phytophagous insects and 15 species of other neutral insects (Zhang et al. 2013). The distribution of functional groups of rice arthropods is depicted in Fig. 8.3. Organic cultivation of paddy is considered as desirable because it prevents 3Rs (resistance, resurgence and pesticide residues). For instance, Kajimura et al. (1993) recorded that the population of the rice brown plant hopper, *Nilaparvata lugens* (Stal), and the white-backed plant hopper (*Sogatella furcifera*) (Hovath) were much less in an organically cultivated field than in chemically fertilized paddy field. A vital basis of integrated pest management is to maximize biocontrol. Therefore, seasonal changes in populations of arthropods, diversity and species richness and community structures are important factors in planning pest management practices. Thus, the study of Zhang et al. (2013) is crucially important. Such studies are needed across other paddy areas, the world over. Ueno (2013) has

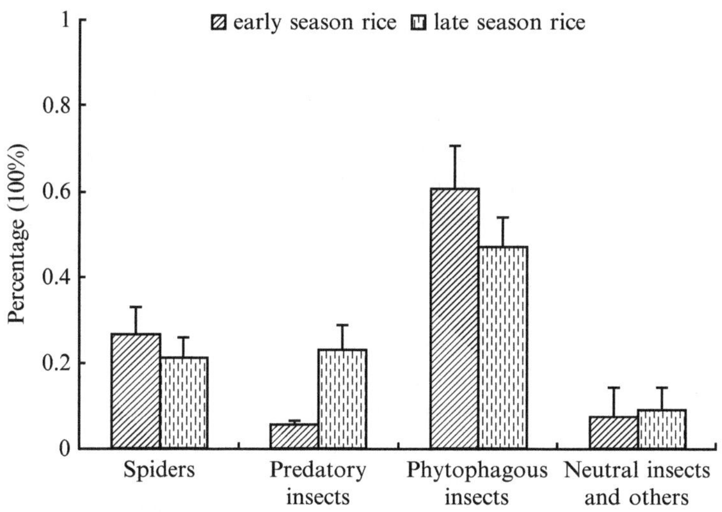

Fig. 8.3 Dominance distribution (±SE) of functional groups of arthropods in the early season (Apr–Jul) and late season (Aug–Nov) crops of double-cropped, organically grown rice at Huizhou, Guangdong Province, China (Zhang et al. 2013)

discussed at length bioindicators of biodiversity in rice fields. According to him, biodiversity of beneficials such as biocontrol agents is a key consideration to faster the productivity and sustainability in paddy fields (Ueno 2010, 2013). However, conserving the diversity of arthropods may be a challenge. The use of insecticides dramatically reduces the incidence of rice pests and the yield loss. Herbicides permit reducing labour required for weed management (Pimentel 1997, Dent 2000). The sustenance of rice cultivation makes the yield higher and the price of rice lower; it causes rice paddies poor in arthropod diversity. In contrast, biodiversity conservation through reduction of agrochemical use reduces labour requirement, helps sustainable use of rice paddies and reduces the incidence of pests. Studies have revealed that diverse species of natural enemies enhance the suppression of pest populations via their complementary functions (Kruess and Tscharntke 1994).

The use of bioindicator organisms is ideal for biodiversity conservation in paddy. More than 5000 species are documented from paddy ecosystems in Japan (Kiritani 2000). In addition, there are many species that are difficult to identify to the species or group level. Thus, indicator organisms that can reflect biological diversity are needed in view of the evaluation of biodiversity in its entirety are not practical (McGeoch 1998; Buchs 2004). Indicator species are useful for connecting biodiversity elements with abiotic environmental sound practices because indicator organisms allow the quantitative assessment of environmental soundness of agricultural practices. Also, such indicators are useful for (1) linking farmers and biodiversity and (2) linking farmers and consumers.

Studies on terrestrial arthropods of paddy-cultivated ecosystems were initiated at Bathalagoda, Sri Lanka. In sum 342 arthropod species were recorded consisting 282 insect species in 90 families and 17 orders and 60 species of arachnids in 14 families. Eight taxa new to Sri Lanka were recorded. Most of the species recorded were hymenopterans. This is interesting and useful because many of the parasitoids and predators belong to hymenoptera. Based on feeding nature, majority of the arthropods were predators (149 species), predominantly spiders. However, in the rice field population, abundance of phytophages was more than parasites and predators. Density fluctuations of predators and parasitoids were positively correlated with density of pest species. Interestingly, higher species richness of arthropods with increase in crop age was recorded but declined following insecticidal applications. Density of terrestrial arthropods was positively correlated with crop age and height of the rice plant and in field bunds with weed cover. Results indicated that a stable relationship was maintained between rice insect pests and their natural enemies through minimal fertilizer and pesticide use (Bambaradeniya and Edirisinghe 2008).

PROSHIKA, a non government organization promoting ecological farming, and its collaborative partners at two locations in Bangladesh conducted studies on rice arthropods. Ecological farming was compared with conventional farming with regard to arthropods. 'Ecological fields' at Koitta supported more arthropods than the conventional fields. At Gabtoli the arthropod load in ecological and conventional fields was similar (Hossain et al. 2002). Leila Luica Fritz et al. (2001) identified arthropods in rice-growing areas of Rio Grande do Sul, Brazil, in three producing regions at different stages of crop development. The study was conducted during 2007–2008 and 2008–2009 at Cachoeira do Sul, Eldorado do Sul and Capivari do Sul. A total of 44, 231 arthropods were collected. Spatial and temporal patterns of arthropods were analysed utilizing 28 principal families and applying the Morisita index, horn coefficient and detrended correspondence analysis (DCA). A dendrogram based on values of Morisita-Horn index in homogeneity of arthropod communities is depicted in Figs. 8.4, 8.5, 8.6, 8.7, 8.8, 8.9, 8.10 and 8.11.

The results indicated that the arthropod communities in Southern Brazilian rice crop agroecosystems are formed of a few families with high abundance and a large number of other smaller families. Among the phytophagous arthropods found, Pentatomidae, Orthoptera and plant hoppers were predominant, while the natural enemies were mainly predatory mites, spiders, Hymenoptera and Odonata.

Arthropod diversity and species abundances were compared with three genotypes of cultivated rice and two genotypes of wild rice, *Oryza rufipogon*, in Southern Luzon, Philippines. Domestication of rice had a small but positive effect on total arthropod diversity. Arthropod species richness was the maximum on IR 64 and minimum on *Oryza rufipogon*. Total arthropod abundance and populations of groups of arthropods did not vary among cultivated and wild paddy. Stem-boring insects and sap-sucking phytophages benefitted from commensalism. Commercialization of rice reduced densities of the wolf spiders.

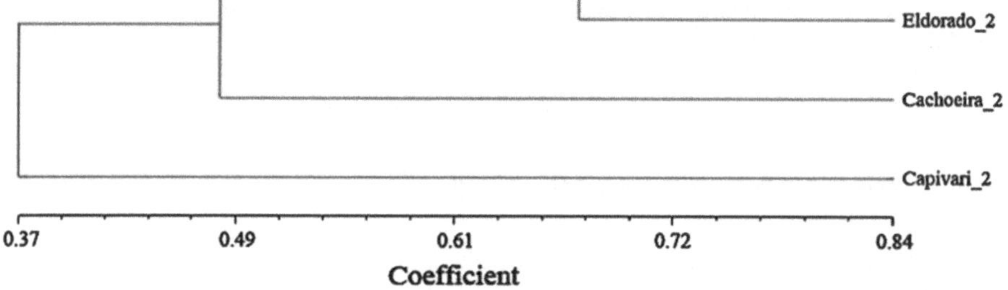

Fig. 8.4 Dendrogram based on values for Morisita-Horn index showing the similarity of the communities of the arthropods in sample sites. Method of cluster analysis: UPGMA (Leila Luica Fritz 2001)

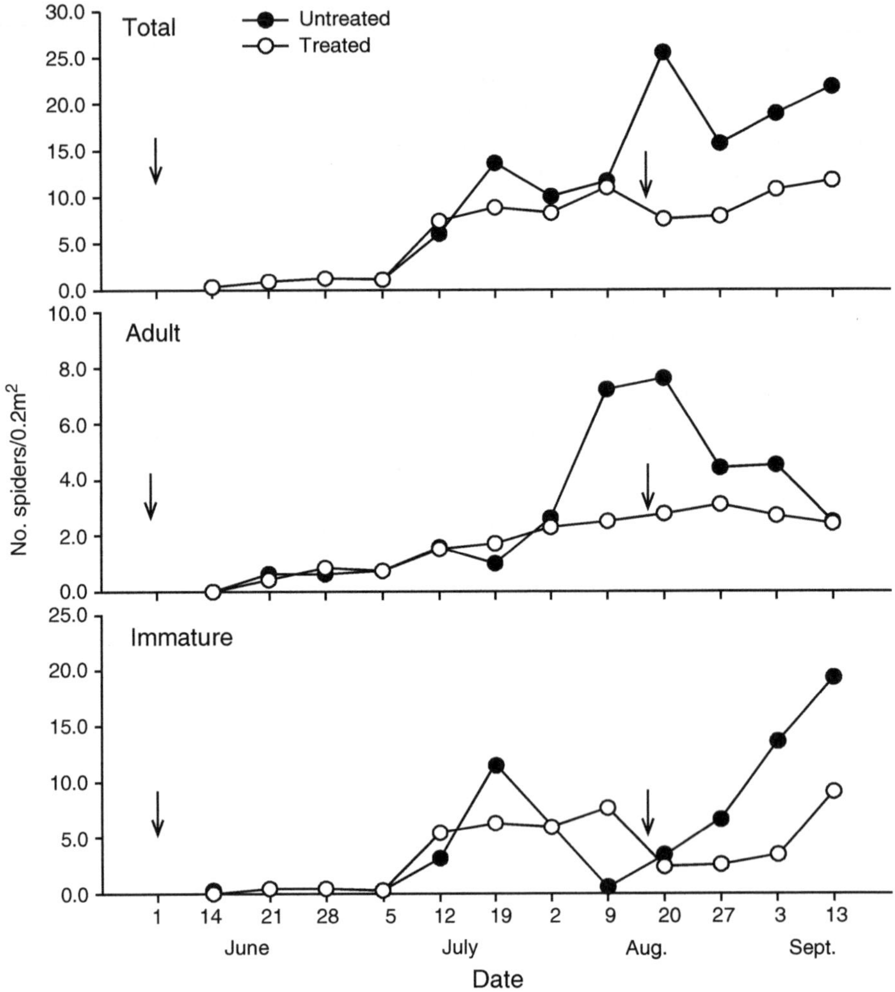

Fig. 8.5 Population patterns of total, adult and immature spiders in untreated and insecticide-treated ricefields, Dangsu-dong, Suwon, Gyeonggi-do in 2001

8.4 Rice Arthropods and Pesticides

Pesticides can suppress many pests, prevent outbreaks and save lives and money. But pesticides can lead to adverse ecological effects. There are several ways pesticides can be altered or manipulated to avoid ecological catastrophes or adverse effects. Some of the methods are a shift from persistent pesticides to less persistent ones; the use of more selective pesticides, systemic or granular pesticides, can be used when possible; and applications of pesticides can be timed so that beneficials

Fig. 8.6 Rice yellow stem borer and their damage

Fig. 8.7 Rice brown plant hopper and their damage

like parasites/predators and pollinators are not affected or the least affected. The dosages of pesticides should be as low as possible so that they do not prove toxic to beneficial. The development of crop-resistant varieties, sterility, pheromones, hormones and cultural tools augment well with biological control (Davis and McMurty 1979).

Recently Ueno (2013) examined the impact of insecticidal applications on natural enemies of rice pests in Japan. For convenience, the bioagents were categorized

Fig. 8.8 Asian rice gall midge and their damage

Fig. 8.9 Rice hispa beetle and their damage

Fig. 8.10 Rice case worm and their damage

Fig. 8.11 Rice leaf folder and their damage

into two main headings based on function, i.e. parasitoids and predators. Obviously, the maximum species richness for parasitoids and predators (Fig. 8.1) were recorded in rice fields where no insecticides had been applied. The application of insecticides decreased the diversity of natural enemies of pests in rice ecosystem.

In recent years, pesticide use has reduced in rice-cultivated areas, not only in Japan but over large landscapes elsewhere. Pesticide adversely affects natural enemies like *Itoplectis naranyae* Ashmead (Ueno 2013), a parasitoid of rice leaf folder, *Cnaphalocrocis medinalis* Guenee and other pests. The percent parasitization exceeds 50 % of pupae in autumn in Japan. Interestingly, there are countries, like Vietnam, where insecticide use has increased on rice (Figs. 8.12a, 8.12b and 8.12c) (Pham and Giang 2006). Species composition of beneficials including natural enemies on rice in Vietnam is presented in Tables 8.1 and 8.2. The relative abundances of the two functional groups were comparable, being almost in balance. Pesticide use increases the infestation of select pests due to a combination of development of insecticide resistance and loss of natural control given by natural enemies (Pimentel 1997; Dent 2000).

Ueno (2013) identified the ichneumonid wasp, *Itoplectis naranyae*, as a bioindicator. It is a solitary parasitoid with a wide biogeographic distribution in Asia (Yasumatsu and Watanale 1965). The wasp can be easily mass reared in laboratory. The natural incidence of this wasp on pests was significantly related to pesticide use (Fig. 8.2). In organic rice fields its incidence on pests was 100 % and only 33 % in rice fields with three insecticide applications. Further, the population of *Itoplectis naranyae* differed significantly among paddy fields varying in insecticide use (Fig. 8.3). *Itoplectis naranyae* is a good environmental indicator of the ecology of rice fields. So farmers may encourage *I. naranyae* by reducing insecticidal applications. Considering the balance between biodiversity conservation and rice yields, one has to determine to what extent reduction in insecticide applications will benefit the wasp parasitoid and, in turn, the rice yields. It is also important that farmers in a region maintain vegetation in surrounding environment to supplement the ecosystem services of natural enemies on rice pests.

In Vietnam 242 species of arthropods were collected from rice fields, of which 36 were pests and 147 natural enemies (Pham and Giang 2006). Observations indicated changes in species composition; population density of herbivore-natural enemy communities in rice was due to human activity via application of broad-spectrum insecticides proving toxic to predaceous and parasitic arthropods. Conservation of arthropods on rice is crucial because production of rice has been a

Trichogramma chilonis **parasite**

Cocoons **of** ***Apanteles*** **spp**

Tachnid fly, ***Chaetomylobia javana***

Netelia **spp**

Fig. 8.12a Natural enemies of rice pests (photos by Rajendra Prasad BS)

Charops bicolor

Mirid, ***Cyrtorhinus lividipennis***

Ceriagrion coromandelianum

Pantala flavescens

Ophionea indica

Chalenius medioguttatus

Fig. 8.12b Natural enemies of rice pests (photos by Rajendra Prasad BS)

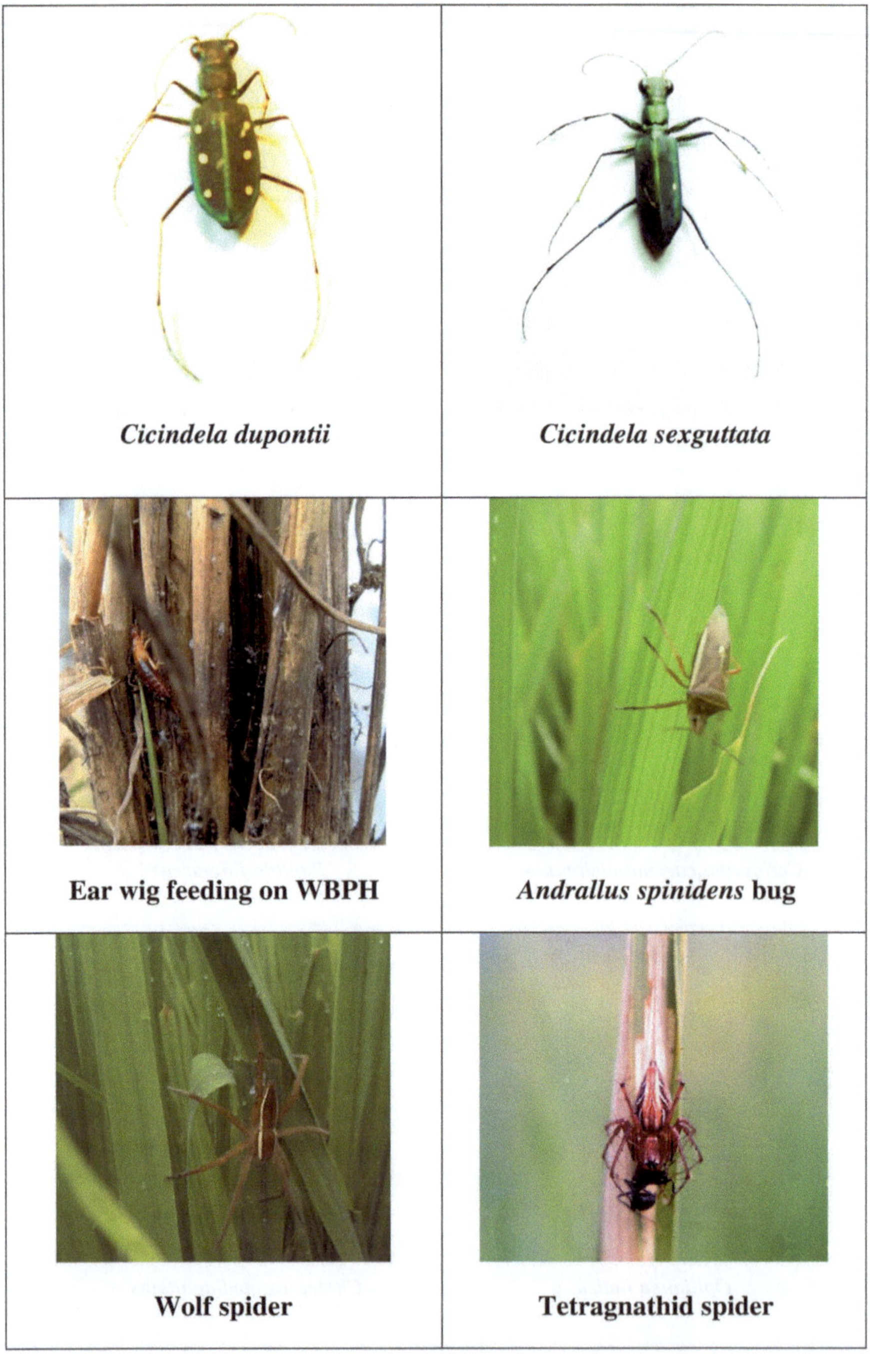

Fig. 8.12c Natural enemies of rice pests (photos by Rajendra Prasad BS)

Table 8.1 Species composition of pests and natural enemies collected in rice fields in 2004–2005

Criteria	Co Nhue (Ha Noi)	Duc Tu (Hu Noi)	Nhat Tan – Van Xa (Ha Nam)	Tien Phong (Vinh Phuc)	Total
Number	159	158	143	106	242
Rice pests	27	29	26	18	36
Natural enemies	97	95	86	60	147
Different species	35	34	31	28	59

Pham and Giang (2006)

Table 8.2 Species composition of pests and natural enemies collected in rice fields in 2004–2005 compared with another

	Ha Noi		Ha Nam		Vinh Phuc	
Criteria	References	2004–2005	References	2004–2005	References	2004–2005
Order	8	7	**7**	**6**	7	6
Family	25	16	**19**	**11**	15	12
Species	67	29	**53**	**26**	65	8

Pham and Giang (2006)

traditional source of income for small-scale farmers in Vietnam, where 75.80 % of the population live and rely on agriculture with rice as the main food crop.

In tropical Asia, the most successful example of rice integrated pest management (IPM) is Indonesia where an estimated 250,000 IPM-trained farmers use varieties and non-chemical methods to prevent yield losses. Insecticides are used as a method of last resort to control pest outbreaks. Several studies have evaluated the effects of insecticides on select crop insect pests and their principal natural enemies, but few studies have taken food webs as a basis to study the effects of pesticides on pest-enemy interactions at the community level (Grigarick et al. 1990; Hurlbert et al. 1972). Schoenly et al. (1996) conducted a study on food webs with insecticides and found that pests increased nearly fourfold in sprayed plots over unsprayed plots. Deltamethrin caused outbreaks in three delphacid pest populations, viz. brown plant hopper, white-backed plant hopper and another plant hopper. This increase was related to reduced mortality of natural enemies killed by sprays, higher reproduction rates of pests and recruitment from surrounding fields (Kenmore et al. 1984; Chellaiah and Heinrichs 1980).

The impact of two insecticides, viz., carbofuran 3 GR and fenobucarb 50 EC which targets rice water weevil, *Lissorhoptrus oryzophilus* Kuschel, and brown plant hopper, *Nilaparvata lugens*, was evaluated in Korean rice ecosystem. Application of insecticides reduced density of total arthropods by 48.40 % compared to control. Both the insecticides reduced filter-feeding chironomids by 50 %. Both web-building and wandering spiders were significantly disturbed by fenobucarb 50 EC although the impact differed according to their behavioural differences.

While the population of web-building spiders decreases over time, that of wandering spiders recovered from the disturbance a few weeks later. The abundances of spiders in treated and untreated plots are shown in Fig. 8.12c. In untreated, control rice fields, 40 taxa represented by five functional groups were collected (Table 8.2). Population patterns of spiders in insecticide-treated and untreated fields lead to the conclusion that pesticides and application methods used in controlling rice pests should be selected carefully to increase the compatibility between pesticide treatments and biological resources in the rice ecosystem.

8.5 Rice fields and Greenhouse Gases

A greenhouse gas (GHG) is a gas that absorbs infrared radiation (IR) and radiates heat in all directions. The more of these gases exist, the more heat is prevented from escaping into space, and, consequently, earth warms up. This increase in heat is called greenhouse effect. Common greenhouse gases are water vapour, carbon dioxide (CO_2), methane, nitrous oxide, ozone and fluorocarbons. Some concentrations of greenhouse gases in our environment are naturally occurring. However, since the industrial age (1750s), carbon dioxide alone has increased by 40 %. Anthropogenic Co_2 emissions come from carbon-based fuels, deforestation and soil erosion (IPCC 2007).

Paddy-farming land use and the wetland changes are attributed to higher methane atmospheric concentrations (Steinfeld et al. 2006). Agricultural activities such as application of fertilizers to paddy fields lead to higher nitrous oxide (N_2O) concentrations. Reducing emission from rice farming is the main thrust in the rice research at the International Rice Research Institute (IRRI), Philippines. Increased methane and Co_2 emissions will affect rice predation. Higher Co_2 levels are typically increase biomass but not necessarily yields. Higher temperatures can decrease rice yields as they can make rice flowers sterile. Higher respiration losses that are linked to higher temperatures also make rice less productive. Integrated Regulatory Review Service (IRRS) in search indicates that a rise in night-time temperature by 1 °C may reduce rice yields by about 10 %. Rise in temperature and Co_2 levels alter sex ratio, longevity, distribution, pest/natural enemy ratio and biology of arthropods. Impact of climate change on arthropods in general is discussed, but briefly in other chapters, in this book.

8.6 Modern Rice Cultivation Practices and Impact on Arthropods

In order to realize higher grain yields, rice breeders and entomologists have evolved high-yielding hybrids. Hybrid rice relies on chemicals and machinery and brings about uniformity. In contrast, traditional rice cultivation maintained ecosystem stability, and rice ecosystems imitated natural aquatic habitats. The system sustained food production by the interaction between biodiversity and biological resources. Modern rice cultivation brought about changes that threatened wild plants and animals. It also caused loss of habitats for several plant and animal species. The paddy biodiversity today has tended to simplify, and the species interaction has been less complex leading to serious damage to the paddy ecosystem biodiversity. This phenomenon in paddy cultivation is widespread as observed in Asia including Japan, Vietnam, Malaysia, the Philippines, Indonesia and China. Discussed below, but briefly, are select cultivation practices that have influenced rice arthropod diversity.

8.6.1 Irrigation

Shortage of quantity and quality water for irrigation has compelled growers to shift to modern methods of irrigating rice paddies. Transitional rice cultivation in most of the rice-growing countries used flooding irrigation. Nowadays, in rice paddies several water-saving irrigation techniques including controlled irrigation (Peng and Xu 2011), intermittent irrigation (Zhu and Gao 1987; Jin et al. 2003), half-dry cultivation (Peng 2011), systems of alternate submergence (Belder et al. 2004) and alternate wetting and drying (Bouman and Tuong 2001) are in use, for instance, in China. Similar trends are observed in other rice-growing countries. These changes inimitably alter the paddy ecological situation and natural balance, affecting rice biodiversity. Aquatic plants/weeds were more common in traditional flooding system of rice cultivation. The species compositions of weeds also vary widely among rice cultivated in different irrigation systems (Luo and Li 2003). Interestingly, under semiarid rice cultivation systems, sheath blight disease incidence decreased by 24 %, plant hoppers decreased by 46 % and rice leaf rollers reduced by 70 %. Thus, this method of irrigation confers multiple advantages to the farmer; Zhu et al. (2000) studied the effects of filmed ground and dry cultivation on pest infestation and on the richness of natural enemies and saprophagous insects. In this irrigation system of rice cultivation, the incidence of sheath blight decreased, while the population densities of parasitic wasps and spiders increased compared to flooded rice cultivation.

8.6.2 Land Use Patterns

Large landscapes of aquatic habitats have undergone changes and are currently undergoing dramatic changes. Fragmentation, canals and other changes in land use have altered the habitat of rice fields, reducing the species richness. Studies in China and India have shown that non-paddy weeds provide a species pool for the rice arthropod diversity, and natural enemies suppress pest populations, if certain weed species are maintained (Liu et al. 2002). The destruction of habitat surrounding paddy fields promotes invasion of *r*-selected, adaptive species, thus increasing the problem of pests on paddy.

The growth of frogs, toads and fish are affected because of the decrease of water bodies due to land levelling and use of irrigation. Similarly, crabs and aquatic birds should inhabit paddy-cultivated ecosystems to increase the biodiversity.

8.6.3 Chemicals

Agrochemicals like pesticides, weedicides and excessive use of fertilizers have adversely affected rice fields' biodiversity (Wu and Chen 2004). Studies have shown, for instance, that natural enemies sensitive to pesticides are rapidly decreased by pesticide application, while insect pests rapidly acquire pesticide resistance. Changes in weed populations result from herbicide use which characterize succession behaviour and promote the emergence of nontarget species as major weeds. To obtain high rice yields, farmers tend to enhance the application of fertilizers, mainly nitrogen to obtain higher rice productivity. Hu et al. (1986) studied the impact of nitrogen fertilizers and irrigation on population variations of serious pests of rice. The physical features of host plant morphology and biochemistry and the ecological adaptability of the arthropods affect the tropic structure, plant-parasite, insect-pest and natural enemy communities. In another study, the number of predators increased significantly with increasing fertilizer application, but the number of egg parasitoids decreased (Lv 2003). Huang et al. (2003) suggests that the abundance and extent of damage of major rice pests can be significantly suppressed by increasing the number of types and proportion of phosphorus.

Developing countries of Asia, Africa and South America which are characterized by tropical environment have poor predicted weather, and farmers have small holdings; here the pest dynamics gets into a flux (Dasgupta and Gangwar 1983). Hence, integrated pest management (IPM) should be planned, and biodiversity in the cultivated ecosystem should be restored. One of the measures could be the conservation of paddy ecosystem habitat. This may be realized by enhancing the diversity of not only rice-cultivated area but also that of the surrounding areas. Multispecies mixed planting may be taken up to fully use the space and biological materials. The amount and frequency of pesticide and fertilizer applications can be reduced to a minimum. Irrigation water also should be limited. Awareness programmers should be launched to educate the farmers, the public and the school children so that rice biodiversity and species richness can be enhanced.

References

Bambaradeniya CNB, Edirisinghe JP (2008) Composition, structure and dynamics of arthropod communities in a rice agro-ecosystem. Ceylon J Sci 37:23–48

Belder P, Bouman B, Cabangon R, Guoan L, Quilang E, Yuanhua L, Spiertz J, Tuong T (2004) Effect of water-saving irrigation on rice yield and water use in typical lowland conditions in Asia. Agric Water Manag 65:193–210

Bouman B, Tuong TP (2001) Field water management to save water and increase its productivity in irrigated low land rice. Agric Water Manag 49:11–30

Buchs W (2004) Biotic indicators for biodiversity and sustainable agriculture-introduction and background. Agric Ecosys Environ 98:1–16

Chellaiah S, Heinrich EA (1980) Factors affecting insecticide-induced resurgence of the brown plant hopper. *Nilaparvata lugens*, on rice. Environ Entomol 9:773–777

IPCC (Intergovernmental Panel on Climate Change) (2007) Summary for policymakers. climate change 2007: synthesis report. http://www.ipcc.ch/pdf/assessment-report/ar4/syr/ar4_syr_spm.pdf. Accessed 1 Mar

Dasgupta MK, Gangwar SK (1983) Integrated management of pest, disease and weeds of rice. Proc 10th Interna Congr Plant Protn, Brighton. Plant Protect Hum Welf 3:934

Davis DW, McMurtry JA (1979) Introduction. In Hoyt SC, McMurtry JA and Ali Niazee MT (eds) Biological control and insect pest management, pp 102

Dent D (2000) Insect pest management, 2nd edn. CABI Willingford, London, p 410

Dudley N, Baldock D, Nasi R, Stolton S (2005) Measuring biodiversity and sustainable management in forests and agricultural landscapes. Phil Trans Royal Soc B 360:457–47

Feng Y (2002) Biodiversity and eco-agriculture. Chin J Eco-Agric 10:5–7

Fritz LL, Heinrichs EA, Machado V, Andreis TF, Pandolfo M, Martins de Salles S, Vargas de Oliveira J, Fiuza LM (2001) Diversity and abundance of arthropods in subtropical rice growing areas in the Brazilian south. Biodivers Conserv. doi:10.1007/s10531-011-0083-3, Springer Science Business Media B.V

Grigarick AA, Webster RL, Meyer RP, Zalom FG, Smith KA (1990) Effect of pesticide treatments on non target organism in California rice paddies. I. Impact of triphenyltin hydroxide, II. Impact of Diflubenzuron and triflumuron. Hilagardia 58:1–36

Hajek A (2004) Natural enemies: an introduction to biological control. Cambridge University Press, Cambridge, p 378

Giang HTT, Pham VL, Truong TL, Nguyen TT, Nguyen VL, Nguyen KH (2009) Study on biodiversity herbivore-natural enemy arthropod fauna in rice fields at red river delta, Viet Nam. Proceedings & Abstract, 2009 Marco symposium Challenges for agro-environmental research in Monsoon Asia. October 5–7, Tsukuba, Japan

Hossain Z, Gurr GN, Wratten SD, Raman A (2002) Habitat manipulation in lucerne *Medicago Sativa*: arthropod population dynamics in harvested and 'refuge' crop strips. J Appl Ecol 39:445–454

Hu J, Lu Q, Yang L, Gao N (1986) Influence of fertilizer level and irrigation on population dynamics of the major insect pests in paddy fields and consequent rice yield. Acta Entomol Sin 29:49–55

Hu Y, Tang Q, Tang J, Hu G (1997) Succession regularity of arthropod community in single-cropping paddy fields. Chin J Rice Sci 12:229–232

Huang Z, He Y, Pi B, Tang H, Liu C (2003) Ecological effects of the fertilizer management on pest population in the high-yield cultivation. Entomol Knowl 37:129–133

Hurlbert SM, Mulla MS, Willson HR (1972) Effects of an organophosphorus insecticide on the phytoplankton, zooplankton and insect populations of freshwater ponds. Ecol Manographs 42:269–299

Jarvis DI, Padach C, Cooper HD (2007) Managing biodiversity in agricultural ecosystems. Columbia University Press, New York

Jin C, Wu Y, Wang D (1990) Diversity of the arthropod communities in paddy fields. Acta Entomol Sin 33:287–295

Jin Q, Ouyang Y, Yu S, Xu D (2003) Water crisis in Agricultural sustainable development and its counter measures in China. Res Agric Mod 1:21–23

Kajimura T, Maeoka Y, Widiarta IN, Sudo T, Hidaka K, Nakasuji F, Nagai K (1993) Effect of organic farming of rice plants on population density of leaf hoppers and plant hoppers. I. Population rate. Jpn J Appl Entomol Zool 37:137–144

Kandibane M, Raguraman S, Mahadevan NR (2007) Diversity and community structure of aquatic arthropods in an irrigated rice ecosystem of Tamil Nadu, India. Asian J Plant Sci 6:741–748

Kenmore PE, Carino FO, Perez CA, Dyyck VA, Gutierrez AP (1984) Population regulation of the rice brown plant hopper (*Nilaparvata lugens Stal*) within rice fields in the Philippines. J Plant Protect Trop 1:19–37

Kenneth Schoenly G, Joel Cohen E, Heong KL, Gertrudo Arida S, Alberto Barrion T, James Litsinger A (1996) Quantifying the impact of insecticides on food web structure of rice-arthropod populations in a Philippine farmer's irrigated field: a case study. In: Polis G, Winemiller K (eds) Food webs: integration of patterns and dynamics. Chapman and Hall, New York, pp 343–351

Kiritani K (2000) Integrated biodiversity management in paddy fields: shift of paradigm from IPM toward IBM. Integr Pest Manag Rev 5:175–183

Koganezawa T (2009) The relationship between rice production and ecosystem services: steps toward sustainable practices. Bull Miyagi Univ Educ 44:15–22

Kruess A, Tscharntke T (1994) Habitat fragmentation, species loss and biological-control. Science (Washington) 264:1581–1584

Kumar A, Khanna V (1983) A review of the taxonomy and ecology of Odonata larvae from India. Orient Insects 17:127–157

Liu Y, Gu D, Zhang G (2002) The community dynamics of predatory arthropods in both weed habitat and paddy field from a double cropping paddy in Guangdong. Acta Entomol Sin 46:591–597

Luo J, Li Y (2003) Theory and technology of water-saving irrigation, 2nd edn. WuHan University Press, Wuhan (In Chinese)

Luo Y, Fu H, Traore S (2014) Biodiversity conservation in rice paddies in China: toward ecological sustainability. Sustainability 6:6107–6124. doi:10.3390/su6096107

Lv Z (2003) Effects of nitrogenous fertilizer on ecological fitness of the brown plant hopper, Nilaparvata lugens, and its relationships with natural enemies and rice. Zhejiang University, Hangzhou (In Chinese)

McGeoch MA (1998) The selection, testing and application of terrestrial insects as bioindicators. Biol Rev 73:181–201

Peng S (2011) Water saving irrigation technologies for rice. Yellow River Water Conservancy Press, Zhengzhou (In Chinese)

Peng S, Xu J (2011) Theory and technology of controlled irrigation for rice. Hohai University Press, Nagjing (In Chinese)

Pham VL, Giang HTT (2006) Study on biodiversity: herbivore-natural enemy arthropod fauna in rice fields at red river delta, Viet Nam. Proceedings & Abstract, 2006 Marco symposium Challenges for agro-environmental research in Monsoon Asia. Tsukuba, Japan. www.niaes.affrc.go.jp/sinfo/sympo/h22/1109/paper_11.pdf

Pimentel D (1997) Techniques for reducing pesticide use: economic and environmental benefits. Wiley, Chichester

Rice Barbosa PR (1998) Conservation biological control. Academic, San Diego

Steinfeld H, Gerber P, Wassenaar T, Castel V, Rosales M, Haan C de (2006) Livestock's long shadow. Environmental issues and options. FAO Agriculture Technical paper, Rome

Ueno T (2010) Biodiversity in rice paddies and use of indicator species to assess environmental friendship of agricultural practices. In International AFAS Joint Symposium between Korea and Japan 7: 126–132

Ueno (2013) Bioindicators of biodiversity and farming practice in rice paddies. Int J Chem Environ Biol Sci (IJCEBS) 1(1):2320–4087

Vinson MR, Hawkins CP (1998) Biodiversity of stream insects: variation at local, basin and regional scales. Ann Rev Entomol 43:271–293

Wang G (2002) Further thoughts on diversity and stability in ecosystems. Biodivers Sci 10:126–134 (In Chinese)

Washitani I (2007) Restoration of biologically-diverse floodplain wetlands including paddy fields. Glob Environ Res 10:135–140

Wu C, Chen X (2004) Impact of pesticides on biodiversity in agricultural areas. J Appl Ecol 15:341–344

Yasumatsu K, Watanabe C (1965) A tentative catalogue of insect natural enemies of injurious insects in Japan, Part 2. Host parasite-predator catalogue. Kyushu Univ Press, Fukuoka, Fac Agric, Entomol Lab

Zhang, Zheng X, Jian H, Qin X, Yian F, Zhang (2013) Institute of biotechnology and germplasm resources. Yunnan Academy of Agricultural Science, Kunming, 650223

Zhu T, Goa P (1987) Shallow-wet irrigation technology for rice. Water Conservancy and Electric Power Press, Beijing (In china)

Zhu Z, Wu L, Wu G, Cheng J (2000) The effects of filmed ground and dry growing conditions on the occurrence of rice pests, natural enemies and saprophagous insects. Acta Phytophylacica Sin 27:295–301

Arthropods on Cotton: A Comparison Between *Bt* and Non-*Bt* Cotton

9

A.K. Chakravarthy, Manja Naik, and T.N. Madhu

Abstract

Adopting monocultures of traditional cotton enhances activity of pest insects and reduces the activity of predatory insects. Cultivating cotton with other crops such as sunflower (*Helianthus annuus*) and sorghum (*Sorghum bicolor*) served as refugia for predators of pests on cotton. Thus, increased habitat diversity by strip cropping in monocultures of cotton increases the population of predators. Transgenic cotton (*Bt*) largely suppressed populations of lepidopteran pests. Insecticidal sprays reduced populations of predators both on non-*Bt* and *Bt* cotton. *Bt* cotton alters the arthropod community by reducing the abundance of *Helicoverpa* populations. *Bt* cotton may also have indirect effects on the abundance of parasitoids and predators that specialize on lepidopteran pests. A 6-year research revealed that the impact of *Bt* cotton on minor pests and non-intended species was of less importance, particularly when compared to insecticides. Cotton ecosystem is uniquely characterized by secondary pest outbreaks, genetically engineered plants, changing arthropod communities and extrafloral (EF) nectaries. Each characteristic influences arthropod communities and crop productivity in turn in different ways. Although reduction in insecticidal use in some regions may alleviate the pest problems, much of the problems can be tackled by adopting integrated pest management (IPM) practices.

A.K. Chakravarthy (✉)
Division of Entomology and Nematology, Indian Institute of Horticultural Research (IIHR), Hessaraghatta Lake Post, Bengaluru 560089, Karnataka, India
e-mail: chakravarthyakshay@gmail.com

M. Naik • T.N. Madhu
Department of Agricultural Entomology, Gandhi Krishi Vignana Kendra (GKVK), University of Agricultural Sciences (UAS), Bengaluru 560065, Karnataka, India

A.K. Chakravarthy, S. Sridhara (eds.), *Economic and Ecological Significance of Arthropods in Diversified Ecosystems*, DOI 10.1007/978-981-10-1524-3_9

Keywords
Arthropods • *Bt* cotton • Insect pests • Conservation • Predators • IPM

9.1 Introduction

Cotton is the soft, staple fibre of the plants of genus *Gossypium* (Malvaceae). The crop is being cultivated both in the Old and the New World. In the Old World, cotton fibre was found as early as 5000 BC in the Indus Valley Civilization in ancient India and also in Mexico. Cotton cultivation accounts for 2.5 % of the world's cultivated tract. China is the largest cotton producer. Four, domesticated, commercially grown species of cotton are:

Gossypium hirsutum – Central America is the origin.
Gossypium barbadense – tropical South America is the origin.
Gossypium arboreum – native of India and Pakistan.
Gossypium herbaceum – Levant cotton, native of South Africa and Arabia.

Between 200 and 100 BC, cotton became widespread across much of India, Iran and many parts of the orient (Stein Burton 1998). Cotton cultivation needs a long, frost-free period, sunshine and moderate rainfall (600–1200 mm). Obviously, favourable ambience conditions are found within the tropics and subtropics in both northern and southern hemispheres. Cotton had been cultivated annually although it is perennial in nature (cottonspinning.com). Early domestication of the crop will provide time for colonization of arthropods that are primary consumers and phytophagous. Subsequently, it would pave way for the development of natural enemies (secondary consumers) on the pests. Cotton contains gossypol, a toxin that makes them inedible. However, scientists have silenced the gene that produces the toxin making it a potential food crop (Bourzac 2006).

In many cotton-growing regions, the major pests in commercial cotton are lepidopteran larvae, which are generally called as bollworms. These include moths of species of *Heliothis*/*Helicoverpa*, *Earias* and *Pectinophora*. Genetically modified (GM) cotton was evolved to lessen the burden of pesticide application. The bacterium *Bacillus thuringiensis* (*Bt*) produces a chemical inimical to larvae of moths and harmless to other forms of life (Mendelsohn et al. 2003). GM cotton eliminated the requirement for large amounts of broad-spectrum insecticides. This spares natural insect enemies and further contributes to non-insecticide pest management. However, *Bt* cotton proved ineffective in suppressing populations of sucking pests like leaf hoppers, bugs and aphids. Based on the situation, the farmers have to use appropriate insecticides. For instance, in South India subsequent to *Bt* cotton introduction, the usage of insecticides increased to suppress populations of leaf hoppers (*Amrasca biguttula biguttula* Ishida). In a study conducted in 2006 by Cornell researchers and the centre for Chinese Academy of Sciences on *Bt* cotton farming in China found that after 7 years the secondary pests generally controlled by pesticides had increased, necessitating similar levels of use of insecticides. This would prove expensive to farmers because of extra cost of seed (Mensah 1999). Investigations by the Chinese Academy of Sciences, Stanford University and Rutgers University declined this trend (Wang et al. 2009). They summarized that the GM cotton provided adequate protection. The secondary pests were mostly Miridae, which attained higher population due to local temperature and rainfall. A 2012 Chinese study concluded that *Bt* cotton reduced the pesticide use by 50 % and increased ladybird beetles, lacewings and spiders (Lu et al. 2012).

9.2 Genetically Modified Cotton

Genetically engineered (GE) cotton to resist specific insect pests has become dominant in several countries worldwide. Between 1996 and 2009, GE crops were grown on 134 million hectares of farmland in 25 countries. GE crop production continues to be dominated (63 % in 2009) by the cultivation of plants tolerant to the herbicides, glyphosate or glufosinate. Insect-resistant crops producing toxins of *Bacillus thuringiensis* (*Bt*) comprise most of the remaining market share (57 %) and 1 % of crops engineered for resistance to viral diseases. Cotton accounts for about 40 % of the world's natural fibre production and is commercially cultivated in 78 countries. Surveys have revealed 1300 species of herbivorous insect inhabit cotton.

Cotton has historically been one of the largest users of insecticides worldwide. In the past two decades, most notable advances in biotechnology have allowed engineering of plants to provide highly effective and selective control of insect pests globally. For example, in Karnataka, South India, 40 % of insecticide usage on cotton was halved. Similar situation of *Bt* cotton is found in many other parts of the world. A few years ago, most *Bt* cotton produced only a single Cry protein (e.g. Cry1Ac in Bollgard), but many countries are now using *Bt* cotton in which two different Cry proteins are produced in the plant (e.g. Bollgard II and Wide Strike). These provided for a broader spectrum of activity against Lepidoptera, enhanced control of caterpillars that were already susceptible to single-toxin transgenic plant and better opportunities for managing insect resistance to Cry proteins.

Biodiversity is threatened by agriculture in general and especially modern methods of agriculture. There was a concern that transgenic crops may affect biodiversity via unintended impacts on nontarget populations of arthropods. On the other hand, GE crops may positively impact agricultural species biodiversity if they enable the targeted management of weeds and insect pests, compared to conventional agriculture. Researchers at the University of Arizona (USA) and McGill University (Canada) conducted a 2-year study to examine whether transgenic *Bacillus thuringiensis* (*Bt*) crops increase agricultural biodiversity while minimizing the environmental impacts of agriculture (Cattaneo 2006). They chose 81 commercial fields in Arizona in which *Bt* cotton represented 48 % and 62 % of the cotton planted in the first and second year of the study, respectively.

The study indicated that cultivation of transgenic and non-Tr cotton had similar effects on arthropod diversity. Growing *Bt* cotton in large commercial fields reduced broad-spectrum insecticide use and increased yields resulting in benefits for higher cultivators, as the yield benefits of transgenic cotton exceeded the cost of transgenic seeds. Since *Bt* cannot control all important cotton pests, the ultimate benefit of *Bt* crops largely depends on whether additional insecticides are required, a factor which cannot be overlooked. The ant and beetle diversity studies indicate that invertebrate taxa may react differently to agricultural practices; however, the authors did not find that transgenic cotton had a greater impact on arthropod diversity than non-transgenic cotton (Janaki Krishna 2006).

Studies in Argentina have shown that the package of restricted pesticide applications halved while significantly enhancing yields of crops. Benefits that accrue can be maximum for small farmers, who are not availing the method. The durability of the advantages is analyzed by using biological models to simulate resistance development in pest populations. It is only after monitoring the populations of the pests for a long time that any conclusions can be drawn about the technology.

9.3 Arthropods on Cotton

Bal and Dhawan (2009) recorded 134 species of arthropods on cotton in Punjab in 2006. Fifty four species were herbivorous, 44 species were natural enemies, 26 species were casual visitors, and 10 species of arthropods were pollinators. The index of diversity of arthropod communities and sub communities was higher on sprayed than on unsprayed cotton. Higher diversity index value was attributed to greater equitability within the arthropod species. A maximum number of species belonged to Hymenoptera (19.40) and Coleoptera (16.40). Earlier workers reported 166 (Khan and Rao 1960; Sohi 1964) species of insects and mite pests (Table 9.1). In Australia, cotton from monoculture fields and the lack of ecological diversity could be the major cause of pest problems because the food, hosts, prey and hibernating of overwintering sites of most of the natural enemies of the pests are reduced, thereby limiting natural control (Beirne 1967; Hagen and Hale 1974). This can lead to outbreaks of pests, because of abundant food (Beirne 1967; Hagen and Hale 1974). *Helicoverpa* spp. are the major pests of cotton crop in Australia. They are highly migratory and can therefore quickly infest large areas of cotton crops and oviposit. If not natural enemies are present and well established in high numbers prior to the pest's establishment, the predators cannot respond rapidly enough to control proliferating pests (Fitt 1989; Mensah and Harris 1995). Major predators of pests on cotton in Australia are given in Table 9.3. The photograph of sucking insects and bollworms in cotton ecosystem is shown in Figs. 9.8 and 9.9.

In regression analyses, the numbers of species that were planted and the number of functional groups planted significantly increased arthropod species richness. The species number planted was lone significant predictor of arthropod species richness. Matrix of analyses between planted species and species of arthropods (Table 9.2) revealed phytophyte diversity was impacted by plant, parasite and predator species richness. Herbivore diversity was closely related with the predator and parasite and number of species and individuals than that of plants. Together with regression analyses, enhancing plant diversity significantly enhanced arthropods; local phytophagous diversity is sustained by a diversity of beneficial insects (Siemann et al. 1998).

Table 9.1 Diversity indices of arthropod communities in RCH 134 *Bt* and non-*Bt* cotton under sprayed and unsprayed conditions

Arthropod community		*Bt* sprayed			Non-*Bt* sprayed			*Bt* unsprayed			Non-*Bt* unsprayed		
		H′	J	D	H′	J	D	H′	J	D	H′	J	D
Total arthropods	Vegetative phase	0.835	0.363	0.637	0.890	0.387	0.613	0.873	0.379	0.621	0.775	0.337	0.663
	Reproductive phase	0.768	0.333	0.667	0.812	0.353	0.647	0.846	0.367	0.633	0.732	0.318	0.682
	Overall	0.978	0.425	0.575	0.963	0.418	0.582	0.954	0.414	0.586	0.940	0.408	0.592
Predatory arthropods	Vegetative phase	0.930	0.374	0.626	1.065	0.428	0.572	0.972	0.391	0.609	1.020	0.410	0.590
	Reproductive phase	0.929	0.374	0.626	0.877	0.353	0.647	0.778	0.313	0.687	0.789	0.318	0.682
	Overall	0.953	0.383	0.617	0.976	0.393	0.607	0.917	0.369	0.631	0.9191	0.370	0.630
Nontarget pests	Vegetative phase	1.044	0.536	0.464	0.995	0.511	0.489	1.023	0.526	0.474	0.894	0.459	0.541
	Reproductive phase	1.343	0.690	0.310	1.169	0.601	0.399	1.264	0.649	0.351	1.167	0.600	0.400
	Overall	1.332	0.685	0.315	1.236	0.635	0.365	1.267	0.651	0.349	1.179	0.606	0.3394

H′ = Shannon and Weaver's diversity index, J = evenness index and D = dominance index
Vegetative phase – observations recorded from 29 June to 10 August 2006
Reproductive phase – observations recorded form 17 August to 28 September 2006
Overall – total observation period from 29 June to 28 September 2006
Source: Bal and Dhawan (2009)

Table 9.2 Simple and partial correlations among number of species planted and arthropod herbivore, parasite and predator species richness

Variable	Log_2 (number of species planted)	Herbivore species richness	Parasite species richness	Predator species richness
Log_2 (number of species planted)	1.00	0.35^{***}	0.27^{***}	0.34^{***}
Herbivore species richness	0.13^{*}	1.00	0.66^{***}	0.54^{***}
Parasite species richness	0.07NS	0.58^{***}	1.00	0.36^{***}
Predator species richness	0.19^{***}	0.39^{***}	−0.01NS	1.00

Number to the right of 1.00 is simple correlations; numbers to the left of 1.00 are partial correlations. $N = 288$; NS means $P > 0.05$

$^{*}0.01 < P < 0.05$

$^{***}P < 0.001$ for significance test from 0 correlations

Plants influence arthropod abundance and diversity in more than one way. Also $\log_2$ (number of species planted) was significantly related to herbivore, parasite and predator species richness (Table 9.2). Additionally, herbivore and species richness of natural enemies were all themselves significantly correlated. Changing plant diversity directly influenced parasite species richness. Plant diversity influenced parasite species richness indirectly through different phytophagous insects; herbivore species richness of phytophagous insects was highly correlated with parasite, predator and plant species number (Fig. 9.1).

Destruction of crop residues and reduction in fallow periods have resulted in large-scale depletion of soil organic matter and degradation of soil fertility in cotton (*Gossypium hirsutum* L.) cropping systems of Cameroon. Soil management ecosystems based on a no till with mulch and intercropped with cereals restore cotton production and boost the biological activity of soil macrofauna. Examination of the soil macrofauna patterns revealed that the abundance and diversity of soil arthropods were significantly higher in patches with grass and legume mulch. Concerning maintenance of ecological functions, herbivores and predators were significantly more abundant in no tillage, grass and legume mulch. The decomposers, predators and herbivores were also considerable in the above plot than in others. Formicidae (53.6 %), Termitidae (24.7 %) and Lumbricidae (9.4 %) were the most abundant detritivores, while Julidae (46.1 %), Coleoptera larvae (22.1 %) and Pyrrhocoridae (11.8 %) were the dominant herbivores (Brévault et al. 2007).

In ecosystems cultivated in the tropics and subtropics of Africa, abundance and biodiversity of soil inhabitants are frequently reduced through habitat structure, disturbances of soil organism communities made by extreme climatic conditions, overgrazing and trampling by cattle, burning of crop residues, ploughing and mechanized seed bed preparation, indiscriminate agrochemical use and monoculture (Lal 1988; Loranger et al. 1999; Brown et al. 2001). Phytophagous were considerably abundant in mulched soils than in non-mulched ones. Detritivores, herbivores and beneficials were abundant in no-tilled soils, especially when mulch was covered in cotton ecosystem (Fig. 9.2).

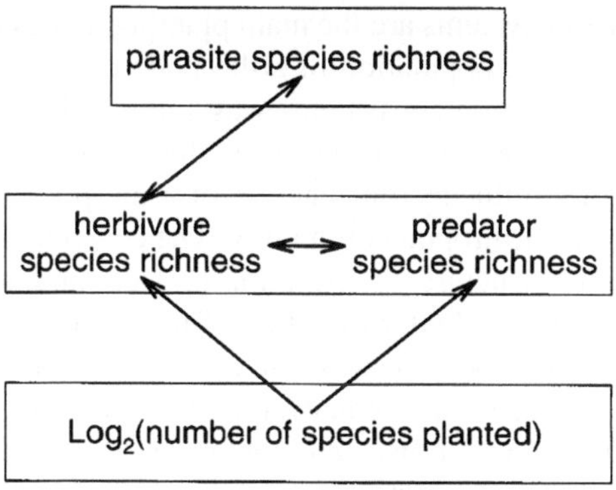

Fig. 9.1 Model consistent with a correlation structure with simple correlations between all four variables and conditional independence of parasite species richness, predator species richness, parasite species richness and $\log_2$ (number of species planted). Following the guidelines of Cox and Wermuth (1993), *arrows* point from explanatory variables to response variables, *lines with two heads* represent correlations among response variables, and *boxes* surround the predictor variable (plant diversity), response variables that respond directly to changes in plant diversity (herbivore and predator diversity) and response variables that respond only indirectly to changes in plant diversity (parasite diversity) (Cox and Wermuth (1993))

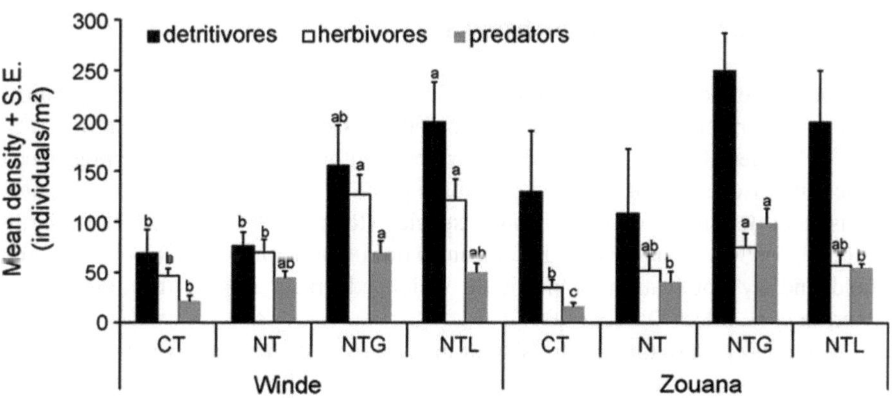

Fig. 9.2 Abundance of soil macrofauna communities as a function of the soil management system and experimental site. Soil macrofauna were sampled by extracting two 30-cm-sided soil cubes (including the litter layer) in the central part of each plot, at the seeding stage and 30 days later (16 and 12 samples per system, from Zouana and Winde sites, respectively) during the 2004 cotton-growing season. *CT* conservation tillage, *NT* no tillage, *NTG* no tillage with grass mulch, *NTL* no tillage with legume mulch. For each individual site, bars of the same colour followed by different letters are significantly different (ANOVA SAS GLM, $P < 0.05$). *SE* standard error (Brévault et al. 2007)

Poplar-cotton agroecosystems are the main planting modes in China. Systematic survey of the diversity and population of arthropod communities in four different combinations of poplar-cotton ecosystems was studied. The main results were as follows: the transgenic poplar-cotton ecosystem had a stronger inhibitory effect on insect pests and had no influence on function of arthropods and, therefore, maintains the diversity of the arthropod community. The character index of the community indicated that the structure of the arthropod community of the transgenic poplar-cotton ecosystem was better than that of the poplar-cotton ecosystem. The transgenic poplar-cotton ecosystem was also better than that of the non-transgenic poplar-cotton ecosystem. The cluster analysis and similarity of arthropod communities between the four different transgenic poplar-cotton ecosystems illustrated that the composition of arthropod community excelled in the small sample of the transgenic poplar-cotton ecosystems (Zhang et al. 2015).

The survey data showed that the transgenic poplar-cotton ecosystem has some effects on the species and tropic structure of the arthropods. In some poplar-cotton ecosystems, a high proportion of pests and a low proportion of natural enemies was observed. Predator quantities in the cotton field next to the transgenic poplar are higher than those in the control field next to the non-transgenic poplar, suggesting that transgenic poplar in the ecosystem contributed partly to the increase in predator numbers in the cotton fields.

9.3.1 Secondary Pest Outbreaks

Secondary pest outbreaks mean a pesticide application to lessen pest populations triggering further population increase of other pests (Ripper 1956; Hardin et al. 1995; Dutcher 2007). They are secondary pest outbreaks, including suppression of beneficial insects, metabolic changes in the plant or nontarget organisms (hormoligosis) and decline in other arthropod species (Ripper 1956; White 1984; Hardin et al. 1995). Secondary pest outbreak can be detrimental to productivity by reducing yield and by application of pesticide which adversely impact the environment (Horton et al. 2005; Dutcher 2007).

Secondary pest outbreaks are also of interest from the perspective of "ecosystem services". This is because quantifying the loss in profit attributable to secondary pest outbreaks may arguably provide a lower bound on the monetary value of regulation of economically injurious pests by natural enemies. For instance, whitefly management of *Lygus* in cotton is thought to provide a prime candidate for secondary pest outbreaks, because cotton harbours a rich community of arthropod herbivores and natural enemies and because, until very recently, only non-selective, broad-spectrum pesticides were available for *Lygus* control (Rao et al. 2003; Dutcher 2007). *Bemisia tabaci* (Genn.) was a minor pest on cotton before the pyrethroid insecticides were applied on cotton in Andhra Pradesh, India. The whitefly became so severe due to injudicious use of insecticides leading to huge economic losses that farmers were driven to the brink of committing suicides.

In California's San Joaquin Valley, judicious use of pesticides was practised so as to achieve maximum pest management services and also to maintain abundant and diverse community of natural enemies (University of California 1996). The threats to cotton production due to phytophagous arthropods change over the course of the growing season. Their populations grow most rapidly under cooler fall temperatures, because their excreta ("honeydew") can contaminate cotton lint, which is exposed once mature cotton fruits ("bolls") start to open as harvest approaches.

9.3.2 Extrafloral Nectaries

Extrafloral nectaries (EF) impact interactions with several species, indicating that nectar attracts diverse arthropods and may enhance the diversity and abundance of arthropods. Experiments conducted in this regard also confirm this evidence on the importance of EF nectar to terrestrial food webs was equivocal. Exploring potential avenues for selection, it was found that several cost-benefit ratios of EF nectary traits have received scarce attention. Some of these aspects include a constraint faced by plants when attracting both pollinators and protectors via nectar and an ecological cost of nectar when herbivores consume EF nectar as adults.

Ant species per EF nectar-bearing plant species range from one to seven ant species ($n = 35$ plant species). The result suggests that EF nectar attracts diverse rather than homogeneous aggregates of ants (Rico-Gray 1993). However, present documentation doesn't allow a similar survey of non-ant arthropods; diverse visitors have been documented, including herbivore, parasite and pollinator in ten arthropod orders (Koptur 1992).

By supplying carbohydrates, plant with EF nectar may assist diversity or abundance of arthropods than nectar-less plants. An analogous example comes from gall-forming aphids (*Pemphigus betae*), which provide honeydew and thereby increase the species richness and abundance of arthropods on cottonwood compared to conspecifics without aphids (Dickson and Whitham 1996). Evidence for similar community-level effects of EF nectar is limited. However, in cultivated cotton, greater abundances of herbivores, parasitoids and non-ant predators were found on plants with EF nectaries compared to near-isogenic (or related) nectar-less lines (Henneberry et al. 1977; Adjei-Maafo and Wilson 1983). Arizona and Mexico indicated that EF nectar may not always promote arthropod abundance or diversity. Conventional wisdom holds that EF nectaries benefit plants by attracting arthropods that reduce herbivores, pathogens or parasites (Bentley 1977; Koptur 1992). EF nectaries can also lure arthropods that deter floral nectar robber (e.g. O'Dowd 1979), although robbing has received less attention than herbivory and pathogen attack.

Arthropod-mediated ecosystem services (AMES) include crop pollination and pest control, which help to maintain agricultural productivity and reduce the need for pesticides. Maximizing survival and reproduction of beneficial arthropods requires provision of pollen and nectar that are often scarce in modern landscapes. Increasingly, native plants are being evaluated for this purpose. Native plants can outperform recommended non-natives and also provide local adaptation, habitat permanency and support for native biodiversity. It is predicted that the success of insect conservation programmes using flowering plants to increase AMES on farmland will depend on landscape context, with the greatest success in landscapes of moderate complexity. Reintegration of native plants into agricultural landscapes has the potential to support multiple conservation goals and requires the collaboration of researchers, conservation educators, native plant experts and farmers.

9.3.3 *Bt* vs. Non-*Bt*

The predator populations between *Bt* and non-*Bt* cotton fields with varying densities of insecticide applications with functional influences on biological control in these fields. Rates of egg and larval predation were significantly higher in *Bt* cotton than on the non-*Bt* cotton in South Carolina, in South Alabama and in Georgia. Differences between *Bt* and non-*Bt* cultivars had minor impact on the arthropod community (Sisterson et al. 2004).

In the transgenic cotton agroecosystem in Coahuila, Mexico, predators regulate pest populations (Whitcomb and Bell 1964). Predators might be exposed to the *Bt* toxin through feeding on lepidopteran larvae that are only partially susceptible to Cry1Ac toxin (Perlak et al. 2001; Stewart et al. 2001). Pest larvae may survive by feeding on other herbivores, such as spider mites and thrips, that may pick up *Bt* toxin from plants (Dutton et al. 2002, 2004).

Most studies to date have addressed species-specific interactions under laboratory conditions, whereas field studies have focused on experiments. This work evaluated the dynamics of major predators in commercial *Bt* and non-*Bt* cotton fields for three consecutive years using assorted sampling plants so that all predators are included (Torres and Ruberson 2005).

Many eggs were attacked by predators with piercing-sucking mouthparts. In addition, several egg sheets were recovered with lacewing larvae (Chrysopidae) on them. The percentage of eggs eaten increased linearly with the number of egg predators in test samples (Fig. 9.3a), which suggests that estimate of egg predator abundance was related to egg predator activity in the field. Treatment did not affect the percentage of eggs eaten or the abundance of egg predators (Fig. 9.3b) (Sisterson et al. 2004).

9.3.4 Intervention by Insecticides

A study was conducted to evaluate the quantitative relationship between insecticide use and predator populations; the implications in the number of insecticide applications on non-*Bt* cotton and *Bt* cotton fields were regressed against the relative size of the ant populations (the most abundant predator in these systems) in each region and year (Fig. 9.4).

The percentage of prey items eaten in a 24-h period in *Bt* cotton fields was consecutively higher than the percentage consumed in non-*Bt* cotton fields. Figure 9.5

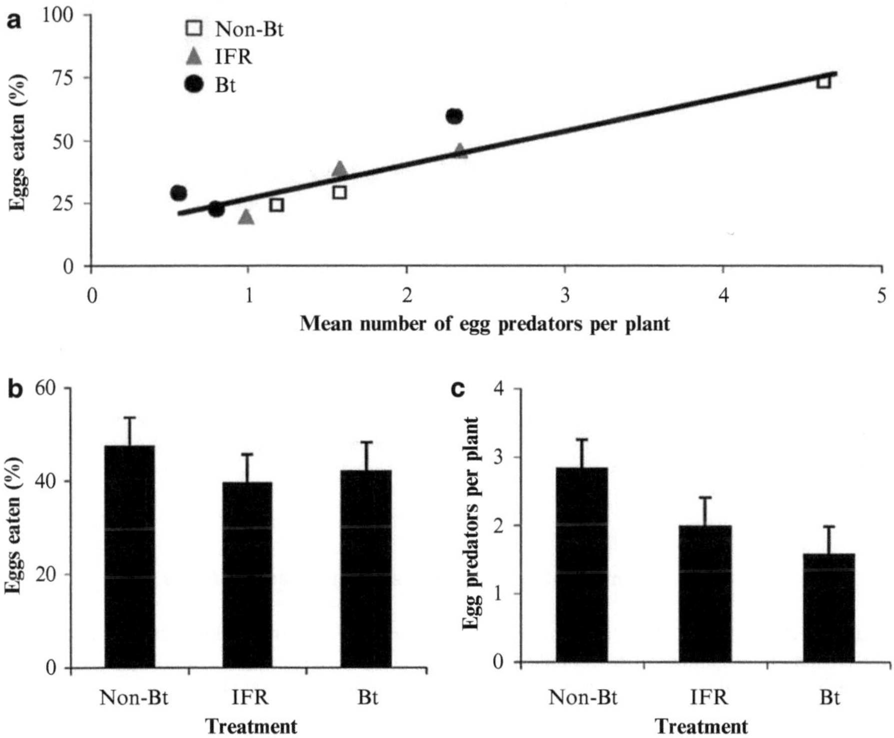

Fig. 9.3 (**a**) Relationship between number of egg predators per plant and percentage of eggs eaten. The relationship was linear and significant (*F* _ 38.0; df _ 1, 7; *P* _ 0.001; *r*2 _ 0.84, y _ 13.44× _ 13.29). (**b**) Percentage of eggs eaten (_SE) in each treatment (*F* _ 0.43; df _ 2, 4; *P* _ 0.68). (**c**) Number of egg predators (_SE) in each treatment (*F*_2.4; df _ 2, 4; *P* _ 0.21) (Source: Sisterson et al. 2004)

shows the results of experiments with sets of *H. zea* eggs in three of the regions in 2002. At all three locations, these experiments were preceded by at least one pyrethroid application for lepidopteran pests on the non-*Bt* cotton fields and two cyhalothrin applications in the preceding 2 weeks at the South Carolina site.

Common and general predators in the cotton ecosystem seem to have been affected more by insecticide use than by *Bt* cotton. Tropic effects play a role under many other conventional pest control methods, including pest resistance traits in cotton plants introduced by conventional breeding approaches (Schuster and Calderon 1986; Cortesero et al. 2000). However, predator populations are considered important for cotton pest management. Results suggest that *Bt* cotton use, coupled with appropriate insecticide selection when economic thresholds are exceeded, has no adverse effect on the predator community. From the data we can observe significant reductions in predator species abundance occurred mid- to late season after broad-spectrum insecticide applications (lambda-cyhalothrin, zeta-cypermethrin and dicrotophos). The results reported here for three successive

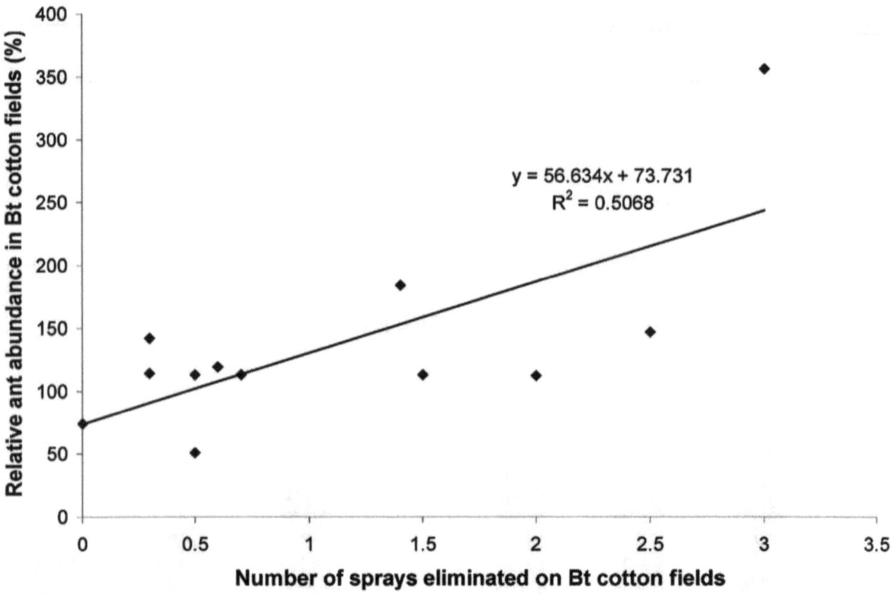

Fig. 9.4 Relative size of ant populations in *Bt* versus non-*Bt* cotton fields regressed against the difference in the number of insecticide applications on non-*Bt* versus *Bt* cotton fields. Values are averages for a region in each of the years 2000–2002 (Source: Head et al. 2005)

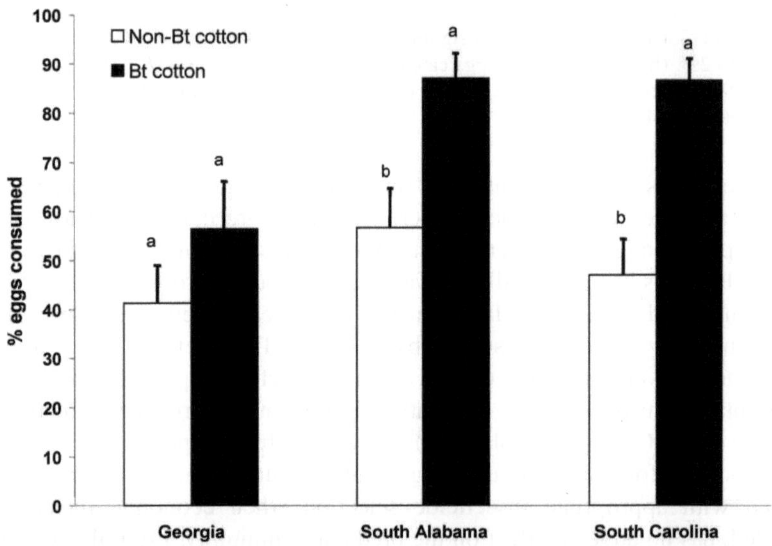

Fig. 9.5 Percentage (mean ± SE) of sentinel eggs consumed within 24 h in non-*Bt* and *Bt* cotton fields in each of three regions (Georgia, South Alabama and South Carolina) in 2002. *Bars* within a region with the same letter above them are not significantly different (*P* _ 0.05) (Source: Head et al. 2005)

seasons conducted in grower fields regulated that differences in relative predator abundance between cotton types could be common among dates for select insects, but these differences largely disappeared when all 3 years of the study were considered.

The analyses indicate no differences in the ground-dwelling arthropod communities between cotton types. One rained species, *Pardosa pauxilla*, comprised 80 % of all araneids, *Labidura riparia* comprised 96 % of all dermapterans, *Megacephala carolina* comprised 97 % of cicindelines, and four carabid species (*Selenophorus palliatus*, *Apristus latens*, *Harpalus gravis* and *Anisodactylus merula*) consisted 80 % of carabid species. *M. carolina* out numbered other species every year. When only predatory carabid species were focused, *A. merula*, *Calosoma sayi*, *Harpalus pensylvanicus* and *Stenolophusochropezus* were dominant, and individuals trapped were almost the same between cotton types. The population of dermapterans, staphylinids, araneids and heteropterans varied among sample dates and across seasons but did not differ between cotton types (Torres and Ruberson 2007).

Transgenic cotton that produces insecticidal proteins of *Bacillus thuringiensis* (*Bt*) was sown in 6.15 million hectares in 11 countries in 2009 and has contributed to a reduction of over 140 million kilogrammes of insecticide active ingredient between 1996 and 2008. As a highly selective form of host plant resistance, *Bt* cotton effectively controls a number of key lepidopteran pests and has become a cornerstone in overall integrated pest management (IPM). *Bt* cotton has resulted in large-scale suppression of pests and benefited non-*Bt* cotton adopters and even producers of other crops affected by polyphagous pests. Although reductions in insecticide use in some regions have elevated the importance of several pest groups, in general it has enhanced natural control. As a result of *Bt* cotton, selective pest suppression alternatives for other key pests in the system along with a complete IPM programme infrastructure permitting for the efficient utilization of all component tactics, insecticide use in Arizona cotton has been driven to low levels (Fig. 9.6).

Stem application of acetamiprid and thiamethoxam was found better not only in suppression of the sucking pests' population, but also the population of predators was significantly less disrupted by the stem application method. The foliar application was, in general, more effective; stem application may be more applicable early in the season when its efficacy was higher and when foliar sprays were particularly destructive to beneficial pests. Systemic neonicotinoids were applied as foliar spray; all the like imidacloprid, clothianidin, admire, thiamethoxam and acetamiprid proved toxic to biocontrol agents compared to spirotetramat, buprofezin and fipronil (Kumar et al. 2012).

Insect-resistant transgenic crops have a significant part in lessening crop losses incurred by pest insects. The commercialization of *Bt* transgenic cotton in India has significantly increased the cotton production from about 10 million bales in 2001–2002 to 34 million bales in 2009–2010. The use of *Bt* cotton in India has considerably lessened the numbers and volume of insecticide preparations, bollworm population, production cost and environmental pollution. However, reduction in insecticide sprays, especially during flowering and boll formation, has resulted in resurgence of minor insect pests such as tobacco caterpillar, mealybugs, thrips,

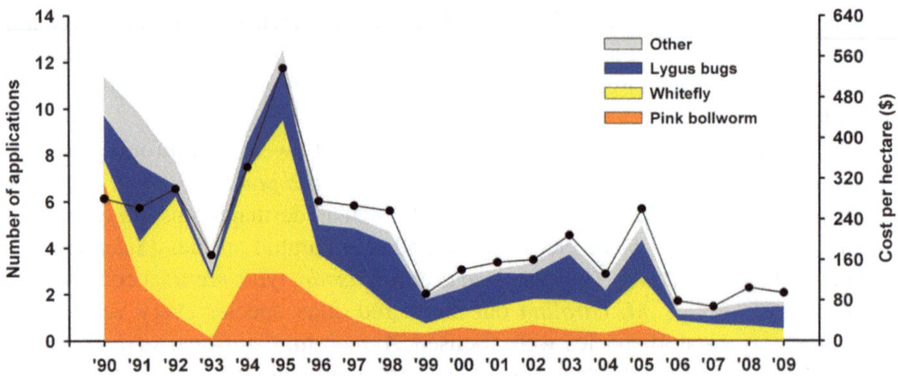

Fig. 9.6 Statewide average foliar insecticide intensity (number of applications per hectare, area graph) and cost (*line* graph) for multiple pest groups in cotton, 1990–2009, Arizona (Compiled from Ellsworth et al. 2007)

aphids, leafhoppers, green stink bug and serpentine leaf miner. The reduced insecticide use and personal experience of farmers of not having adverse effect of *Bt* cotton on the population build-up of bee colonies have encouraged beekeepers to keep beehives in *Bt* cotton in Haryana, Rajasthan and Punjab. Studies have shown that there are no bad effects of *Bt* cotton on the arthropods and natural enemy and their functions in cotton ecosystem and if any are much lower than that of insecticides and have negligible ecological impact (Dhillon et al. 2011). Indian experience with *Bt* cotton cultivation showed that the development of resistance in target pests to *Bt* can't be ignored, and the resistance monitoring and management strategies are essential and need greater attention to sustain *Bt* technology.

On ground-dwelling arthropods, surveys have indicated that in commercial *Bt* and non-*Bt* cotton fields, 65 species of ground-dwelling arthropods (carabids, cicindelines, staphylinids, dermapterans, heteropterans and araneids) of importance for cotton were documented. The analyses demonstrated no differences in the ground-dwelling arthropod communities between cotton types. The frequent trapping of *M. carolina*, *S. palliatus* and *P. pauxilla* in all fields and seasons in both cottons suggests that these species may be important for monitoring further changes in local communities in view of cultivation practices (Torres and Ruberson 2007).

Figure 9.7 shows the linkages between insecticide use, *Bt* and yields. The curves shown are based on the econometric estimates of the damage control function. When insecticides are less used in conventional cotton, adoption of *Bt* causes a significant yield effect, as actually observed in Argentina. Yet, the distance between the curves diminishes gradually with increasing pesticide use; this is why yield effects are smaller in the USA and China. In these countries, yield losses in conventional cotton are low, so that *Bt* is mainly pesticide reducing at constant output levels. These relationships support Qaim and Zilberman's (2003) hypothesis that *Bt* yield effects be higher in situations where crop damage is not effectively controlled through chemical pesticides. Similar results were also obtained by Thirtle et al. (2003) from South Africa and by Qaim and Zilberman (2003) from India.

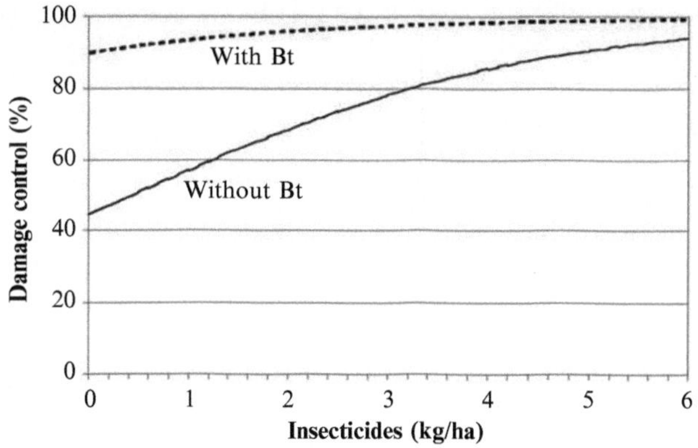

Fig. 9.7 Estimated relationship between insecticide use and damage control with and without *Bt* (Source: Qaim and Janvry (2005)

9.3.5 Conservation Strategies

Generally, agroecosystems disfavour natural enemies. Habitat management, a type of conservation biological control, is an ecological approach focusing natural enemies in agricultural systems. The aim of habitat management is to create a suitable ecological infrastructure in landscape to facilitate resources such as nectar for adult natural enemies, alternative prey as well as shelter. These resources should be favourably integrated into the landscape in a way that is practical for growers (Mensah 1996).

Diversity in agroecosystems may lessen pest population and enhance activity of natural enemies (Altieri 1991; Ryszkowski et al. 1993; Stamps and Linit 1998). However, several researchers have noted that to increase natural enemies, the important components of diversity should be indentified and provided rather than encouraging diversity per se (Goller et al. 1997; Speight 1983; Van Emden 1990; Van Emden and Williams 1974; Way 1966). In fact, simply increasing diversity can exacerbate pest problems (Andow and Risch 1985; Baggen and Gurrm 1998; Collins and Johnson 1985; Gurr et al. 1998). Identifying the key elements of diversity may be a difficult process, but natural enemies can be identified much easily (Wratten and Van Emden 1995). While some parasitoids are able to obtain needed resources from hosts (Jervis and Kidd 1986), others require access to non-host food. Floral nectar is taken by many species (Jervis et al. 1993) and can result in increased rates of parasitism (Powell 1986). Extrafloral nectar is produced by plants such as faba bean (*Vicia faba* L.) and cotton (*Gossypium hirsutum* L.) and is an important food source for adult parasitoids (Bugg et al. 1989; Treacy et al. 1987). The honeydew-producing insect has been suggested as desirable for select parasitoids (England and Evans 1997). Most habitat management attempts with alternative food sources have involved hymenopteran parasitoids (Topham and Beardsley 1975).

Cotton thrips, *Thrips tabaci*

Cotton leaf hopper, *Amrasca biguttula Biguttula*

Aphid, *Aphis gossypii*

Whitefly, *Bemisia tabaci*

Red cotton bug, *Dysdercus cingulatus*

Dusky cotton bug, *Oxycarenus Hyalinipennis*

Fig. 9.8 Sucking insects on cotton (Photos by: Harish Badigere)

Fig. 9.9 Bollworms on cotton (Photos by: Harish Badigere)

Ground covers or intercrops influence natural enemy density, like carabids in maize (Brust et al. 1985), parasitoids in cabbage (*Brassica oleracea capitata* L.) (Theunissen et al. 1995), natural enemies in peas (*Carya illinoensis* Koch) (Peng et al. 1998) and cotton (Xia 1997). The photograph of common natural enemies found in cotton ecosystem was provided in Fig. 9.10.

Successful implementation of natural enemy conservation involves assessing levels of disturbance in agricultural systems. Practices such as cover cropping, intercropping and reduced tillage relax the overall disturbance regime, although they may require some new disturbances (i.e. herbicides) in order to manage weeds. Alternatively, some new technologies such as transgenic cotton expressing *Bt* toxins may appear to reduce disturbance by eliminating pesticides. Habitat management will normally be complemented by other methods and should not be promoted as a stand-alone method. Commonly these will employ biological control agent that has been released in classical or augmentative manners.

Campoletis chloridae | *Bracon lefroyi*

Apanteles anguleti

Photos by: Vishlesh Shankar Nagrare, V Chinna Babu Naik, Prabhulinga T.

Fig. 9.10 Natural enemies on insect pests of cotton

Agricultural ecosystems, which occupy large areas of land, are critical in maintaining biodiversity (McIntyre 1994). Thus the encouragement of natural enemies by strategic increases in habitat diversity offers potential to align the goals of agriculture with those of nature conservation (Gillespie and New 1998). The benefits arise partly from the lessened need for synthetic pesticides and the attendant direct and indirect off-target impacts on organisms such as butterflies (Longley and Sotherton 1997), birds (Rands 1985) and small mammals (Tew et al. 1992) and partly from the introduction or maintenance of structural heterogeneity.

The improvements in conservation and maximization of the abundance and efficacy of *Helicoverpa* spp. natural enemies in cotton crop will be dire essential to increase the control of these pests. There are many instances to show that increased habitat diversity in crops increases population densities of indigenous predators to enhance biological control of pests on different crops (Southwood and Way 1970; Pimentel 1961; Van Emden and Williams 1974; Risch et al. 1983; Wetzler and

Table 9.3 Major predators identified on cotton in Australia (1993–1996)

Order	Family	Species	Group
Coleoptera	Coccinellidae	*Coccinella transversalis* (Fabricius)	Predatory beetles
		Adalia bipunctata (Linnaeus)	
	Melyridae	*Dicranolaius bellulus* (Guerin-Meneville)	
Hemiptera	Nabidae	*Nabis capsiformis* (Germar)	Predatory bugs
	Lygaeidae	*Geocoris lubra* (Kirkaldy)	
	Pentatomidae	*Cermatulus nasalis* (Westwood)	
		Oechalia schellenbergii (Guerin-Meneville)	
	Reduviidae	*Coranus trisbratus* (Horvath)	
Neuroptera	Chrysopidae	*Chrysopa* spp.	Predatory lacewings
	Hemerobiidae	*Micromus tasmaniae* (Walker)	
Araneida	Lycosidae	*Lycosa* spp.	Spiders
	Oxyopidae	*Oxyopes* spp.	
	Salticidae	*Salticidae* spp.	
	Araneidae	*Araneus* spp.	

Source: Mensah (1999)

Risch 1984; Andow and Risch 1985; Bugg et al. 1989; Way and Heong 1994; Mensah and Khan 1997).

The predatory insects of *Helicoverpa* spp. in cotton that have been identified from the study plots in Australia are given in Table 9.3. Significant differences ($P < 0.001$) in numbers of predators were found among fields with cotton-lucerne intercropping. Significantly higher ($P < 0.001$) numbers of predatory beetles were found on lucerne strips than any of the crops. Significant difference ($P < 0.01$), with the exclusion of spiders, was found between the numbers of predatory beetles, bugs and lacewings recorded on cotton with and without lucerne strips. The maximum numbers of predatory beetles, bug and lacewings were recorded on the lucerne strips, followed by cotton with lucerne strips with the least on cotton without lucerne strips. In contrast, a number of spiders recorded from the lucerne strips and cotton with and without lucerne strips were not significantly different ($P > 0.05$) (Mensah 1999).

The use of synthetic food sprays to increase the abundance and impact of natural enemies of arthropod pests has been recognized for over 40 years. Nevertheless, artificial food sprays are applied in relatively few conservation biological control programmes, possibly because of inconsistent performance. The Wade et al. (2008) quantitatively reviewed 234 trials from 77 publications. The levels of assessment of food sprays most commonly found in the publications were the densities of arthropod pests (59 trials) and their natural enemies (124). Although the density of natural enemies increased in 108 of the 124 trials (or 87 % of cases) and pest populations declined in 28 of the 59 trials (or 47 %), increased profit was not demonstrated in the five trials where it was examined. The most commonly studied natural enemies belonged to the order Neuroptera (104 trials). Nevertheless, the parasitic Hymenoptera had the highest proportion of positive successes (56 of the 69 trials or 81 %).

Food sprays can also benefit predators and parasitoids, but this approach may be economically viable only in relatively high-value crops. In cotton, field testing of the product "Envirofeast" showed treated areas to be attractive to natural enemies including Coccinellidae and Melyridae (Coleoptera), Lygaeidae and Nabidae (Hemiptera) and Chrysopidae (Neuroptera) (Mensah 1996).

9.4 Resistant Genotypes

The cotton genotype FBRN 2–6, a combination of frego bract, red stem and nectaries, showed the least preface for oviposition by *Earias vittella* Fab. (0–1 egg/plant). Similarly experiments conducted at Central Institute for Cotton Research, Coimbatore, Tamil Nadu, India revealed that LK 861 and JGL 14515 showed resistance to whitefly, *B. tabaci* (5–6 adults/plant) (Natarajan and Sundaramurthy 1988). Such cotton cultivars can be cultivated or be used in field-tolerant/field-resistant cultivars that obviate the need for insecticide applications and will contribute to the conservation of natural enemies and other beneficials. Under Indian conditions, it has been observed that the egg-larval parasitoid, *Chelonus blackburni* Cameron, is effective against cotton bollworms and it should be conserved. Conservation biological control (CBC) aims at improving the efficacy of natural enemies and can contribute to safer and more effective biological control practices. Considerable progress in this field has been made during recent years, and it is therefore justified to review key findings in a special issue of biological control. The following topics, with primary emphasis on CBC of arthropods by arthropods, are (1) honeydew as a food source for natural enemies, (2) artificial food sprays, (3) shelter habitats, (4) natural enemy diversity and CBC and (5) CBC as provider of multiple ecosystem services.

Acknowledgements The authors thank Nagaraja T. and Raghava T. for select review and Prabhulinga T., Harish Badigere, Dr. Vishlesh Shankar Nagrare and Dr. V Chinna Babu Naik for lending select photos.

References

Abney RM, Sorenson EC, Bradley JR (2007) Alternate crop hosts as resistance management refuges for tobacco budworm, Heliothis virescens, (Lepidoptera: Noctuidae) in North Carolina. J Cotton Sci 11:35–39

Adjei-Maafo IK, Wilson LT (1983) Factors affecting the relative abundance of arthropods on nectaried and nectariless cotton. Environ Entomol 12:349–352

Altieri MA (1991) Increasing biodiversity to improve insect pest management in agro-ecosystems. In: Hawksworth DL (ed) Biodiversity of microorganisms and invertebrates: it's role in sustainable agriculture. CAB Int, Wallingford, p 302

Andow DA, Risch SJ (1985) Predation in diversified agroecosystems: relations between a coccinellid predator *Coleomegilla maculata* and its food. J Appl Ecol 22:357–372

Baggen LR, Gurrm GM (1998) The influence of food on *Copidosoma koehleri*, and the use of flowering plants as a habitat management tool to enhance biological control of potato moth. *Phthorimaea operculella.* Biol Control 11(1):9–17

Bal HK, Dhawan AK (2009) Effect of transgenic cotton on arthropod diversity under sprayed and unsprayed conditions in cotton agro-ecosystem. Pestic Res J 21(2):126–132

Beirne BP (1967) Pest management. CRC Press, Cleveland, p 123

Bentley BL (1977) Extrafloral nectaries and protection by pugnacious bodyguards. Ann Rev Ecol Syst 8:407–427

Bourzac Katherine (2006) Edible cotton. MIT Technology Review

Brévault T, Bikay S, Maldes JM, Naudin K (2007) Impact of a no-till with mulch soil management strategy on soil macrofauna communities in a cotton cropping system. Soil Tillage Res 97:140–149

Brown G, Pasini A, Benito NP, Aquino AM, Correia E (2001) Diversity and functional role of soil macrofauna communities in Brazilian no tillage agroecosytems. In: Lal R (ed) International symposium on managing biodiversity in agricultural ecosystems, Montreal. p 19

Brust GE, Stinner BR, McCartney DA (1985) Tillage and soil insecticide effects on predator-black cutworm (Lepidoptera: Noctuidae) interactions in corn agroecosystems. J Econ Entomol 78:1389–1392

Bugg RL, Ellis RT, Carlson RW (1989) Ichneumonidae (Hymenoptera) using extrafloral nectar of faba bean (*Vicia faba* L.: Fabaceae) in Massachusetts. Biol Agric Hortic 6:107–114

Burton S (1998) A history of India. Blackwell Publishing, Oxford, p 47

Cattaneo MG (2006) Farm scale evaluation of the impacts of transgenic cotton on biodiversity, pesticide use and yield. PNAS 103(20):7571–7576

Collins FL, Johnson SJ (1985) Reproductive response of caged adult velvetbean caterpillar and soybean looper to the presence of weeds. Agric Ecosyst Environ 14:139–149

Cortesero AM, Stapel JO, Lewis WJ (2000) Understanding and manipulating plant attributes to enhance biological control. Biol Control 17:35–49

Cox RD, Wermuth N (1993) Linear dependencies represented by chain graphs. Stat Sci Inst Math Stat 8(3):204 218

Dhillon KM, Gujar TG, Kalia V (2011) Impact of *Bt* cotton on insect biodiversity in cotton ecosystem in India. Pak Entomol 33(2):161–165

Dickson LL, Whitham TG (1996) Genetically-based plant resistance traits affect arthropods, fungi, and birds. Oecologia 106:400–406

Dutcher JD (2007) A review of resurgence and replacement causing pest outbreaks in IPM. In: Ciancio A, Mukerji KG (eds) General concepts in integrated pest and disease management. Springer, Dordrecht, pp 27–43

Dutton A, Klein H, Romeis J, Bigler F (2002) Uptake of *Bt*-toxin by herbivores feeding on transgenic maize and consequences for the predator *Chrysoperla carnea.* Ecol Entomol 27:441–447

Dutton AL, Obrist, D'Alessandro M, Diener L, Mu J, Romeis M, Bigler F (2004) Tracking *Bt*-toxin in transgenic maize to assess the risks on non-target arthropods. IOBC WPRS Bull 27:57–63

Ellsworth PC, Fournier A, Smith TD (2007) Arizona cotton insect losses, publication AZ1183. College of Agriculture and Life Sciences, Cooperative Extension, University of Arizona, Tucson

England S, Evans EW (1997) Effects of pea aphid (Homoptera: Aphididae) honeydew on longevity and fecundity of the alfalfa weevil (Coleoptera: Curculionidae) parasitoid, *Bathyplectes curculionis* (Hymenoptera: Ichneumonidae). Environ Entomol 26:1437–1441

Fitt GP (1989) The ecology of Heliothis in relation to agroecosystems. Ann Rev Entomol 34:17–52

Gillespie RG, New TR (1998) Compatibility of pest management and conservation strategies. In: Pest management—Future challenges: Proceedings of 6th Australian applied entomology research conference 29 Sept–2 Oct. vol 2, ed. MP Zaluki, RAI Drew, GG White, Brisbane: University of Queensland. pp 356

Goller E, Nunnenmacher L, Goldbach HE (1997) Faba beans as a cover crop in organically grown hops: influence on aphids and aphid antagonists. Biol Agric Hortic 15:279–284

Gurr GM, Van Emden HF, Wratten SD (1998) Habitat manipulation and natural enemy efficiency: implications for the control of pests. In: Barbosa P (ed) Conservation biological control. Academic, San Diego, pp 155–183

Hagen KS, Hale R (1974) Increasing natural enemies through use of supplementary feeding and non-target prey. Proceedings Summer Inst. for Biological control of plant insects and diseases. Mississippi Univ. Press, Jackson, p 647

Hardin MR, Benrey B, Coll M, Lamp WO, Roderick GK, Barbosa P (1995) Arthropod pest resurgence: an overview of potential mechanisms. Crop Prot 14:3–18

Head G, Moar W, Eubanks M, Freeman B, Ruberson J, Hagerty A, Turnipseed S (2005) A multi-year, large-scale comparison of arthropod populations on commercially managed *Bt* and Non-*Bt* cotton fields. Environ Entomol 34:1257–1266

Henneberry TJ, Bariola LA, Kittock DL (1977) Nectariles cotton: effect on cotton leafperforator and other cotton insects in Arizona. J Econ Entomol 70:797–799

Horton D, Bellinger B, Pettis GV, Brannen PM, Mitchum WE (2005) Pest management strategic plan for eastern peaches. USDA Agricultural Research Service/CSREES [Cooperative State Research, Education, and Extension Services]. http://www.ipmcenters.org/pmsp/pdf/EasyPeach.pdf; http://www.cottonspinning.com/grow/plantingcottonseeds.html "Planting Cotton Seeds"

Janaki Krishna PS (2006) Effects of transgenic cotton on biodiversity, pesticide use and yield. ISB News Report

Jervis MA, Kidd NAC (1986) Host feeding strategies in hymenopteran parasitoids. Biol Rev 61:395–434

Jervis MA, Kidd NAC, Fitton MG, Huddleston T, Dawah HA (1993) Flower visiting by hymenopteran parasitoids. J Nat Hist 27:67–105

Khan MQ, Rao VP (1960) Insect and mite pest on cotton in India. Indian Cent Cotton Common, Bombay, pp 207–301

Koptur S (1992) Extrafloral nectary-mediated interactions between insects and plants. In: Bernays E (ed) Insect–plant interactions, vol IV. CRC Press, Boca Raton, pp 81–129

Kumar R, Kranthi S, Nitharwal M, Jat SL, Monga D (2012) Influence of pesticides and application methods on pest and predatory arthropods associated with cotton. Phytoparasitica 40:417–424

Lal R (1988) Effects of macrofauna on soil properties in tropical ecosystems. Agric Ecosyst Environ 24:101–116

Longley M, Sotherton NW (1997) Factors determining the effects of pesticides upon butterflies inhabiting arable farmland. Agric Ecosyst Environ 61:1–12

Loranger G, Ponge JF, Blanchart E, Lavelle P (1999) Influence of agricultural practices on arthropod communities in a vertisol (Martinique). Eur J Soil Biol 34:157–165

Lu Y, Wu K, Jiang Y, Guo Y, Desneux N (2012) Widespread adoption of *Bt* cotton and insecticide decrease promotes biocontrol services. Nature 487(7407):362–365

McIntyre S (1994) Integrating agricultural land-use and management for the conservation of a native grassland flora in a variegated landscape. Pac Conserv Biol 1:236–244
Mendelsohn M, Kough J, Vaituzis Z, Matthews K (2003) Nat Biotechnol 21(9):1003–1009
Mensah RK (1996) Suppression of *Helicoverpa* spp. (Lepidoptera: Noctuidae) oviposition by use of the natural enemy food supplement Envirofeast (R). Aust J Entomol 35:323–329
Mensah RK (1999) Habitat diversity: implications for the conservation and use of predatory insects of Helicoverpa spp. in cotton systems in, Australia. Int J Pest Manag 45(2):91–100
Mensah RK, Harris WE (1995) Using Envirofeast spray and refugia technology for cotton pest control. Aust Cotton Grow 16:30–33
Mensah RK, Khan M (1997) Use of *Medicago sativa* (L.) interplantings/trap crops in the management of the green mirid, *Creontiades dilutus* (Stål) in commercial cotton in Australia. Int J Pest Manag 43:197–202
Natarajan K, Sundaramurthy VT (1988) Management of *Bemisia tabaci* in cotton system through host plants resistance. In: National symposium on integrated pest control- progress and perspectives. pp 112–115
O'Dowd D (1979) Foliar nectar production and ant activity on a neo-tropical tree, *Ochroma pyramidale*. Oecologia 43:233–248
Peng RK, Christian K, Gibb K (1998) The effect of non-crop vegetation on the insect pests and their natural enemies in cashew (*Anacardium occidentale* L.) plantations. Plant Prot Q 13:16–20
Perlak FJ, Oppenhuizen M, Gustafson K, Voth R, Sivasupramaniam S, Heering D, Carey R, Ihrig RA, Roberts JK (2001) Development and commercial use of Bollgard cotton in the USA nearly promises versus today's reality. Plant J 27:489–501
Pimentel D (1961) Species diversity and insect population outbreaks. Ann Entomol Soc Am 54:76–86
Powell W (1986) Enhancing parasite activity within crops. In: Waage JK, Greathead D (eds) Insect parasitoids. London Academic Press, London, p 389
Qaim M, de Janvry A (2005) *Bt* cotton and pesticide use in Argentina: economic and environmental effects. Environ Dev Econ 10:179–200
Qaim M, Zilberman D (2003) Yield effects of genetically modified crops in developing countries. Science 299:900–902
Rands MRW (1985) Pesticide use on cereals and survival of grey partridge chicks: a field experiment. J Appl Ecol 22:49–54
Rao CN, Shivankar VJ, Singh S (2003) Secondary pest resurgence. Encycl Pest Manag 1:1–5
Rico-Gray V (1993) Use of plant-derived food resources by ants in the dry tropical lowlands of coastal Veracruz, Mexico. Biotropica 25:301–315
Ripper WE (1956) Effect of pesticides on balance of arthropod populations. Annu Rev Entomol 1:403–408
Risch SJ, Andow D, Altieri MA (1983) Agroecosystem diversity and pest control: data, tentative conclusions, and new research directions. Environ Entomol 12(3):625–629
Ryszkowski L, Karg J, Margalit G, Paoletti MG, Zlotin R (1993) Above-ground insect biomass in agricultural landscapes of Europe. In: Bunce RG, Hl R, Paoletti MG (eds) Landscape ecology and agroecosystems. Lewis, Boca Raton, pp 71–82
Schuster MF, Calderon M (1986) Interaction of host plant resistant genotypes and beneficial insects in cotton ecosystems. In: Boethel DJ, Eikenbary RD (eds) Interactions of plant resistance and parasitoids and predators of insects. Ellis Horwood Limited, Chichester, pp 84–97
Siemann E, Tilman D, Haarstad J, Ritchie M (1998) Experimental tests of the dependence of arthropod diversity on plant diversity. Am Nat 152(5):738–750
Sisterson SM, Biggs WR, Olson C, Carrière Y, Dennehy JT, Bruce ET (2004) Arthropod abundance and diversity in *Bt* and non-*Bt* cotton fields. Environ Entomol 33(4):921–929
Sohi GS (1964) Pests of cotton. In: Entomology in India. Entomology society of India, pp 111–148

Southwood TRE, Way MJ (1970) Ecological background to pest management. In: Rabb RL, Guthrie FE (eds) Concepts of pest management. North Carolina State University Press, Raleigh, pp 6–29

Speight MR (1983) The potential of ecosystem management for pest control. Agric Ecosyst Environ 10:183–199

Stamps WT, Linit MJ (1998) Plant diversity and arthropod communities: implications for temperate agroforestry. Agrofor Syst 39:73–89

Stewart SD, Adamczyck JJ, Knighten KS, Davis FM (2001) Impact of *Bt* cotton expressing one or two insecticidal proteins of *Bacillus thuringiensis* Berliner on growth and survival of noctuid (Lepidoptera) larvae. J Econ Entomol 94:752–760

Tew TE, MacDonald DW, Rands MRW (1992) Herbicide application affects microhabitat use by arable wood mice (*Apodemus sylvaticus*). J Appl Ecol 29:532–539

Theunissen J, Booji CJH, Lotz LAP (1995) Effects of intercropping white cabbage with clovers on pest infestation and yield. Entomol Exp Appl 74:7–16

Thirtle C, Beyers L, Isma¨el Y, Piesse J (2003) Can GM-technologies help the poor? The impact of *Bt* cotton in Makhathini Flats, KwaZulu-Natal. World Dev 31:717–732

Topham M, Beardsley JW Jr (1975) Influence of nectar source plants on the New Guinea sugarcane weevil parasite, *Lixophaga sphenophori* (Villeneuve). Proc Hawaii Entomol Soc 22:145–155

Torres BJ, Ruberson RJ (2005) Canopy- and ground-dwelling predatory arthropods in commercial *Bt* and non-*Bt* cotton fields. Patterns Mech Environ Entomol 34:1242–1256

Torres JB, Ruberson JR (2007) Abundance and diversity of ground-dwelling arthropods of pest management importance in commercial *Bt* and non-*Bt* cotton fields. Ann Appl Biol 150(1):27–39

Treacy MF, Benedict JH, Walmsley MH, Lopez JD, Morrison RK (1987) Parasitism of bollworm (Lepidoptera: Noctuidae) eggs on nectaried and nectariless cotton. Environ Entomol 16:420–423

University of California (1996) Integrated pest management for cotton in the western region of the United States. University of California, Division of Agriculture and Natural Resources, Oakland, Publication 3305

Van Emden HF (1990) Plant diversity and natural enemy efficiency in agroecosystems. In: Mackauer M, Ehler LE, Roland J (eds) Critical issues in biological control. Intercept, Andover, p 330

Van Emden HF, Williams GF (1974) Insect stability and diversity in agroecosystems. Annu Rev Entomol 19:455–475

Wade MR, Zalucki MP, Wratten SD, Robinson KA (2008) Conservation biological control of arthropods using artificial food sprays: current status and future challenges. Biol Control 45(2):185–199

Wang Z, Lin H, Huang J, Hu R, Rozelle S, Pray C (2009) *Bt* cotton in China: are secondary insect infestations offsetting the benefits in farmer fields? Agric Sci China 8:83–90

Way MJ (1966) The natural environment and integrated methods of pest control. J Appl Ecol 3:29–32

Way MJ, Heong KL (1994) The role of biodiversity in the dynamics and management of insect pests of tropical irrigated rice – a review. Bull Entomol Res 84:567–587

Wetzler RE, Risch SJ (1984) Experimental studies of beetle diffusion in simple and complex crop habitats. J Anim Ecol 53:1–9

Whitcomb WH, Bell K (1964) Predaceous insects, spiders, and mites of Arkansas cotton Þelds. Agric. Exp. Sta., Univ of Arkansas, Fayetteville, AR Bull pp 690

White TCR (1984) The abundance of invertebrate herbivores in relation to the availability of nitrogen in stressed food plants. Oecologia 63:90–105

Wratten SD, Van Emden HF (1995) Habitat management for enhanced activity of natural enemies of insect pests. In: Glen DM, Greaves MP, Anderson HM (eds) Ecology and integrated farming systems. Wiley, Chichester, p 329

Xia J (1997) Biological control of cotton aphid (Aphis gossypii Glover) in cotton (inter) cropping systems in China; a simulation study. PhD diss. Landbouwuniversiteit, Wageningen, p 173

Zhang DJ, Lu ZY, Liu JX, Li CL, Yang MS (2015) Diversity of arthropod community in transgenic poplar-cotton ecosystems. Genet Mol Res 14(4):15713–15729

Arthropod Biodiversity on Jute and Allied Fibre Crops

10

K. Selvaraj, B.S. Gotyal, S.P. Gawande, S. Satpathy, and S.K. Sarkar

Abstract

Biotic stresses, particularly insect attacks, adversely affect the yield potential and the fibre quality of jute and allied fibre crops. Insect and mite pests attack these plants at seedling, growth and fibre development stages. Moreover, fibre crops have witnessed the effect of the gradual shift in the climatic pattern in terms of increased diversity and intensity of biotic stresses. The pest outbreaks in these crops have become more frequent which elevated the minor pests to the status of major pests. Considering the occasional, low-input management strategy adopted in these crops, the enhanced pest status of the existing pests, greater severity and the report of many new insect pests indicates the role of various biotic and abiotic stresses in triggering this dynamism in the pest scenario. The report of cotton mealybug, gram pod borer and safflower caterpillar in jute indicates the emerging new pests which may cause economic damage in future. The infestation of yellow mite, *Polyphagotarsonemus latus* in jute, and Bihar hairy caterpillar, *Spilosoma obliqua* in all fibre crops, is more regular and so severe that the situation may reach the status of outbreaks. The scenario is more or less same in the allied fibre crops. Published information on arthropods of jute and allied fibre crops is meagre. A few natural enemies and pollinators have been listed on fibre crops, but the attempt to utilize them is wanting. Considering the cost of cultivation and profitability, it is important to develop a low-cost, easily adoptable integrated pest management (IPM) technology for these crops including natural enemies.

K. Selvaraj (✉) • B.S. Gotyal • S. Satpathy • S.K. Sarkar
Division of Crop Protection, Central Research Institute for Jute and Allied Fibre Crops (CRIJAF), Barrackpore, Kolkata 700120, West Bengal, India
e-mail: selvaentomo@yahoo.co.in

S.P. Gawande
Ramie Research Station, Central Research Institute for Jute and Allied Fibre Crops (CRIJAF), Sorbhog 781317, Assam, India

A.K. Chakravarthy, S. Sridhara (eds.), *Economic and Ecological Significance of Arthropods in Diversified Ecosystems*, DOI 10.1007/978-981-10-1524-3_10

Keywords
Bihar hairy caterpillar • Climatic pattern • Jute • Ramie and yellow mite

10.1 Introduction

Jute (*Corchorus* spp.) and allied fibres such as tossa jute (*Corchorus olitorius* L.) and white jute (*C. capsularis* L.), kenaf (*Hibiscus cannabinus* L.), roselle (*H. sabdariffa* L.), sunnhemp (*Crotalaria juncea* L.), sisal (*Agave sisalana* Perr.) and ramie (*Boehmeria nivea* Gaud.) are called natural bast fibres. Fibre is extracted from the stem of jute, mesta, sunnhemp and ramie and from leaf of sisal. Except sisal and ramie, all others are annual crop whereas sisal and ramie are plantation crops of 6–10 years duration (Das and Mait 1998). The jute-producing countries are India, Bangladesh, China, Uzbekistan, Nepal, Vietnam, Zimbabwe, Thailand and Egypt, with India and Bangladesh holding first two positions (FAO 2014). In India, current area under these crops is 0.79 m ha with production 1.98 m tonnes of fibres during 2012–2013 (IJSG 2013) providing livelihood to 5 million people in farming, industry and trade involving 4 million farm families, 0.25 million industrial workers and 0.5 million traders in India. Raw jute farming alone supports about 10 million man-days of employment. The jute and mesta productivity has increased by twofolds during the last 50 years which was possible due to the concerted effort to develop high-yielding, premature flowering resistance, shorter duration jute varieties fitting well into the existing cropping system and appropriate policy interventions (Tripathi and Rama 1971; Mahapatra et al. 2009). There are several constraints in increasing the productivity of jute and allied fibre crops, of which the loss due to insect pests and diseases is of major concern. The infestation pattern and damage caused by these insect pests have indicated gradual shift. The status of pest and disease management, the study on biodiversity associated with agroecosystems has grown. So also the importance in ecology and conservation because maintenance of biodiversity is essential for ecological sustainability in agriculture (Pimentel et al. 1992; Perfecto et al. 1997). Secondly, arthropod community is important in cultivation, like in fibre crops, as it shows differences in composition, species richness and abundance. Several studies on arthropod diversity from tropical areas on cotton are available, but not on other fibre crops. Eleven insect and spider species were recorded on jute in West Bengal, India. Among the predators, there were coccinellids, staphylinid beetle, mirid bug, reduviid bug, dragonfly and damselfly (Rahman and Khan. 2009). But the task of collection, sampling and identification of arthropods is not easy and time consuming. Relevant experts for identification of arthropods are difficult to get. To start with, the use of the taxonomic family level for monitoring and biodiversity studies is considered acceptable (Basset et al. 2004). For the purpose of comparison, arthropods can be organised into functional groups, viz., phytophagous insects, natural enemies and other arthropods. Other arthropods include neither pests nor natural enemies. Although intervention in jute and allied fibre crops is marginal, recent outbreaks of few existing and newly emerging pests and diseases in jute, mesta, sunnhemp and ramie in different states have signalled gradual dynamism in the pest scenario in the traditional growing belts effecting fresh research impetus for sustainable management (Satpathy et al. 2014a, b).

10.2 Insect Pests

10.2.1 Jute (*Corchorus* spp.)

Jute is one of the most important commercial crops of eastern Indian states of West Bengal, Assam, Bihar, Orissa and eastern Uttar Pradesh. It is an annual plant belonging to the family Malvaceae under the genus *Corchorus*. Only two species of jutes are cultivated commercially, viz., *Corchorus olitorius* L. (tossa jute) and *C. capsularis* L. (white jute). From jute plants, cellulose fibres are extracted from the phloem tissues in the stem, after harvesting and retting. Fibre yield is directly correlated with plant height and basal girth. Pest problem is one of the major constraints responsible for low productivity of jute because the crop is damaged by more than 40 species of pests including insects and mites from seedling stage to harvesting of the crop which results in decline in yield as well as the quality of the fibres. About a dozen important pests are reported, of which jute stem weevil (*Apion corchori*), jute semilooper (*Anomis sabulifera*, *A. involuta*), yellow mite (*Polyphagotarsonemus latus*), Bihar hairy caterpillar (*Spilosoma obliqua*) and indigo caterpillar (*Spodoptera exigua*) are the most important pests regularly occurring everywhere (Das et al. 1999). Besides, Gram caterpillar (*Helicoverpa armigera*), safflower caterpillar (*Condica capensis*), green semilooper (*Amyna octa*), leaf webber (*Homona* sp. Tortricidae) and leaf miner (*Trachys pacifica*) have also emerged as insect pests in jute recently (Table 10.1). Tossa jute occupies 80 % of the jute-growing area as against 20 % by the white jute, unfortunately the incidence of major pests except stem weevil is more on tossa jute than on white jute (Saha 2000). It is estimated that under certain conditions, the loss of jute production due to pest damage alone may be as high as 12 %. However, the intensity of damage varies in different areas and in different years. The incidence of pest infestation largely depends upon weather conditions, variety of jute, method of cultivation and the presence of predatory and parasitic organisms.

10.2.1.1 Semilooper, *Anomis sabulifera* and *A. involuta* Guenee (Noctuidae: Lepidoptera)

Anomis sabulifera is one of the most important foliage and specific pest of jute and occurs in all the jute-growing tracts. Crop loss due to this pest was estimated up to 22–42 % (Sheikh 2012). The magnitude of loss in fibre depends on the age of the crop and number of infestations during crop growth. Damage starts, in all cases, from unopened leaves in upper part of the plant which represents the most susceptible portion. Newly hatched larvae are very active and start voraciously feeding on the epidermal membrane of one side of the mesophyll, leaving the epidermal membrane intact. As the larvae grow bigger, holes become evident and the edges of the

Table 10.1 List of insect and mite pests' scenario in jute-based cropping system in India

Common name	Scientific name	Family	Order	Status
Insect pest				
Bihar hairy caterpillar	*Spilosoma obliqua*	Arctiinae	Lepidoptera	Major
Jute semilooper	*Anomis sabulifera*, *A. involuta*	Noctuidae	Lepidoptera	Major
Jute stem weevil	*Apion corchori*	Curculionidae	Coleoptera	Major
Mealybug	*Phenacoccus solenopsis*, *P. hirsutus*, *Ferrisia virgata*	Pseudococcidae	Homoptera	Major
Stem girdler	*Nupserha bicolor*	Lamiidae	Coleoptera	Minor
Grey weevil	*Myllocerus discolor*	Curculionidae	Coleoptera	Minor
Indigo caterpillar	*Spodoptera exigua*	Noctuidae	Lepidoptera	Minor
Leaf miner	*Trachys pacifica*	Buprestidae	Coleoptera	Minor
Non-insect pest				
Yellow mite	*Polyphagotarsonemus latus*	Tarsonemidae	Acarina	Major
Red mite	*Tetranychus bioculatus*	Tarsonemidae	Acarina	Minor
New and emerging pest				
Gram caterpillar	*Helicoverpa armigera*	Noctuidae	Lepidoptera	Minor
Safflower caterpillar	*Condica capensis*	Noctuidae	Lepidoptera	Minor
Green semilooper	*Amyna octa*	Noctuidae	Lepidoptera	Minor
Hairy caterpillar,	*Dasychira mendosa*, *Euproctis chrysorrhoea*	Lymantriidae	Lepidoptera	Minor
Leaf-folding caterpillar/leaf webber	*Homona* sp.	Tortricidae	Lepidoptera	Minor

tender leaves are eaten, and serrated, diagonal cuts appear in apical leaves (Fig. 10.1). In severe attack the growing points are eaten and destroyed, the stems are totally defoliated and profuse branching occurs, and internodes are shortened resulting in reduction of fibre yield and quality. Pods and unripe seeds are also damaged by semilooper. The *C. olitorius* pods are more susceptible than *C. capsularis*. High forenoon RH for 15 days followed by drizzling during night hours and bright sunshine during daytime is more conducive for the outbreak of this insect during June to September months.

10.2.1.2 Stem Weevil, *Apion corchori* Marshall (Curculionidae: Coleoptera)

The jute stem weevil occurs in all the jute-growing tracts of India. This is the most harmful pest and causes loss at about 18 %. *C. capsularis* is more susceptible to stem weevil attack starting from seedling stage to harvesting than *C. olitorius*. Crops sown early are more susceptible than those sown comparatively late. The adults, both male and female, feed on jute foliage; small holes in the lamina due to such feeding indicate their presence. The nature of damage is one or more punctures

Fig. 10.1 Typical semilooper damage of serrated leaves symptom and larva of *Anomis sabulifera*

Fig. 10.2 Apion grub infestation: drying of top shoot and 'knot' symptom

by the female at the top nodes, where knot is formed, and there are corresponding numbers of grubs inside the plant. The destruction of the tissues by the grub results in withering and drying of the crown leaves just above the seat where the insect is concealed, and the leaves below remain unaffected (Fig. 10.2). Loss of apical meristem checks the vertical growth and thus affects the fibre both in terms of quality and quantity. In older plants mucilaginous substances produced around the tissues damaged by the grub bind the fibres together; the fibres break at these points during fibre extraction and results in 'knotty fibre' (Fig. 10.2). Cloudy damp weather associated with low daytime temperatures of both soil and air is congenial for incidence and multiplication during April to May. Early sown crops with more nitrogenous fertilisers suffer most.

10.2.1.3 Bihar Hairy Caterpillar, *Spilosoma obliqua* Walker (Noctuidae: Lepidoptera)

It is a polyphagous pest attacking several crops including jute, mesta, ramie and sunnhemp. However, jute is the preferred host than mesta. It was once considered as a sporadic and irregular pest on jute, but nowadays in high rainfall areas, it is a regular and major pest on jute. Young larvae feed gregariously and scrap the chlorophyll

Fig. 10.3 Bihar hairy caterpillar infested jute plants

content mostly under the surface of the leaves (Fig. 10.3). Later they disperse to the entire field and prefer to defoliate the older leaves; particularly, the third and fourth stages feed voraciously on the jute leaves and may completely skeletonise the jute plant. In later stages, the larvae eat the leaves from the margin. The leaves of the plant give an appearance of net or web, and under severe condition, complete defoliation may occur. *Tossa* jute is somewhat tolerant to the hairy caterpillar as compared to white jute. This pest attacks jute during June and continues till mid-September. They overwinter as hibernating pupae.

10.2.1.4 Mealybug, *Phenacoccus solenopsis* Tinsley and *P. hirsutus* (Pseudococcidae: Homoptera)

These mealybugs are highly polyphagous and occur almost on all the plants, and no plant is free from these pests as they have greater survival, high reproductive potential and shorter generation time. Earlier three species of mealybugs, i.e., *Maconellicoccus hirsutus*, *Ferrisia virgata* and *Pseudococcus filamentosus*, were reported to infest jute. Recently, for the first time cotton mealybug, *Phenacoccus solenopsis*, has been recorded on jute (*C. olitorius* and *C. capsularis*) and kenaf (*Hibiscus cannabinus* L. and *H. sabdariffa* L) in West Bengal (Satpathy et al. 2013). Apical meristem is the most susceptible part of the plant and presence of red/black ant for honeydew secretion is seen (Fig. 10.4). Plant infested during vegetative phase exhibits symptoms of distorted and bushy shoots, crinkled and/or twisted bunchy leaves and stunted plants that dry completely in severe cases (Fig. 10.4). The damage is mostly caused by the immature stages of mealybug which suck the sap. The vertical growth of plant is arrested with shortened internodes and the plant gives bushy appearance. Repeated attacks on the stem cause the development of crust due to which fibre bundles resist separation at the time of retting, resulting in the formation of 'barky fibre'. Late season infestation during reproductive stage of the crop results in reduced plant vigour and early crop senescence. The infestation of mealybug on tossa jute, particularly in the early crop growth stage of the plant during intermittent stretches of dry period, has been witnessed in many parts of South Bengal.

Fig. 10.4 Mealybug infested jute plants, severely infested jute field

Fig. 10.5 Grey weevil infested jute plant and adult weevil

10.2.1.5 Grey Weevil, *Myllocerus discolor* Boheman (Curculionidae: Coleoptera)

This is a highly polyphagous pest attacking several cultivated crops. Earlier this pest was reported as a minor pest of jute, but in recent days, it emerged as one of the major insect pests of jute. But in jute crop, it attacks *olitorius* jute; *capsularis* jute is resistant to this pest. The grey weevil (*Myllocerus discolor*) became important recently which attacks only *C. olitorius* jute. Adult weevils attack 30–45-day-old plants and causes considerable damage to *tossa* jute by defoliating the crop. Unopened tender apical leaves are mostly preferred. The adult feeds on leaves by making irregular holes on the apical leaves (Fig. 10.5). Grubs are voracious feeder of roots causing stunted plant growth. White jute varieties are immune to the attack of grey weevil. One or two pre-monsoon showers, temperature range of 35–39 °C and RH 85–94 % are conducive for the maximum incidence of grey weevil. April to May is active period of attack.

Fig. 10.6 Indigo caterpillar infested jute plants and skeletonised leaf

10.2.1.6 Indigo Caterpillar, *Spodoptera exigua* Hubner (Noctuidae: Lepidoptera)

Popularly known as indigo caterpillar, highly polyphagous pest sporadically assumes destructive nature in the early sown jute crop where its activity is confined to seedling or young jute plants. Occurrence is more prevalent in jute seed crop, active throughout the crop growth stages. Their infestation causes twisting of top leaves and stunted plant growth. The extent of damage may go up to 20 % yield loss. The young larvae after hatching web, either at the margin of the same leaf or two top leaves, shelter inside and start voraciously feeding on the green matters. The damage is noticed even from distance especially border plants (Fig. 10.6). Within these webs the young larvae live gregariously only for 2 or 3 days and thereafter they separate and spread out. The feeding activity of a grown-up larva is generally confined to a few morning hours and late evening. They are very voracious and quite large patches of foliage are quickly stripped and top plants are webbed together. The leaves are skeletonized; the older caterpillars often devour the entire lamina (Fig. 10.6). March to April is the peak period of infestation.

10.2.1.7 Stem-Girdling Beetle, *Nupserha bicolor* Dutt (Lamiidae: Coleoptera)

About 30 % loss is caused by this pest in young plants and grown-up plants suffer least. The adult beetle girdles the stem at two levels, 1–1.4 cm apart, before it starts oviposition. Then three punctures or slits are made within the two girdles, and the middle slit is used to lay a single egg. Because of this, nutritional transfer is arrested, the stem above the lower girdle dies and dries up (Fig. 10.7). Branching takes place below the lower girdle of such stem. After hatching the larva feeds on pith tissue, moves downwards along the central hollow and pupates within stem. Ovipositional damage by female is more than the larval feeding. High humidity or rainfall breaks the larval diapause during March–April.

Fig. 10.7 Stem girdling beetle infested jute plants and dried jute plant

Fig. 10.8 *Yellow* mite infested jute plant

10.2.1.8 Yellow Mite, *Polyphagotarsonemus latus* Banks (Tarsonemidae: Acarina)

Yellow mite is one of the important destructive pests of jute. It is widely distributed in all the jute-growing tracts of India. The yield loss due to this pest has been estimated between 10 % and 42 % depending upon the level of infestation may reach 20–90 %. *Olitorius* jute suffers more than *capsularis* jute by mite. The fibre strength is reduced to 12.13 g/tex in infested plants as against 15.43 g/tex. Both nymphs and adults suck the sap from the ventral surface of young leaves even before they are unfolded. The midrib curves downwards and the lamina roll inwards from two sides. The secondary veins wrinkle and give the leaf a rough and crumpled look and do not grow to their full size. The infested leaves turning deep green with coppery brown shades and drop prematurely (Fig. 10.8). The vertical vegetative growth of the crop is arrested, internodes become shortened, and significant yield loss occurs regularly. The fibre quality and strength also deteriorates. *Tossa* jute suffers more due to yellow mite infestation than the white jute. A crop exposed to prolonged periods of high relative humidity with occasional drizzle favour incidence. These pests remain active throughout the year on different crops. High temperature and humidity with poor sunshine during May to August favour multiplication.

10.3 Emerging Insect Pests of Jute

10.3.1 Gram Caterpillar, *Helicoverpa armigera* (Noctuidae: Lepidoptera)

The gram caterpillar is cosmopolitan and widely distributed in India. It is reported on *tossa* jute (Selvaraj et al. 2013a, b). The pest was found to be defoliating, feeding and cutting the terminal succulent portion of the stem of about 65–70-day-old crop which resulted in drooping and drying of the stem that eventually reduced the yield (Fig. 10.9). In severe cases, larvae are found scooping the succulent stem, which results in wilting and drying (Fig. 10.9). The caterpillars when full grown are 3.5 cm in length, being greenish with dark broken grey lines along the sides of the body.

10.3.2 Safflower Caterpillar, *Condica capensis* (Noctuidae: Lepidoptera)

Safflower caterpillars occasionally feed on *tossa* jute. This caterpillar is often confused with gram pod borer, *Helicoverpa armigera*. This is one of the important insect pests of safflower in India. The larvae in the early stages bite holes in the leaves and they feed voraciously as they grow. The fully grown larvae are brown in colour and smooth, tapering towards the posterior region with a network of brown lines on head and slightly hampered anal segment, and have creamy white line running on dorsolateral side (Fig. 10.10), and it measures about 25 mm.

Fig. 10.9 *Helicoverpa armigera* damaged jute field and scooping symptom

Fig. 10.10 *C. capensis* larva and their dorsolateral view

Fig. 10.11 *Green* semilooper larva feed on jute plant and prepupal larvae

10.3.3 Green Semilooper, *Amyna octa* (Noctuidae: Lepidoptera)

Synonymously called *Amyna axis*—unlike the jute semilooper (*Anomis sabulifera*), this green semilooper differs in polyphagous nature. This pest is reported to feed on other host plants such as jute, sweet potato, sunnhemp, mesta, *Urena lobata* and ramie (Pradhan and Chatterji 1978). The larvae feed voraciously making large irregular holes on the jute leaves. The *A. octa* larvae are green in colour, about 18–20 mm in length having setae on the body and white stripes on dorsolateral from anterior to posterior region (Fig. 10.11). When alarmed the larva essentially jumps from the host and continues to wreathe and wriggle wildly. Prior to pupa formation, larvae turn pinkish from green colour and body length contracts (Fig. 10.11). In this stage, they make leaf folds by joining two leaves with their webs and pupate there in.

10.3.4 Hairy Caterpillar, *Dasychira mendosa* and *Euproctis scintillans* (Lymantriidae: Lepidoptera)

The tussock caterpillar, *Dasychira mendosa*, and mesta hairy caterpillar, *Euproctis scintillans*, are polyphagous pests of wild and cultivated plants which is reported for the first time on jute (*Corchorus* spp.) from West Bengal (Selvaraj et al. 2015). It is

Fig. 10.12 (**a**) Tussock caterpillar on jute plant and its adult moth. (**b**) Mesta hairy caterpillar on jute and its adult moth

a sporadic pest of mesta crop in India. The pest is active throughout the year, but its activity is reduced in winter. Larvae are gregarious feeders and cause defoliation of jute leaves. The larvae have densely urticating hair and often have tufts of hairs on the dorsal aspects of certain segments (Fig. 10.12a). Adult is yellowish-brown moth. *E. scintillans* (Fig. 10.12a) larva has yellowish-brown head, a yellow dorsal stripe with a central red line on the body and tufts of black hairs dorsally on the first three abdominal segments (Fig. 10.12b). Adult is yellowish with spots on the edges of forewings and apart from a tuft of brown hairs at the end of the abdomen (Fig. 10.12b).

10.3.5 Leaf Webber, *Homona* sp. (Tortricidae: Lepidoptera)

This leaf webber is reported for the first time on jute (*Corchorus* spp.) from West Bengal (Ramesh Babu et al. 2015). The young larvae fold the leaf in such a manner that the margin of leaf blade comes together and then starts feeding from its edge (Fig. 10.13). When mature, they bind several leaves together to make a nest. The mature larvae are voracious feeders on the leaves, often leaving partly fed or dead leaves on plants. This feeding activity causes distortion of the leaves and young shoots and also defoliation, which can be seen from a distance. The adult moth is

Fig. 10.13 Leaf webber infested jute plant

brownish-yellow, small in size with a wing expanse of about 25–27 mm. The wings are held roof like over the body when the insect is at rest. The fore wings bear an oblique band and few transverse wavy lines.

10.3.5.1 Mesta (*Hibiscus* spp.)

Mesta fibre is obtained from two cultivated species, *Hibiscus sabdariffa* (roselle) and *H. cannabinus* (kenaf), and is used for blending with jute in the manufacture of jute goods. Leaves are used for preparing pickles, etc., and fleshy calyx for preparing natural dyes, soft drinks, jelly, etc. It is also a very important raw material for newsprint production. In India, both species are attacked by a number of major insect pests and diseases which were reviewed by Dempsey (1975). Although more than a dozen insect pests attack mesta in different regions, the spiral borer (*Agrilus acutus*), mealybug (*Phenacoccus solenopsis*, *Maconellicoccus hirsutus*), jassids (*Amrasca biguttula biguttula*), flea beetle (*Nisotra orbiculata*), leaf roller (*Haritalodes derogata*), semilooper (*Cosmophila erosa*), Bihar hairy caterpillar (*Spilosoma obliqua*), mesta hairy caterpillar (*Euproctis scintillans*) and aphid (*Myzus persicae*) are the most important ones, occurring regularly almost everywhere (Pandit and Pathak 2000). The major pests, however, have limited geographical distribution and do not occur in all the mesta-growing states in India.

10.3.6 Spiral Borer, *Agrilus acutus* Thumb. (Buprestidae: Coleoptera)

In India, *Agrilus acutus* is a serious pest of *Hibiscus cannabinus* and it also infests *H. sabdariffa*. The insect bores through the main stem of the plant. It forms rings on the main stem and the stem breaks at that portion. The fibre obtained from the infected plants become useless. The adult lays its eggs on the stem most frequently at the nodal region below the leaf base, and the larva on hatching burrows its way beneath the cambium layer and starts feeding upon the woody tissues, making a spiral around the stem beneath the bark. During feeding, the larva travels spirally throughout the entire length of the stem. A portion of the infested region swells up considerably to form an elongated gall (Fig. 10.14). The lignification of the cell wall of the sclerenchymatous tissue is affected by which stem at the level of gall becomes

Fig. 10.14 Spiral borer infested jute plant

very weak and breaks by a strong gust of wind. The portion of the stem above the region of the gall dies and dries up after the break. The galls are usually 9–15 cm long. The gall on *H. cannabinus* is usually 9–15 cm. long, the range being 3–30 cm, and it is initiated by the late-stage larva before it tunnels into the woody tissue to form the pupal chamber. It was found that crops sown during April are usually the most infested.

10.3.7 Flea Beetle, *Nisotra orbiculata* Mots (Curculionidae: Coleoptera)

The adult beetle is shiny black having the habit of quick jumping rather than flying (Fig. 10.15). This is one of the most important early season pests of mesta. Adult beetle's attack commences with week-old seedlings which are highly susceptible and ceases just before harvest. Resowing is warranted in case of heavy infestation. Eggs in clusters are found in the lower surface of the leaves. The larvae feed on roots. The shiny black adults are injurious to plants as they feed on the tender stem and leaves. Irregular cut and holes on the leaves are the typical damage symptoms. Intermittent showers followed by dry spell with high humidity were observed to be conducive for its multiplication.

Fig. 10.15 Flea beetle adult on leaf (Courtesy for adult photo: http://insecta.idv.tw)

Fig. 10.16 *Sylepta derogata* damaged mesta plants and the larva on leaf

10.3.8 Leaf Roller, *Sylepta derogata* (Pyralidae: Lepidoptera)

It is primarily a sporadic pest of cotton, jute (*Corchorus* spp.) and major pest of Congo jute (*Urena lobata*) in India. Besides, it was reported to feed on bhendi, eggplant, several *Hibiscus* spp. (including *H. cannabinus*, *H. esculentus*, *H. columnaris* and *H. rosa-sinensis*) and several other malvaceous plants. The larvae feed on the lower surface of leaves when they are young, and as they grow, they feed on the edges of leaves and roll inwards up to the midrib into a trumpet fastened by silken threads (Fig. 10.16), and a marginal portion of leaves are eaten away. The larvae remain inside the roll and feed outside the marginal portion of the leaves. They are seen in groups amidst faecal materials inside the folds and infestation spreads to neighbouring plants and hence the symptoms of the pests are patchy. The plants are defoliated in severe attack and plants along the field borders are more vulnerable for the attack. In the presence of a large number of leaf rolls, the plants become stunted ultimately. The larvae are glistening green in colour and semitranslucent with a dark brown head in early instars, but later becoming dark pink before pupation (Fig. 10.16).

Fig. 10.17 *Myzus persicae* damaged mesta plant and their nymphs

Full-grown larvae attain a length of about 15 mm, bright green (glistening) in colour with dark head and prothoracic shield. Pupation is reddish brown in colour and takes place inside the rolled leaf. Adult moths are medium sized with yellow wings having series of brown wavy markings.

10.3.9 Aphid, *Myzus persicae* (Sulzer) (Aphididae: Hemiptera)

Myzus persicae is known as the peach tree green aphid and is the most harmful aphid species associated with vegetables such as cabbage, beet and cauliflower including mesta. Greenish-yellow aphids are seen in colonies on the underside of tender leaves, stem and the pods (Fig. 10.17). Sometimes it is a serious problem especially during prolonged dry spells. Due to de sapping the plant loses its vitality. In severe cases curling of leaves, stunted growth and drying and death of the plants occur. Nymphs initially are greenish, but soon turn yellowish, greatly resembling viviparous (parthenogenetic) adults. Development can be rapid, often 10–12 days for a complete generation.

10.3.9.1 Ramie

Ramie fibre is one of the strongest, natural fine textile fibres in the world, obtained from the bark of the plant *Boehmeria nivea* L. Gaud. The strength, lustre and absorbance capacity of the fibre makes it a special one among the natural fibres. Ramie fibre is primarily used for blending with cotton and silk for its unique strength and absorbance. China is the major producer of ramie fibre contributing to 96.3 % of the global production (Mitra et al. 2013). In India, ramie cultivation is restricted to some pockets of Assam and North Bengal covering about 100 ha area. The infestation of insect pests and diseases in these crops has not been so alarming in India at present (Singh 1998). Among the insects, Bihar hairy caterpillar (*Spilosoma obliqua*), leaf roller (*Sylepta derogata*) and leaf-eating caterpillar (*Spodoptera exigua*) are observed from time to time (Mustafee 1977). However, incidence and severity of insect pests increased in the recent past due to continuous availability of host plants and change in climatic conditions which results in losses in fibre yield. There are six new insect pests, viz., Indian red admiral caterpillar (*Vanessa indica*), leaf

Fig. 10.18 Indian red admiral caterpillar damaged ramie plant (Courtesy for larvae: http://www.ifoundbutterflies.org)

folder (*Pleuroptya* sp.), leaf-eating beetle (*Pachnephorus bretinghami*), white grub (*Lepidiota* sp.), termite (*Microtermes* sp.) and mealybug (*Maconellicoccus hirsutus*) (Gawande et al. 2014, 2015). Details on Bihar hairy caterpillar, leaf-eating caterpillar and leaf roller were discussed on jute and mesta crop section.

10.3.10 Indian Red Admiral Caterpillar, *Vanessa indica* Herbst (Nymphalidae: Lepidoptera)

The *V. indica* was reported for the first time in India in 2011 at Assam as a pest on a ramie, and it is causing considerable damage (up to 10–50 %) due to feeding on young tender leaves (Gawande et al. 2014). *V. indica* is found in higher-altitude regions (above 2000 ft) of India including the Nilgiri Hills in southern India. Adult lays the eggs on tender leaves; upon hatching larvae feed on young leaves preferably at neck region, where the stalk is attached to the leaf, and fold them in such a way that both the margins get attached by a silky web (Fig. 10.18). They cut trenches at the base of host plant leaves and construct leaf-fold shelters. The function of trenching is to facilitate leaf folding, then larvae would cut trenches before folding leaves. Pupation takes place inside the folded leaf. The adult emerge as butterfly. The level of infestation is maximum during the month of December and January; later on it declines.

10.3.11 Leaf Folder, *Pleuroptya* sp. (Crambidae: Lepidoptera)

This insect species is also called the mother of pearl moth, leaf folder, bean webworm and pearl caterpillar. This pest was first reported feeding on ramie crop in India by Gawande et al. (2015). Initially larvae roll up leaf midrib around themselves as a shelter and attach the leaf margins together with silk strands (Fig. 10.19). The larvae

Fig. 10.19 Leaf folder damaged ramie field and their larva

are greenish yellow with dark spots along the body, very small and delicate (Fig. 10.19). They feed on the inner layers of leaf, and at later stage, preferably after third instars, the larva moves up and rolls another leaf. It pupates inside the folded/rolled leaf. During severe infestation leaf margins dry completely which affects plant photosynthesis and results in reduction in yield. Adults are medium-sized yellowish moth with numerous wavy lines on both wings. Maximum plant damage was observed during the months of October and November.

10.3.12 Leaf-Eating Beetle, *Pachnephorus bretinghami* Baly. (Chrysomelidae: Coleoptera)

The chrysomelid *P. bretinghami* is recorded for the first time as a pest of ramie in Assam, India (Gawande et al. 2013). The damage was recorded up to 30–40 %, and it was highest during the months of May to August. Nowadays this is the most important insect pest of ramie crop, causing considerable damage to ramie crop in Assam. Adults feed on the soft part of ramie leaves at night, with holes appearing on the leaves by feeding on plant tissue between leaf veins and leaving behind a lace-like pattern (Fig. 10.20). In severe cases most leaves are skeletonised and completely devoured. The adults are sturdy, pale golden-brown beetles, and very small in size. Its body is covered with fine white hairs that can give the beetle a greyish appearance (Fig. 10.20).

10.3.13 White Grub, *Lepidiota* sp. (Scarabaeidae: Coleoptera)

The outbreak of *Lepidiota* sp. was experienced in ICAR-CRIJAF Ramie Research Station (RRS), Sorbhog, Assam, since 2012. *Lepidiota* sp. is credited as the second species of white grub belonging to the genus '*Lepidiota*' reported from Assam after

Fig. 10.20 Leaf eating beetle damaged ramie plant and adult

Lepidiota mansueta which had appeared as an extremely serious key pest of many field crops in Majuli, Assam. Endemism of white grub species belonging to the genera '*Lepidiota*' is chiefly governed by the presence of river/rivulet/large water-bodies and light soil with high organic carbon content. Beetles were also found to feed on leaves of ramie, litchi, rubber, black pepper and areca nut in field conditions.

10.3.13.1 Sunnhemp

Sunnhemp crop is affected by a number of insect pests in the field starting from seedling to harvesting stage. Top shoot borer (*Laspeyresia tricentra* Meyr.), sunnhemp hairy caterpillar (*Utetheisa pulchella* Linn), sunnhemp weevil (*Alcidodes leopardus*), flea beetle (*Longitarsus belgaumensis* Jac) and Bihar hairy caterpillar (*Spilosoma obliqua*) are the major ones. Besides, there are a number of minor pests, which also cause considerable damage of the crop under certain specific conditions (Chaudhury et al. 1997; Sarkar et al. 2015). Details on Bihar hairy caterpillar have already been discussed on jute section.

10.3.14 Top Shoot Borer, *Laspeyresia tricentra* Meyr (Eucosmidae: Lepidoptera)

It is the most serious pest of sunnhemp. Late sown crop in June and July suffers most, while their incidence in early sown crop during April and May is significantly less resulting in higher fibre yield. The loss in fibre yield due to this pest varied from 11.5 % to 20.6 %. Further, it affects the pod and reduces the seed yield in seed crop up to 20 % (Prakash 2003). Apart from the direct loss, the insects also create the avenues for the infection of seed deteriorating mycoflora (Sarkar and Tripathi 2003; Sarkar 2007). The larvae initially bore into the top portion of the stem of young plants resulting in formation of characteristic gall, which affects further growth of the plant, and consequently, side branches appear. In addition to apical portion, larvae also bore the stem near the node of the plant and feed inside, and as a result, the affected stems swell up into a gall. The fibre obtained from such plant is short, coarse and specks, which are highly undesirable qualities. The attack of top shoot borer greatly reduces the quality of the fibre by making gall at the apical region of the stem, which stops further growth and induces branching. Due to gall formation, the fibre continuity is disrupted and the lateral branches produce fibre of shorter length and weaker in strength. Finally when pod formation takes place in seed crop, the larvae bore into the pod and feed upon the seeds. Thus due to its triple nature of damages, this pest is considered to be the most serious.

10.3.15 Sunnhemp Hairy Caterpillar, *Utetheisa pulchella* (Arctiinae: Lepidoptera)

It is also one of the serious pests of sunnhemp occurring all over India and it generally attacks sporadically, but sometimes in hot summer, it becomes a serious problem in North India. The young caterpillars are gregarious in nature and feed on upper foliage of the plants. The caterpillars cause two types of damage. Initially, they feed voraciously on the foliage and skeletonise them completely (Fig. 10.21), but at later stage, during pod formation, they start to migrate upwards and bore into pods and eat away the unripe seeds. As the crop matures, the caterpillars feed by thrusting the head in and leaving the rest of the body exposed. Pupation takes place either in the leaves' fold or in the soil. In this way it causes considerable loss in fibre yield as well as seed production.

10.3.16 Sunnhemp Flea Beetle, *Longitarsus belgaumensis* Jac (Chrysomelidae: Coleoptera)

Sunnhemp flea beetle is a minor pest of sunnhemp. The adult generally feeds on tender leaves at apical parts of plants, and as a result, several small elongated holes

Fig. 10.21 *U. pulchella* larva feed on leaves and on pod (Courtesy for larva on pod photo: NBAIR)

Fig. 10.22 Sunnhemp flea beetle damaged leaves and adult

of 2–4 mm in length appear on the leaves (Fig. 10.22). The larva after hatching enters into the root and cotyledon of germinated seed and thereby feeds by making a tunnel.

10.4 Predators and Parasitoids

10.4.1 Jute Crop-Based Ecosystem

The immense value of natural enemies in pest suppression has been understood, and a renewed interest is seen in pest management through biological agents in different parts of India (Kalita and Borah 1993; Rahman and Khan 2009; Sarma et al. 2010). Besides, occurrence of two braconid parasitoids, i.e., *Protapanteles obliquae* and *Meteorus spilosomae* on *Spilosoma obliqua*, was reported by Selvaraj et al. (2012, 2013a, b). The relative abundance of natural enemies is determined on the basis of the extent of parasitisation and predation by predators (Table 10.2).

Table 10.2 Natural enemies of jute pests in jute-based ecosystem in India

Parasitoid name	Host insect/s	Taxonomic position	Host stage	Remarks
Jute				
Protapanteles obliquae	*Spilosoma obliqua*	Braconidae: Hymenoptera	Larva	24–38 %
Meteorus spilosomae	*Spilosoma obliqua*	Braconidae: Hymenoptera	Larva	28–25 %
Blepharalla lateralis	*Spilosoma obliqua*	Tachinidae: Diptera	Larva	Moderately abundant
Aenasius bambawalei	*Phenacoccus solenopsis, P. hirsutus,*	Encyrtidae: Hymenoptera	Nymph	15–32 %
Parachremylus sp.	*Trachys pacifica*	Braconidae: Hymenoptera	Grub	2–10 %
Sisiropa formosa, Trichologa sorbilans	*Anomis sabulifera, A. involuta*	Hymenoptera	Larvae	–
Entedon manilensis, E. urozonus	*Apion corchori*	Eulophidae: Hymenoptera		50–94 %
Apanteles spp.	*Anomies* sp.	Braconidae: Hymenoptera	Larvae	70 %
Trichogramma minutum, Tetrastichus Howardi	*Anomies* sp.	–	Eggs, pupae	Moderately abundant
Predators				
Harmonia octomaculata	*Polyphagotarsonemus latus*	Coccinellidae: Coleoptera	Egg, nymph and adult	Less abundant
Menochilus sexmaculatus	*Polyphagotarsonemus latus*	Coccinellidae: Coleoptera	Egg, nymph and adult	Less abundant
Coccinella septempunctata	*Polyphagotarsonemus latus*	Coccinellidae: Coleoptera	Egg, nymph and adult	Less abundant
Holobus kashmiricus	*Polyphagotarsonemus latus*	Coccinellidae: Coleoptera	Egg, nymph and adult	Less abundant
Trombidium holosericeum	*Polyphagous*	Trombidiidae: Trombidiformes	–	Less abundant

Larval parasitoid, *Protapanteles obliquae***:** The full-grown parasitic larvae emerged out through the ventrolateral body region of the host insect larva (mostly second to third instars) (Fig. 10.23a, b). Immediately after exiting from the host, each grub started spinning a cocoon and soon a compact mass of milky white cylindrical cocoons appeared on the side of the host larva. It was a gregarious endoparasitoid, specific to *S. obliqua*. The activity of this parasitoid was noticed from mid-May to mid-July during the cropping season. The early instars (up to third

Fig. 10.23 *Spilosoma obliqua* larvae parasitised by *Protapanteles obliquae* (**a, b**), *Meteorus spilosomae* (**c**), *Phenacoccus solenopsis* parasitised by *Aenasius bambawalei* (**d, e**), *T. pacifica* parasitised by *Parachremylus* sp. (**f, g**) and some Coccinellidae adult and grub (**h–j**) (Courtesy for (**e**) Moazzem Khan; for **g** Photographer Direct.com)

instars) of *S. obliqua* were more vulnerable to this parasitoid. Hence, insecticidal interference may be avoided during the early instar stages (Selvaraj et al. 2013a, b).

Larval parasitoid *Meteorus spilosomae***:** Potential larval endoparasitoid and very specific to *S. oblique* (Fig. 10.23c). The apodous grub that emerges from the host insect body (mostly third to fourth instars) followed by forms pupal cocoons. The pupal cocoon measures about 5–6 mm in length with a maximum width of 2.2 mm. The silken thread that attaches the cocoon with the host insect is 4–6 cm in length. Fully developed cocoons are honey brown in colour and the adult parasitoid emerges in 5–7 days. It is solitary, koinobiont in nature (Selvaraj et al. 2012).

Mealybug parasitoid, *Aenasius bambawalei***:** It is a potential nymphal parasitoid of mealybug and the extent of average parasitisation ranges from 15 % to 32 % with peak activity from the second fortnight of June to late August (Fig. 10.23d, e) (Satpathy et al. 2014a, b).

Fig. 10.24 Photo (1) Xylocopa fenistroides. Photo (2) *Megachile lanata*

Parasitoid, ***Parachremylus*** **sp. (Braconidae: Hymenoptera):** The genus *Parachremylus* sp. is a larval parasitoid on jute leaf miner, *Trachys pacifica* (Selvaraj et al. 2014). The complete host range of *Parachremylus* has not been known yet. However, the members of related genera of the tribe Avgini (*Parahormius*, *Avga*, *Allobracon*) are recorded as parasitoids of the leaf rollers or leaf miners of the families (Fig. 10.23f, g).

Sunnhemp crop-based ecosystem: In India the parasites of top shoot borer were reported by Reddy (1956) and Ram (1968). Five parasites, namely, *Apanteles teragamae* V, *Cremastus* (*Trathala*) sp., *Goniozus* sp., *Elasmus homonae* F. and *Sphyracephala hearciana* W, were reported from Uttar Pradesh (Ram 1968). The extent of parasitisation was 3.0–57.5 % in different months. David and Kumarswami (1960) recorded a tachinid parasite, *Drino* (*Prosturmia*) *inconspicua* M., on this caterpillar and Ayyar (1963) reported a braconid, *Bracon brevicornis* W, and a tachinid parasite, *Padomyia setosa* D., from South India.

Pollinators: Sunnhemp is propagated only through the seed. It is an obligatory cross-pollinated crop although self-compatible strains were also reported. Different techniques including the normal selfing were followed for effective autogamy, but there was no pod formation in any one of them. Few pods, which may set by brushing method, are due to contamination of foreign pollen during the operation of the technique. A large number of insect visitors are noticed during the flowering season including honeybees and bumblebees. But three insect species, namely, *Xylocopa fenistroides*, *X. latipies* and *Megachile lanata*, are largely responsible for pollination of sunnhemp flower. *Apis florea* and *A. indica*, although they visit the flower, are not effective pollinators because of their lower body weight. A refugium with bee forage crops is required to sustain the activities of generalist pollinators.

10.4.2 Soil Scavengers

The field cricket and mole cricket are more predominant in jute ecosystem. However, both are reported to feed on jute plants.

Field cricket, ***Brachytrupes portentosus*****:** Also known as burrowing cricket. It is not only destructive for jute by digging the soil at root zone but also plays a major role in delivering ecosystem services.

Mole cricket, ***Gryllotalpa orientalis*** **Burmeister:** These insects make holes in the land and live inside the holes at day. At night, they come out from the holes, cut the base of the jute seedlings and keep the cut seedlings inside the holes. Consequently, the jute field sometimes becomes vacant. The infestation becomes higher when there is no rainfall, and the infestation becomes lower after rainfall. The adult insects eat the root and the base of the stem of jute plants. Thus jute and allied fibres suffer due to the ravages caused by a number of insect pests. The changing cropping sequence, presence of favourable alternate hosts during and prior to the cropping season and persistence of slightly elevated temperature are the reasons for such erratic and increasing pest status in these crops. Since there is no attempt on the utility of natural enemies and other taxa for pest management, research must be initiated in their direction. The recurrent infestation of cotton mealybug, *Phenacoccus solenopsis*, in jute and mesta crop in South Bengal and Northern Andhra Pradesh has confirmed the increasing pest status of mealybug particularly during the dry, hot period in the early crop growth stage. The elevated pest status of Bihar hairy caterpillar (BHC), *Spilosoma obliqua,* is evident from few outbreaks in jute and sunnhemp, and hence there may be a chance of these getting established on these crops or may even spread to other economically important crops. So the surveillance, development and validation of integrated pest management strategies are most urgent.

References

Ayyar TVR (1963) Hand book of economic entomology for South India. Govt of Madras, Chennai, pp 235–237

Basset Y, Mavoungou JF, Mikissa JB, Missa O, Miller SE, Kitching RL, Alonso A (2004) Discriminatory power of different arthropod data sets for the biological monitoring of anthropogenic disturbance in tropical forests. Biodivers Conserv 13:709–732

Chaudhury J, Singh DP, Hazra SK (1997) Sunnhemp (*Crotalaria juncea*, L). CRIJAF (ICAR) Technical Bulletin. No. 5 pp 1–50

Das BB, Mait SN (1998) Jute (*Corchorus* species) and allied fibres research in India. Indian J Agric Sci 68(8):484–493

Das LK, Laha SK, Pandit NC (1999) Entomology. In: Fifty years of research on jute and allied fibres agriculture. Published by Central Research Institute for Jute and Allied Fibres. pp 142–164

David BV, Kumarswami T (1960) *Drino (Prosturnia) inconspicua* Mg (Tachinid: Diptera), a parasite of *Utethasia pulchella* Linn in South India. Madras Agric J 47:481

Dempsey JM (1975) Fiber crops. The University Press of Florida, Gainesville

FAO (2014) World food and agriculture. In: FAO statistical yearbook 2014. Food and Agriculture Organization of the United Nations. p 307

Gawande SP, Sharma AK, Selvaraj K, Satpathy S (2013) Record of *Pachnephorus bretinghami* Baly: a new insect pest of ramie in Assam. JafNews 11(2):15

Gawande SP, Sharma AK, Satpathy S (2014) New record of Indian red admiral caterpillar (Vanessa indica Herbst.) as a pest of ramie (*Boehmeria nivea* L. Gaud) from Assam. Curr Biotica 8(1):93–96

Gawande SP, Sharma AK, Selvaraj K, Gotyal BS, Satpathy S (2015) Leaf folder, *Pleuroptya* sp. (Lepidoptera: Crambidae): a new insect pest of ramie, *Boehmeria nivea* L. Gaud. Curr Biotica 9(1):86–87

IJSG (2013) Annual report 2012–13 of International Jute Study group (IJSG). p 181

Kalita DN, Borah DC (1993) Parasitoids and predators of jute pests in certain localities of Central Brahmaputra Valley zone of Assam. J Agric Sci Soc Northeast India 6:19–23

Mahapatra BS, Mitra S, Ramasubramanian T, Sinha MK (2009) Research on jute (*Corchorus olitorius* and *C. capsularis*) and kenaf (*Hibiscus cannabinus* and *H. sabdariffa*): present status and future perspective. Indian J Agric Sci 79(12):951–967

Mitra S, Saha S, Guha B, Chakrabarti K, Satya P, Sharma AK, Gawande SP, Kumar M, Saha M (2013) Ramie: the strongest bast fibre of nature, Technical Bulletin No. 8. Central Research Institute for Jute and Allied Fibres, ICAR, Barrackpore, p 38, 120

Mustafee TP (1977) *Spodoptera Litura* Fab. causing damage to ramie in Assam. Curr Sci 46(10):350

Pandit NC, Pathak S (2000) Management of insect pests in mesta. Central Research Institute for Jute and Allied Fibres, Barrackpore, p 39

Perfecto I, Vandermeer J, Hanson P, Cartı´n V (1997) Arthropod biodiversity loss and the transformation of a tropical agroecosystem. Biodivers Conserv 6:935–945

Pimentel D, Stachow U, Takacs DA, Brubaker HW, Dumas AR, Meaney JJ, O'Neil J, Onsi DE, Corzilius DB (1992) Conserving biological diversity in agricultural/forestry systems. Bioscience 42:354–362

Pradhan SK, Chatterji SM (1978) Bionomics of the green semilooper, *Ilattia* (*Amyna*) *octa* G. (Lepidoptera: Noctuidae), a new pest of 'tossa' jute (*Corchorus olitorius* L.). J Entomol Res 2(1):116–119

Prakash S (2003) Seed yield loss in sunnhemp (*Crotalaria juncea* L) due to pod borer complex. Legum Res 24(1):48–50

Rahman S, Khan MR (2009) Natural enemies of insect and mite pests of jute ecosystem. Ann Plant Prot Sci 17(2):466–467

Ram S (1968) Record of parasites of sunnhemp top shoot borer, *Laspeyresia tricentra* Meyr. (Tortricidae/Lepidoptera) in Uttar Pradesh. Indian J Entomol 30(4):254

Ramesh Babu V, Selvaraj K, Gotyal BS, Satpathy S (2015) Record of hairy caterpillars on jute crop in West Bengal. JafNews 13(2):15

Reddy DB (1956) Sunnhemp and its insect fauna. Proc 10th Int Congr Entomol 3:439–440

Saha SN (2000) Improved varieties of jute for maximization of fibre yield. In: Pathak S (ed) Workshop cum training on adoptive research on improved varieties of jute and allied fibres and their utilization for enhanced income generation of farmers. Central Research Institute for Jute and Allied Fibres, ICAR, pp 5–6. 48p

Sarkar SK (2007) Quality assessment of sunnhemp seed collected from different districts of eastern Uttar Pradesh. J Mycol Plant Pathol 37(3):488–490

Sarkar SK, Tripathi MK (2003) Summer crop of sunnhemp escape major pests and diseases. ICAR News 9(4):16

Sarkar SK, Hazra SK, Sen HS, Karmakar PG, Tripathi MK (2015) Sunnhemp in India. ICAR-Central Research Institute for Jute and Allied Fibres (ICAR), Barrackpore, p 140

Sarma KK, Borah BK, Debnath MC, Das B (2010) Natural enemies of jute pests in Nagaon, Assam. Insect Environ 15(4):157–158

Satpathy S, Gotyal BS, Ramasubramanian T, Selvaraj K (2013) Mealybug, *Phenacoccus solenopsis* Tinsley infestation on jute (*Corchorus olitorius*) and mesta (*Hibiscus sabdariffa*). Insect Environ 19(3):187–188

Satpathy S, Gotyal BS, Selvaraj K (2014a) Record of *Aenasius bambawalei* Hayat on *Phenacoccus solenopsis* Tinsley in jute ecosystem. JafNews 12(1):15

Satpathy S, Selvaraj K, Gotyal BS, Biswas C, Gawande SP, Sarkar SK, De RK, Tripathi AN, Ramesh Babu V, Mandal K, Meena PN (2014b) Problems and prospects of pest management in jute and allied fibre crops. In: International conference on natural fibres

Selvaraj K, Satpathy S, Gotyal BS, Ramesh Babu V (2012) *Meteorus spilosomae*: a potential larval parasitoid of *Spilosoma obliqua*. Jaf News 10(2):10

Selvaraj K, Satpathy S, Gotyal BS, Ramesh Babu V (2013a) First record of *Protapanteles obliquae* on *Spilosoma obliqua*. Jaf News 11(1):15

Selvaraj K, Satpathy S, Gotyal BS, Ramesh Babu V (2013b) *Helicoverpa armigera* (Hubner): a new pest of *tossa* jute, *Corchorus olitorius* L. Insect Environ 19(3):166–167

Selvaraj K, Gotyal BS, Ramesh Babu V, Satpathy S (2014) Record of parasitoid, *Parachremylus* sp. (Braconidae: Hymenoptera) on leaf mining beetle in jute. JafNews 12(2):29

Selvaraj K, Satpathy S, Gotyal BS, Ramesh Babu V (2015) Record of hairy caterpillars on jute crop in West Bengal. Jaf News 13(2):14

Sheikh MS (2012) Studies on life cycle and population structure of jute semilooper (*Anomis sabulifera* guenee, Lepidoptera, Noctuidae) on tossa jute (*Corchorus olitorius* L.) in the district of Barpeta, Assam, India. Ecoscan 6(3&4):129–131

Singh DP (1998) Ramie (*Boehmeria nivea*). CRIJAF (ICAR) Technical Bulletin. No. 6, pp 1–52

Tripathi RL, Rama S (1971) Review of Entomological Researches on Jute, Mesta, Sunhemp and Allied Fibres. Indian Institute of Agricultural Research (ICAR). Technical Bulletin (AGRIC.) No. 36 p 42

Arthropod Diversity and Management in Legume-Based Cropping Systems in the Tropics

11

V. Sridhar and L.S. Vinesh

Abstract

Despite the wealth of arthropod diversity in legumes, most species are not considered economically important. In this chapter, information on various arthropod communities associated with leguminous crop-based agroecosystems/agroforestry systems as pests (sucking and chewing herbivores) and beneficials like pollinators, natural enemies like parasites and predators, and their ecological roles in these cultivated ecosystems is discussed.

Keywords

Arthropods • Ecology • Leguminous crops • Natural enemies • Tropics

11.1 Introduction

Legumes (Family: Fabaceae) are one of the largest families of flowering plants, comprising of around 630 genera and 18,860 species. They are primarily grown for food grain seeds (beans, lentils), vegetables (cowpea, beans), livestock forage (alfalfa), silage (lucerne, red clover), and soil fertilizers (mucuna). Other than as sole crops, legumes are also grown as cover crops as a part of cropping systems thus affecting the arthropod fauna associated with them. Several arthropods are associated with legumes as pests or as beneficials. An array of arthropod species can be found in fields used for legume-based cropping systems promoting species diversity.

V. Sridhar (✉) • L.S. Vinesh
Division of Entomology and Nematology, Indian Institute of Horticultural Research (IIHR), Hessaraghatta Lake Post, Bengaluru 560089, Karnataka, India
e-mail: vsridhar@iihr.res.in

A.K. Chakravarthy, S. Sridhara (eds.), *Economic and Ecological Significance of Arthropods in Diversified Ecosystems*, DOI 10.1007/978-981-10-1524-3_11

11.2 Arthropods and Legumes

Legumes are a primary source of nitrogen in cropping systems and provide food for humans and animals. The ability of legumes to fix atmospheric nitrogen has resulted in their worldwide use in crop-soil enrichment. Legumes are found in all kinds of habitats like deserts, rain forests, arctic, alpine, etc. (Schrire et al. 2005). Legumes like green gram, pigeon pea, and black gram are mainly grown in semiarid and lower humid tropics of Southeast Asia, Africa, and Central America (30° N to 30° S), whereas chickpea and lentil are cultivated in extensively low altitudes ranging from 15° N up to 40° N. In many cultivated ecosystems, legumes are grown as cover crops primarily to manage soil erosion, fertility, quality, water, weeds, pests, diseases, biodiversity, and wildlife (Lu et al. 2000). Cover crops harbor both harmful and beneficial arthropods (Altieri and Letourneau 1982; Andow 1988). Though cover crops harbor low incidence of pests, these pests may disperse to adjoining cash crops (Kennedy and Margolies 1985). Thus cover crops play a role in the life cycles, and population dynamics of various arthropods can be considered as management tools, to minimize pest problems while maximizing beneficial arthropod activity thereby reducing the dependence on insecticides and ill effects of insecticides in the environment and their residues.

Insects cause an average economic loss of 30–100 % in India (Dhaliwal and Arora 1994). In Africa, the damage can be up to 100 % in cowpea (Singh and Jackai 1985). In Pakistan, nearly 10 % of legume seed loss is attributed to bruchids in storage (Aslam 2004). *Helicoverpa armigera* is reported to cause loss of US$ 317 million in pigeon pea and US$ 328 million in chickpea (ICRISAT 1992). Economic loss up to US$ 2 billion annually is reported from this pest globally with over US$ 1 billion worth of insecticides used for its control (Sharma 2005). In the tropics, the extent of damage caused by insect pests is 50–100 % when compared to 5–10 % in temperate regions (van Emden et al. 1988).

11.3 Arthropod Pests

In Bangladesh, insect pests like ants, leaf and stem feeders, aphids, beetles, leaf binders, caterpillars, mites, borers, and storage pests were recorded on grain legumes like hyacinth bean, cowpea, black gram, soybean, pigeon pea, and lentils (Kabir 1978). In Sri Lanka, nearly 40 insect species were reported on legume crops of which only few were considered economically important. On cowpea, green gram, and black gram, the bean fly *Ophiomyia phaseoli* was reported as major pest; likewise in groundnut, *Aphis craccivora*, *Riptortus* spp., *Spodoptera littoralis*, *Helicoverpa armigera*, and *Callosobruchus* spp. were considered as pests. In pigeon pea, the most common insect pests are *Maruca vitrata*, *Etiella zinckenella*, *H. armigera*, *Lampides boeticus*, *Sphenarches anisodactylus*, *Exelastis atomosa*, *Melanagromyza obtusa*, and *Mylabris* spp. (Fig. 11.1) (Suba Singhe and Fellowes 1978).

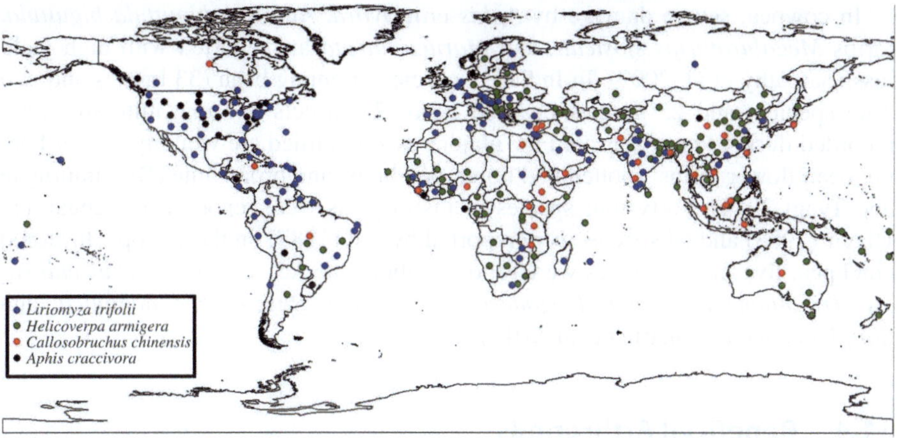

Fig. 11.1 Distribution of major pests of legumes

In Thailand, *Ophiomyia phaseoli*, *Stomopteryx subsecivella*, *Lamprosema* spp., *Archips micacaeana*, *Aphis glycines*, *Empoasca* spp., *Agrotis ipsilon*, *Spodoptera littoralis*, *H. armigera*, and termites were recorded on soybean crop (Arunin 1978). In the Philippines, eight insect species were recorded as major pests and around eight to ten as minor pests on legume crops. Among the major pests, *Ophiomyia phaseoli*, *Hedylepta indicate*, *Homona coffearia*, *H. armigera*, *Euchrysops cnejus*, *Empoasca sp*, *S. littoralis*, *Chrysodeixis chalcites*, and *Maruca testulalis* were reported on soybean, mung bean, cowpea, asparagus bean, etc. In Japan, 30 major insect pests attack legume crops like soybean, adzuki bean, common bean, cowpea, and broad bean. Few major insect pests reported are *Leguminivora glycinivorella*, *Matsumuraeses phaseoli*, *Etiella zinckenella*, *Nezara antennata*, etc. Major pests reported from Taiwan are agromyzid flies, spider mites, bruchid beetles, pod borers, leaf feeders, leafhoppers, stink bugs, plant bugs, and aphids which were reported to cause less damage on legume crops.

In India, 60 species of insect pests and four species of mites are recorded on green beans (FAO 2007; Sharma et al. 2010; Parvatha Reddy 2014). Several insect pests, like aphids, thrips, beetles, caterpillars, borers, etc., are recorded in mung bean and urd bean from India (Sharma et al. 2011). In chickpea, major insect pests recorded are borers, termites, aphids, loopers, and cutworms (Chandrashekar et al. 2014). On groundnut aphids, white grubs, thrips, cutworms, leaf miners, jassids, bruchids, borers, beetles, wire worms, etc., are reported. Stem fly, cutworms, semi-loopers, beetles, borers, whitefly, leaf miner, and weevils are reported from India on soybean (Singh et al. 2014; Whitman and Ranga Rao 1993). About 40 insect species attack green gram in India (Swaminathan et al. 2012). Jhansi Rani and Sridhar (2004) have reported incidence of several insect pests on cow's itch, *Mucuna pruriens*, from Bangalore, India.

In cowpea, severe damage by *Aphis craccivora*, *Amrasca biguttula biguttula*, thrips *Megalurothrips sjostedti*, and *Maruca vitrata* are reported with high yield loss (Satpathy et al. 2009). In India, 35 species on mung bean (33 insects and two mite species) and 25 species on urad bean (24 insects and one mite sp.) were recorded during kharif season. Few major pests identified are whitefly, blister beetle, bean flower thrips, spotted pod borer, pod bugs, and broad mite (Duraimurugan and Tyagi 2014). Forty-four species of insect pests were reported by Sehgal and Ujagir (1988) and 64 species were reported by Lal (1987) on these crops. In stored chickpea, five insect species were reported abundant, i.e., *Callosobruchus chinensis*, *Tribolium castaneum*, *Trogoderma granarium*, *Rhyzopertha dominica*, and *Sitophilus oryzae* (Sharma et al. 2013).

11.4 Beneficial Arthropods

Eighteen species of wasps and several species of predatory mites and bugs were reported on insect pests associated with green beans (Sharma et al. 2010; FAO 2007). In chickpea, 16 parasitoids and 17 predators were recorded as beneficial arthropods on many insect pests (Chandrashekar et al. 2014). Twenty-five parasitoids and 29 predators were recorded on insect pests of groundnut in India (Singh et al. 2014). In soybean, nine parasitoids and few predators were recorded on *Helicoverpa armigera*, *Spodoptera litura*, and *Bemisia tabaci* in India (Sharma et al. 2014). *Coccinella septempunctata* was recorded as the abundant predator on cowpea aphid (Arvind and Akhilesh 2015).

11.5 Management

11.5.1 Cultural Practices

In India, Weigand et al. (1994) and Dahiya et al. (1999) have demonstrated that early planting of legume crops can lower the incidence of *H. armigera*. Intercropping is a multiple cropping system wherein two or more crops are grown in proximity to enhance the yield and to provide a barrier to the spread of a pest or disease of the main crop (Willey et al. 1983). Bud worm infestation in sole maize was greater than in maize intercropped with soybean (Brown 1935; Seran and Brintha 2010).

The infestation of corn borer in maize was less when intercropped with soybean (Sastrawinata 1976). Soybean and groundnut are more effective in suppressing termite attack than common beans (Sekamatte et al. 2003). Maize stalk borer infestation was significantly greater in maize monocrop (70 %) than in maize-soybean intercrop (Martin 1990).

Degri et al. (2014) studied the incidence of stem borer on pearl millet intercropped with groundnut and inferred that the intercrop pattern of a millet and groundnut in the ratio of 1:2 and 1:1 recorded less stem borer infestation in pearl millet and resulted in high panicle weight and grain yield and also natural enemies were encouraged. Intercropping of groundnut with pearl millet (*Pennisetum glaucum*) has increased the population of *Goniozus* sp., a parasitoid species against leaf miner in groundnut (*Arachis hypogaea*) (Dhaliwal and Arora 1996).

Rao (1982) reviewed several improved resistant varieties of legume crops, viz., pigeon pea, urad bean, mung bean, cowpea, chickpea, pea, lentil, and groundnut. Leafhopper-resistant varieties of cowpea were developed at the International Institute of Tropical Agriculture, Nigeria, viz., Tvu59, Tvu123, VITA-1, and VITA-3 (Ofuya and Fayape 1999). *Bt* gene used in *Maruca*-resistant cowpeas is widely used in many tropical countries like China, Brazil, Korea, Japan, South Africa, the Philippines, Taiwan, etc. (Oyewale and Bamaiyi 2013). In Taiwan, common bean varieties resistant to aphids have been developed (Rose et al. 1978).

11.6 Methods of Influencing Arthropod Pest Management in Legumes

The use of fertilizers in legumes can be an innovative method of pest management. Application of 20 kg P_2O_5/ha has reduced the *Aphis craccivora* damage on a resistant cowpea variety (Annan et al. 1997). Higher dosage of phosphorus (30 kg P_2O_5/ha) reduced the incidence of *Mylothris sjostedti* through increased nutrition enabling the plant to overcome pest damage. In cowpea, damage by *A. craccivora*, *M. sjostedti*, and *M. vitrata* was reduced, coupled with higher grain yields at 30 and 45 kg P_2O_5/ha (Asiwe 2009).

11.6.1 Host-Arthropod Interactions

Lopez et al. (2010) studied the effect of the farming system on the arthropod community through its effects on plant community characteristics by comparing organic and conventional winter wheat fields in the Mediterranean region. Arthropods were classified into seven feeding groups (chewing herbivores, flower consumers, omnivores, saprophages, sucking herbivores, parasitoids, and predators), and plant species were classified into three functional groups (grasses, forbs, and legumes) representing highly distinct resources for the arthropods. Legumes enhance the richness of saprophages, parasitoids, and predators. The plant community through farming system affects the entire community of arthropods.

The numbers of chewing and sucking herbivores across fields were similar regardless of vegetation parameters; the family richness of both groups was closely associated to plant characteristics. However, the plant species richness and cover of grasses and legumes proved to affect the family richness of sucking herbivores, while the richness of chewing herbivores was positively associated to a greater legume cover. Thus the diversity of plant feeders was related to the diversity of resources (Knops et al. 1999; Murdoch et al. 1972; Siemann et al. 1998).

11.6.2 Saprophages

With the application of organic fertilization in organic fields, the taxa involved in decomposition are likely to be benefitted more in such fields (Mader et al. 2002; Moreby et al. 1994). However, saprophage community was richer and abundant where the grass cover was greater, which usually occurred in conventional fields where higher productivity was reported.

11.6.3 Flower Consumers

Flower consumers and legume cover showed positive relationship, where flower consumers were directly enhanced by floral food resources, such as nectar and pollen provided by the plant community (Bianchi and Wäckers 2008).

11.6.4 Predators, Parasitoids, and Their Interactions

The abundance of potential preys results in a greater abundance of flower consumers, and sucking herbivores lead to higher predator and parasitoid abundances (Haddad et al. 2001; Koricheva et al. 2000). The number of predators also positively relates to the abundance of saprophages (Haddad et al. 2001; Wardle et al. 1999). In addition, higher cover of grasses and legumes favors higher predator and parasitoid richness. The positive response of parasitoids and predators to grass cover increase can be attributed to greater prey diversity. However, the positive correlation shown between parasitoids and predators and legume cover also indicated a direct enhancement to alternative resources such as floral resources (Bianchi and Wackers 2008; Norris and Kogan 2000).

Rhizobia increased the abundance of arthropod herbivores on the plants. Increased plant biomass can increase the abundance of a wide variety of arthropod herbivores (Siemann et al. 1998; Forkner and Hunter 2000; Fonseca et al. 2005). Rhizobia increased plant biomass in terms of size and leaf number. However, rhizobia need not affect the abundance of arthropod herbivores with increase in plant size alone but through modifying other plant traits. In this context, nitrogen is an essential limiting element for survival and/or growth of many herbivorous arthropods (Mattson and Scriber 1987; White 1993). Several studies have proved this observation. The abundance of herbivorous arthropods was significantly greater on new willow leaves with high nitrogen content than on mature leaves with low nitrogen content (Nakamura et al. 2006; Utsumi and Ohgushi 2009). Leaf phenolics are defensive substances against arthropod herbivores (Feeny 1970; Larson and Berry 1984; Dudt and Shure 1994). Herbivore abundance is negatively correlated with concentration of foliar phenolics (tannin) of oak species (Feeny 1970; Forkner et al. 2004). Rhizobia positively affected above ground herbivores via changes in plant quality, because rhizobia-associated soybeans increased leaf nitrogen by 50 % and decreased phenolics by 12 % (Katayama et al. 2011). The species richness of herbivores on R+ plants was significantly greater than that on R− plants. Increased plant biomass and improved nutrient conditions of plants can increase species richness of herbivores (Siemann et al. 1998; Fonseca et al. 2005).

In the presence of rhizobia, the taxonomic richness, diversity, and abundance of predators increased, but community evenness decreased, although the community composition did not change suggesting the bottom-up effects initiated by rhizobia can extend beyond trophic levels. First, plants support increased herbivore abundance, which in turn increase species richness of predators (Siemann et al. 1998; Knops et al. 1999; Forkner and Hunter 2000). This is because a variety of predator species can aggregate when prey becomes abundant (Ives et al. 1993; Cardinale et al. 2006). Second, plants support increased species richness of herbivores, which may provide a wider range of prey items for generalist predators (Hunter and Price 1992). In addition, changes in plant size or architecture may directly affect the abundance or diversity of predators by providing shelter and foraging and/or oviposition sites (Langellotto and Denno 2004; Denno et al. 2005).

Symbiotic microbes like mycorrhizae and endophytes associated with terrestrial plants can have strong impacts on plants and their consumer diversity (van der Heijden et al. 2008) and can influence multi-trophic interactions of arthropods on host plants (Omacini et al. 2001; Gange et al. 2003; Chaneton and Omacini 2007; Hartley and Gange 2009; Koricheva et al. 2009). In this regard, several studies have illustrated the strong impacts on biodiversity and the abundance of higher trophic levels (Omacini et al. 2001; Finkes et al. 2006; Rudgers and Clay 2008). For example, Rudgers and Clay (2008) showed the important role of a symbiotic grass endophyte on arthropod communities of *Lolium arundinaceum*. The presence of the endophyte reduced abundance and species diversity of arthropods. Finkes et al. (2006) also documented that the species richness of spiders on tall fescue grass without a fungal endophyte was greater than endophyte-infected grass. The endophyte may have decreased spider species richness by reducing prey abundance. Total herbivore abundance declined 25–55 % in the presence of the endophyte, which could indicate a reduction in prey. Jani et al. (2010) examined how endophyte alkaloids affect the abundance and species richness of arthropod communities on sleepy grass and found that endophyte-produced alkaloids were associated with increased herbivore and natural enemy abundance and herbivore species richness.

11.7 Quarantine Importance of Arthropods

The quarantine risk in exchange of germplasm is mainly due to the hidden nature of infestation caused by pests. This group includes several species of bruchids (Coleoptera: Bruchidae) belonging to the genera *Aeanthoseelides*, *Bruchidius*, *Bruchus*, *Callosobruchus*, *Caryedes*, *Caryedon*, *Conieobruchus*, *Kytorhinus*, *Megabruchidius*, *Mimosestes*, *Pseudopachymerina*, *Specularius*, *Spermophagus*, and *Stator*. These beetles infest legume seeds in several countries (Udayagiri and Wadhi 1989; Bhalla et al. 2006). The grain legumes are infested by various native species of bruchids and spread to various continents. Some of the examples includes:

Acanthoscelides obtectus indigenous to North and South America got introduced into Asia, Africa, Europe, and Australia; *Callosobruchus analis* indigenous to Asia got introduced into Africa; C. *chinensis*, C. *maculatus*, and *Caryedon serratus* indigenous to Asia and Africa got introduced into Europe, North and South America, and Australia; and *Zabrotes subfasciatus* indigenous to North and South America got introduced into Asia, Africa, and Europe (Southgate 1978; Bhalla et al. 2006).

There are 142 insects and mite pests associated with seed or can be introduced as a contaminant along with soil or plant debris accompanying seeds of grain legumes. These include 134 insects belonging to six different orders of the class Insecta, viz., Coleoptera (104), Hemiptera (6), Hymenoptera (2), Lepidoptera (19), Orthoptera (1), and Thysanoptera (2), and eight mite pests belonging to class Acarina. Of these, 140 are exotic pests not yet reported from India or have restricted distribution. The major order infesting grain legumes is Coleoptera which includes pulse beetles, weevils, and scolytids that infest developing seeds in the field as well as the seeds in storage (Kavita et al. 2012).

11.8 Legumes as Pollinator Enhancers

Over 20 bee species have been recorded on alfalfa, berseem, white clover, red clover, pigeon pea, sun hemp, and pea in India (Abrol 2012). Honeybees were responsible for 65–70 % cross-pollination in pigeon pea (Saxena et al. 1993). Several pollinators are recorded on pigeon pea in India; among them major pollinators are *A. florae*, *Aegachile spp.*, and *A. dorsata*. Foraging activities of bees are considered to have direct effect on yield in Pusa 33 variety of pigeon pea (early maturing variety) (Upadhyay et al. 1997). In Hyderabad, *A. dorsata* and *Chalicodoma* spp. were frequent visitors to pigeon pea. In Ludhiana, five species of bees, *A. mellifera* (Fig. 11.2), *A. dorsata*, *Xylocopa* spp., *Megachile lanata*, and *Ceratina binghami*, were recorded by William (1977). Deodikar and Suryanarayana (1977) and Partap (1997) have reported many legume crops, viz., red gram, Bengal gram, horse gram, cowpea, lucerne, black gram, green gram, peas, sesbania, dhaincha, berseme, etc., having high density of bees.

Intensive agricultural management has increased productivity detriment of floral resources vital for insect pollinators like bees, butterflies, and hover flies across tropical and subtropical places. While the creation of wild flower habitats has been widely used to reestablish such resources into arable ecosystems (e.g., sown into field margins), comparable low-cost methods for enhancing floristic diversity in production grasslands are lacking. Woodcock et al. (2014) investigated how simple and cheap seed mixtures based around three plant functional groups (grasses, legumes, and non-leguminous herbs) could be used to enhance flowering resources to benefit insect pollinator communities. The abundance and species richness of pollinators were correlated with the increased availability of legume and non-legume flowers. Though flowering resources provided by agricultural cultivars of legumes declined rapidly once sown, inclusion of a forb component within seed mixtures was effective in increasing the long-term persistence of these resources.

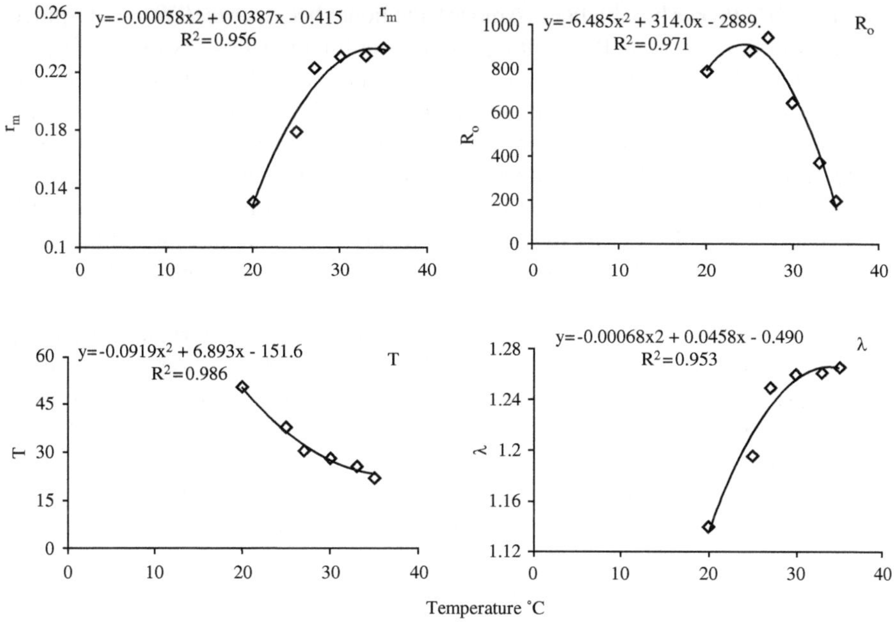

Fig. 11.2 Relation between life table parameters of *S. litura* and temperature on peanut at *e*CO_2

This will result in the abundance and species richness of insect pollinators over 4 years with greater stability. Management also played a role in the persistence of floral resources, with grazing more likely to maintain legume cover than cutting. In conclusion, it was demonstrated that low-cost seed mixtures can be used to enhance floristic diversity to benefit pollinators. In Brazil, several pollinators belonging to families Andrenidae (*Psaenythia* sp.), Apidae (*Ancyloscelis* sp., *Centris analis*, *Exomalopsis analis*, *Florilegus* sp., *Melitomella grisescens*), Halictidae (*Augochloropsis* sp., *Augochlorella* sp., *Augochlora* sp., *Dialictus* sp.), Megachilidae (*Megachile* sp.), and Syrphidae were recorded on soybean. Pollinators play a very important role in increasing the yield of soybeans. In the presence of pollinators, an extra 179.2 kg seeds/ha was obtained.

Tamarind flowers are mainly bee-pollinated (Radhamani et al. 1993; Nagarajan et al. 1997). In Brazil, 33 species of bees were observed pollinating tamarind, among which the major pollinators were *Centris* (*Heterocentris*) *analis*, *Centris* (*Centris*) *aenea*, *Xylocopa* (*Neoxylocopa*) *suspecta*, *Centris* (*Paremisia*) *fuscata*, *Partamona cupira*, *Apis mellifera scutellata*, *Trigona spinipes*, *Centris* (*Hemisiella*) *tarsata*, and *Ceratina* (*Crewella*) *madeirae* (Castro and Oliveira 1998). In Sudan, *Xylocopa olivacea*, *Megachile* sp., *Apis mellifera*, *Trigona* sp., *Syrphus* sp., and *Bombylius* sp. were recorded on tamarind (Diallo et al. 2014).

Majority of the legumes including alfalfa are primarily insect pollinated. Very early in northern Saskatchewan, ten species of native *Megachile* and three species of *Coelioxys* leafcutting bees (Megachilidae), 13 species of *Bombus* and three species of *Psithyrus* bumblebees (Apidae), the honeybee *Apis mellifera* L. (Apidae), and *Anthophora furcata* Pz. (Anthophoridae) pollinating flowers of alfalfa (Peck and Bolton 1946) are reported, before the introduction of the domesticated leafcutting bee *Megachile rotundata* Fab. Later, Harper (1988) recorded 15 species of pollinating Hymenoptera present in alfalfa fields in Alberta, i.e., six species of leafcutting *Megachile* bees, eight species of bumblebees, and the honeybee.

Honeybees are not well suited to extracting nectar from long, tubular corollas typical of alfalfa flowers as they have relatively short tongues (Waddington and Herbst 1987). Reinhardt (1952) anthropomorphically described the alfalfa flower as a trap for honeybees trying to extract nectar from the nectaries within the corolla. When a honeybee is caught in a tripped flower, considerable effort is needed on the part of the bee to extricate itself. Naive honeybees will occasionally trip alfalfa flowers, but experienced bees avoid doing so, indicating a learning ability in the species (Reinhardt 1952). Bumblebees are fairly efficient pollinators of alfalfa flowers (Peck and Bolton 1946; Plowright and Lavetry 1984), but their numbers in comparison to other pollinators in Saskatchewan alfalfa fields have been low (Knowles 1943) and may be getting lower throughout western North America (Grixti et al. 2009).

Megachilid bees are among the world's most efficient pollinators, tripping an average of 17.3 alfalfa flowers per minute (Knowles 1943). This is in part because their long tongues allow them to reach the nectar in tubular flowers with ease, in part because the "swimming" or vibrating movements in flowers facilitate effective pollen transfer, and in part because of their high frequency of flower visits. However, leaf cutting bees are inefficient nest provisioners and require 15–27 flower visits to gather sufficient resources to provision a single brood cell (Richards 1989).

Several features of alfalfa leaf cutting bees make them amenable to domestication. In addition to being efficient pollinators, they prefer to visit alfalfa or other legume blossoms and restrict their foraging to a small area of a few hectares. In contrast, honeybees and bumblebees forage on many plants over great distances. Megachilids are less aggressive than honeybees and do not defend their nests and their venom is less toxic. They are solitary yet gregarious and will live their separate lives in close proximity to other leaf cutting bees. They live in man-made structures, and their life cycle can be synchronized with pollination requirements. Unlike native leaf cutting bees, *M. rotundata* stay where placed. Although less cold tolerant than bumblebees, they will visit flowers at colder temperatures than honeybees.

11.9 Tritrophic Interactions

A tritrophic interaction study was conducted between pigeon pea genotypes, *H. armigera*, and natural enemies during 2008–2012 at ICRISAT, Hyderabad, India (Hugar 2012). Under no, dual, and multi-choice conditions, *H. armigera* egg laying on ICPW 125 was minimum due to high density of type D trichomes and was maximum on ICPL 87. The percent parasitization by *Trichogramma chilonis* Ishii was higher on ICPB 2042, and by *Campoletis chlorideae* Uchida, it was greater on

ICPL 87 and ICPL 87091 due to longer pods and clustering type of inflorescence and ICPL 87119 due to higher pod wall thickness. Odors from the flowers of ICPL 84060 and ICP 7035 attracted *T. chilonis* and *C. chlorideae*. *C. chlorideae* performed better on LRG 41, ICP 7035, and ICPL 84060 (Hugar 2012).

Ozawa et al. (2012) assessed the effect of temperature on a tritrophic system of lima bean, the herbivorous spider mite, *Tetranychus urticae*, and the predatory mite, *Phytoseiulus persimilis*, and found that these relationships are temperature dependent. The plant defended itself against *T. urticae* by emitting volatiles that attract *P. persimilis*. Over 20–40 °C, the emission of volatiles by infested plants and the subsequent attraction of *P. persimilis* peaked at 30 °C, but the number of eggs laid by *T. urticae* adults and the number of eggs consumed by *P. persimilis* peaked at 35 °C indicating that the spider and predatory mites performed best at a higher temperature than that at which most volatile attractants were produced. Transcriptome pyrosequencing of the mites revealed *P. persimilis* upregulated gene families for heat shock proteins (HSPs) and ubiquitin-associated proteins, whereas *T. urticae* did not. RNA interference-mediated gene suppression in *P. persimilis* revealed differences in temperature responses. Predation on *T. urticae* eggs by *P. persimilis* which fed PpHsp70-1 dsRNA was low at 35 °C but not at 25 °C when PpHsp70-1 expression was very high.

In pigeon pea and chickpea ecosystem, tritrophic interaction was assessed by Bisane et al. (2008). It was observed that early instars of *H. armigera* larvae were parasitized by *Eriborus argenteopilosus*, *Bracon* sp., *C. chlorideae*, and tachinid fly (Bisane et al. 2008).

11.10 Climate Change and Arthropods on Legumes

Climate change poses a threat to the control of insect pests. Due to climate change, new pests may invade areas previously uninhabitable for the pests. Climate change is predicted to cause an increase in global temperatures, alterations in rainfall, and insect distribution and spread. Increase in temperature due to climate change has resulted in poleward migration of many insect species (Parmesan 2006). Some times, drought situation alters the species composition of pests of legumes and can cause the out break of the pests (Mattson and Haack 1987).The elevated CO_2 concentrations and rising temperatures associated with climate change will have substantial impacts on plant-insect interactions, integrated pest management (IPM) programs, and the movement of non native insect species (Trumble and Butler 2009). Research in California has demonstrated that lima beans (*Phaseolus lunatus*) photosynthesized better and grow more rapidly under elevated CO_2 concentration with its primary pest, the cabbage looper (*Trichoplusia ni*), consuming 20 % more leaf area, as the leaves contained about 28 % less nitrogen in comparison to plants grown in ambient levels of carbon dioxide. As nitrogen is essential for the development of cabbage loopers, crops grown under elevated carbon dioxide levels contains less of nitrogen resulting in higher consumption of cabbage leaf in order to get the same amount of 'N'. This effect of increased feeding was observed in many insect groups such as butterflies, beetles, moths, and grasshoppers (Coviella and Trumble 1999).

Elevated CO_2 positively promoted number and size of root nodules in white clover (*Trifolium repens*) and also larger populations of *Sitona lepidus* larvae that developed at a much faster rate. However, more root nodules were damaged and consequently nitrogen concentrations in the roots declined. Thus increase in such pests could therefore compromise the perceived benefits of legumes in arable rotations under future climate scenarios (Staley and Johnson 2008).

Sharma (2010), Babasaheb et al. (2012), and Sridhar and Reddy (2013) gave an elaborate account on the impact of climate change and global warming on arthropod biodiversity, pest management, food security and prediction of pest and beneficial activities. Climate change will trigger major changes in geographical distribution of insect pests, population dynamics, insect biotypes, herbivore-plant interactions, diversity and abundance of arthropods including natural enemies, species extinction, and efficacy of crop protection technologies. Changes in geographical range and insect abundance will increase the extent of crop losses and thus will have a major bearing on crop production and food security. Climate change-triggered cropping patterns will also influence the distribution of insect pests. A shift in the distribution of major insect pests, viz., pod borers (*Helicoverpa*, *Maruca*, and *Spodoptera*), aphids, and white flies, to the temperate regions will ultimately lead to greater damage in grain legumes and vegetables. The effectiveness of host plant resistance, transgenic plants, natural enemies, biopesticides, and synthetic chemicals for pest management may be highly affected by climate change. Newer technologies that will be effective under climate change situations need to be developed to enhance the production of legume crops by decreasing the incidence of insect pest by adopting innovative pest management techniques. For knowing the information on potential distribution of legume pests, there is a need to generate information on the likely effects of climate change on insect pests and natural enemies to develop robust technologies that will be effective in the future under global warming and climate change. In this volume, a separate chapter on climate change and arthropods is included.

Rao et al. (2014) under eCO_2 condition developed nonlinear models for *S. litura* on peanut for four life table parameters, viz., "r_m," "R_o," "T," and "λ," where "r_m" is the intrinsic rate of increase, "R_o" is the net reproductive rate, "T" is the mean generation time, and "λ" is the finite rate of increase (Fig. 11.2), along with equations. The relationship between "r_m" and temperature followed the quadratic form and was best fit with higher R^2 (0.95) at eCO_2 and the other parameters, viz., "R_o," "T," and "λ," followed the similar trend ($R^2 = 0.95–0.98$) (Fig. 11.3).

Fig. 11.3 Major insect pests and predators associated with legumes

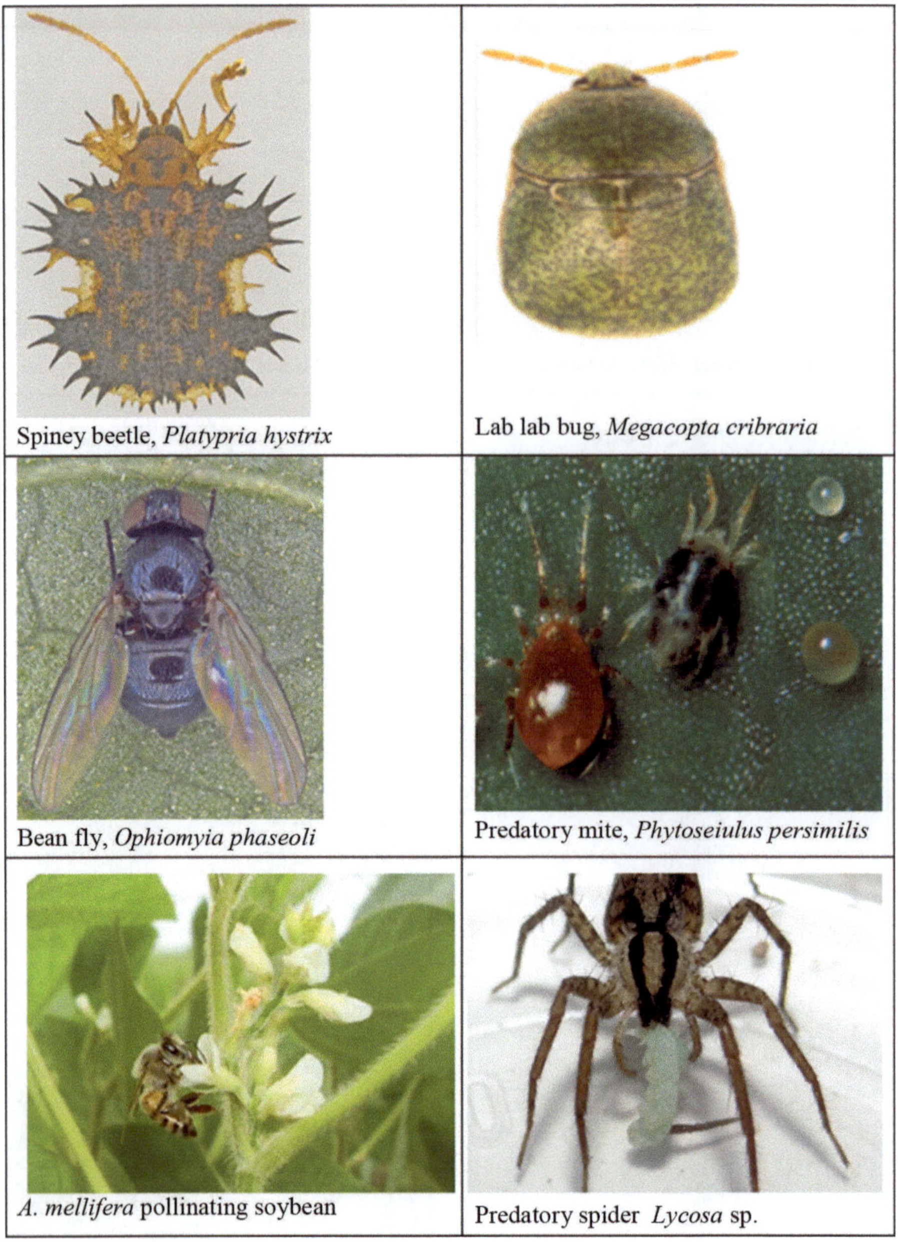

Fig. 11.3 (continued)

References

Abrol DP (2012) Pollination biology: biodiversity conservation and agricultural production. Springer, Dordrecht, ISBN 978-94-007-1941-5

Altieri MA, Letourneau DK (1982) Vegetation management and biological control in agroecosystems. Crop Prot 1:405–430

Andow DA (1988) Management of weeds for insect manipulation in agroecosystems. In: Altieri MA, Liebman M (eds) Weed management in agroecosystems: ecological approaches. CRC Press, Boca Raton, pp 265–301

Annan IB, Ampong-Nyarko K, Tingey WM, Schaefers GA (1997) Interactions of fertilizer, cultivar selection, and infestation by cowpea aphid (Aphididae) on growth and yield of cowpeas. Int J Pest Manag 43(4):307–312

Arunin A (1978) Pests of soybean and their control in Thailand. In: Singh SR, Van Emden HF, Ajibola Taylor T (eds) Pests of grain legumes: ecology and control. Academic, London, pp 43–46

Arvind K, Akhilesh K (2015) Effect of abiotic and biotic factors on incidence of pests and predator in cowpea [*Vigna unguiculata* (L.) Walp.]. Legum Res 38(1):121–125

Asiwe JAN (2009) The impact of phosphate fertilizer as a pest management tactic in four cowpea varieties. Afr J Biotechnol 8:7182–7186

Aslam M (2004) Pest status of stored chickpea beetle, *Callosobruchus chinensis* L. on chickpea. J Entomol 1(1):28–33

Babasaheb BF, Ankush LK, Kumar M (2012) Will climate change pose serious threat to crop pest management: a critical review? Inter J Sci Res Publ 2(11):1–14

Bhalla S, Kapur ML, Gupta K, Lal B, Khetarpal RK (2006) Checklist of bruchids associated with seeds of different plant genera. NBPGR, New Delhi, p 38

Bianchi FJJA, Wäckers FL (2008) Effects of flower attractiveness and nectar availability in field margins on biological control by parasitoids. Biol Control 46:400–408

Bisane KD, Borkar SL, Lande GK, Bhalkare SK (2008) Parasitization of *Helicoverpa armigera* (Hubner) on pigeon pea. J Biol Control 22(2):455–461

Brown HB (1935) Effect of soybean on corn-yields Louisiana. Exp Agric 265:31–31

Cardinale BJ, Weis JJ, Forbes AE, Tilmon KJ, Ives AR (2006) Biodiversity as both a cause and consequence of resource availability: a study of reciprocal causality in a predator–prey system. J Anim Ecol 75:497–505

Castro MS, Oliveira CM (1998) Abelhas visitantes do tamarindo *Tamarindus indica* L. em duas áreas da Bahia: caatinga (Milagres) e sub-úmida (Conceição do Almeida). Anais do XVII Congresso Brasileiro de Entomologia; Rio de Janeiro; Brasil. Rio de Janeiro

Chandrashekar K, Om Gupta, Suhas Yelshetty, Sharma OP, Someshwar Bhagat, Chattopadhyay C, Mukesh Sehgal, Arpana Kumari, Amaresan N, Sushil SN, Sinha AK, Ram Asre, Kapoor KS, Satyagopal K, Jeyakumar P (2014) Integrated pest management for chickpea, pp 43

Chaneton EJ, Omacini M (2007) Bottom-up cascades induced by fungal endophytes in multitrophic systems. Ecological communities. In: Ohgushi T, Craig TP, Price PW (eds) Plant mediation in indirect interaction webs. Cambridge University Press, Cambridge, UK, pp 164–187

Coviella C, Trumble JT (1999) Elevated atmospheric CO_2 and insect-plant interactions: implications for insect conservation. Conserv Biol 13:700–712

Dahiya SS, Chauhan YS, Johansen C, Shanower TG (1999) Adjusting pigeon pea sowing time to manage pod borer infestation. Int Chickpea Pigeonpea Newsl 6:44–45

Degri MM, Mailafiya DM, Mshelia JS (2014) Effect of intercropping pattern on stem borer infestation in pearl millet (*Pennisetum glaucum* L.) grown in the Nigerian Sudan Savannah. Adv Entomol 2:81–86

Denno RF, Finke DL, Langellotto GA (2005) Direct and indirect effects of vegetation structure and habitat complexity on predator–prey and predator–predator interactions. In: Barbosa P, Castellanos I (eds) Ecology of predator–prey interactions. Oxford University Press, Oxford, pp 211–239

Deodikar GB, Suryanarayana MC (1977) Pollination in the service of increasing farm production in India. In: Nair PKK (ed) Advance in pollen-spore research, vol II. Maharashtra Association for the Cultivation of Science, New Delhi, pp 60–82

Dhaliwal GS, Arora R (1994) Components of insect pest management: a critique. In: Dhaliwal GS, Arora R (eds) Trends in agricultural insect pest management. Commonwealth Publishers, New Delhi, pp 1–55

Dhaliwal GS, Arora R (1996) Principles of insect pest management. National Agricultural Technology Information Centre, Ludhiana, p 374

Diallo BO, Ouedraogo M, Chevallier M-H, Joly HI, Hossaert-Mckey M, Mckey D (2014) Potential pollinators of *Tamarindus indica* L. (Caesalpinioideae) in Sudanian region of Burkina Faso. Afr J Plant Sci 8(12):528–536

Dudt JF, Shure DJ (1994) The influence of light and nutrients on foliar phenolics and insect herbivory. Ecology 75:86–98

Duraimurugan P, Tyagi K (2014) Pest spectra, succession and its yield losses in mung bean and urad bean under changing climatic scenario. Legum Res 37(2):212–222

FAO (2007) Green beans integrated pest management: an ecological guide. FAO Regional Vegetable IPM Programme, pp 82

Feeny P (1970) Seasonal changes in oak leaf tannins and nutrients as a cause of spring feeding by winter moth caterpillars. Ecology 51:565–581

Finkes LK, Cady AB, Mulroy JC, Clay K, Rudgers JA (2006) Plant – fungus mutualism affects spider composition in successional fields. Ecol Lett 9:347–356

Fonseca CR, Prado PI, Almeida-Neto M, Kubota U, Lewinsohn TM (2005) Flower-heads, herbivores, and their parasitoids: food web structure along a fertility gradient. Ecol Entomol 30:36–46

Forkner RE, Hunter MD (2000) What goes up must come down? Nutrient addition and predation pressure on oak herbivores. Ecology 81:1588–1600

Forkner RE, Marquis RJ, Lill JT (2004) Feeny revisited: condensed tannins as anti-herbivore defences in leaf-chewing herbivore communities of Quercus. Ecol Entomol 29:174–187

Gange AC, Brown VK, Aplin DM (2003) Multitrophic links between arbuscular mycorrhizal fungi and insect parasitoids. Ecol Lett 6:1051–1055

Grixti JC, Wong LT, Cameron SA, Favret C (2009) Decline of bumble bees (*Bombus*) in the North American Midwest. Biol Cons 142:75–84

Haddad NM, Tilman D, Haarstad J, Ritchie M, Knops MH (2001) Contrasting effects of plant richness and composition on insect communities: a field experiment. Am Nat 158:17–35

Harper AM (1988) Insects and mites on Alfalfa in Alberta, Technical Bulletin 1988-3E. Research Branch, Agriculture Canada, Ottawa, p 38

Hartley SE, Gange AC (2009) Impacts of plant symbiotic fungi on insect herbivores: mutualism in a multitrophic context. Annu Rev Entomol 54:323–342

Hugar SV (2012) Tritrophic interaction between pigeonpea genotypes, *Helicoverpa armigera* (Hübner) and natural enemies. PhD thesis, University of Agricultural Sciences, Dharwad, India

Hunter MD, Price PW (1992) Playing chutes and ladders: heterogeneity and the relative roles of bottom-up and top-down forces in natural community. Ecology 73:724–732

ICRISAT (1992) Medium term plan. International Crops Research Institute for the Semi-Arid Tropics, Patancheru, Hyderabad, India.

Ives AR, Kareiva P, Perry R (1993) Response of a predator to variation in prey density at 3 hierarchical scales – lady beetles feeding on aphids. Ecology 74:1929–1938

Jani AJ, Faeth SH, Gardner D (2010) Asexual endophytes and associated alkaloids alter arthropod community structure and increase herbivore abundances on a native grass. Ecol Lett 13:106–117

Jhansi Rani B, Sridhar V (2004) Record of arthropod pests on velvet bean, *Mucuna pruriens* var. *utilis* under Bangalore conditions. J Med Arom Plant Sci 26(3):505–506

Kabir AKMF (1978) Pests of grain legumes and their control in Bangladesh. In: Singh SR, van Emden HF, Taylor TA (eds) Pests of grain legumes: ecology and control. Academic, London

Katayama N, Zhang Z, Ohgushi T (2011) Community-wide effects of below-ground rhizobia on above-ground arthropods. Ecol Entomol 36:43–51

Kavita G, Bhalla S, Manju LK, Beche L, Charan Singh (2012) Insects and mites of quarantine significance in grain legumes. In: Chalam VC, Bhalla S, Singh B, Rajan (eds) Potential quarantine pests for India in grain legume. National Bureau of Plant Genetic Resources, New Delhi, pp 1–108

Kennedy GG, Margolies DC (1985) Mobile arthropod pests: management in diversified agroecosystems. Bull Entomol SOCA 31(3):21–27

Knops JMH, Tilman D, Haddad NM, Naeem S, Mitchell CE, Haarstad J, Ritchie ME, Howe KM, Reich PB, Siemann E, Groth J (1999) Effects of plant species richness on invasion dynamics, disease outbreaks, insect abundances and diversity. Ecol Lett 2:286–293

Knowles RP (1943) The role of insects, weather conditions, and plant character in seed setting of Alfalfa. Sci Agric 24:29–50

Koricheva J, Mulder CPH, Schmid B, Joshi J, Huss-Danell K (2000) Numerical responses of different trophic groups of invertebrates to manipulation of plant diversity in grasslands. Oecologia 125:271–282

Koricheva J, Gange AC, Jones T (2009) Effects of mycorrhizal fungi on insect herbivores: a meta-analysis. Ecology 90:2088–2097

Lal SS (1987) Insect pests of mung, urad, cowpea and pea and their management. In: Rao MV, Sithanantham S (eds) Plant protection in field crops. Plant Protection Association of India, Hyderabad, pp 185–201

Langellotto GA, Denno RF (2004) Responses of invertebrate natural enemies to complex-structured habitats: a meta-analytical synthesis. Oecologia 139(1):1–10

Larson KC, Berry RE (1984) Influence of peppermint phenolics and monoterpenes on two spotted spider mite (Acari: Tetranychidae). Environ Entomol 13:282–285

López CB, Blanco-Moreno JM, Pérez N, Pujade-Villar J, Ventura D, Oliva F, Sans FX (2010) A functional approach to assessing plant–arthropod interaction in winter wheat. Agric Ecosyst Environ 137:288–293

Lu YC, Watkins KB, Teasdale JR, Abdul-Baki AA (2000) Cover crops in sustainable food production. Food Rev Int 16:121–157

Mäder P, Fliessbach A, Dubois D, Gunst L, Fried P, Niggli U, Masters G (2002) Soil fertility and biodiversity in organic farming. Science 296:1694–1697

Martin RC (1990) Intercropping corn and soybean for high-protein silage in a cool temperate region. PhD thesis, McGill University, Montreal, Quebec, pp 185

Mattson WJ, Haack RA (1987) The role of drought in outbreaks of plant-eating insects. Biogeosciences 37:110–118

Mattson WJ, Scriber JM (1987) Feeding ecology of insect folivores of woody plants: water, nitrogen, fiber, and mineral considerations. In: Slansky F Jr, Rodriguez JG (eds) The nutritional ecology of insects, mites and spiders. Wiley, New York, pp 105–146

Moreby SJ, Aebischer NJ, Southway SE, Sotherton NW (1994) A comparison of the flora and arthropod fauna of organically and conventionally grown winter-wheat in southern England. Ann Appl Biol 125:13–27

Murdoch WW, Evans FC, Peterson CH (1972) Diversity and pattern in plants and insects. Ecology 53:819–829

Nagarajan B, Nicodemus A, Mandal AK, Verma RK, Gireesan K, Mahadevan NP (1997) Phenology and controlled pollination study in tamarind. Silvae Genet 47:237–241

Nakamura T, Ogawa H, Maripi DK, Uematsu M (2006) Contribution of water soluble organic nitrogen to total nitrogen in marine aerosols over the East China Sea and Western North Pacific. Atmos Environ 40:7259–7264

Norris RF, Kogan M (2000) Interaction between weeds, arthropod pests, and their natural enemies in managed ecosystems. Weed Sci 48:94–158

Ofuya TI, Fayape OE (1999) Evaluation of cashew nut shell liquid and powder for the control of *Callosobruchus maculatus* (Fabricus) (Coleoptera: Bruchidae) infesting stored seeds of cowpea, *Vigna unguiculata* (L.) Walpers. Appl Trop Agric 4:72–77

Omacini M, Chaneton EJ, Ghersa CM, Muller CB (2001) Symbiotic fungal endophytes control insect host – parasite interaction webs. Nature 409:78–81

Oyewale RO, Bamaiyi LJ (2013) Management of cowpea insect pests. Sch Acad J Biosci 1:217–226

Ozawa R, Nishimura O, Yazawa S, Muroi A, Takabayashi J, Arimura G (2012) Temperature-dependent, behavioural, and transcriptional variability of a tritrophic interaction consisting of bean, herbivorous mite, and predator. Mol Ecol 21(22):5624–5635

Parmesan C (2006) Ecological and evolutionary responses to recent climate change. Ann Rev Ecol Evol Syst 37:637–669

Partap U (1997) Bee flora of the Hindu Kush-Himalayas: inventory and management. International Centre for Integrated Mountain Development (ICIMOD), Kathmandu, p 297

Parvatha Reddy (2014) Leguminous vegetable crops. In: Parvatha Reddy (ed) Biointensive integrated pest management in horticultural ecosystems, vol XVI. Springer, pp 277, ISBN 978-81-322-1844-9

Peck O, Bolton JL (1946) Alfalfa seed production in Northern Saskatchewan as affected by bees, with a report on means of increasing the populations of native bees. Sci Agric 26:388–418

Plowright RC, Laverty TM (1984) The ecology and sociobiology of bumble bees. Annu Rev Entomol 29:175–199

Radhamani A, Nicodemus A, Narajan B, Subramanian KN (1993) Reproductive biology and breeding system studies in *Tamarindus indica* L. In: Veeresh GK, Shaanker RU, Ganeshaiah KN (eds) Pollination in tropics. International Union for the Study of Social Insects, India Chapter, Bangalore, pp 33–35

Rao MR (1982) Legume production in traditional and improved cropping systems in India. In: Symposium on grain legumes production, Chiang Mai, Thailand, November 1980

Rao MS, Manimanjari D, Rama Rao AC, Swathi P, Maheswari M (2014) Effect of climate change on *Spodoptera litura* Fab. on peanut: a life table approach. Crop Prot 66:98–106

Reinhardt JF (1952) Some responses of honey bees to alfalfa. Am Nat 86:257–275

Richards KW (1989) Alfalfa leafcutter bee management in Western Canada. Agriculture Canada, Ottawa, Publication 1495/E

Rose RI, Chiang HS, Harnoto I (1978) Pests of grain legumes and their control in Taiwan. In: Singh SR, van Emden HF, Taylor TA (eds) Pests of grain legumes: ecology and control. Academic, London, pp 67–71

Rudgers JA, Clay K (2008) An invasive plant – fungal mutualism reduces arthropod diversity. Ecol Lett 11:831–840

Sastrawinata SE (1976) Nutrient uptake, insect, disease, labor use, and productivity characteristics of selected traditional inter cropping patterns which together affect their continued use by farmers. Ph.D. Thesis, University of Philippines, p 130

Satpathy S, Shivalingaswamy TM, Akhilesh K, Rai AB, Rai M (2009) Efficacy of biopesticides and new insecticides for management of cowpea pod borer, *Maruca vitrata*, Symposium on international conference on Grain Legumes: Quality improvement value addition and trade. IIPR, Kanpur, pp 292–293

Saxena NP, Johansen C, Saxena MC, Silim SN (1993) Selection for drought and salinity tolerance in cool – season food legumes. In: Singh KB, Saxena MC (eds) Breeding for tolerance in cool season food legumes. Wiley, Chichester, pp 245–270

Schrire BD, Lavin M, Lewis GP (2005) Global distribution patterns of the Leguminosae: insights from recent phylogenies. Biol Skr 55:375–422

Sehgal VK, Ujagir R (1988) Insect and pest management of mungbean in India. Mung bean: proceedings of the Second International Symposium. Asian Vegetable Research and Development Center, Shanhua, pp 315–328

Sekamatte BM, Ogenga-Latigo M, Russell-Smith A (2003) Effects of maize-legume intercrops on termite damage to maize, activity of predatory ants and maize yields in Uganda. Crop Prot 22:87–93

Seran TH, Brintha I (2010) Review on maize based intercropping. J Agron 9(3):135–145

Sharma HC (ed) (2005) Heliothis/helicoverpa management: emerging trends and strategies for future research. Oxford & IBH, and Science Publishers, USA, New Delhi, p 469

Sharma HC (2010) Global warming and climate change: impact on arthropod biodiversity, pest management and food security. In: National symposium on perspectives and challenges of integrated pest management for sustainable agriculture. Solan, India

Sharma OP, Gopali JB, Suhas Yelshetty OM, BaFmbawale DK, Garg BB, Bhosle (2010) Pests of Pigeon pea and their management. NCIPM, LBS Building, IARI Campus, New Delhi

Sharma OP, Bambawale OM, Gopali JB, Someshwar B, Suhas Y, Singh SK, Rajesh A, Om Pal S (2011) Field guide-mung bean and urad bean. National Centre for Integrated Pest Management (Indian Council of Agricultural Research) LBS Building, IARI Campus, New Delhi

Sharma S, Upadhyaya HD, Roorkiwal M, Varshney RK, Gowda CLL (2013) Chickpea. In: Singh M, Upadhyaya HD, Bisht IS (eds) Genetic and genomic resources of grain legume improvement. Elsevier, London, pp 81–104

Sharma AN, Gupta GK, Verma RK, Sharma OP, Someshwar Bhagat, Amaresan N, Saini MR, Chattopadhyay C, Sushil SN, Ram Asre, Kapoor KS, Satyagopal K, Jeyakumar P (2014) Integrated pest management for soybean, p 41

Siemann E, Tilman D, Haarstad J, Ritchie M (1998) Experimental test of the dependence of arthropod diversity on plant diversity. Am Nat 152:738–750

Singh KR, Jackai LEN (1985) Insect pests of cowpeas in Africa: their life cycle, economic importance and potential for control. In: Singh SR, Rachie KO (eds) Cowpea research, production and utilisation. Wiley, London, pp 217–231

Singh S, Thirumalaisamy PP, Harish G, Datta Ram, Sushil SN, Sinha AK, Ram Asre, Kapoor KS, Satyagopal K, Jeyakumar P, Ajanta Birah, Sharma OP, Someshwar Bhagat, Verma PV, Sunil Kumar, Chattopadhyay C, Yadav MS (2014) Integrated pest management package for groundnut, p 49

Southgate BJ (1978) The importance of Bruchidae as pests of grain legumes, their distribution and control. In: Pests of grain legumes: ecol control, pp 219–229

Sridhar V, Reddy PVR (2013) Use of degree days and plant phenology: a reliable tool for predicting insect pest activity under climate change conditions. In: Singh HP, Rao NKS, Shivashankar KS (eds) Climate resilient horticulture: adaptation and mitigation strategies. Springer Publisher, New Delhi, pp 287–294

Staley JT, Johnson SN (2008) Climate change impacts on root herbivores. In: Johnson SN, Murray PJ (eds) Root feeders: an ecosystem perspective. CABI, Wallingford, pp 192–213

Suba Singhe SMC, Fellowes RW (1978) Recent trends in grain legume pest research in Sri Lanka. In: Singh SR, van Emden HF, Taylor TA (eds) Pests in grain legumes. Ecology and control. Academic, London, pp 37–41

Swaminathan R, Singh K, Nepalia V (2012) Insect pests of green gram *Vigna radiata* (L.) Wilczek and their management. INTECH 51(1):978–953

Trumble JT, Butler CD (2009) Climate change will exacerbate California's insect pest problem. Calif Agric 63(2):73

Udayagiri S, Wadhi SR (1989) Catalogue of Bruchidae. Mem. Amer. Ent. Inst (45):300

Upadhyay SK, Singh RP, Rizvi SMA (1997) Impact of bee pollination on the yield components of early variety of pigeon pea. National symposium on management of biotic and abiotic stresses in pulse crops. IIPR, Kanpur

Utsumi S, Ohgushi T (2009) Community-wide impacts of herbivore-induced plant regrowth on arthropods in a multi-willow species system. Oikos 118(12):1805–1815

Van der Heijden MGA, Bardgett RD, van Straalen NM (2008) The unseen majority: soil microbes as drivers of plant diversity and productivity in terrestrial ecosystems. Ecol Lett 11:296–301

Van Emden HF, Ball SL, Rao MR (1988) Pest, disease and weed problems in pea lentil and faba bean and chickpea. In: World crops: cool season food legumes. Kluwer Academic Publishers, Dordrecht, pp 519–534

Waddington KD, Herbst LH (1987) Body size and the functional length of the proboscis of honey bees. Fla Entomol 70:124–128

Wardle DA, Nicholson KS, Bonner KI, Yeates GW (1999) Effects of agricultural intensification on soil-associated arthropod population dynamics, community structure, diversity and temporal variability over a seven-year period. Soil Biol Biochem 31:1691–1706

Weigand S, Lateef SS, El Din Sharaf N, Mahmoud SF, Ahmed K, Ali K (1994) Integrated control of insect pests of cool season food legumes. In: Muehlbauer EJ, Kaiser WJ (eds) Expanding the production and use of cool season food legumes. Kluwer Academic Publishers, Dordrecht, pp 679–694

White TCR (1993) The inadequate environment: nitrogen and the abundance of animals. Springer, New York, pp 397–343

Whiteman JA, Ranga Rao GV (1993) A groundnut insect identification handbook for India, Information Bulletin No. 39. International Crops Research Institute for the Semi-Arid Tropics, Patancheru, Hyderabad, India p 64

Willey RW, Natarajan M, Reddy MS, Rao MR, Nambiar PTCM, Kannaiyan J, Bhatnagar VS (1983) Intercropping studies with annual crops. In: Nugent J, O'Connor M (eds) Better crop for food. Pitman Co, London

Williams IH (1977) Behaviour of insects foraging on pigeon pea *Cajanus cajan* (L.) Millsp. in India. Trop Agric (Trinidad) 54(4):353–363

Woodcock BA, Savage J, Bullock JM, Nowakoski M, Orr R, Tallowin JRB, Pywell RF (2014) Enhancing floral resources for pollinators in productive agricultural grasslands. Biol Conserv 171:44–51

Arthropod Diversity in Non leguminous Vegetable Crops

12

N.R. Prasannakumar, K.P. Kumar, and A.T. Rani

Abstract

Arthropods are important in maintaining the ecological balance in nonleguminous vegetable ecosystems. They provide natural services to human welfare and act as the pollinators, natural enemies, scavengers, leaf-litter sweepers, garbage collectors, soil conditioners and natural fertiliser producers in nature. The stable food chain and food webs are being maintained due to the presence of arthropod diversity. Conserving biodiversity of arthropods within agroecosytems enhances the processes of plant and soil enrichment which in turn improves crop yields and provides new sources of crop germplasm and cultivars. Biodiversity of arthropod fauna consists of major insect pests and non insect pests, which are causing significant crop loss and their natural enemies. In this chapter, arthropod diversity on different nonleguminous vegetable crops such as vegetables of Brassicaceae, Solanaceae, Malvaceae and Cucurbitaceae are discussed.

Keywords

Arthropod fauna • Biodiversity • Conservation • Nonleguminous vegetables

N.R. Prasannakumar (✉)
Division of Entomology and Nematology, Indian Institute of Horticultural Research (IIHR), Hessaraghatta Lake Post, Bengaluru 560089, Karnataka, India
e-mail: prasannaent@gmail.com

K.P. Kumar • A.T. Rani
Department of Agricultural Entomology, University of Agricultural Sciences (UAS), Gandhi Krishi Vignana Kendra (GKVK), Bengaluru 560065, Karnataka, India

A.K. Chakravarthy, S. Sridhara (eds.), *Economic and Ecological Significance of Arthropods in Diversified Ecosystems*, DOI 10.1007/978-981-10-1524-3_12

12.1 Introduction

Arthropods play a crucial role in vegetable crops, which includes arthropod species that are involved in cycling of nutrients, yield enhancers and facilitators of better yield, natural enemies and pollinators. They are elaborated below:

- *Ecological role*
  - In soil, arthropods play multiple roles as consumers, scavengers and decomposers and facilitate biogeochemical cycling of nutrients. Soil aeration, its capacity to retain water and tilth improvement are improved by the activities of arthropods. Their burrowing and nesting activities in the soil result in turnover of soils and redistribution of nutrients much more than earthworm. Flies and dung beetles prevent the build-up of manure from large animals by decomposing it and speed up its further degradation by fungi and bacteria. Without such scavengers, the gradual accumulation of waste products from herbivores would render much of the landscape unsuitable for living.
- *Natural Enemies*
  - Parasites and predatory arthropods play a vital role in maintaining the equilibrium of pests which unchecked lead to population explosions and exhaust food resources. So far, over 6000 insect species have been evaluated and released as biological control agents to manage pests like insects and weeds. But there are also many unidentified species that act as natural enemies in population regulation, often unnoticed until they become extinct by a natural disaster or anthropogenic intervention. Anthropogenic disruption of natural ecosystems is a common cause of pest outbreaks.
- *Pollination*
  - Arthropods play crucial role in the pollination of many flowering plants (angiosperms). Many plants produce nectar and pollen, a wide range of colours and odours, rewards to attract pollinators. In fact, there is a close symbiotic relationship between flowers and pollinators. Every flower has a distinct pattern of floral structure that facilitates adhesion of pollen from one flower and its deposition on stigma of another flower nearby. For instance, wild ginger produces an unobtrusive brown flower that smells like a dead rat. Many fungal gnats are attracted to these flowers, and the flower gets pollinated by fungus gnats, flesh flies and other saprophytic insects. Cleistogamy in brinjal makes it completely dependent on insect pollination. Cabbage, cauliflower, broccoli, knol khol, radish, turnip, cucurbits, etc. depend on insect pollination.

12.2 Impact of Agriculture on Arthropods

About 25–30 % of the world's land area is under agriculture which invariably affects biological diversity due to various farm activities. Since civilisation, man has started manipulating the natural resources such as land, forest, river, etc. to cultivate food crops in a wide area. While simplifying the landscape of the environment over wide areas, replacement of many diversified flora and fauna has happened. Worldwide, there are 12 species of grain crops, 23 vegetable species and about 35 fruit and nut crop species (Fowler and Mooney 1990). Continuous cultivation of the same crop in each season, i.e. monoculture, resulted in the extinction of many flora and fauna by replacing the existing biodiversity. Modern agriculture is mostly by monoculture which has reduced plant diversity which in turn affected the composition and abundance of associated biota such as wildlife, pollinators, insect pests and their natural enemies, soil invertebrates and microorganisms (Matson et al. 1997). The pest outbreaks are linked to the expanding crop monocultures which alter natural vegetation, thus decreasing local habitat diversity (Altieri and Letourneau 1982; Flint and Roberts 1988). Monocultures are often more sensitive to pest and disease outbreaks and therefore require higher inputs of pesticides (Power and Flecker 1996). Excessive use of associated intensive agrochemicals, fertilisers and mechanical technology to improve crop production also affects biodiversity. For instance, about 17.8 mt of fertilisers and about 500 mt of pesticides are used annually in the United States. Although these measures have increased crop yields, in the long run, sustainability of agriculture is affected due to their harmful impact on the environment (Altieri 1994). In the Philippines, pesticides are the main, if not the only, control measure used on major vegetable crops such as cabbage, green beans, eggplant and tomato (Sumalde 1995).

In practical terms, pest and disease management in major vegetable-growing belts worldwide is only through chemical control. The World Resources Institute (WRI) (1992) identified five fundamental causes of biodiversity loss:

1. Increase in human population growth and consumption of natural resources for their livelihood
2. Economic systems and policy issues that fail to value the environment and its resources
3. Disproportion in the consumption of resources, conservation and flow of benefits
4. Lack of knowledge, awareness and its proper application
5. Legal and institutional systems that promote unsustainable exploitation

The six major causes of biodiversity loss are habitat loss and fragmentation of lands; introduction of new species from one place to another; over-exploitation of plant and animal resources; pollution of soil, water and atmosphere; global climate change; urbanisation; industrial agriculture; and forestry (Fig. 12.1). The loss of biodiversity has a range of negative effects on the entire ecosystem. More immediately, the loss of biodiversity can have significant impacts on the disruption of food chain and food webs in the ecosystem that results in the extinction of certain species and reduced returns from agriculture by another species replacement.

Fig. 12.1 Larvae of *Pieris brassicae* feeding on cabbage

12.3 Enhancing Arthropod Diversity

Conserving and augmenting diverse flora and fauna provide several benefits to sustainable vegetable production. Uncultivated species, including wild relatives of crops, are important sources of germplasm for developing new crops and cultivars which are resistant to pests and diseases. Undisturbed natural areas adjacent to agricultural systems provide source of pollen and nectar for pollinators and natural pest enemies to flourish. Diversification of agroecosystem through polycultures, intercropping and border cropping can augment the resources available to pests, pollinators and natural enemies resulting in balanced arthropod diversity (Andow 1991). Minimising the use of spurious, substandard chemicals can also help in conserving beneficial organisms and functional processes such as decomposition and nutrient recycling. Thus crop productivity and sustainability of ecosystem can be maintained (Matson et al. 1997).

Alternative methods of pest management instead of pesticides specially non-chemical control techniques are widely used in modern agriculture. Ecologically and economically sound pest control procedures should involve the following:

- The use of cultural, physical, mechanical and biological methods is encouraged to conserve and maintain sustainable ecosystems.
- The use of semiochemical-based pest management such as pheromones, e.g. lures for melon fly management.

- Intercropping such as growing of African tall marigold in tomato crop in the ratio of 16 (tomato):2 (marigold) to attract oviposition of *Helicoverpa armigera*, in cabbage 21 (cabbage):2 (bold-seeded mustard) to trap diamondback moth, etc.
- Augmentation of natural enemies in the cropping system.
- Stage-specific and time-specific insecticide application and discontinuing fixed-schedule sprays can reduce the use of hazardous chemicals and the improvement of the pest's natural enemies' dynamics (Gonzales 1976).
- The reduced use of chemical pesticides to eradicate a key pest from a wide area will enhance the biological diversity (Thomas 1996).

Arthropods are very important in maintaining the ecological balance in nonleguminous vegetable ecosystems. The stable food chain and food webs are being maintained mainly due to the presence of arthropod diversity. Disturbances from natural and anthropogenic factors lead to the collapse of the entire food chain and food web of the system. Arthropods of different nonleguminous vegetable crops such as vegetables of Brassicaceae, Solanaceae, Malvaceae, cucurbits and gourds are discussed below.

12.4 Arthropods Diversity in Vegetable Crops

12.4.1 Cruciferous

The biodiversity of arthropod fauna in cruciferous ecosystems consists of three species of aphids, viz. *Lipaphis erysimi*, *Brevicoryne brassicae* and *Myzus persicae*. These along with the large white cabbage butterfly (*Pieris brassicae*) and diamondback moth (DBM), *Plutella xylostella*, are found to be prominent pests of cruciferous plants, while pentatomid bugs (*Nezara viridula*), flea beetle (*Phyllotreta cruciferae*), leaf beetle (*Monolepta signata*), sawfly (*Athalia lugens proxima*) and small white cabbage butterfly (*Pieris rapae*) appeared to be minor pests. Natural enemies, especially predators of aphid, viz. coccinellid beetles and syrphid flies, include spiders (Table 12.1). Parasitoids include nymphal-adult parasitoid of aphids, *Diaeretiella rapae*, and larval parasitoids of cabbage butterfly, *Hyposoter ebeninus* and *Cotesia glomerata*. Besides, activities of several pollinators are also recorded during the flowering stage. The major predators of aphids are *Coccinella septempunctata* complex and *C. transversalis* complex (Firake et al. 2012). Adults of syrphids, bees, sepsids, ants, halactids, some wasps and Muscidae will act as pollinators and help in the pollination of cabbage, cauliflower, mustard and broccoli. The DBM has developed resistance to insecticides beginning with DDT (dichlorodiphenyltrichloroethane) in 1953. Since then, the DBM has become resistant to every new class of insecticide whenever they were used indiscriminately and repeatedly. If a newer insecticide is more effective, then it is likely that more rapidly selection for resistance occurs. If multiple insecticides have similar modes of action (MoAs), the DBM tends to develop resistance to all of them. These MoAs are categorised by the

Table 12.1 Select arthropod fauna on cruciferous vegetable crops

Arthropod	Scientific name	Order	Family	Crop	Status
Diamondback moth	*Plutella xylostella*	Lepidoptera	Plutellidae	Crops	Major
Cabbage butterfly	*Pieris brassicae*	Lepidoptera	Pieridae	Oilseed and vegetable brassica	Major
Aphids	*Brevicoryne brassicae*, *Lipaphis erysimi*, *Myzus persicae*	Hemiptera	Aphididae	Oilseed and vegetable brassica	Major
Honeybee	*Apis dorsata*, *A. mellifera*, *A. florae*, *A. cerana indica*	Hymenoptera	Apidae	Oilseed and vegetable brassica	Major
Mustard sawfly	*Athalia lugens proxima*	Hymenoptera	Tenthredinidae	Oilseed brassica	Minor
Leaf webber	*Crocidolomia binotalis*	Lepidoptera	Pyralidae	Vegetable brassica	Major
Ichneumonid wasp	*Diadegma semiclausum*, *Hyposoter ebeninus*	Hymenoptera	Ichneumonidae	Oilseed and vegetable brassica	Major
Egg parasitoid	*Trichogrammatoidea bactrae*	Hymenoptera	Trichogrammatidae	Oilseed and vegetable brassica	Major
Spider	*Pardosa altitudis*	Araneae	Lycosidae	*Brassicae*	Frequent occurrence
	Leucage celebesiana		Tetragnathidae		
	Neoscona rumpfi		Araneidae		
	Theridion manjithar		Theridiidae		

Insecticide Resistance Action Committee. If the DBM population is not controlled by an insecticide, it should not be used with another insecticide in the same IRAC group.

12.4.2 Okra

Arthropods of okra include lepidopteran pests which bore inside the shoot and fruits, defoliators, leaf rollers and sucking pests such as aphid, whitefly, mealybug and cotton bug. There are pheromonal components available to manage okra shoot and fruit borers like *Earias* spp. and *Helicoverpa armigera* as an eco-friendly pest management technique. For sucking pests, sticky traps are being used to trap and kill the pests. Whitefly, *Bemisia tabaci*, has not only acted as a pest but also as a vector of yellow vein mosaic virus. Downward cupping of leaves in okra is always due to mite, *T. urticae*, whereas upward curling is due to leafhopper, *Amrasca biguttula biguttula*.

12.4.3 Cucurbits

Select arthropod fauna of cucurbits and gourd vegetables are presented in Table 12.2. The button-shaped extrafloral nectaries (EFNs) located on the bracts, bracteoles, calyces and leaves of cultivated sponge gourd plant, *Luffa cylindrica*, were visited by insects belonging to five different orders: Hemiptera, Diptera, Coleoptera, Hymenoptera and Lepidoptera (Agrawal and Rastogi 2010). Insects with biological control potential recorded at the EFNs included ants, wasps and ladybird beetles. Ten species of ants constituted by far the most abundant group (84.44 ± 4.34 % of individuals) at the EFNs and were found in large numbers on EFN-bearing plant parts, particularly and calyces. Insect pollinators included honeybees, butterflies and wasps, which, while visiting the floral nectaries (FNs), also visited the EFNs. Ant species visiting the plants included *Camponotus compressus*, *C. paria* and *Pheidole* sp. The major insect pest was the red pumpkin beetle, *Raphidopalpa foveicollis* Lucas (Fig. 12.2), which fed predominantly on the corolla of the plants and, to a lesser extent, on each of the EFN-bearing vegetative parts. The low incidence of insect pests on vegetative parts indicates that the ants and, to a lesser extent, wasps, bees and ladybirds visiting the EFN-bearing plant structures may be protecting the plant from the herbivores. Thus, though the EFNs of sponge gourd plants attract predators, pollinators and also extrafloral nectar thieves such as flies, ants are the major insect group involved in the facultative association with the plant. Such studies may aid in an environment-friendly management approach involving the natural enemies of insect pests of EFN-bearing annual crops (Table 12.3).

Table 12.2 Select arthropod fauna of cucurbits

Arthropod	Scientific name	Order	Family	Crop	Status
Cucumber	Pumpkin beetle	*Aulacophora foveicollis* and *A. lewisii*	Coleoptera	Chrysomelidae	Major
	Fruit fly	*Bactrocera cucurbitae*, *B. ciliates*, *B. zonata*	Diptera	Tephritidae	Major
	Peach aphid, melon aphid	*Myzus persicae* *Aphis gossypii*	Hemiptera	Aphididae	Major
	Red spider mite	*Tetranychus urticae*	Trombidiformes	Tetranychidae	Major
	Red spider mite	*Tetranychus urticae*	Trombidiformes	Tetranychidae	Major
	Seed-corn maggot	*Delia platura*	Diptera	Anthomyiidae	Minor
	Sweet potato whitefly	*Bemisia tabaci*	Hemiptera	Aleyrodidae	Minor
	Red pumpkin beetle	*Aulacophora foveicollis*, *A. lewisii*	Coleoptera	Chrysomelidae	Major
	Mirid bug	*Nesidiocoris cruentatus*, *N. tenuis*	Hemiptera	Miridae	Major

Fig. 12.2 Red pumpkin beetle

Table 12.3 Select arthropod fauna of other vegetable crops

Arthropod	Scientific name	Order	Family	Crop	Status
Parsley	Green peach Aphid	*Myzus persicae*	Hemiptera	Aphididae	Major
	Aster leafhopper	*Macrosteles quadrilineatus*	Hemiptera	Cicadellidae	Minor
	Potato leafhopper	*Empoasca fabae*	Hemiptera	Cicadellidae	Minor
	Carrot weevil	*Listronotus oregonensis*	Coleoptera	Curculionidae	Minor
Leek	Onion maggot	*Delia antique*	Diptera	Anthomyiidae	Major
	Onion thrips	*Thrips tabaci*	Thysanoptera	Thripidae	Major
Beetroot	Beet webworm	*Spoladea recurvalis*	Lepidoptera	Crambidae	Major
Amaranthus	Amaranthus weevil	*Hypolixus truncatulus*	Coleoptera	Curculionidae	Major
	Amaranthus caterpillar	*Hymenia recurvalis*	Lepidoptera	Pyraustidae	Major
	Leaf webber	*Psara basalis*	Lepidoptera	Pyraustidae	Major
	Leaf webber	*Eretmocera impactella*	Lepidoptera	Heliodinidae	Minor
Okra	Shoot and fruit borer	*Earias vitella, E. insulana*	Lepidoptera	Noctuidae	Major
	Bhendi fruit borer	*Helicoverpa armigera*	Lepidoptera	Noctuidae	Major
	Shoot weevil	*Alcidodes affaber*	Colleoptera	Curculionidae	Major
	Stem weevil	*Pempherulus affinis*	Coleoptera	Curculionidae	Minor
	Leaf roller	*Sylepta derogata*	Lepidoptera	Pyralidae	Major
	Semilooper	*Anomis flava*	Lepidoptera	Noctuidae	Minor
	Whitefly	*Bemisia tabaci*	Hemiptera	Aleyrodidae	Major
	Leaf hopper	*Amrasca devastans*	Hemiptera	Cicadellidae	Major

12.4.4 Solanaceous Crops

Arthropods of solanaceous crops include pests, parasites, parasitoids, scavengers and mites (Table 12.4). Among the pests, *Leucinodes orbonalis* (Lepidoptera: Pyralidae) is a monophagous pest which feeds only on eggplant (Fig. 12.3). It attacks the shoot during the early stage of the plant and the flowers, flower buds and fruits during the later stage. However, some polyphagous pests like *Helicoverpa armigera*, *Spodoptera litura* and aphids are also found in solanaceous crop

Table 12.4 Select arthropod fauna of solanaceous crops

Arthropod	Scientific name	Order	Family	Crop	Status
Fruit borer	*Helicoverpa armigera*	Lepidoptera	Noctuidae	Tomato, chilli, capsicum	Major
Leaf miner	*Liriomyza trifolii*	Diptera	Agromyzidae	Tomato	Major
Pin worm	*Tuta absoluta*	Lepidoptera	Gelechiidae	Tomato	Major
Cutworm/leaf-eating caterpillar	*Spodoptera litura*	Lepidoptera	Noctuidae	Tomato, chilli, capsicum	Major
Whitefly	*Bemisia tabaci*	Hemiptera	Aleyrodidae	Brinjal, tomato, chilli, capsicum	Major
Sucking pest	*Thrips tabaci*, *Frankliniella schultzei*	Thysanoptera	Thripidae	Tomato	Minor
Striped mealybug	*Ferrisia virgata*	Hemiptera	Pseudococcidae	Tomato	Minor
Red spider mite	*Tetranychus* spp.	Trombidiformes Class: Arachnida	Tetranychidae	Brinjal, tomato, chilli, capsicum	Major
Mirid bug	*Nesidiocoris tenuis*	Hemiptera	Miridae	Brinjal, tomato, chilli, capsicum	Major
Shoot and fruit borer	*Leucinodes orbonalis*	Lepidoptera	Pyralidae	Brinjal	Major
Budworm	*Scrobipalpa blapsigona*	Lepidoptera	Gelechiidae	Tomato, chilli, brinjal	Minor
Hadda beetle	*Henosepilachna vigintioctopunctata*	Coleoptera	Coccinellidae	Tomato, brinjal	Minor
Gall midge	*Asphondylia capparis*, *A. capsici*	Diptera	Cecidomyiidae	Brinjal, tomato, chilli, capsicum	Major
Yellow mite	*Polyphagotarsonemus latus*	Trombidiformes	Tarsonemidae	Brinjal tomato chilli capsicum	Major
Spider mite	*Tetranychus urticae*	Class: Arachnida			

Fig. 12.3 Brinjal shoot and fruit borer (adult) and its damage on brinjal fruit

ecosystems. Phytophagous arthropods cause huge damage to solanaceous vegetables resulting in huge losses. Among the Acarina, *Tetranychus* spp. popularly known as red spider mite feed on the ventral surface of the leaves by forming a small web, and *Polyphagotarsonemus latus* sucks sap from the plants.

A zoophytophagus mirid, *Nesidiocoris tenuis*, is reported in the tomato ecosystem, which feeds on plant as well as on some of the insect pests such as *Tuta absoluta*. Pollinators such as *Apis* spp. are reported from eggplant ecosystem as the plant has cleistogamous flowers and needs external pollination for fruit setting. Parasitoids like broconids and egg parasitoid, Trichogrammatidae, keep the phytophagous arthropods in balance.

Latif et al. (2009) identified 20 species of harmful arthropods under 17 families belonging to 6 different orders. The brinjal shoot and fruit borer (*Leucinodes orbonalis*), jassid (*Amrasca bigutulla bigutulla*), epilachna beetle (*Epilachna* sp.), whitefly (*Bemisia tabaci*) and aphid (*Aphis gossypii*) are found to be the most common and major arthropod pests of brinjal. Ten plant-dwelling predaceous arthropod families were found in field; among them 42–44 % belong to three families under coleopteran insects. Spiders and Lycosidae family formed 30.23 %. Surface-dwelling arthropods caught in pitfall traps belonged to 17 families; among them 7 families were predators. Formicidae constituted 67 % of the total surface-dwelling predaceous arthropods. Insecticide resistance in brinjal shoot and fruit borer especially to pyrethroids is now widespread in many brinjal-producing countries. Continuous spraying of pyrethroids resulted in increased selection pressure of the BSFB to get resistance. In order to avoid the deleterious effect of pesticides, the *Bt* came in to existence.

12.5 Conclusions

Increasing plant diversity considerably enhanced the arthropod diversity mainly via effects of diversified plant structure on herbivores and predators. The low arthropod diversity in monocultures may affect the energy flow at different trophic structures resulting in weak associations between the plant and arthropod diversity. This emphasises the richness of the flora in determining local arthropod diversity. Further, increase in biotic potential and decrease in environmental resistance result in outbreak of arthropod pests. Decrease in biotic potential and increase in environmental resistance result in extinction of the species. So, both biotic potential and environment resistance keep the arthropod diversity in equilibrium. Any anomalies in these two factors due to natural or anthropogenic influences result in boom and bust in the life cycle of an organism, including pest.

Arthropods are good indicators of the habitat fragmentation. Fragmented ecosystems isolate the populations and impose barriers to dispersal. These barriers limit gene flow and preclude migration as a response to environmental change (Ledig 1992). Fragmented populations contain only a part of the original gene pool and often are subject to substantial genetic drift and loss of genetic biodiversity (Brown 1992). Geographically circumscribed species with little genetic diversity have proven highly prone to extinction (Ehrlich 1992). Genetic diversity of arthropod populations in fragmented ecosystems can be measured, and the rate of genetic drift can be assessed with respect to non-fragmented populations. In this way, an advance warning of ecosystem changes due to fragmentation can be obtained. Consequently, policy and management practices can be modified to reduce their impact (BSC 1996). Fossil remains demonstrate that arthropod species are robust over long periods, and that given the opportunity, they migrate with changing conditions rather than evolve as new species.

References

Agrawal VM, Rostogi M (2010) Ants as dominant insect visitors of the extrafloral nectaries of sponge gourd plant, *Luffa cylindrica* (L.) (Cucurbitaceae). Asian Myrmecol 3:45–54

Altieri MA (1994) Biodiversity and pest management in agroecosystem. Food Products Press/The Haworth Press, Binghamton

Altieri MA, Letourneau DK (1982) Vegetation diversity and insect pest outbreaks. CRC, Critic Rev Plant Sci 2:131–169

Andow DA (1991) Vegetational diversity and arthropods population response. Ann Rev Entomol 36:561–586

Brown AHD (1992) Human impact on gene pools and sampling for their conservation. Oikos 63:109–118

BSc (Biological Survey of Canada) (1996) The advantages of using arthropods in ecosystem management. January

Ehrlich PR (1992) Population biology of checkerspot butterflies and the preservation of global biodiversity. Oikos 63:6–12

Firake DM, Lytan D, Behere GT (2012) Bio-diversity and seasonal activity of Arthropod Fauna in Brassicaceous crop ecosystems of Meghalaya, North East India. Mol Entomol 3(4):18–22

Flint ML, Roberts PA (1988) Using crop diversity to manage pest problems: some California examples. Am J Alternat Agric 3:164–167
Fowler C, Mooney P (1990) Shattering: food, politics and the loss of genetic diversity. University of Arizona Press, Tucson
Gonzales RH (1976) Crop protection in Latin America with special reference to integrated pest control. FAO PH Prot Bull 24:3
Latif MA, Rahman MM, Islam MR, Nuruddin MM (2009) Survey of Arthropod diversity in the brinjal field. J Entomol 6(1):28–34
Ledig FT (1992) Human impacts on genetic diversity in forest ecosystems. Oikos 63:87–108
Matson PA, Parton WJ, Power AG, Swift MJ (1997) Agricultural intensification and ecosystem properties. Science 277:504–509
Power AG, Flecker AS (1996) The role of biodiversity in tropical managed ecosystems. In: Orians GH, Dirzo R, Cushman JH (eds) Biodiversity and ecosystem process in tropical forests. Springer, New York, pp 173–194
Sumalde AC (1995) Environmental impact of pesticides and other concerns related to their use in vegetable production in the Philippines. In: Magallona ED, Sumalde AC, Pola AC, MacQuarrie P (eds) Selected papers on pesticides management and regulation in the Philippines. Environment and Resource Management Project (ERMP), Philippines
Thomas JD (1996) Using marine invertebrates to establish research and conservation priorities. In: Reaka ML, Kudla D, Wilson E (eds) Biodiversity II: understanding and protecting our biological resources. National Press, Washington, DC
World Resources Institute (WRI), World Conservation Union (WCU), United Nations Environment Programme (UNEP) (1992) Global biodiversity strategy guidelines for action to save, study and use Earth's biotic wealth sustainably and equitably

[illegible]

Diversity of Mites on Vegetable Crops, Kerala, South India: Documentation for Conserving Predatory and Other Beneficial Mites on Vegetables

13

K.V. Binisha, Haseena Bhaskar, and Sosamma Jacob

Abstract

Spider mites (*Tetranychus* spp.) were the dominant species of phytophagous mites on six vegetable crops in Kerala, South India. Predatory mites belonging to five families were identified on the vegetables. Phytoseiid mites were found effective against phytophagous mites. The database generated on mite fauna will assist in developing management strategies against not only mites but insect pests too. The results will also help in conserving beneficial mites in cultivated vegetable crop ecosystems.

Keywords

Vegetable crops • Conservation • Phytoseiid • Phytophagous • Spider mites

13.1 Introduction

Phytophagous mites are important pests of crops and some are quite injurious causing heavy loss to farmers. Contrary to this, predatory mites help in the natural control of mite pests to some extent. In India, 2350 species of mites belonging to 725 genera under 190 families are reported (Gupta and Gupta 1999). This forms only a small percentage of the world's known biodiversity. However, information on the diversity of phytophagous mites affecting major vegetable crops in Kerala in

K.V. Binisha (✉)
Plant Quarantine Station, Bengaluru 560017, Karnataka, India
e-mail: binishabaskar@gmail.com

H. Bhaskar • S. Jacob
Department of Agricultural Entomology, College of Horticulture,
Kerala Agricultural University, Vellanikkara, Thrissur, Kerala, India
e-mail: bhaskarhaseena@yahoo.co.in

A.K. Chakravarthy, S. Sridhara (eds.), *Economic and Ecological Significance of Arthropods in Diversified Ecosystems*, DOI 10.1007/978-981-10-1524-3_13

South India is limited. The proposed study was made to provide an overview of the important acarinae species, phytophagous and predatory, associated with major vegetable crops grown in Thrissur district, Kerala. Surveys were conducted in different vegetable-growing tracts of Thrissur to collect mite-infested leaf samples from selected vegetable plants, and mite fauna up to generic/species level were identified.

13.2 Materials and Methods

The work was carried out during 2011–2012 to explore the faunal composition of mites associated with six major vegetable crops of Thrissur district, Kerala. Random roving surveys were carried out in the farmers' fields from three vegetable-growing tracts, namely, Pazhayannur, Kannara, and Vellanikkara of Thrissur districts of Kerala to collect phytophagous and predatory mites associated with six vegetable crops, viz., *Solanum melongena* L., *Abelmoschus esculentus* L., *Amaranthus* sp., *Vigna unguiculata* L., *Momordica charantia* L., and *Capsicum annum* L. Mite-infested leaf samples (three leaves per plant) were collected from ten randomly selected vegetable plants separately in polythene bags from each locality and brought to the laboratory. In the laboratory, the leaves were observed under a stereomicroscope, and mite specimens were collected using camel hair brush and preserved in 70 % ethyl alcohol with a few drops of glycerol taken in glass vials.

The mites collected in the survey were mounted in Hoyer's media to prepare permanent slides. Single specimen representing both male and female mites of the same species was mounted separately on different slides. The male tetranychid mites were mounted in the lateral position to ensure better orientation of the genital structures which are important for species determination. The mounted specimens were kept in an oven at 40 °C for 7–10 days, and dried specimens were then labeled and numbered serially for identification. The permanent slides prepared were observed under phase-contrast microscope with image analyzer software to study the taxonomic characters. Identification of the slide-mounted mite specimens was made using appropriate literature and also with the help of mite taxonomists at the University of Agricultural Sciences, Bangalore.

13.3 Results

A total of 19 species of phytophagous and predatory mites (Fig. 13.1) belonging to eight families were identified (Table 13.1). The phytophagous mite families recorded were Tetranychidae, Tenuipalpidae, and Tarsonemidae represented by the genera *Tetranychus*, *Eutetranychus*, *Brevipalpus*, and *Polyphagotarsonemus* (Fig. 13.1). The predatory mite families included Phytoseiidae, Stigmaeidae, Cunaxidae, Bdellidae, and Tydeidae. The acarinae faunal diversity in different vegetable crop environments is detailed below (Table 13.2).

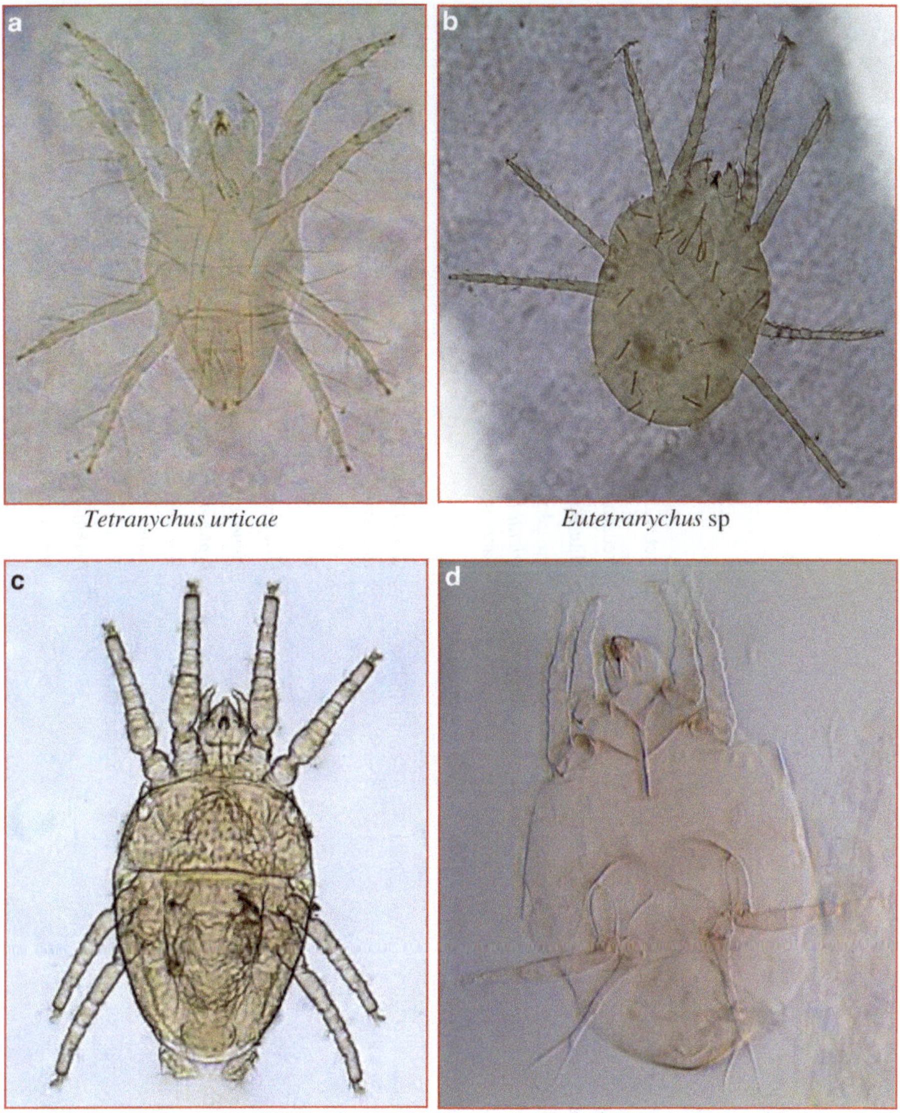

Tetranychus urticae *Eutetranychus* sp

Brevipalpus phoenicis *Polyphagotarsonemus latus*

Fig. 13.1 Phytophagous mites

13.3.1 Mite Fauna on Brinjal (*Solanum melongena* L.)

Ten species of mites belonging to three phytophagous and seven predatory mites were observed in brinjal. The two-spotted spider mite, *Tetranychus urticae* Koch, was the predominant phytophagous mite. *Polyphagotarsonemus latus* (Banks) and *Brevipalpus phoenicis* (Geij.) occurred in very minor form. The predatory mite

Table 13.1 Phytophagous and predatory mites associated with vegetable crops in Thrissur district, Kerala

	Phytophagous mites			Predatory mites		
Host plant	Mite genus/species	Suborder	Family	Mite genus/species	Suborder	Family
Brinjal (*Solanum melongena* L.)	*Tetranychus urticae* Koch	Prostigmata	Tetranychidae	*Paraphytoseius orientalis* Narayanan	Mesostigmata	Phytoseiidae
	Polyphagotarsonemus latus (Banks)	Prostigmata	Tarsonemidae		Mesostigmata	Phytoseiidae
	Brevipalpus phoenicis (Geij.)	Prostigmata	Tenuipalpidae	*Neoseiulus longispinosus* (Evans)	Mesostigmata	Phytoseiidae
					Mesostigmata	Phytoseiidae
				Amblyseius sp.	Mesostigmata	Phytoseiidae
				Euseius sp.	Mesostigmata	Phytoseiidae
				Phytoseius sp.	Prostigmata	Stigmaeidae
				Typhlodromips sp.		
				Agistemus gamblei Gupta		
Bhindi (*Abelmoschus esculentus* L.)	*T. urticae*	Prostigmata	Tetranychidae	*N. longispinosus*	Mesostigmata	Phytoseiidae
				Typhlodromips sp.	Mesostigmata	Phytoseiidae
				Amblyseius sp.	Mesostigmata	Phytoseiidae
				Agistemus sp.	Prostigmata	Stigmaeidae

Chili (*Capsicum annum* L.)	*P. latus*	Prostigmata	Tarsonemidae	*Amblyseius paraaerialis* Muma	Mesostigmata	Phytoseiidae
					Mesostigmata	Phytoseiidae
						Phytoseiidae
						Tydeidae
				Euseius sp.	Mesostigmata	Stigmaeidae
						Cunaxidae
					Prostigmata	Bdellidae
					Prostigmata	
				Typhlodromips sp.	Prostigmata	
				Tydeus sp.	Prostigmata	
				Agistemus sp.		
				Cunaxa sp.		
				Bdella sp.		

Table 13.2 Phytophagous and predatory mites associated with vegetables in Thrissur district

	Mite genus/species	Suborder	Family	Mite genus/species	Suborder	Family
Amaranthus (*Amaranthus* sp.)	*Tetranychus sp.*	Prostigmata	Tetranychidae	*N. longispinosus*	Mesostigmata	Phytoseiidae
	B. phoenicis	Prostigmata	Tenuipalpidae	*Amblyseius* sp.	Mesostigmata	Phytoseiidae
				Euseius sp.	Mesostigmata	Phytoseiidae
				Scapulaseius sp.	Mesostigmata	Phytoseiidae
				Typhlodromips sp.	Mesostigmata	Phytoseiidae
Cowpea (*Vigna unguiculata* L.)	*Tetranychus* sp.	Prostigmata	Tetranychidae	*N. longispinosus*	Mesostigmata	Phytoseiidae
	Eutetranychus sp.	Prostigmata	Tetranychidae	*Typhlodromips* sp.	Mesostigmata	Phytoseiidae
	P. latus	Prostigmata	Tarsonemidae	*Amblyseius* sp.	Mesostigmata	Phytoseiidae
Bitter gourd (*Momordica charantia* L.)	*P. latus*	Prostigmata	Tarsonemidae	*Typhlodromips* sp.	Mesostigmata	Phytoseiidae
				Euseius macrospatulatus Gupta	Mesostigmata	Phytoseiidae

fauna includes *Neoseiulus longispinosus* Evans, *Paraphytoseius orientalis* Narayanan, *Amblyseius* sp., *Phytoseius* sp., *Typhlodromips* sp., and *Euseius* sp. all belonging to the family Phytoseiidae and *Agistemus gamblei* Gupta of the family Stigmaeidae.

13.3.2 Mite Fauna on Bhindi (*Abelmoschus esculentus* L.)

Five different mites were observed from bhindi which included one phytophagous and four predatory mites. *T. urticae* was the only phytophagous mite recorded from this crop, and the predatory mites include *N. longispinosus*, *Typhlodromips* sp., and *Amblyseius* sp. of the family Phytoseiidae and *Agistemus* sp. belonging to Stigmaeidae.

13.3.3 Mite Fauna on Amaranthus (*Amaranthus* sp.)

On amaranthus, two species of phytophagous mites recorded were *Tetranychus* sp. and *B. phoenicis* (Geij.) of which *Tetranychus sp.* was the predominant one. Predatory mites included *N. longispinosus* Evans, *Amblyseius* sp., *Euseius* sp., *Scapulaseius* sp., and *Typhlodromips* sp. all belonging to the family Phytoseiidae.

13.3.4 Mite Fauna on Cowpea (*Vigna unguiculata* L.)

On cowpea, three phytophagous and three predatory mites were collected. The phytophagous mites included *Tetranychus* sp. and *Eutetranychus* sp. both belonging to the family Tetranychidae and *P. latus* (Banks) of the family Tarsonemidae. Predatory mites recorded were *N. longispinosus* Evans, *Typhlodromips* sp., and *Amblyseius* sp.

13.3.5 Mite Fauna on Chili (*Capsicum annum* L.)

P. latus was the only phytophagous mite recorded from chili. However, seven different species of predatory mites were collected during the study which included *Amblyseius paraaerialis* Muma, *Euseius* sp., and *Typhlodromips* sp. of the family Phytoseiidae, *Tydeus* sp. of the family Tydeidae, *Agistemus* sp. of Stigmaeidae, *Cunaxa* sp. belonging to Cunaxidae, and *Bdella* sp. of Bdellidae.

13.3.6 Mite Fauna on Bitter gourd (*Momordica charantia* L.)

One phytophagous mite, *P. latus*, and two predatory mites, namely, *Euseius macrospatulatus* Gupta and *Typhlodromips* sp., were recorded on bitter gourd.

13.4 Discussion

Faunal studies of mites in six vegetable crops revealed highest diversity of mites on brinjal and the least diversity on bitter gourd. Spider mites belonging to the genus *Tetranychus* were dominant phytophagous mites on brinjal, bhindi, amaranthus, and cowpea, whereas in chili and bitter gourd, the tarsonemid mite, *P. latus*, was the phytophagous mite. *B. phoenicis* was the only tenuipalpid mite observed during the study, and it was reported in amaranthus and brinjal. These mites were reported as important mite pests of vegetable crops from different parts of India (Gupta 1991; Gulati 2004; Rai and Indrajeet 2011). Karmakar (1997) reported the broad mite, *P. latus*, as one of the most destructive pests and a major contributing agent of the devastating "Murda" complex on chili.

The predatory mites found associated with vegetables in the present study belonged to five families, viz., Phytoseiidae, Stigmaeidae, Tydeidae, Cunaxidae, and Bdellidae among which Phytoseiidae predominates. Several species of phytoseiid mites were reported as effective predators of plant-feeding mites all over the world in many diverse crop ecosystems (Abhilash 2001; Sadanandan and Ramani 2006; Karmakar and Gupta 2010).

Predatory mites *Cunaxa* sp. of the family Cunaxidae, *Bdella* sp. of the family Bdellidae, and *Tydeus* sp. of the family Tydeidae (Fig. 13.3) were found in association with *P. latus* on chili. The Stigmaeid mites of the genus *Agistemus* (Fig. 13.3) were found to be associated with phytophagous mites on brinjal, bhindi, and chili. *Agistemus* sp. has gained a great economic importance as a biocontrol agent and plays a pivotal role in controlling phytophagous mites and soft-bodied insects on different vegetables (Khan et al. 2008).

The study has helped to develop a database on the major phytophagous and predatory mite fauna associated with important vegetable crops of Kerala. Additional studies have to be conducted to explore the host range and extent of damage caused by the mite pests in different vegetable crops. Further, extensive studies have to be carried out to assess the efficacy of various predatory mites for the identification of potent species and to standardize the mass production techniques for utilization in successful biological control programs.

13.5 Predatory Mites

From the past several years, predatory mites are being used to effectively control spider mites infesting European ornamental crops like chrysanthemum, rose, etc., as well as vegetables in greenhouses. Not only species selection but also release rates considerably vary with respect to the plant species and abiotic factors such as temperature and humidity which affect the growth rate for both the predator and prey. An excellent spider mite predator, *P. persimilis* is found on small plants in humid greenhouses with moderate temperatures. There are also some limitations with respect to *P. persimilis* as they cannot be used on crops like tomato because the mites become trapped on glandular hairs on the leaf petioles and stems and are also affected by toxic compounds in the tomato leaf. Similarly, *P. persimilis* slips off the stems and leaves of carnations. *P. macropilis* is the most effective biocontrol agent on ornamental plants, like parlor palm, dieffenbachia, schefflera, and dracena, under warm, humid conditions. *M. longipes* is often used in controlling spider mites in warm, dry greenhouses on taller plants as it can tolerate low humidity better than its counterpart *P. persimilis*. *N. californicus* performs best on several potted plants in greenhouses with ambient temperatures and average humidity. *G. occidentalis* and *N. californicus* are best suited for application on semipermanent greenhouse crops such as rose or gardenia than on short-term vegetable crops. A combination of predators released at regular intervals works best in greenhouses or interior plant capes with a variety of plant species and growing conditions.

Spider mites' distribution on a particular plant species is influenced by plant density and plant architecture and the ease with which the predators can find patches of prey. For example, *P. persimilis* is very efficient on cucumbers with large leaves which intermingle, but less effective on peppers with small leaves that don't touch each other. *P. persimilis* also possesses lower efficacy on cut rose varieties with fewer leaves because the movement for the mites around these plants is not so easy. *N. californicus* is a better option to control spider mites on roses, if early introduced. Pruning or Trailing of plants so that the leaves intermingle or touch may improve the efficacy of biological control on some plant species.

Cultural practices can have a significant impact on spider mites. Mites outbreak under dusty conditions. Water-stressed trees and plants are more susceptible to spider mite damage. Watering pathways as well as other dusty areas at regular intervals and providing adequate irrigation reduce spider mite infestation. Mid season washing of trees and vines with water to remove dust may help prevent serious late-season mite infestations. In gardens and on small fruit trees, regular, forceful spraying of plants with water often reduces spider mite numbers adequately. If more control is essential, the use of insecticidal soap or oil in spray is advised. But the product has to be tested on two or more plants to be sure it isn't damaging them.

Spider mites quite often become problematic after application of insecticides. Such outbreaks are common as a result of killing mite's natural enemies, and also certain insecticides stimulate mite reproduction. For example, spider mites exposed to carbaryl (Sevin) in the laboratory have shown faster reproduction than untreated populations. Carbaryl, some organophosphates, and some pyrethroids apparently favor spider mites by increasing the level of nitrogen in leaves. Pesticides applied during summer usually appear to have the greatest effect, causing dramatic spider mite outbreaks within few days of application.

If a treatment for mites is essential, the use of selective materials, preferably insecticidal soap or insecticidal oil, is recommended. Both petroleum-based horticultural oils and phenolic oils such as canola, neem, or cottonseed oils are acceptable. There are also several plant extracts formulated as acaricides that exert an effect on spider mites, which include clove oil, garlic extract, rosemary oil, mint oils, cinnamon oil, and others. Application of soaps or oils on water-stressed plants or when temperatures exceed 90 °F is not recommended. These materials inflict injury in some plants; testing them out on a portion of the foliage several days before applying a full treatment is advised. Oils and soaps must effectively come in contact with mites to kill them, so excellent coverage, especially on the undersides of leaves, is dire essential, and several repeat applications may be required.

Natural enemies play a significant role in managing pest populations. Utilization of predatory mites in strawberries against spider mites is a good example where farmers take advantage of the potential of the natural enemies. Multiple species of predatory mites are commercially available, and several California strawberry growers use biological control as a complementary tool to chemical control.

13.6 Predatory Mites Belong to Four Categories

Type I: These predatory mites are specialists feeding specifically on spider mites (family Tetranychidae) that produce webbing. *Phytoseiulus persimilis* Athias-Henriot is a Type I specialist which exclusively feeds on spider mites. This mite is bright orange, is tear drop-shaped, and moves rapidly. It prefers cooler climate and is sensitive to hot and dry conditions. So, it is effective in the initial stage of the crops before temperatures increase.

Type II: These are also specialist predators, but they feed on spider mites and other species of mites. They also feed on pollen and in some cases on thrips and other species of predatory mites. *Neoseiulus fallacis* (Garman) is a Type II specialist that primarily feeds on spider mites (Fig. 13.2b). It is translucent to peach or orange and appears to have a flatter body compared to spider mites or *P. persimilis*. It is also sensitive to hot and dry conditions. *Galendromus occidentalis* (Nesbitt) also known as western predatory mite is a Type II specialist that primarily feeds on spider mites. It prefers warm temperatures and tolerates dry conditions below 30 % relative humidity. It is sensitive to cooler temperatures.

Type III: These are generalist predators that feed on several species of mites including spider mites, eriophyid mites, and tarsonemid mites and insects such as thrips and whiteflies. *Neoseiulus californicus* (McGregor) is a Type II specialist that primarily feeds on spider mites but also has Type III generalist characters. It appears similar to *N. fallacis*. It can withstand warmer conditions better than *P. persimilis* and *N. fallacis*. It can withstand cold temperatures for short periods and tolerates relative humidity range from 40 % to 80 %. *Amblyseius andersoni* (Chant) is a Type III generalist predator. It can tolerate high temperatures when relative humidity is high.

Type IV: These mites primarily feed on pollen and can also feed on pest mites.

There are five species of predatory phytoseiid mites (Acari: Phytoseiidae) that are available commercially for spider mite control.

Live predator mites are generally shipped mixed in vermiculite, bran, or a similar material to cushion them in transit. The carrier-mite mixture can be applied directly onto the foliage of infested plants and thus the mites will disperse on their own. Predator mites can be applied uniformly throughout the greenhouse or can be concentrated in infested patches. Uniform distribution of predators throughout the greenhouse is the most common method of introduction. It provides predictable levels of control. However, application in patches of mite damage will often result in better control than uniform distribution.

The purchase and application of predatory mites are useful in the establishment of populations in orchards or large plantings, but the best results are obtained by the creation of favorable conditions for naturally occurring predators, like avoiding dusty conditions and insecticide sprays. In a heavily infested orchard or garden that has few predators, the use of soap spray or selective miticide brings pest mites to a lower level, and then predatory mites should be released. A good guideline is that for every ten spider mites, one predator is needed in order to effectively control them. More than one application of predatory mites may be required if pest populations have to be reduced rapidly. Applications in hot spots are suggested where spider mite densities are highest.

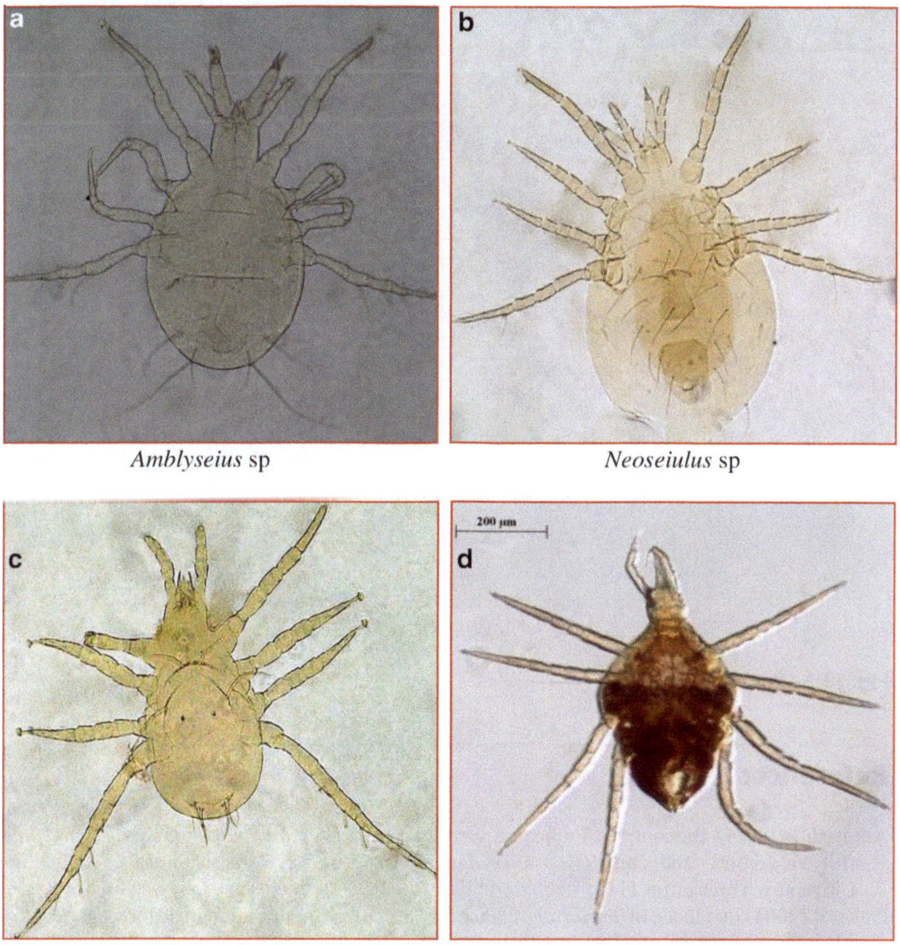

Amblyseius sp — *Neoseiulus* sp — *Paraphytoseius* sp — *Cunaxa* sp

Fig. 13.2 Predatory mites

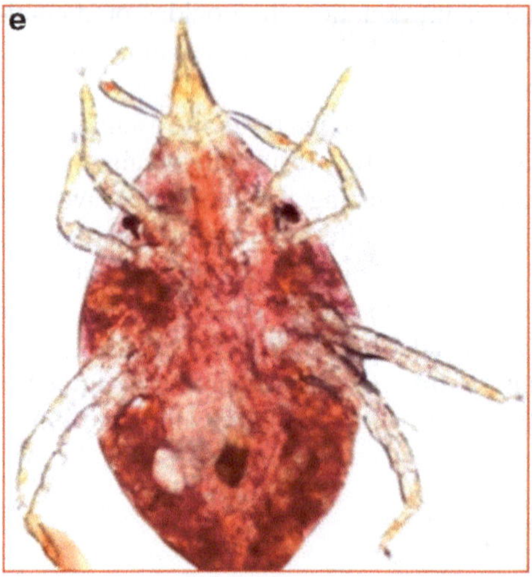

Bdella sp

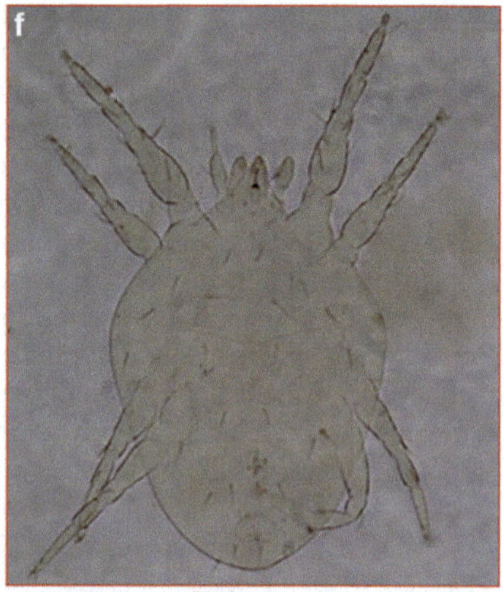

Tydeus sp

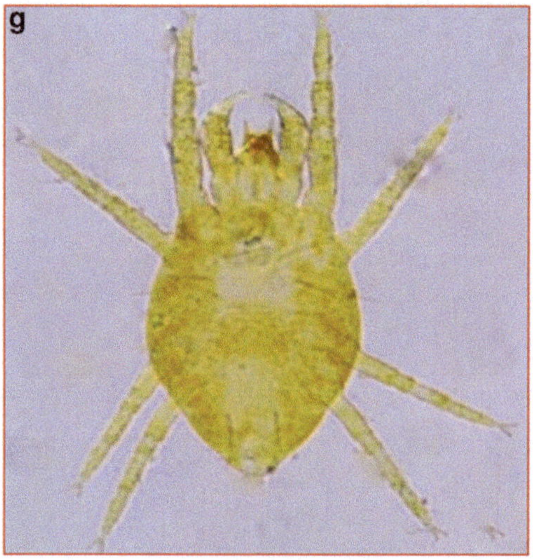

Agistemus sp

Fig. 13.3 Predatory mites

References

Abhilash B (2001) Biocontrol of mites on yard long bean (*Vigna unguiculata* sp. *Sesquipedalis* (L.) Verdecourt) and chilli (*Capsicum annum* (L.) MSc. (Ag) thesis, Kerala Agricultural University, Thrissur pp 114

Gulati R (2004) Incidence of *Tetranychus cinnabarinus* (Boisd.) infestation in different varieties of *Abelmoschus esculentus* (L.). Ann Plant Prot Sci 12:45–47

Gupta SK (1991) Mites of agricultural importance in India and their management. All India Coordinated Research Project on Agricultural Acarology. Tech Bull Indian Council of Agriculture Research, New Delhi

Gupta SK, Gupta A (1999) Progress of taxonomic research on Indian mites up to the end of twentieth century and prospects of research in the next millennium. J Acarol 15:80–83

Karmakar K (1997) Chilli mite *Polyphagotarsonemus latus* (Banks) a serious pest. Madras Agric J 84(8):218–220

Karmakar K, Gupta SK (2010) Diversity of predatory mites associated with agri-horticultural crops and weeds from Gangetic plains of West Bengal, India [Abstract]. In: International Congress of Acarology, 23–27August 2010, Recife-PE, Brazil pp 119

Khan BS, Afzal M, Bashir MH (2008) Effects of some morphological leaf characters of some vegetables with incidence of predatory mites of the genus *Agistemus* (Stigmaeidae: acarina). Pak J Bot 40(3):1113–1119

Rai SN, Indrajeet (2011) Note on phytophagous mites associated with common vegetables in Varanasi and Azamgarh districts of eastern Uttar Pradesh. J Insect Sci 24(2):199–200

Sadanandan MA, Ramani N (2006) Two new species of predatory mites acarina: phytoseiidae from Kerala, India. Zoos' Print J 21(6):2267–2269

Arthropod Communities Associated with Mango (*Mangifera indica* L.): Diversity and Interactions

14

Poluru Venkata Rami Reddy and Kolla Sreedevi

Abstract

Arthropod abundance and diversity are valuable indicators of the impact of agricultural practices on biodiversity. Mango (*Mangifera indica* L.), being an evergreen perennial tree, harbours diverse arthropod fauna at different phenological stages. The fauna associated with mango consists of pests, natural enemies, pollinators, millipedes and centipedes. Insects are the dominant arthropods both as harmful pests and beneficial organisms. Mites, though in small proportion, form an important group as pests. About 400 species of insect and mite pests are reported on mango. Of them only a few like leafhoppers, fruit flies, stone weevil, stem and shoot borers, mealybugs and leaf webbers are of economic importance. Among pestiferous arthropods, majority (about 45 % of total species) are foliage feeders followed by fruit feeders (32 %). Predator arthropods like spiders, ladybird beetles, mantids and ants help to maintain the general equilibrium in the ecosystem. Since mango is a cross-pollinated crop, insect pollinators form a crucial component of mango ecosystem and help to sustain the genetic diversity. Flies of families Calliphoridae and Syrphidae and honeybees constitute the major chunk of pollinators. Intensive orchard system of mango involving large-scale use of pesticides, clean cultivation and dominance of a few varieties resulting in narrowed genetic diversity has adversely impacted the species richness of arthropods. Adoption of ecologically sustainable crop management practices is essential to conserve the diversity of beneficials.

P.V.R. Reddy (✉)
Division of Entomology and Nematology, Indian Institute of Horticultural Research (IIHR), Hessaraghatta Lake Post, Bengaluru 560089, Karnataka, India
e-mail: pvreddy2011@gmail.com

K. Sreedevi
Division of Entomology, Indian Agricultural Research Institute (IARI), New Delhi 110012, India

A.K. Chakravarthy, S. Sridhara (eds.), *Economic and Ecological Significance of Arthropods in Diversified Ecosystems*, DOI 10.1007/978-981-10-1524-3_14

Keywords
Arthropods • Bioecology • *Mangifera indica* • Species richness

14.1 Introduction

Mango (*Mangifera indica* L.) belonging to the family Anacardiaceae, aptly called the King of Fruits, is the most important commercially grown fruit crop in India and other tropical countries. It originated in the Indo-Burma region and has become naturalized and adapted throughout the tropics and subtropics (Mukherjee 1953). There are more than 1000 named mango varieties all over the world, and India has the richest collection of them. Besides commercial value, the cultural and religious importance of mango is well acknowledged and recorded (Mukherjee 1997). Major mango-producing countries are India, China, Thailand, Mexico, Pakistan, the Philippines, Indonesia, Brazil, Nigeria and Egypt. India ranks first and accounts for about 50 % of the world's mango production with an area of 2.5 million hectares producing annually 18.0 million tons. Significant contributors to the national mango pool are Andhra Pradesh, Bihar, Gujarat, Karnataka, Maharashtra, Odisha, Tamil Nadu, Uttar Pradesh and West Bengal (Indian Horticulture Database 2013).

Mango, at both individual tree level and orchard level, sustains several kinds of arthropods. Their diversity, abundance and interactions play a significant role in shaping up the production potential of the mango crop. Pests, predators, parasitoids and pollinators represent the wide diversity of arthropods in an agricultural ecosystem. Besides them, certain species of centipedes and millipedes, though in small proportion, add to the richness of arthropod diversity.

14.2 Arthropod Pests

Mango trees are vulnerable to attack by a number of insect and mite pests. About 400 species of insect pests are known to occur on mango in different parts of the world (Peña et al. 1998). Worldwide lists of pests of mango have been compiled by de Laroussilhe (1980), Tandon and Verghese (1985) and Veeresh (1989). In a precise division of pests according to the part they attack, majority (about 45 % of total species) are foliage feeders followed by fruit feeders (32 %), and the rest feed on the inflorescence, branches and the trunk. The three or four key pests, including fruit flies, seed weevils, tree borers and leafhoppers, require annual control measures. Secondary pests may become serious pests as a result of changes in cultural practices or climate and or cultivars or indiscriminate use of insecticides against a key pest. Mohyuddin and Mahmood (1993) reported that scale insects became serious pests following non-judicious use of insecticides against fruit flies. Similarly, mites, considered as minor pests, may become serious because of human intervention. Occasional or incidental pests also can cause economic damage only in localized areas at certain times. An account of the major species of pestiferous arthropods on mango in presented below.

14.2.1 Hemiptera

14.2.1.1 Leafhoppers (Family: Cicadellidae)

Diversity and Distribution

Among arthropod fauna associated with mango, leafhoppers are the most widely studied group considering their economic importance, with a potential to cause even complete crop loss. Thirty-seven species of Auchenorrhyncha in seven families are associated with mango in the world, and the group forms a major pest taxon of mango in the oriental region (Viraktamath 1989).

The most predominant and widespread are *Idioscopus clypealis* Lethierry, *I. nitidulus* (Walker), *I. nagpurensis* Pruthi and *Amritodus atkinsoni* Lethierry (Veereh 1989; Waite 2002) of which *I. nitidulus* is considered more destructive (Sohi and Sohi 1990). This hopper breeds on both shoots and inflorescence unlike *I. clypealis* and *I. nagpurensis*, which breed only on inflorescence (Verghese and Devi Thangam 2011). However, the intensity of the occurrence of different species varies across places. For example, all species occur together in Bihar and South India, or one particular species may dominate as in the case of *A. atkinsoni* and *I. clypealis* in Punjab and *I. niveosparsus* in Gujarat (Veereh 1989). Dalvi et al. (1992) reported species composition of hoppers in the Konkan region of Maharashtra. Viraktamath and Viraktamath (1985) reported three new species of mango hoppers, viz. *Busoniominus manjunathi*, *Idioscopus anasuya* and *I. jayshriae*, in Karnataka. Outside India, *I. clypealis* was reported to constitute >95 % of hopper population in the Philippines (Alam 1994), and *I. incertus* Baker was found to be a serious pest in China (Waite 2002). In Australia, *I. niveosparsus* was recorded for the first time in 1998 in Queensland in a remote sea port but did not spread to commercial plantations (Waite 2002).

Bioecology and Host Plant Association

Leafhoppers are monophagous and feed only on mango. Both nymphs and adults congregate on and suck the sap from tender shoots (Fig. 14.1), inflorescence and occasionally fruits. The continued feeding results in withering and dropping of florets thus leading to failure of fruit setting. Leafhoppers excrete honeydew which attracts sooty mould and affects photosynthetic efficiency (Butani 1979). The honeydew also has potential to distract honeybees from flowers which are low in nectar thus affecting the pollination (Pena and Mohyuddin 1997).

Several workers studied the biology of hoppers in different countries (Patel et al. 1973; Pena et al. 1998). The number of generations varies with place. The insect overwinters as adult. Patel et al. (1973) noted these hoppers to rest in the cracks and crevices of the bark of mango during hot noon and rainy days. The population reaches a peak during March–April, and maximum and minimum temperature and

Fig. 14.1 Congregation of nymphs of mango leafhoppers. Honeybees feeding on the honeydew excreted by leafhoppers

relative humidity were major abiotic factors contributing to population fluctuations (Tandon et al. 1983).

The spacing of mango trees in orchards also plays an important role in breeding of the hoppers. The orchards with closer spacing and varieties of dense inflorescence attract high hopper population (Reddy and Dinesh 2005). The distribution of hopper nymphs on a tree across four directions and at different canopy levels was found uniform (Tandon et al. 1983). Nachiappan and Bhaskaran (1983) categorized the varieties *Baneshan*, *Chinnarasam*, *Totapuri* and *Khadar* as resistant varieties, *Gaddemar* and *Rumani* as moderately resistant, *Himayuddin* as susceptible and *Padri*, *Neelam*, *Mulgoa*, *Peter* and *Sindura* as highly susceptible to hopper infestation. The resistance was found to be correlated with the biochemical composition of leaves. While higher nitrogen content has rendered the plant more susceptible, more of phosphorus, potassium, calcium and phenols conferred resistance. Khaire et al. (1987) screened 19 varieties and found *Rajmana* and *Vanraj* least susceptible to *I. clypealis*. Srivastava et al. (1982) reported *Langra*, *Bombay Green* and *Neelum* as highly susceptible to mango hoppers, while *Baneshan*, *Ratna* and *Mallika* were found to be moderately resistant. Of 392 accessions evaluated by Devi Thangam et al. (2013), 32 were identified as least susceptible to leaf hoppers based on vegetative-phase screening.

14.2.1.2 Mango Mealybugs (Family: Pseudococcidae)

Diversity and Distribution

More than 20 species of mealybugs attack mango. Of them, three species, viz. *Drosicha mangiferae*, *D. stebbingi* and *Rastrococcus iceryoides*, are serious in nature and are found in India, Nepal, Bhutan, China, Pakistan and Bangladesh, while *Rastrococcus iceryoides* is reported from Malaysia (Tandon and Verghese 1985).

Fig. 14.2 Mealybug colonies tended by black ants on a mango panicle

Host Plant Association, Bioecology and Symptomatology

Damage is caused by feeding of nymphs and female adults on the terminal parts of the panicles and shoots. They remain stationary and adhere to the panicles and shoots. Affected panicles shrivel and get dried resulting in size reduction and premature dropping of fruits (Singh and Mukherjee 1989).

The gravid females crawl down the trees during March to May and enter the soil (80–150 mm deep) wherein they excrete whitish foam containing 400–500 eggs and die soon after the oviposition. Eggs remain in diapause till the winter sets in. Soon after hatching, the majority of nymphs start crawling up the tree trunks, and clusters of these may be seen on young shoots and panicles, sucking the cell sap (Fig. 14.2). A few nymphs crawl away to neighbouring trees as well during December–January. Mealybugs are more active on bright sunny days. The population peak is observed during March–April on inflorescence (Butani 1979).

14.2.1.3 Mango Shoot Gall Psyllid (Family: Psyllidae)

Diversity and Distribution

Mango shoot galls were first recorded from Dehra Dun, North India, and were described first as *Psylla cistellata*. The species was later changed to *Apsylla cistellata* Buckton. The psyllid is a monophagous pest of mango and is distributed in plains of North India, Nepal and Bangladesh (Tandon and Verghese 1985) and in north-eastern states (Singh 1978).

Bioecology and Host Plant Association

The psyllid induces apical and axillary buds to form cone-shaped green galls on mango shoots directly interfering with the formation of inflorescence and ultimately affect the yield of mango crop (Singh 1978). About 140–145 eggs are laid in the midrib of the tender leaves in two parallel rows during March–April (Fig. 14.3). Nymphs start feeding in situ starting from mid-August and, while feeding, secrete chemicals like phenyl amino acids which initiate the buds to grow and convert into galls. They have a single generation per year. The number of galls is proportional to the number of embryos feeding on the leaves. Prasad (1971) worked out a relationship between the number of galls and nymphs on a shoot.

14.2.1.4 Scales (Family: Coccidae)

Diversity and Distribution

Three types of scales, viz. fluted (Monophlebids), armoured (Diaspids) and soft (Lecanids), have been reported infesting mango all over India. Though about 15 species are recorded, the armoured scale, *Aspidiotus destructor*, is the most economically important one. Besides India, this insect has been reported from Sri Lanka, China, Taiwan, Fiji Island, Mexico, West Indies, British Guiana, Africa, Mauritius, etc. In India, it is found throughout the plains and foot of the hills (Butani 1979). Besides mango, it affects banana, guava, *jamun*, papaya, etc. In Australia and South Africa, the mango scale, *Aulacaspis tubercularis* (Newstead), is considered a serious pest as its infestation on fruits and leaves blemishes resulting in loss market value (Joubert 1997). Swirski et al. (1997) listed 63 species of soft scales of which *Ceroplastes pseudoceriferus* Green is a key pest.

Bioecology and Host Plant Association

Nymphs and adults of scales suck the sap from leaves and fruits. Though scales are not considered as major pests, severe infestation may affect the growth and fruit

Fig. 14.3 Mango leaf galls

setting capacity of the tree. Female scale is circular, semi-transparent and pale brown and reproduces oviparously. Total life cycle lasts for 32–34 days. Nymphs are small (2 mm long), oval, translucent, yellowish-brown crawlers which colonize underside of leaves and tender shoots. They are covered with waxy material and become immobile. A number of parasitoids are recorded to exercise natural control of the scales in India (Sankaran 1955), Bangladesh (Ali 1978) and Taiwan (Wen and Lee 1986). Branches with high density of scales showed more decline in mango yield than branches with low scale density. Nevertheless the role of scales in mango decline is yet to be established (Pena 1993).

14.2.2 Coleoptera

14.2.2.1 Stone Weevils (Family: Curculionidae)

Diversity and Distribution

Three species of curculionids, viz. *Sternochetus mangiferae* (Fabricius), *S. gravis* and *S. frigidus*, have been reported as pests inhabiting mango fruits (Tandon and Verghese 1985), of which *S. mangiferae* is the most widely distributed and serious and specific pest of mango. It is distributed in almost all mango-growing areas except the Canary Islands, Italy, Israel and Egypt (IIE 1995; Tandon and Verghese 1985). The other species, *S. gravis*, is found in India, Indonesia, Malaysia, Myanmar and Pakistan, while *S. frigidus* is restricted to Malaysia, Pakistan and Thailand. Mango stone weevil was not reported from America until it was found in the southern Caribbean region (Johnson 1989). In India, *S. mangiferae* is distributed across the country but is serious in southern states, while *S. gravis* is confined to north-eastern parts.

Bioecology and Host Plant Association

Eggs are laid singly in depressions along the fruit surface. On hatching, the grubs enter the nut or stone. Initially damage is caused by feeding on the outer coat of the stone in a zigzag fashion. After hatching, the larvae burrow through the pulp to the young, developing seed (Fig. 14.4). Consequently, complete stone is destroyed leaving behind a black mass. Damaged stones lose their viability and the fruits become unsuitable for consumption and processing. Generally, only a single larva completes development in each fruit, but as many as five larvae have been found (Hansen et al. 1989). Extent of damage can be up to 60–65 % in susceptible varieties like *Neelum*, *Totapuri* and *Banganpalli*. Weevils hibernate in cracks and crevices of the tree trunk and under the fallen leaves (Shukla and Tandon 1985). Verghese et al. (2005) reported an association between stone weevil infestation and fruit drop. Similarly in Hawaii, Follett (2002) found that stone weevil infestation could increase fruit drop in mango during early fruit development.

Fig. 14.4 Mango stone weevil infestation on fruits

14.2.2.2 Mango Stem Borers (Family: Cerambycidae, Scolytidae and Buprestidae)

Diversity and Distribution

The larval stages of certain species of the cerambycid longicorn beetles inhabit mango trunk and stem for prolonged periods of 6–8 months. The beetles of the genus *Batocera* are considered a serious problem to mango in India (Veereh 1989). The species recorded in India include *Batocera rufomaculata* (Geer), *B. rubus* (Linnaeus), *B. roylei* (Hope), *B. numitor* (Newmann) and *B. titana* (Thomson) (Butani 1979). Of them, *Batocera rufomaculata* De Geer is the most destructive and frequently found borer in mango orchards (Fig. 14.5). Besides mango, beetles attack fig, jackfruit, mango, mulberry, papaya, apple, etc.

Other longicorn beetles attacking mango trees in India are *Acanthophorus serraticornis* (Oliver), *Anoplophora versteegi* (Ritseema), *Epepeotes luscus* (Fabricius), *Rhytidodera bowringi*, *R. simulans* (White) and *Stromatium barbatum* (Fabricius). A few scolytids (shot-hole beetles) like *Hypocryphalus mangiferae* (Stebbibg), *Xyleborus keraatzi* (Eichhoff) and *X. semigranosus* (Blandford) have also been reported infesting the mango stems, though not of much economic importance (Butani 1979). In a recent study, Reddy et al. (2014) reported that at Bengaluru in South India, besides *B. rufomaculata*, mango is attacked by two other cerambycids, viz. *Glenea multiguttata* Guerin-Meneville and *Coptops aedificator* (Fabricius) and one buprestid. In Pakistan, the scolytid *H. mangiferae* was reported to be associated with sudden death disease of mango (Masood et al. 2009).

Bioecology and Host Plant Association

Generally the older trees of more than 15 years old or those already weakened from other causes, either pathological or environmental, are more vulnerable to attack by stem borers (Waite 2002). Female beetle lays eggs singly on the main trunk of relatively older mango trees. A single beetle lays up to 200 eggs. After hatching from

Fig. 14.5 Mango stem borer, *Batocera rufomaculata*

the egg, the neonate larva initially feeds under the bark and then tunnels through the sapwood (about 2–3 cm width) which interferes with sap flow and affects foliage and production. A hole with dripping sap and frass on the bark are symptoms visible in advanced stages of infestation. The damage results in the yellowing of branches followed by drying and die-back of terminal shoots and branches ultimately leading to the death of the whole tree. The beetle emerges in July–August and there is only one generation per year. The life span of the beetle is about 8 months (Krishnamoorthy et al. 2014). Varietal preference of borer is evident with 'Alphonso', 'Langra' and 'Jehangir' being the most susceptible (25–50 % damage) and 'Himayuddin' and 'Banganapalli' being susceptible (Reddy et al. 2015). Rootstock and spacing influence the borer infestation levels (Reddy et al. 2015).

14.2.2.3 Leaf-Cutting Weevils (Family: Curculionidae) and Defoliating Beetles (Family: Scarabaeidae)

A few weevil species feed on tender foliage and cause extensive defoliation. Of them, mango leaf-cutting weevil, *Deporaus marginatus* (Pascal), is important. It is found all over India, Bangladesh and Burma and is a specific pest of mango. The pest is active from August to October. Young trees suffer comparatively more than the older ones. Eggs are laid singly on either side of the midribs on lower surfaces of tender leaves. The leaves are then cut by the weevils, near the base. Upon hatching, the grubs mine between the two epidermal layers of the leaf and feed within. When fully grown, the grubs come out of the mines to pupate in the soil. There are three generations in a year (Butani 1979).

Fig. 14.6 Leaf-mining weevil, *Rhynchaenus mangiferae*

Mango leaf-twisting weevil, *Apoderus transquebaricus* (Fabricius), is another minor pest of mango reported from South India. Its other main host is almond, *Terminalia catappa* Linnaeus. Eggs are laid singly on leaf tips. These leaves are then rolled tip downwards into neat thimble-shaped structures, and the earlier stages of the pest are passed within these rolled leaves. The adults come out by making a hole at the side of these rolled leaves.

Mango leaf-mining weevil, *Rhynchaenus mangiferae* Marshall (Fig. 14.6), is widely distributed in South India, where it is active during March to July. As many as 20–30 grubs may be found in a single leaf. Affected leaves turn reddish-brown, crumple, dry and fall off. 'Langra' variety of mango is most susceptible to this weevil. Other curculionid weevils that feed on the mango leaves include *Alcidodes* spp., *Myllocerus discolor* Boheman, *M. laetivirens* Marshall, *M. undecimpustulatus maculosus* Desbrocher and *M. sabulosus* Marshall. These are all pests of minor importance reported occasionally from various parts of India.

The adults of white grubs (Scarabaeidae), viz. *Anomola* sp. and *Holotrichia* sp., attack leaves in groups during the night in monsoon season (June–July). These are polyphagous and, besides mango, are recorded feeding on several fruit and avenue trees (Butani 1979).

14.2.3 Lepidoptera

14.2.3.1 Mango Leaf Webbers (Family: Pyralidae)

Diversity and Distribution

Two leaf webber species, viz. *Orthaga euadrusalis* Walker and *O. exvinacea* Hampson, have been recorded from North and South India, respectively (Butani 1979). Hampson (1896) recorded *Orthaga euadrusalis* from India, Sri Lanka and

Indonesia, and *Orthaga chionalis*, *O. melanoperalis*, *O. icarusalis*, *O. leucatma* and *O. vitialis* were reported from Sri Lanka (Rajapakse and Kulasekera 1982).

Bioecology and Host Plant Association

Eggs are laid singly or in clusters within silken webbings or on leaves. Upon hatching, the larvae feed gregariously by scraping the leaf surface (Fig. 14.7). Soon they web together tender shoots and leaves and feed within. Several caterpillars may be found in a single webbed cluster of leaves, and pupation takes place inside these webs in silken cocoons. As a consequence of severe feeding, clusters of webbed leaves become dry and brown. Affected trees present sickly appearance and can be observed from a distance due to brown, dry, clustered leaves. Though precise data on crop losses is lacking, *O. euadrusalis* has been reported to cause 25–80 % damage (Srivastava et al. 1982). The pest completes several overlapping generations from July to December on mango trees. Three distinct peaks can be observed in first fortnight of August, September and October. Verma and Singh (2010) found no significant correlation between webber infestation and weather parameters.

14.2.3.2 Mango Shoot Borers (Family: Noctuidae)

Diversity and Distribution

Mango shoots are infested by two species of borers, viz. *Chlumetia transversa* Walker and *C. alternans* Moore. Of them, the former causes extensive damage to young plants. Besides mango, it also attacks litchi leaves. It is found in India, Sri Lanka, Malaysia, the East Indies, the Philippines and Indonesia (Butani 1979).

Fig. 14.7 Leaf webber infestation on mango

Bioecology and Host Plant Association

Oval, pale yellow eggs are laid singly on tender shoot or flower panicle which hatch in 2–3 days. The caterpillars first bore into the midribs for a few days and later tunnel into the shoot downwards (Chahal and Singh 1977).

14.2.3.3 Bark-Eating Caterpillars (Family: Cossidae)

Diversity and Distribution

Bark-eating caterpillars are found all over India and subcontinents including Bangladesh, Burma, Sri Lanka and Pakistan damaging *aonla*, ber, citrus, guava, *jamun*, litchi, loquat, mango, mulberry, pomegranate and a number of forest and ornamental trees. *Indarbela quadrinotata* (Walker) is the most commonly found species on mango, and *I. tetraonis* (Moore), normally associated with guava, is also found attacking mango especially when guava trees are around. *Indarbela dea* (Swinhoe) and *I. theivora* (Hampson) are other minor species reported from India.

Bioecology and Host Plant Association

Adults are pale brown moths. Freshly hatched larvae nibble the tree trunk and after 2–3 days bore into the stem and feed within. This interrupts the transportation of cell sap (Butani 1977). The caterpillars spin silken webs consisting of their excreta and chewed wood particles which are seen hanging loosely on the bark of affected trees, more commonly at the junction of two stems or main branches. Pupal period varies between 21 and 31 days. Moth emergence continues till June and their longevity is not more than 3 days. There is only one generation in a year. The older trees and neglected orchards are more prone to pest incidence than the clean and well-maintained orchards (Butani 1979).

14.2.3.4 Fruit Borers (Family: Pyralidae)

The major borer pest of mango fruit is *Deanolis albizonalis* (Hampson) which is synonymous with *Autocharis albizonalis* (Hampson) and *Noorda albizonalis* Hampson. Commonly called red-banded mango caterpillar, or mango seed borer, this pest was reported for the first time in India as early as 1955 (Sengupta and Behura 1955), and the severe incidence of *D. albizonalis* was later reported from coastal districts of Andhra Pradesh (Sujatha and Zaheeruddin 2002). It is widely distributed in mango-growing areas of the West Bengal and east coast of Andhra Pradesh causing 10–52 % damage of fruits from pin-headed stage to full maturity. All major belts in South India are free of the borer. It is likely to spread across major mango-growing areas, unless strict stringent domestic quarantine is put into regulation (Krishnamoorthy et al. 2014). Upon hatching, larvae enter the fruit by boring holes on the apex or narrow tip of the fruit. It tunnels through flesh and skin and then feeds on seed causing fruit spoilage and premature fruit drop. Larvae feed on mango fruit at all stages of development. The first sign of infestation is the presence of a sap stain running from the caterpillar's entry hole and collecting on the drip point at the fruit apex. The sap darkens over time and becomes very noticeable (Fenner 1997). The larvae usually enter through one hole, typically laid in the lower half of the fruit

(Krull 2004). First and second instar caterpillars feed just beneath the skin surface, tunnelling towards the seed. Later instars feed on the seed itself (Kalshoven 1981; Waterhouse 1998). Krull (2004) observed in Papua New Guinea that mango fruits of all sizes were attacked, but marble-sized fruits were preferred sites for oviposition. *Banganapalli*, *Langra*, *Fazil*, etc., are most susceptible cultivars. Though there are a couple of studies on the biology of this borer, the bioecology is not completely understood (Singh and Kishore 2014).

Citripestis eutraphera (Meyrick) is another mango fruit borer. It is geographically distributed in Java, Indonesia, India and Northern Territory in Australia. However, in India, the only official record of *C. eutraphera* is from the Andaman Islands reported by Bhumannavar (1991) on local endemic mango species, *Mangifera andamanica* L. It was also reported as a major pest on cashew, *Anacardium occidentale*, another member of the Anacardiaceae family from the Andaman Islands (Jacob et al. 2004). However, Jayanthi et al. (2014) recorded its occurrence in Karnataka and Kerala indicating the geographical expansion of the pest. Like a typical pyralid, larval stage bores the fruit and feeds on pulp.

Castor capsule borers, *Conogethes punctiferalis* (Guenee) and *Hyapsila leuconeurella* (Ragonot), are occasionally found boring mango fruits and are considered to be of minor importance (Butani 1979).

14.2.4 Diptera

14.2.4.1 Fruit Flies (Family: Tephritidae)

Diversity and Distribution

Fruit flies are serious pests of mango in most parts of the world and cause economic losses (Veereh 1989; Verghese et al. 2011). They are also the major constraint in the export of fresh mango fruits to foreign countries. There are several species of Tephritidae associated with mango across the globe. White and Elson-Harris (1992) revised the taxonomy of fruit flies and reported 48 species of fruit flies attacking mango across the globe. They belong to genera *Anastrepha* (eight spp.), *Bactrocera* (30 spp.), *Ceratitis* (seven spp.), *Dirioxa* (two spp.) and *Toxotrypana* (one sp.) (Wharton and Marsh 1978). All *Dacus* species attacking mango have recently been placed under the genus *Bactrocera*. Kapoor (1970) listed 128 species of fruit flies, and out of these eight are found infesting mango fruits in India. They are *Bactrocera dorsalis*, *B. zonata*, *B. correcta*, *B. caryeae*, *B. diversa*, *B. cucurbitae*, *B. hageni* and *B. tau*. In India, *Bactrocera dorsalis* (Hendel), commonly called the Oriental fruit fly, which earlier was considered to be a species complex, is credited as the most important and dominant species. It is reported from India, Sri Lanka, Myanmar, Nepal, Bhutan, Thailand, Vietnam and Cambodia in Asia. The insect is distributed throughout India; in the north it overwinters in pupal dormancy, but in the south it is active throughout the year. *B. dorsalis* occurs on a wide range of fruit crops including guava, custard apple, banana, papaya, peaches and plums (Tandon 1995). *Anastrepha* species are endemic to the Western Hemisphere and their range extends from the Southern USA to Northern Argentina and includes the Caribbean islands. *A. oblique* (Macquarat) is the most common fruit fly pest in the Americas (Aluja 1994).

Bioecology and Host Plant Association

Female fly punctures the skin of mature fruits with ovipositor and inserts white banana-shaped eggs (six to ten/batch) in clusters into the mesocarp. Upon hatching (after 1–2 days), the maggots tunnel into the fruit and feed on the pulp. It is hypothesized that a female fly chooses a fruit with a higher biomass (pulp) that will suffice for the development of all the maggots that hatch from the eggs it can potentially lay at a point of time (Verghese et al. 2011). All commercial varieties of mango are susceptible. However, *Langra*, *Dashehari* and Bombay Green are least susceptible (Jothi et al. 1994). Early harvesting would bring down the infestation.

14.2.4.2 Gall Midges (Family: Cecidomyiidae)

Diversity and Distribution

A midge is a tiny dipteran fly, which as adult is harmless and is short lived and dies within 24 h of emergence after copulation and oviposition. Mango is attacked by about 16 species of midges in Asia and in the Caribbean region. Two genera, *Procontarinia* Kieffer and Cecconi and *Erosomya* Felt, are particularly associated with mango (Harris and Schreiner 1992). In India, 12 species of midges representing three genera are known to produce different types of galls on mango leaves. Five species including *Erosomyia indica* Grover and Prasad are reported to attack mango flowers (Kulkarni 1955; Prasad 1972; Butani 1979). The mango gall midge or blister midge, *Erosomya mangiferae* Felt, is a major pest on flowers (Fig. 14.8) and reduces up to 70 % of fruit set (Verhgese et al. 1988a). Similarly the leaf gall midge, *Procontarinia matteiana*, is a serious pest of mango in Oman. This pest is distributed in India, Indonesia, Kenya, Mauritius, Oman, Reunion, South Africa and the United Arab Emirates.

Fig. 14.8 Mango blossom midge

Bioecology and Host Plant Association

The midge infests and damages the crop at the floral bud burst stage and young fruiting stage and on foliage. Infested panicles have a characteristic right-angled bend, with an exit hole, from which last instar maggots emerge to pupate in the soil. The second generation then infests on very young fruits, which eventually drop before the marble stage (Prasad 1971). Harris and Schreiner (1992) reported that the rainfall and temperature are the most influential abiotic factors that affect the populations of the midge. Mature larvae undergo diapause in the soil and break diapause on the arrival of favourable conditions in the following January. Verghese et al. (1988) studied the spatial distribution of blister midge and developed a sampling plan. They reported that infestation in east lower, south lower and south upper sections of canopy correlated significantly with total population.

14.2.5 Thysanoptera

14.2.5.1 Thrips (Family: Thripidae)

Diversity and Distribution

Thrips are among an emerging group of sucking pests that infest leaves, flowers and fruits of mango. From India, the species reported on mango are *Aeolothrips collaris* (Priesner), *Anaphothrips sudanensis* (Trybom), *Caliothrips indicus* (Bagnall), *Rhipiphorothrips cruentatus* (Hood), *Selenothrips rubrocinctus* (Giard), *Haplothrips ganglbaueri* (Schmutz), *Neoheegeria mangiferae* (Priesner), *Ramaswamiahiella subnudula* (Karny) and *Scirtothrips dorsalis* (Hood). Of these, the first four species feed on the leaves and the last four infest the inflorescence (Butani 1979). Tandon and Verghese (1987) recorded for the first time *Thrips palmi* Karny infesting mango flowers in India, which caused scab-like feeding marks and retarded fruit development. Outside the subcontinent, *Frankliniella bispinosa* (Morgan) and *F. kelliae* (Sakimura) were reported infesting mango blossoms and feeding on the nectaries and anthers in Florida (Pena 1993). The western flower thrips, *Frankliniella occidentalis* (Pergande), was found to damage flowers and fruits in Israel (Wysoki et al. 1993), while *Scirtothrips dorsalis* (Hood) is regarded as an important pest of mango flowers in Thailand and the plague thrips, *Thrips imaginis* Bagnall, in Australia (Waite 2002). *Thrips hawaiiensis* (Morgan), *Scirtothrips dorsalis* (Hood), *Frankliniella schultzei* (Trybom) and *Megalurothrips usitatus* (Bagnall) were recorded as pests of mango in Malaysia (Aliakbarpour and Che Salmah 2010).

Bioecology and Host Plant Association

Thrips colonize the leaves, inflorescence, fruit and new flush and suck the sap by lacerating the tissues (Higgins 1992; Pena et al. 2002). Apart from weakening the inflorescence and reducing fruit set, thrips cause serious bronzing of the fruit surface due to the presence of air in emptied cell cavities which is more pronounced in mature fruits (Lewis 1973). In case of severe infestation, the leaf tips turn brown and get curled (Aliakbarpour and Che Salmah 2010). *Thrips palmi* showed preference to lower canopy over upper canopy (Verghese et al. 1988b). Populations reach a peak during hot dry weather.

14.2.6 Isoptera

14.2.6.1 Termites (Family: Termitidae)

Termites or white ants are social insects that live in colonies called the termitaria. The species on mango trees in India include *Coptotermes heimi* (Wasmann), *Neotermes bosei* (Snyder) (*gardneri* Snyder), *N. mangiferae* (Roonwal and Sen-Sarma), *Heterotermes indicola* (Wasmann), *Microtermes obesi* (Holmgren) (*anandi* Holmgren), *Odontotermes assmuthi* (Holmgren), *O. feae* (Wasmann), *O. obesus* (Rambur), *O. lokanandi* (Chatterjee and Thakur), *O. gurdaspurensis* (Holmgren and Holmgren), *O. wallonensis* (Wassman) and *O. horai* (Roonwal and Chotani) along with *Microtermes obesi* (Holmgren). Of these, the most destructive one, found all over India, is *Odontotermes obesus* (Butani 1979; Pena and Mohyuddin 1997).

The insects remain underground and feed on roots and then move upwards making the trunks completely hollow, or they construct mud galleries mostly during night on tree trunks. Under the protection of these galleries, termites feed on the bark of the trunks. They are active all year round, though the incidence during monsoon months is rather low.

14.2.7 Hymenoptera

14.2.7.1 Ants (Family: Formicidae)

Ants are important components of ecosystems with high ecological value and function as 'ecosystem engineers'. They are omnivores and play an important role in linking the food web (Folgarait 1998). Mango trees are often found harbouring some ant species which are not considered pests but thought to influence the orchard production dynamics. Of them, *Oecophylla smaragdina* (Fabricius), called the red ant, green ant or weaver ant, is the most widespread and frequently encountered species (Fig. 14.9). It is reported from the entire oriental region extending from Australia to Africa (Atwal 1963). The ants web and stitch together a few leaves usually at the top of the branches and build their nests on trees. These nests are waterproof and the leaves remain green as they are not detached from the trees. The ants are carnivorous and considered effective predators of a range of soft-bodied insect pests like thrips, fruit-spotting bugs, scales, seed weevil, etc. (Butani 1979). However, ants attain indirect pest status, as they deter the predators of scales, aphids, mealybug, etc. They are also considered nuisance to orchardists and pose problem while harvesting. Verghese et al. (2013) developed a quick and non destructive sampling plan to estimate the number and biomass of *O. smaragdina* using brood nest as index in fruit crop ecosystem.

Fig. 14.9 Red ant nest on a mango tree

14.2.8 Arachnida: Acari

14.2.8.1 Mites

Diversity and Distribution

In any agricultural ecosystem, mites constitute the major non insect arthropod pest group. Phytophagous mites belong mainly to families Tetranychidae, Eriophyidae, Tarsonemidae and Tenuipalpidae. They cause damage either directly by feeding or spreading diseases as vectors. On mango, mites are known to infest leaves, buds and fruits. The bud mite, *Eriophyes mangiferae* (Sayed), is a major pest in the northern states of India (Singh and Mukherjee 1989). The infestation starts from April and gradually reaches a peak in June. The mango bud mite attack results in proliferation of shoots on the terminal, giving rise to a witches' broom effect. In association with the fungus, *Fusarium* sp., mite infestation results in floral and foliar galls resembling witches' broom (Ochoa et al. 1994). In Florida, *E. mangiferae* is reported to be associated with malformed mango flowers (Pena 1993), and hence there is an impression that it may be vectoring the diseases that could be the real cause of the malformation.

Spider mites, belonging to the genus *Oligonychus* of the family Tetranychidae, are the other group of mites that feed on the upper surface of mango foliage. The mango mite, *Oligonychus mangiferae* Rahman and Sapra, is a common pest in India, Egypt, Mauritius, Peru, Israel and some parts of Asia. In other countries like Australia and Central America, the tea red spider mite, *Oligonychus coffeae*

(Nietner), and the avocado brown mite, *Oligonychus punicae* (Hirst), are reported be minor pests of mango (Cunningham 1989). Besides these mites, broad mite, *Polyphagotarsonemus lotus* (Banks), of the family Tarsonemidae was reported to occasionally infest the nursery seedlings causing stunting and crinkling of new leaves and rolling of leaf margins.

14.3 Sustainable Pest Management

Considering the economic significance of mango crop, it is essential to undertake measures to keep pest populations under check. However, in the present era of 'input-intensive cultivation', orchardists are relying more on chemical means, which has triggered a series of ecological consequences of undesirable nature. Large-scale and indiscriminate use of insecticides has almost eliminated natural enemies and pollinators. Increased machanization of mango orchards had a negative impact on the environment and arthropod diversity (Cabreara-Mireles et al. 2011). Hence, pest management options should take the arthropod diversity and its relevance to the ecosystem into consideration. The following are some of management practices that could protect the crop from pests without affecting the beneficial biota.

Adequate attention should be given to cultural practices like deep ploughing and orchard sanitation which would help in managing leaf hoppers, fruit flies, mealybugs and stone weevil. Crop habitat management, like timely pruning ensures percolation of optimum light and discourages proliferation of sucking pests and reducing the load of pests like shoot borers, webbers, etc. Maintaining refugee flora helps in sustaining the beneficial arthropods during off season. Exploiting the locally available natural enemies serves the long-term goal of food safety and ecological sustainability. Several species of natural enemies like *Promuscidea unfasciaventris* and *Anagyrus pseudococci* were recorded on scales of *R. iceyroide* (Tandon and Lal 1978).

For instance, entomopathogens like *Metarhizium anisopliae* can drastically bring down requirement of insecticides to manage leaf hoppers and thrips. Jayanthi et al. (2015) reported that infection of entomopathogenic fungus, *Aspergillus flavus*, disabled the antioxidative enzyme system of *S. mangifera*. Plant-originated pesticides like azadirachtin, lemon grass oil (0.125 %) and citronella oil (0.25 %) were found to be quite effective in controlling the hopper population at lower densities (Verghese 2000). Other innovative practices like male annihilation technique and bait splash for fruit fly management, sticky banding around the stem to prevent mealybugs from climbing up the tree (Tandon 1995) and stem wrapping to collect and kill the emerging adults of stem borers (Reddy et al. 2014) have a lot of potential as effective pest management strategies without involving blanket application of toxic chemicals. Irradiation is suggested to be a useful disinfestation treatment for fruit flies and stone weevils (Heather and Corcoran 1992).

14.4 Natural Enemies

Besides the above-mentioned long list of arthropod pests, there are several insect and non insect arthropods which either predate or parasitize the pests and thus play a crucial role in maintaining the general equilibrium of pest populations in mango ecosystem. Natural enemies include predators and parasitoids.

14.4.1 Predators

Many predators are generalist in nature and prey on a variety of other arthropods. Major groups of predators include ladybird beetles (feed on mealybugs, aphids, scales, leaf hoppers, etc.), praying mantids (on thrips, hoppers, scales, caterpillars), predatory thrips and red ants. Spiders and predatory mites represent non insect predators. In addition to 12 species of spiders belonging to eight families, *Coccinella septempunctata*, *C. transversalis* and *Menochilus sexmaculatus* preyed upon leaf hopper, *I. clypealis*. Two species of mantids and two species of Neuroptera, i.e. *Mallada boninensis* and *Chrysopa lacciperda*, were found preying upon nymphs and adults of *I. nitidulus* besides *Bochartia* sp., a mite predating the nymphs (Srivastava and Tandon 1981).

14.4.2 Spiders (Arachnida: Araneae)

Spiders are the seventh most diverse group of organisms. Approximately 120,000 species of spiders occur worldwide, but till now only 38,432 species have been described (Platnick 2004). They are entirely carnivorous and feed mainly on insects and rarely on other arthropods including other spiders (Nyffeler and Benz 1987). Spiders play an important predatory role in orchard ecosystems due to their high abundance and predominantly insectivorous feeding habits. In Maharashtra, India, 12 species belonging to six families and ten genera were recorded from mango fields. The most dominant family was Eresidae, forming 76 % of the whole collection, followed by Lycosidae (10 %). Guild structure analysis of spiders revealed five feeding guilds, namely, web weavers, ground runners, stalkers, sheet web weaver and foliage runners. Species diversity is affected by environmental factors: spatial heterogeneity, competition, predation and habitat type (Phartale et al. 2014). In an apple ecosystem of Kashmir, India, the proportion of visual hunters was highest (44.97 %), followed by web builders (31.91 %) and tactile hunters (23.12 %) (Khan 2012). Srivastava and Tandon (1980) recorded 12 species of spiders belonging to eight families, namely, *Phidippus* sp., *Rhene indicus*, *Marpisa* sp., *Oxyopes shweta*, *Cyrtophora* sp., *C. cicatrosa*, *Araneus sinhagadensis*, *Cheiracanthium donicli*, *Linylia* sp. *Stegodyphus sarasinorum*, *Uloborus* sp., *Hersilia sarigryi* and *Theridion indica* in mango orchards. Diversity of spiders is much higher in inter cropped and undisturbed orchards than those with monoculture in Pakistan (Maqsood 2011). Implementing non-chemical and environment-friendly management practices helps in conserving the spider fauna which in turn brings down the pest populations thus minimizing the usage of pesticides.

14.4.3 Parasitoids

Insect parasitoids mainly belong to the order Hymenoptera. Two species of parasitoids, *Pipunculus annulifemer* Brun and *Pyrilloxenos paracompactus* Pierce were found parasitizing the adults of leaf hoppers of all the species. Larvae of *Epipyrops fuliginosa* Toms were also reported to parasitize the hoppers. Egg parasitoids, viz. *Aprostocetus* sp., *Gonatocerus* sp., *Polynema* sp. and *Tetrastichus* sp., were recorded on hoppers, *I. nitidulus* and *A. atkinsoni*. The extent of parasitization was 5–10 % in Lucknow, India (Srivastava and Tandon 1981). The parasitoids associated with fruit fly are *Opius compensates* Silvestri, *O. persulcatus* Silvestri, *Biosteres arisanus* (Sonam), *O. incises* Silvestri and *O. manii* (Braconidae); *Spalangia philippinensis* Mill., *S. afra*, *S. stomoxysine* Gir. and *S. grotiuse* Gir. (Pteromalidae); *Dirhinus giffardi* Silvestri (Chalcididae); and *Pachycrepoideus dubiers* Ashmead and *Trybliographa daci* Weld (Eucoilidae) (Tandon 1995).

14.5 Pollinators

Mango tree produces both male and bisexual (hermaphrodite or perfect) flowers on branched terminal panicles. The number of panicles ranges from 200 to 3000 per tree with 500 to 10,000 flowers per panicle. The proportion of bisexual flowers is very low compared to male and varies from 1 % to 35 % depending on the variety. In the perfect flower, there is a fleshy disc around the ovary which secretes nectar. The stamen is on the outer margin of this disc. Since the pistil and stamen are of the same length, the insects that feed on either nectar or pollen are likely to transfer pollen from the anther to the stigma (Mukherjee 1953; Free and Williams 1976).

There are diverse opinions on the essentiality of insects for mango pollination. For example, Free and Williams (1976) reported that mangoes were able to set fruit even though insects had been excluded by bagging, thus suggesting that at least some pollination is assisted by wind or gravity. However much before this observation, Popenoe (1920) pointed out that the mango flowers had none of the characteristics of a wind-pollinated flower, and he considered the mango to be an insect-pollinated plant. Hermaphrodite flowers are self-pollinated, but the incompatibility of some pollen and stigmas causes failure in mango fruit set (Mukherjee et al. 1968; Sharma and Singh 1970; Dag et al. 2006). The contribution of insect foragers to the pollination and fruit set in mango was evidently proven through caging experiments by different workers (Bhatia et al. 1995; Singh 1997). Galan Sala et al. (1997) found in the Canary Islands that when all insects were excluded, no fruit was set in 'Tommy Atkins' mangoes, but when bees were introduced, there was a significant increase in fruit set.

14.5.1 Diversity of Insect Foragers on Mango

Several reports documented the insect fauna attending mango flowers in different countries. Bhatia et al. (1995) and Singh (1997) found that of insect species visiting mango flowers, the highest number (17) of species was from the order Diptera (flies), followed by Coleoptera (beetles and weevils), Hymenoptera (ants, bees, wasps, sawflies), Heteroptera (bugs) and Lepidoptera (butterflies, moths). Verghese and Tandon (1990) studied the pollination behaviour of *Apis florea* on mango and found it to be an important pollinator. Reddy et al. (2012) recorded four species, viz. *Apis florea* (Hymenoptera: Apidae), *Chrysomya megacephala* and *Stomorhina discolor* (Diptera: Calliphoridae) and *Eristalinus arvorum* (Diptera: Syrphidae) (Fig. 14.10), as the dominant foragers significantly contributing to mango pollination in Bengaluru, while the stingless bee *Trigona iridipennis* was the most abundant in the *Konkan* region in India. Even thrips and leaf hoppers are credited as beneficial to mango pollination.

In Israel, 46 distinct species or types of pollinators were found, and most belonged to the orders Diptera (26), Hymenoptera (12) and Coleoptera (6). Of them two blowflies (*Chrysomya albiceps* and *Lucilia sericata*), the honeybee (*Apis mellifera*) and the housefly (*Musca domestica*) played a significant role in mango pollination in most orchards (Dag and Gazit 2000). In Taiwan, 126 individual insects

Fig. 14.10 Pollinator diversity on mango

belonging to 39 species in 23 families and five orders were recorded as a visitors or pollinators on mango (Sung et al. 2006). Hymenoptera, especially honeybees, were found to be more prevalent in terms of species and were observed collecting extra-floral nectar from the bases of mango panicles in Australia (Anderson et al. 1982) and South Africa (Eardley and Mansell 1994). Since mango flowers are generally considered to be unattractive to honeybees, *Apis mellifera* Linnaeus (Free and Williams 1976), Anderson et al. (1982) suggested that *Trigona* sp. might be used to augment the pollinating fauna, since it is common and prevalent on mango blossom and can be hived.

Pollinator abundance seems to be influenced by varietal differences. In terms of total pollinator activity, 'Ratna' attracted maximum numbers (3.24/10 panicles), while it was lowest in 'Sindhu' (0.63). Dipterans and bees showed varied preferences to varieties as was evident in case of 'Alphonso' which recorded the highest number of bees (2.6), while dipteran activity was lowest (0.02). Across the varieties, *A. florea* was the most dominant forager followed by *E. arvorum* (Reddy et al. 2013b).

14.5.2 Pollinator Conservation and Enhancement

In recent years, there has been a concern about declines in both wild and domesticated pollinators. In different countries, 35–50 % decline in honeybee populations was estimated within a span of 15–20 years (Table 14.1) (Gallai et al. 2009). Besides other factors like habitat loss and insecticides, climate change is reported to significantly impact pollinators, especially bees at various levels, including their pollination efficiency (Fig. 14.11) (Hegland et al. 2009; Reddy et al. 2012a, b). Since mango flowers open in large numbers at a designated span of time, it necessitates the visitation of a large number of insects. Hence, augmentation of pollinators may be helpful to obtain maximum fruit set. The positive impact of the introduction of honeybee and the bumblebee (*Bombus terrestris*) on mango yield was reported in Israel (Dag and Gazit 2000). Besides honeybees, the activity of other pollinators, especially that of *C. megacephala*, can be enhanced by placing simple fish

Table 14.1 Population decline of honeybees in world scenario

Country	Decline (%)	Duration
Germany	57	Last 15 years
UK	61	Last 10 years
USA	>50	Last 20 years
Poland	>35	Last 15 years
India	>40	Last 25 years
Brazil	>53	Last 15 years
Netherlands	58–65	Last 25 years
China	>50	Last 20 years

Source: Gallai et al. (2009)

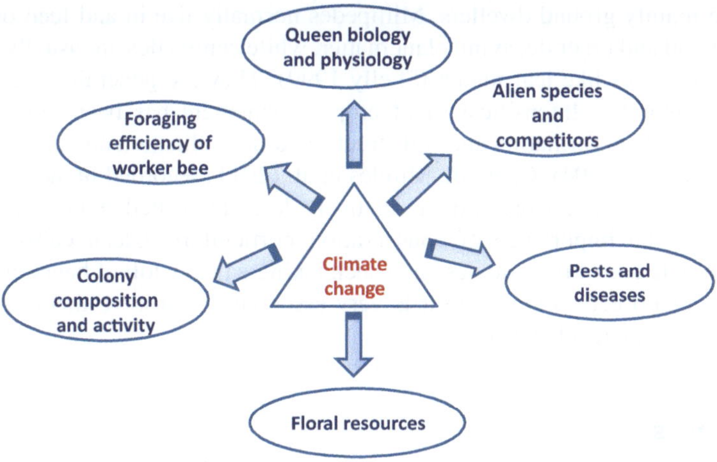

Fig. 14.11 Multilevel impacts of climate change on honeybees

meal-based traps which attracted adult flies to breed on fish-based diet. These traps should be erected just before flowering so that population can be enhanced by the time full blossom occurs (Reddy et al. 2013a). Conservation and management of native insect pollinators are very important for sustaining crop productivity as well as insect species diversity in the long run. Efforts must be made to identify, assess and develop techniques/methodology to rear and manage them for enhancing crop pollination. Even though both the need and the potential exist, the practice of rearing and managing natural pollinators for crop pollination has not been given due attention. Another area that deserves to be seriously looked into is the enhancement of non-*Apis* pollinators like stingless bees, syrphid and calliphorid flies. Enhancing the population of wild pollinators such as *Eristalinus*, *Chrysomya*, *Stomorhina*, *Sarcophaga* and *Camponotus* especially those of big size and hairy body helps to improve pollination service in the mango orchards (Nurul Huda et al. 2015).

Some of the measures which would be of great help to conserve and sustain the pollinator populations in orchards include providing sufficient flora for off-season sustenance of pollinators; protecting and conserving nest sites of natural pollinators; providing more non-crop flowering resources in fields, such as cover crops, strip crops or hedgerows; avoiding insecticide applications during blossom period; and most importantly, spreading the awareness among growers on the importance of pollinator conservation (FAO 2009; Reddy and Sreedevi 2014). Organic horticulture favours establishment of diversified arthropod communities on mango.

14.6 Centipedes and Millipedes

In addition to pests, pollinators and natural enemies, orchard ecosystem supports other arthropods like centipedes and millipedes. They may not be of direct economic significance, but their presence strengthens the ecological balance of the crop habitat.

They are mainly ground dwellers. Millipedes normally live in and feed on rotting leaves, wood and other decaying plant matter, while centipedes are usually found in damp dark places like leaf mulch (Shelly 1999). They are generalist predators on other invertebrates. Intensification of crop management has been affecting processes like soil conformation and soil structure, crop pollination and natural control of pests (Donald 2004). Cabreara-Mireles et al. (2011) recorded higher number of millipedes in a mango orchard of relatively less intensified management. This underscores the importance of conservation horticulture. Clean cultivation and indiscriminate use of insecticides have been taking a heavy toll of beneficial organisms in orchard ecosystems, and it is very essential to conserve them in order to sustain the ecological balance.

References

Alam SN (1994) Population dynamics, varietal reactions and microbial control of different species of mango leafhoppers. MSc thesis, University of the Philippines, Los Banos

Ali M (1978) A report on the wax scales, *Ceroplastes ceriferus* Green and *Chloropulvinaria polygonata* (Ckll.) (Homoptera: Coccidae) on mango and their natural enemies. Bangladesh J Zool 6:69–70

Aliakbarpour H, Che Salmah MR (2010) Diurnal activity of four species of thrips (Thysanoptera: Thripidae) and efficiencies of three non destructive sampling techniques for thrips in mango inflorescences. J Econ Ento 103:631–640

Aluja M (1994) Bionomics and management of *Anastrepha*. Annu Rev Entomol 39:155–178

Anderson DL, Sedgeley M, Short JRT, Allwood AJ (1982) Insect pollination in mango in northern Australia. Aust J Biol Sci 33:541–548

Atwal AS (1963) Insect pests of mango and their control. Punjab Hortic J 3:235–258

Bhatia R, Gupta D, Chandel JS, Sharma NK (1995) Relative abundance of insect visitors on flowers of major subtropical fruits in Himachal Pradesh and their effect on fruit set. Indian J Agric Sci 65(12):907–912

Bhumannavar BS (1991) Record of *Citripestis eutraphera* Meyrick (Pyralidae: Lepidoptera) on *Mangifera andamanica* in India. J Bombay Nat Hist Soc 88(2):299

Butani DK (1977) *Indarbela quadrinotata* (Wlk.). In: Paul P (ed) Diseases, pests and weeds in tropical crops. Verlag, Paul Parey, Berlin, pp 438–439

Butani DK (1979) Insects and fruits. Periodical Expert Book Agency, Delhi

Cabreara-Mireles H, Murillo-Cuevas FD, Ortega-Zaleta DA, Villanueva-Jiminez JA, Esobar-Dominguez AA (2011) Impact of mango manila management systems on arthropods in foliage and weeds. Trop Subtrop Agroecosyst 13:317–326

Chahal BS, Singh D (1977) Bionomics and control of mango shoot borer, *Chlumetia transversa* Walk. (Noctuidae: Lepidoptera). Indian J Hortic 34:188–193

Cunningham I (1989) Pests. In: Bagshaw J (ed) Mango pests and disorders. Department of Primary Industries Information Series, Q189007, pp 10–21

Dag A, Gazit S (2000) Mango pollinators in Israel. J Appl Hortic 2(1):39–43

Dag A, Degani C, Gazit S (2006) Gene flow in mango orchards and its impact on yield. Acta Hortic 820:347–350

Dalvi CS, Dumbre RB, Khanvilkar VG (1992) Record of mango hopper species and their composition in the Konkan region. J Maharashtra Agric Univ 17(3):404–406

de Laroussilhe F (1980) Le Manguier. G. P Masoneuve and Larose, Paris

Devi Thangam S, Verghese A, Dinesh MR, Vasugi C, Kamala Jayanthi PD (2013) Germplasm evaluation of mango for preference of the mango hopper, *Idioscopus nitidulus* (Walker)

(Hemiptera: Cicadellidae): the first step in understanding the host plant resistance. Pest Manag Hortic Ecosyst 19:10–16

Donald PF (2004) Biodiversity impacts of some agricultural commodity production system. Conserv Biol 18:17–37

Eardley CD, Mansell MW (1994) Report on the natural occurrence of insect pollinators in a mango orchard. Year Book S Afr Mango Growers Assoc 14:65–66

FAO (2009) Global action on pollination services for sustainable agriculture. Food and Agriculture Organization of the United Nations. Vialedelle Terme di Caracalla, 00153 Rome, Italy

Fenner T (1997) Red-banded mango caterpillar: biology and control Prospects. Northern Territory Department of Primary Industry and Fisheries, pp 2

Folgarait PJ (1998) Ant biodiversity and its relationship to ecosystem functioning: a review. Biodivers Conserv 7:1221–1244

Follett PA (2002) Mango seed weevil (Coleoptera: Curculionidae) and premature fruit drop in mangoes. J Econ Entomol 95(2):336–339

Free JB, Williams IH (1976) Insect pollination of *Anacardium occidentale* L., *Mangifera indica* L., *Blighiasapida* Koenig

Galan-sala V, Fernandezg D, Hemandez Conde JC, Morales NA (1997) Preliminary studies on fruit-set of mango cultivar Tommy Atkins under greenhouse cultivation in the Canary Islands. Acta Hortic 455:530–537

Gallai N, Salles JM, Settele J, Vaissi BE (2009) Economic valuation of the vulnerability of world agriculture confronted with pollinator decline. Ecol Econ 68:810–821

Hampson GF (1896) The Fauna of British India including Ceylon and Burma, Moths 4. Taylor and Francies, London, p 494

Hansen JD, Armstrong JW, Brown S (1989) The distribution of the mango weevil, *Cryptorhynchus mangiferae* (Coleoptera: Curculionidae), in Hawaii. Proc Hawaii Entomol Soc 29:31–39

Harris KM, Schreiner H (1992) A new species of the gall midge (Diptera: Cecidomyiidae) attacking mango foliage in Guam, with observations on its pest status and biology. Bull Entomol Res 82:41–48

Heather NW, Corcoran RJ (1992) Effects of ionizing energy on fruit flies and seed weevil in Australian mangoes. In: Panel proceedings of the final research coordination meeting on 'Use of Irradiation as a Quarantine Treatment of Food and Agricultural Commodities'. Kuala Lumpur, Malaysia, August 1990, IAEA, Vienna

Hegland SJ, Nielsen A, Lázaro A, Bjerknes AL, Totland (2009) How does climate warming affect plant-pollinator interactions? Ecol Lett 12:184–195

Higgins CJ (1992) Western flower thrips (Thysanoptera: Thripidae) in greenhouses: population dynamics, distribution on plants and association with predators. J Econ Ento 85:1891–1903

IIE (1995) Distribution maps of plant pests. CAB International, Wallingford, p 180

Indian Horticulture Database (2013) Statewise area, production and productivity of Mango. National Horticulture Board, Government of India, pp 96

Jacob TK, Veenakumari K, Bhumannavar BS (2004) Insect pests of cashew in the Andaman Islands. Cashew 18(4):25–28

Jayanthi PDK, Verghese A, Shashank PR, Kempraj V (2014) Spread of indigenous restricted fruit borer, *Citripestis eutraphera* (Meyrick) (Pyralidae: Lepidoptera) in mango-Time for domestic quarantine regulatory reforms. Pest Manag Hortic Ecosyst 20(2):227–230

Jayanthi PDK, Ayyasamy A, Kemparaj V, Aurade RM, Govindan S, Verghesze A (2015) Aspergillus flavus impairs antioxidative enzymes of *Sternochetus mangiferae* during mycosis. J Invertebr Pathol 124:73–77

Johnson PJ (1989) Mango seed weevil (*Sternochetus mangiferae* F.): identification, biology and management. Univ Idaho Pip Tip 6:3–7

Jothi DB, Tandon PL, Verghese A (1994) Hot water immersion as a quarantine treatment for Indian mangoes infested with the oriental fruit fly, *Bactrocera dorsalis* (Hendel) (Diptera: Tephritidae). FAO Plant Prot Bull 42(3):158–159

Joubert P (1997) Pest management – mango. In: *Biennial Report,* April 1995–March 1997. Institute for Tropical and Subtropical Crops, Nelspruit, South Africa, p 25

Kalshoven LGE (1981) The pests of crops in Indonesia. PT Ichtiar Baru – Van Hoeve, Jakarta, p 701

Kapoor VV (1970) Indian Tephritidae with their recorded hosts. Orient Insects 4(2):207–251

Khaire VA, Kolhe DS, Patil JD (1987) Relative susceptibility of mango varieties to mango hoppers and powdery mildew. Haryana J Hortic Sci 16(3):214–217

Khan AA (2012) Spider Fauna (arachnida: Araneae) in Temperate Fruit Orchards of Kashmir. J Biol Control 26(2):112–115

Krishnamoorthy A, Reddy PVR, Verghese A (2014) Borer pests: status and challenges. Indian Institute of Horticultural Research, Bengaluru, pp 14–15

Krull SME (2004) Studies on the mango-ecosystem in Papua New Guinea with special reference to the ecology of *Deanolis sublimbalis* Snellen (Lepidoptera: Pyralidae) and to the biological control of *Ceroplastes rubens* Maskell (Homoptera, Coccidae). PhD thesis, Justus-Liebig-Universitat Gieben, p 190

Kulkarni HL (1955) Incidence of mango flower galls in Bombay and Karnataka. J Bombay Nat Hist Soc 53:147–148

Lewis T (1973) Thrips, their biology, ecology and economic importance. Academic, New York

Maqsood I (2011) Taxonomical and ecological studies of spiders from citrus and fruit gardens of district Faisalabad, Pakistan, M.Phil. thesis, Department of Wildlife and Fisheries, Government College University of Faisalabad, Pakistan

Masood A, Saeed S, Sajjad A, Ali M (2009) Life cycle and biology of mango bark beetle, Hypocryphalus mangiferae (Stebbing), a possible vector of mango sudden disease in Pakistan. Pak J Zool 41(4):281–288

Mohyuddin AI, Mahmood R (1993) Integrated control of mango pests in Pakistan. Acta Hortic 341:467–483

Mukherjee SK (1953) The mango, its botany, cultivation, uses and future improvement. Econ Bot 7(2):130–162

Mukherjee SK (1997) Introduction: botany and importance. In: Litz RE (ed) The Mango: botany, production and uses. CAB International, Wallingford, pp 1–19

Mukherjee SK, Singh RN, Majumder PK, Sharma DK (1968) Present position regarding breeding of mango in India. Euphytica 17:462–467

Nachiappan RM, Bhaskaran R (1983) Biochemical constitution of inflorescence of certain varieties of mango in relation to mango leafhoppers. S Indian Hortic 31:160–165

Nurul Huda A, Che Salmah MR, Abu Hassan A, Hamdan A, Abdul Razak MN (2015) Pollination services of mango flower pollinators. J Insect Sci 15(1):113

Nyffeler M, Benz G (1987) Spiders in natural pest control: a review. J Appl Entomol 103:321–339

Ochoa R, Aguilar H, Vargas C (1994) Phytophagous mites of Central America: an illustrated guide. Centro Agronomico deInvestigacio y Enseilanza, Technical Series 6. Turrialba, Costa Rica

Patel AK, Patel SR, Shah AH (1973) Studies on sex ratio, longevity and seasonal incidence of mango leafhopper, *Amritodus atkinsoni* in South Gujarat. Indian J Ent 37:150–153

Pena JE (1993) Pests of mango in Florida. Acta Hortic 341:395–406

Pena JE, Mohyuddin AI (1997) Insect pests. In: Litz RE (ed) The mango: botany, production and uses. CAB International, Wallingford, pp 327–362

Pena JE, Mohyuddin AI, Wysoki M (1998) A review of the pest management situation in mango agroecosystems. Phytoparasitica 26:129–148

Pena JE, Sharp JL, Wysoki M (2002) Tropical fruit pests and pollinators: biology, economic importance, natural enemies and control. CABI Publishing, New York

Phartale NN, Shinde A, Gyananath G, Kadam JA (2014) Diversity and abundance of spiders in mango fields of Latur district, Maharashtra state. J Ent Res 38(4):295–299

Platnick NI (2004) The world spider catalog, version 5.0. American Museum of Natural History

Popenoe W (1920) Manual of tropical and subtropical fruits. The Macmillan Co, New York, p 474

Prasad SN (1971) The Mango midge pests. Cecidological Society of India, Allahabad

Prasad SN (1972) On the control of mango midge pests. Cecidologia Indic 7:51–63

Rajapakse RHS, Kulasekera VL (1982) Some observations on insect pests of cinnamon in Sri Lanka. Entomon 7(2):221–223

Reddy PVR, Dinesh MR (2005) Evaluation of mango exotic collections for resistance to hopper, *Idioscopusniveosparsus* Lethierry. Indian J Plant Genet Resour 18(1):69–70

Reddy PVR, Sreedevi K (2014) Role of insects in fruit crop pollination. In: Dinesh MR (ed) Fruit crop breeding. New India Publishing Agency, New Delhi, pp 67–86

Reddy PVR, Verghese A, Varun Rajan V (2012a) Potential impact of climate change on honeybees (*Apis* spp.) and their pollination services. Pest Manag Hortic Ecosyst 18(2):121–127

Reddy PVR, Verghese A, VarunRajan V, Rashmi T, Kavitha (2012b) Diversity and foraging ecology of pollinators in mango (*Mangifera indica*): an Indian perspective. International Congress of Entomology, Daegu, South Korea, 19–24 August 2012

Reddy PVR, VarunRajan V,Verghese A (2013a) A semi-synthetic diet based trap to attract and monitor the calliphorid pollinator, *Chrysomya megacephala* (F.). 100th Indian Science Congress, Kolkata, 3–7 January 2013

Reddy PVR, Varun Rajan V, Vasugi C, Verghese A (2013b) Pollinator activity in relation to varietal differences in mango (*Mangifera indica* L.). International conference on Insect Science, Bangalore, 14–17 February 2013

Reddy PVR, Chakravarthy AK, Sudhagar S, Kurian R (2014) A simple technique to capture, contain and monitor the fresh-emerging beetles of tree borers. Curr Biot 8(2):191–194

Reddy PVR, Varun Rajan V, Thangam D, Chakravarthy AK (2015) Stem borers in mango: species diversity and damage patterns. 4th congress on Insect Science, Punjab Agricultural University, Ludhiana, 16–17 April 2015, pp 150–151

Sankaran T (1955) The natural enemies of *Ceroplastes pseudoceriferus* Green (Hemiptera:Coccidae). J Sci Res Banaras Hindu Univ 5:100–119

Sengupta GC, Behura BK (1955) Some new records of crop pests from India. Indian J Entomol 17:283–285

Sharma DK, Singh RN (1970) Self incompatibility in mango (*Mangifera indica* L.). Hortic Res 10:108–118

Shelly RM (1999) Centipedes and millipedes with emphasis on North American fauna. Kan School Nat 45(3):1–16

Shukla RP, Tandon PL (1985) Bio-ecology and management of the mango weevil, *Sternochetus mangiferae* (Fabricius) (Coleoptera: Curculionidae). Int J Trop Agric 3:293–303

Singh RN (1978) Mango. Indian Council of Agricultural Research, New Delhi, p 99

Singh G (1997) Pollination, pollinators and fruit setting in mango. Proceedings of the 5th International Mango Symposium. Acta Hortic 455:116–123

Singh HS, Kishore K (2014) Status and strategies for red banded mango caterpillar, *Deanolis albizonalis* (Hampson) – an emerging pest of in eastern India. J Appl Zool Res 25(1):11–20

Singh J, Mukherjee IN (1989) Pest status of phytophagous mites in some northern states of India. In: Proceedings of The first Asia-Pacific conference of entomology, Chiang Mai, Thailand, pp 192–203

Sohi A, Sohi AS (1990) Mango leafhoppers (Homoptera: Cicadellidae) – a review. J Insect Sci 3:1–12

Srivastava RP, Tandon PL (1980) New records of parasites and predators of important insect pests of mango. Entomon 5(3):243–244

Srivastava RP, Tandon PL (1981) Bioagents of insect pests of mango in India. In: Proceedings of horticultural seminar, Directorate of Horticulture, Government of Uttar Pradesh, Lucknow

Srivastava RP, Tandon PL, Verghese A (1982) Evaluation of insecticides for the control of mango shoot gall psyla, *Apsylla cistelata* (Buckton) (Psyllidae, Homoptera). Entomon 1:281–284

Sujatha A, Zaheeruddin SM (2002) Biology of pyralid fruit borer, *Deanolis abizonalis*: a new pest of mango. J Appl Zool Res 13:1–5

Sung IH, Lin M-Y, Chang C-H, Cheng A-S, Chen W-S (2006) Pollinators and their behaviors on mango flowers in Southern Taiwan. Formosan Entomol 26:161–170

Swirski E, Ben -Dov Y, Wysoki M (1997) Mango. In: Ben-Dov Y, Hodgson CJ (eds) Soft scale insects – their biology. Natural Enemies and Control (7B), pp 241–254

Tandon PL (1995) Integrated insect pest management in mango. In: Proceedings National Symposium on Integrated Pest Management and Environment organised by Plant Protection Association of India and FAO at Madras, 2–4 February, 1995, India

Tandon PL, Lal B (1978) The mango coccid, *Rastrococcus iceryoides* (Green) (Homoptera:Coccidae) and its natural enemies. Curr Sci 47:467–468

Tandon PL, Verghese A (1985) World list of insect, mite and other pests of mango. Technical Documentation Indian Institute Horticulture Research 5

Tandon PL, Verghese A (1987) New insect pests of certain fruit crops. Indian J Hortic 44:120–121

Tandon PL, Lal B, Rao GSP (1983) Prediction of mango hopper *Idioscopus clypealis* (Leth.) population in relation to physical environmental factors. Entomon 8:257–261

Veeresh GK (1989) Pest problems in mango-world situation. Acta Hortic 231:551–565

Verghese A (2000) Status and management of the mango stone weevil *Sternochetus mangiferae* (Fab.) (Coleoptera: Curculionidae) in India. Pest Manag Hortic Ecosyst 6(1):15–21

Verghese A, Devi Thangam S (2011) Mango hoppers and their management. Extension folder No: 71–11, ATIC Series: 31–11, Indian Institute of Horticultural Research, Hessaraghatta Lake post, Bangalore, India

Verghese A, Tandon, PL (1990) Pollination behavior of *Apis florea* on mango. In: Social insects – an Indian perspective. IUSSI, Bangalore, India, pp 60–63

Verghese A, Tandon PL, Rao GSP (1988a) Spatial distribution pattern and sampling plan for the blister midge, *Erosomyia indica* Grover (Cecidomyiidae: Diptera) in India. Insect Sci Appl 9(4):515–518

Verghese A, Tandon PL, Rao GSP (1988b) Ecological studies relevant to the management of *Thrips palmi* Karny on mango in India. Trop Pest Manag 34(1):55–58

Verghese A, Nagaraju DK, Kamala Jayanthi PD, Madhura HS (2005) Association of mango stone weevil, *Sternochetus mangiferae* (Fabricius) (Coleoptera: Curculionidae) with fruit drop in mango. Crop Prot 24:479–481

Verghese A, Uma S, Jayanthi PD, Mouly R, Helen M (2011) Evidence of a random ovipositional strategy by female fruit fly *Bactrocera dorsalis* (Tephritidae: Diptera) with reference to host quantum. Curr Sci 100(2):246–248

Verghese A, Kamala Jayanthi PD, Sreedevi K, Sudha Devi K, Pinto V (2013) A quick and non-destructive population estimate for the weaver ant Oecophylla smaragdina Fab (Hymenoptera: Formicidae). Curr Sci 104(5):641–646

Verma R, Singh S (2010) Seasonal activity of mango leaf webber, *Orthaga eudrasalis* walker. Insect Environ 16(1):22

Viraktamath CA (1989) Auchenorrhyncha (Homoptera) associated with mango, *Mangifera indica* L. Trop Pest Manag 35(4):431–434

Viraktamath S, Viraktamath CA (1985) New species of *Busoniomimus* and *Idioscopus* (Homoptera: Cicadellidae: Idiocerinae) breeding on mango in south India. Entomon 10:305–311

Waite GK (2002) Pests and pollinators of mango. In: Pena JE (ed) Tropical fruit pests and pollinators: economic importance, natural enemies and control. CABI Publishers, Wallingford, p 103

Waterhouse DF (1998) *Deanolis sublimbalis* (chapter 4.6). In: Biological control of insect pests: south-east Asian prospects, ACIAR monograph no. 51. Australian Centre for International Agricultural Research, Canberra, p 548

Wen HC, Lee HS (1986) Seasonal abundance of the ceriferous wax scale (*Ceroplastes pseudoceriferus) in* southern Taiwan and its control. J Agric Res China 35:216–221

Wharton RA, Marsh PM (1978) New World Opiinae (Hymenoptera: Braconidae) parasitic on Tephritidae (Diptera). J Wash Acad Sci 68:147–165

White IM, Elson-Harris M (1992) Fruit flies of economic significance. CABI, Wallingford, p 768

Wysoki M, Ben-Dov Y, Swirski E, Izhar Y (1993) The arthropod pests of mango in Israel. Acta Hortic 341:452–466

Arthropod Communities in Cashew: A Perennial Reservoir of Species Assemblages

15

P.S. Bhat, K. Vanitha, T.N. Raviprasad, and K.K. Srikumar

Abstract

Over 200 arthropod species are associated with cashew, some of which are common the world over. Depending on the climate, location and age of the plantation, each geographic region has its own distinctive pest complex. Tea mosquito bug and cashew stem and root borer are the two major pests of cashew in most of the cashew-growing tracts of the world. In addition, shoot tip caterpillars, leaf miners, hairy caterpillars, leaf thrips, leaf beetles and inflorescence feeders are capable of causing economic damage during cropping season. Cashew serves as perennial reservoir of arthropod communities, and it is vital to make it balanced to sustain yields and maintain diverse arthropod communities.

Keywords

Cashew • Integrated pest management • Tea mosquito bug • Stem and root borer

P.S. Bhat (✉)
Division of Entomology and Nematology, Indian Institute of Horticultural Research (IIHR), Hessaraghatta Lake Post, Bengaluru 560089, Karnataka, India
e-mail: pshivarama59@gmail.com

K. Vanitha • T.N. Raviprasad
Division of Entomology, ICAR-Directorate of Cashew Research, Puttur 574202, Karnataka, India

K.K. Srikumar
United Planters Association of Southern India (UPASI), Tea Research Foundation, Tea Research Institute, Valparai 642 127, Coimbatore, Tamil Nadu, India

A.K. Chakravarthy, S. Sridhara (eds.), *Economic and Ecological Significance of Arthropods in Diversified Ecosystems*, DOI 10.1007/978-981-10-1524-3_15

15.1 Introduction

Cashew, *Anacardium occidentale* L. (Anacardiaceae), is an economically important plantation crop native to South eastern Brazil. India was the first country in the world to exploit international trade in cashew kernels in the early part of the twentieth century. Cashew occupies 53.2 lakh ha with a production of 41.5 lakh tonnes. India has the largest area (16.8 %) under cashew followed by Ivory Coast, Brazil, Indonesia, Benin, Tanzania, Nigeria, Guinea-Bissau, Kenya, Vietnam and the Philippines. The highest raw cashew nut production is from Vietnam followed by Nigeria, India, Ivory Coast, Benin, the Philippines, Guinea-Bissau, Tanzania, Indonesia, Brazil and Kenya. There is an ever-increasing demand for cashew kernel both in international and domestic markets. All parts of the plant are fed upon by at least one pest species, resulting in huge yield loss if left unchecked.

Cashew is a commercial nut crop that thrives in hot humid regions and hence is distributed in countries near the equatorial region. The long-lived cashew plantations provide a relatively steady microclimate and food supply for arthropod communities thus serving as perennial reservoir of species assemblage. Cashew plantations nearly resemble a single-species forest, and insect pest species coexist by way of intra-tree distribution or well-defined stratification/ecological niche formation similar to rubber or tea plantations. Weeds are a major component of the cashew plantations and serve as alternative hosts for pests as well as a refuge for their natural enemies and other arthropods (Sundararaju and Bakthavatsalam 1994).

15.2 Arthropod Communities on Cashew

Every part of the cashew tree is economically useful to humans in one or the other way; every part of the tree is damaged by one or other pests. Globally, more than 200 arthropod species are associated with cashew. The production loss is estimated to be about 20–30 % by tea mosquito bug alone and death of 5–10 % of productive trees every year by cashew stem and root borer (CSRB) (Rai 1984). Hence, insect pests pose a severe constraint for cashew production. There are significant contributions in the field of cashew entomology by Ayyar (1942), Abraham (1958), Pillai et al. (1976), Ohler (1979), Stonedahl (1991), Sundararaju (1984, 2000a) and Sundararaju et al. (2006). A meticulous knowledge about the pests is one of the prerequisites in evolving suitable management approach against pests.

15.2.1 Tea Mosquito Bug (TMB): *Helopeltis* spp.

Tea mosquito bug is a low-density pest incurring 36–75 % damage at its mean population level of 0.15–0.36 nymphs and adults per shoot/panicle. Information on distribution, nature and extent of damage, biology, seasonal abundance, natural enemies, alternate host plants and control measures have been reviewed by Devasahayam and Nair (1986) and Stonedahl (1991). The genus *Helopeltis* has a palaeotropical distribution extending from West Africa to New Guinea and Northern Australia. Of the recognised species, 26 are restricted to Africa; 4 to Oriental Region; 4 are endemic to the Philippine Islands; 2 are distributed throughout the Malay Peninsula, Sumatra and Java; and 1 species is endemic to South India, Sri Lanka, Laut Island (southeast coast of Borneo), Sulawesi, New Guinea and Northern Australia. Most of the earlier reports pertaining to *Helopeltis antonii* Signoret may also be of *Helopeltis bradyi* Waterhouse since *H. bradyi* has close resemblance with *H. antonii* (Stonedahl 1991). *H. antonii* existed on varied host plants, viz. neem, cashew, guava, ber, drumstick, Indian gooseberry, cotton, *Ailanthus excelsa* Roxb. and cow pea, whereas in the Western Ghats, *H. antonii* coexisted as a dominant species along with *H. bradyi* and *Helopeltis theivora* Waterhouse on cashew and cocoa. *H. antonii* also coexisted as dominant species on guava along with *H. bradyi*; only *H. theivora* exists on *Chromolaena odorata* (L.) and tea (Sundararaju and Sundarababu 1999a). Venkata (2009) recorded the activity of *Helopeltis* spp. on *Annona* spp., while Srikumar and Bhat (2013a) recorded its activity on Singapore cherry (*Muntingia calabura* L.). Apart from *H. antonii*, *H. theivora*, *H. bradyi* and *Pachypeltis measarum* Kerk were also recorded on cashew causing similar damage in certain areas. Rebijith et al. (2012a, b) used molecular biology tools for the identification of *Helopeltis* spp. and *Pachypeltis measarum*.

Helopeltis antonii

Helopeltis bradyi

Helopeltis theivora

Pachypeltis measarum

The egg and nymphal period last 6–12 and 10–14 days, respectively. The adults can survive for more than a month, and the female bug lays up to 259 eggs during its lifetime (Desai et al. 1977; Abraham and Nair 1981). It generally spreads from neem trees to guava, ber, drumstick and cashew. The eggs are inserted in the tender parts of shoots, petioles and midribs of leaves during flushing and on flower buds, flowering panicles, peduncle, rachis and immature fruits and nuts resulting in severe yield losses.

TMB damage on shoot

TMB damage on developing cashew apple /nut

Cashew plantation damaged by TMB

The typical feeding damage results in the formation of brownish or darker necrotic lesions. Under vulnerable stage, its population increases very rapidly within a month leading to severe loss in yield. Each nymph or adult can cause more than 100 lesions on fruit buds or immature fruits, and feeding lesions coalesce, and the tender shoots and the entire panicles having tender immature apples and nuts dry subsequently resulting in a burnt-up appearance.

The life table analysis carried out in Indonesia (Siswanto et al. 2008a, b) and in India (Sundararaju and Sundarababu 1998; Srikumar and Bhat 2013b) indicated that the survivorship of the *H. antonii* population was a type II with a high hatchability and bulk death occurring during early nymphal stages followed with a relatively lower death rate throughout the older stages (Sundararaju and Sundarababu 1998). Build-up of *H. antonii* population and its damage commenced from October to November onwards synchronising with the emergence of new flushes/panicles after the cessation of monsoon rains. Maximum shoot damage of 49.5 % during November and high panicle damage of 72.1–73.9 % from December to February with a peak pest population during February were recorded by Sundararaju (1984). In young plantations, the pest was noticed continuously with a higher intensity during February and March (Sathiamma 1978). When weather parameters were related to weekly mean TMB population, only minimum temperature had shown consistently negative relationship. Thus the prevalence of low minimum temperature (i.e. below 20.0 °C) continuously for more than 1 week during the months of December and January provides a clue for further surveillance to decide on management option (Sundararaju 2007).

15.2.2 Cashew Stem and Root Borer (CSRB)

Plocaederus ferrugineus L. (Coleoptera: Cerambycidae) is the most important species that infests cashew in most of cashew-growing areas and two other species, viz. *Plocaederus obesus* G. and *Batocera rufomaculata* Deg., were also reported in association with this species (Abraham 1958; Rai 1984). The pest is capable of causing death of 1–10 % of the productive trees annually if left unnoticed (Pillai et al. 1976; Ramadevi and Krishnamurthy 1983; Jena et al. 1985a; Godse et al. 1990; Samiyyan et al. 1991; Mohapatra and Mohapatra 2004; Mohapatra et al. 2007). *P. ferrugineus* was recorded as an emerging serious pest of cashew in Nigeria (Anikwe et al. 2007; Asogwa et al. 2008). Management of this pest is a tough job as the borer remains in a concealed condition in the interface of bark and hardwood which facilitates its escape from the attack of the natural enemies. Curculionid weevil borers *Marshallius multisignatus* (Boheman), *M. anacardi* Lima and *M. bondari* Rosado-Neto were reported from Brazil and French Guiana (Bleicher et al. 2010) and were responsible for death of trees especially dwarf types.

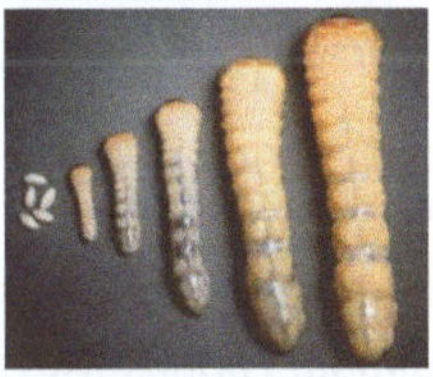		
Different instars of CSRB grubs	Adults of *P. ferrugineus*	Adults of *P. obesus*

Adults of *B. rufomaculata* are greyish, measuring 50 mm in length, and have yellowish or orange spots on the forewings. *P. obesus* are chestnut-coloured, longicorn beetles, measuring about 40 mm in length and with slight pubescence. Adults of *P. ferrugineus* are dark reddish brown, medium-sized beetles (25–40 mm in length). Eggs are usually deposited in the crevices of the bark of the main trunk up to one metre in height from ground level and also on the exposed roots. The nascent first instar grubs feed on the tissue near the site of oviposition, and extrusion of fine dusty frass is noticed within few days of hatching. After hatching, the grubs bore into the fresh tissues of the bark and feed on the subsequent subepidermal and sapwood tissues and make tunnels in irregular directions (Bhat et al. 2002). The larval period continues for 6–7 months. The fully grown grub measuring about 100 mm in length enters into heartwood for pupation and makes a circular exit hole of 1.5 cm width for adult emergence. The pupation takes place inside a calcareous cocoon (Pillai et al. 1976; Godse et al. 1990; Bhat and Raviprasad 1996).

Symptoms of damage include extrusion of frass, occurrence of gummosis, premature shedding of leaves, drying of twigs and, finally, death of the tree (Misra and Basu choudhuri 1985). Its infestation is severe in unattended plantation, and infested trees act as source of inoculum (Jena 1990).

Accumulation of frass at the base of cashew tree	Tunnelling by CSRB grub

Even though the occurrence of the pest is noticed throughout year in both East and West Coast, a relatively large population of grubs and severe infestation could be seen in coastal Karnataka and Andhra Pradesh during March–May and May–July, respectively (Abraham 1958; Ramadevi and Krishnamurthy 1983; Jena et al. 1985a, b).

15.2.3 Shoot Tip Caterpillar

Lepidopteran caterpillars are known to infest cashew shoot tips during flushing period and cause considerable damage. Gelechid caterpillar, *Anarsia epotias* Meyr, is pale yellowish-green with black head and bores into the terminal shoots and tunnel inside. A gummy substance oozes out from the infected tips and finally the attacked shoots dry up (Remamony 1965; Subba Rao et al. 2006). Similarly, the tiny, yellowish to greenish-brown larvae of the moth *Hypotima* (=*Chelaria*) *haligramma* M. (Lepidoptera: Gelechidae) also damage shoot tips by folding the fresh leaves and feeding within (Pillai et al. 1976; Mohapatra et al. 1998).

Damage by shoot tip caterpillar

15.2.4 Leaf Miner

The leaf miner, *Acrocercops syngramma* M. (Lepidoptera: Gracillariidae), is one of the serious pests of cashew (Sundararaju 1984; Jena and Satpathy 1989; Jacob and Belvadi 1990). The caterpillars mine and feed below the epidermal layer of the tender leaves causing extensive leaf blisters which later dry up, causing distortion, browning and curling of the leaves. The freshly hatched larvae and younger larvae are pale white in colour, while full-grown caterpillars are reddish-brown and feed by scraping the mesophyll below the epidermis. Abraham (1958) estimated the leaf miner damage to be 26 % in severely infested leaves, while 70–80 %, 60 %, 6–20 % and 18–20 % leaf damage was reported by Basu Choudhuri (1962), Rai (1984), Ayyanna et al. (1985) and Chatterjee (1997), respectively. Besides cashew, jamun and mango serve as additional hosts for this pest (Butani 1979; Sundararaju 1984).

Damage by leaf miner

Leaf miner caterpillars inside the blotch

15.2.5 Hairy Caterpillars

The hairy caterpillars of *Euproctis* spp. (Lepidoptera: Lymantriidae) feed in groups on the inflorescence and tender nuts of cashew. They scrape the green tissues on the inflorescence branches and feed on the shell of the nut in the tender green stage. *Metanastria hyrtaca* Cram. (Lasiocampidae) and *Lymantria obfuscata* Wlk. (Lymantriidae) cause severe sporadic defoliation in cashew (Arjuna Rao et al. 1976; Ramaseshaiah and Bali 1987). Early instar caterpillars are gregarious feeders on tender foliage, and the full-grown caterpillars fed voraciously on mature leaves. They congregate in large numbers on the ground under dry leaves near the base of the tree in crevices of bark or on lower parts of well-shaded branches.

15.2.6 Leaf Thrips

Occurrence of foliage thrips, viz. *Selenothrips rubrocinctus* Giard, *Rhipiphorothrips cruentatus* Hood and *Retithrips syriacus* (Mayet), has been reported on cashew (Ananthakrishnan 1984; Ayyanna et al. 1985; Jena et al. 1985a, b). The red-banded thrip, *S. rubrocinctus* is a tropical–subtropical species thought to have originated in northern South America (Chin and Brown 2008) and is found in parts of Asia, Africa, Australia, South America and the West Indies. *S. rubrocinctus* and *R. cruentatus* cause severe damage to young plantations, particularly during summer, and the adults and immature stages of thrips colonise the lower surface of leaves. As a result of its rasping and sucking activity, the leaves become pale brown and slightly crinkled with roughening of the upper surface.

15.2.7 Leaf Beetles and Weevils

During rainy season (June–August), the chrysomelid leaf beetles and weevils defoliate cashew. The chrysomelid beetle *Monolepta longitarsus* Jacoby is an important regular pest in the West Coast regions during the southwest monsoon. These appear in young trees and skeletonize the leaves which gradually dry up. An ash-coloured chrysomelid, *Neculla pollinaria* Baly also attacks the postharvest flushes and also the upcoming tender shoots and buds. *Microserica quadrinotata* Moser (Melolonthinae) was recorded as another defoliator which skeletonizes the leaf by scrapping chlorophyll (Jena et al. 1985b).

Leaf beetles on cashew shoot

15.2.8 Pests of Cashew Apples and Nuts

Thylocoptila paurosema Meyrick and *Hyalospila leuconeurella* R. (Lepidoptera) and *Nephopteryx* sp. (Lepidoptera) attack tender apples and nuts. Damaged nuts get deformed and dry away (Rai 1984; Ayyanna et al. 1985; Dharmaraju et al. 1974, 1976). Besides leaves and shoot, *Orthaga exvinacea* (Hampson) also damages the apple. Similarly, *Hyalospila leuconeurella* Ragnot (Pyralidae) and *Anarsia epotias* Meyr. (Gelechidae) were recorded as apple and nut borers (Basu Choudhuri and Misra 1973). The larvae of *H. leuconeurella* bore through the apple and remained inside the apple till the fruit droped, and when nuts are attacked they get deformed.

Aphids [*Toxoptera aurantii* (Boyer de Fonscolombe), mealy bugs *Planococcus citri* (Risso) and *Ferrisia virgata* (Cockerell)] suck saps of immature apples and nuts. Flower thrips such as *Rhynchothrips raoensis* G. and *Scirtothrips dorsalis* H., besides flowers, scrap immature apples and nuts resulting in the malformation of nuts and immature fruit drop (Bhat et al. 2002). A pentatomid bug, *Catacanthus incarnatus* Dru Drury) also occurs as cashew apple pest (Bhat and Srikumar 2013a). *Drosophila melanogaster* Meigen is a very serious apple-feeding insect during fruiting stages followed by *Bactrocera* spp. Under coleopteran pests, *Carpophilus* sp. was recorded in India and *Macrodactylus pumilio* Burm. from Brazil feeding on ripe apples (Ohler 1979).

Catacanthus incarnates on cashew apple

15.2.9 Inflorescence Feeders

Cashew shoots bearing fresh flushes and flowers are attacked by two species of leaf- and shoot-webbing caterpillars, *Lamida (=Macalla) moncusalis* Wlk. (Lepidoptera: Pyralidae) and *Orthaga exvinacea* Hamps. (Lepidoptera: Noctuidae). Symptoms of infestation are presence of webs on terminal portions, with clumped appearance and drying of webbed shoot/inflorescences (George et al. 1984). This pest was sporadic in certain pockets, and maximum infestation of 26 % was noticed in one of the affected areas. During post-monsoon period, the caterpillars feed on the terminal leaves of new shoots and blossoms after webbing them.

Damage due to leaf and blossom webber

Flower thrips such as *Rhynchothrips raoensis* Giard, *Haplothrips ganglbauer* (Schmutz), *Thrips hawaiiensis* (Morgan), *H. ceylonicus* Schmutz, *Frankliniella schultzei* (Trybom) and *Scirtothrips dorsalis* Hood cause premature shedding of flowers, scabs on floral branches, apples and nuts (Thirumalaraju et al. 1990; Bhat et al. 2002). The occurrence of damage, extent of damage and seasonal incidence were reported for *R. roanensis* (Abraham 1958; Ayyanna et al. 1985; Patnaik et al. 1986; Thirumalaraju et al. 1990), *S. dorsalis*, *H. ganglbauer*, *T. hawaiiensis* (Ayyanna et al. 1985), *H. ceylonicus* and *F. schultzei* (Patnaik et al. 1987).

15.3 Factors Influencing Arthropod Population in Cashew

Insect population always fluctuates according to the dynamic condition of its environment. Climatic factors such as rainfall and humidity have been known to greatly influence the population change of *Helopeltis* spp. (Pillai et al. 1979; Muhamad and Chung 1993; Karmawat et al. 1999). Knowledge of the seasonal abundance and trends in the population build-up of pest have become important for effective control schedules. Population fluctuation study conducted in Indonesia provides good information that rainfall increased the number of shoots and inflorescence which indirectly influenced the number of *H. antonii* population (Siswanto et al. 2008a).

Other factors include natural enemies (Giesberger 1983; Karmawat et al. 1999; Peng et al. 1999a, b), temperature (Pillai et al. 1979) and food supply (Pillai et al. 1979). Less population of tea mosquito bug in the older plantation during monsoon period was due to existence of resistant phenological (matured flush) stage of cashew (Sundararaju and Sundararbabu 1999a). Population growth of TMB was estimated by obtaining the difference between average TMB population recorded during particular week and that recorded during the preceding week. Minimum temperature between 15 and 20 °C was reported optimum for triggering the population build-up of *H. antonii* (Rao et al. 2002; Sundararaju 2005).

15.4 Tactics for Integrated Pest Management

Habitat Management It is quite obligatory to keep proper surveillance at vulnerable habitats (young cashew plantations or neem groves). Neem trees existing on the border or fence side of cashew fields act as reservoir for tea mosquito bug throughout the year (Sundararaju and Sundarababu 1999b). Therefore, it is quite obligatory to eliminate the growth of neem in the vicinity of these fields, and thereby the spread of tea mosquito bug from neem to all horticultural crops can be curtailed. Severely infested trees and dead trees should be uprooted before and after monsoon season as a main phytosanitary measure to manage cashew stem and root borer.

Monitoring The population build-up of tea mosquito bug can be monitored by using single virgin (unmated) adult TMB female as bait insect. It is possible to detect the male population at 20 m distance in few minutes during day time by this pheromone-based technology (Sundararaju and Sundarababu 1999b).

Soil Fertility In general, the optimal physical, chemical and biological properties of soils determine the capability of a crop to resist or tolerate insect pests. These properties can be manipulated through soil fertility management by way of application of organic amendments/manures. Balanced N, P and K levels induce tolerance to many of the pests and besides indirectly induce resistance upon any pest attack.

Weed Management/Phytosanitation The weeds serve as host plants to important pests of cashew. For example, *Helopeltis theivora* is capable of completing its life cycle and multiplies in a very common weed, *Chromolaena odorata*, which is present in cashew plantations (Srikumar and Bhat 2013c). Fourteen weed species belonging to ten different families were found as alternate hosts of TMB during flushing period of cashew (September–October). Hence, weed management is very important during vulnerable period (Vanitha et al. 2014). Phytosanitation activities involving removal of infested dead trees to achieve reduction of pest population in a given location are very essential for the management of cashew stem and root borer (Misra and Basu Choudhuri 1985; Raviprasad and Bhat 1998; Raviprasad et al. 2009).

Host Plant Resistance Histopathological investigations made in the tea mosquito bug-infested tender cashew shoot revealed that cashew is inherently (genetically) provided with very active phenol–phenolase system (Sundararaju and Sundarababu 1999b). Any feeding injury will result in rapid hypersensitive reactions leading to necroses, blighting and drying of affected parts especially tender shoot, panicle and fruits. The matured shoots of cashew irrespective of varieties exhibited highest oviposition and feeding deterrence. Every year, this type of resistant phenological stage brings down the population build-up during non-flushing period (June–September) on older plantation. Mid-season/late-season flowering cashew varieties are able to escape from the severity of the pest infestation. One such variety, Goa 11/6, showed consistent performance with yield of 2 t/ha under unsprayed situations, under moderate level of pest incidence which was later released as ‘Bhaskara’ (Sundararaju et al. 2006). Even though incidence of shoot tip caterpillars and apple and nut borers was observed in all the recommended varieties of cashew, sometimes the fruit set was partially affected in the varieties which were having early mixed phase of flowering with male and hermaphrodite flowers, whereas in varieties which were having early male phase, the damage was severe resulting in poor fruit set.

Pheromones and Kairomones The use of attractants in pest management systems can be a precise, specific and ecologically sound pest management approach. Kairomonal effect of cashew bark and frass extracts towards cashew stem and root borer adults have been confirmed. The components accountable for their kairomonal activity were studied. Similarly, in the case of tea mosquito bug, presence of sex pheromone activity is confirmed for *H. antonii*. A confined virgin female TMB can attract more than 30 male insects in a single day under field condition. The whole body extract of virgin females of *H. antonii* was analysed through GC-MS, and 17 components were identified including pinene, 9-hexadecenoic acid and 9-octadecenoic acid, but none of them could be implicated as sex pheromone (Sundararaju and Sundarababu 1999b). The volatiles collected from virgin female and field-collected female were analysed, and methyl butyrate, a compound exhibiting pheromone activity in other insects of family Miridae, was one of the compounds detected in the analyses (Bhat and Raviprasad 2008).

Biological Control Natural enemy (NE) diversity in the cashew ecosystem has a significant role in biological control of various cashew pests.

(A) Parasitoids A parasitoid *Erythmelus helopeltidis* Gahan (Mymaridae: Hymenoptera) was recorded to parasitise the eggs of *H. antonii* (Devasahayam and Nair 1986; Devasahayam 1989). Subsequently, two hymenopteran egg parasitoids, namely, *Telenomus* sp. (*laricis* group) (Scelionidae) and *Chaetostricha* sp. (Trichogrammatidae) were reported (Sundararaju 1993a). *E. helopeltidis* Gahan (Hymenoptera: Mymaridae), *Telenomus* sp. (*laricis* group) (Scelionidae), *Chaetostricha* sp. (Trichogrammatidae) and *Gonatocerus* sp. nr. *bialbifuniculatus* Subba Rao were the egg parasitoids reported on this pest from West Coast regions, while *Ufens* sp. was an egg parasitoid reported from the East Coast (Vridhachalam). *E. helopeltidis* Gahan (Hymenoptera: Mymaridae) was recorded as an egg parasitoid from *Pachypeltis maesarum* (Heteroptera: Miridae) (Bhat and Srikumar 2012). The build-up of TMB was naturally regulated through these egg parasitoids (Devasahayam 1989; Sundararaju 1993a, 1996). Two eulophid larval parasitoids, viz. *Sympiesis* sp. and *Cirrospilus* sp. (Hymenoptera: Eulophidae), were recorded on leaf miners. *Panerotoma* sp. (Braconidae) and *Trathala* sp. (Ichneumonidae) were recorded as hymenopteran larval parasitoids of *T. paurosema*.

(B) Predators An array of ants, spiders, mantids, reduviids, coccinellids and a few wasps were identified as the main natural enemies of various cashew pests during different parts of the season. Many pests were preyed upon by a large assortment of natural predators such as spiders and mites, lacewings, predatory thrips and predatory bugs (Sundararaju 1993b; Chin and Brown 2008).

(C) Spiders Spiders are potential biological control agents in agroecosystems including cashew (Riechert and Bishop 1990). Several species of spiders, *Hyllus* sp., *Oxyopes sehireta*, *Phidippus patch* and *Matidia* sp. *Sycanus collaris* (Fab.), *Sphedanolestes signatus* Dist. and *Endochus inornatus* Stal., *Irantha armipes* Stal. and *Occamus typicus* Dist. have been recorded as predators (Sundararaju 1993b). The mean spider population varied from 0.22 to 0.31 per panicle, and the spider population had influence on arthropod complex during fruiting season (Sundararaju 2004). Bhat et al. (2013a, 2013b) reported spiders as indigenous natural enemies of tea mosquito bug (TMB), and the study revealed occurrence of 117 species of spiders belonging to 18 families, viz. Araneidae, Clubionidae, Corinnidae, Gnaphosidae, Hersiliidae, Linyphiidae, Lycosidae, Miturgidae, Nephilidae, Oxyopidae, Pholcidae, Pisauridae, Salticidae, Sparassidae, Tetragnathidae, Theridiiae, Thomisidae and Uloboridae. Salticids were predominant (30 %) followed by Araneidae (22 %). Field observation revealed that *Telamonia dimidiata* and *Oxyopes shweta* as the major predators of *Helopeltis* spp. The spiders, viz. *Argiope pulchella*, *Cyclosa fissicauda*, *Eriovixia laglazei*, *Neoscona mukerjei*, *Nephila pilipes*, *Oxyopes sunandae*, *Bavia kairali*, *Carrhotus viduus*, *Epocilla aurantiaca*, *Hyllus semicupreus*, *Achaearanea mundula*, *Camaricus formosus* and *Thomisus lobosus*, were also superior with respect to their predatory activity. This rich diversity of spiders is indicative of overall insect biodiversity of cashew plantation since spiders are considered to be useful indicators of species richness and health of terrestrial ecosystem.

Telamonia dimidiate

Oxyopes shweta

Oxyopes sunandae

Agriope pulchella

Telamonia dimidiate feeding on TMB

Oxyopes shweta feeding on TMB

(D) Ants The green ant, *Oecophylla smaragdina* (Fabricius), is an effective predator, and it can significantly reduce the numbers of over 30 important insect pest species of many tropical crops (Way and Khoo 1992). The green ant can significantly reduce the damage levels of the main cashew insect pests, such as the tea mosquito bug, *Helopeltis pernicialis* (Stonedahl, Malipatil and Houston); the mango tip borer, *Penicillaria jocosatrix* (Guenee); the fruit-spotting bug, *Amblypelta lutescens* (Distant); the leaf roller, *Anigraea ochrobasis* (Hampson); and the green vegetable bug, *Nezara viridula* (Fabricius) (Peng et al. 1995, 1997a, b, c, 1998). In the cashew ecosystem of the west coast of India, inter colony rivalry and death of queen due to infection by broad-spectrum mycopathogen (*Beauveria bassiana*) were commonly observed in the case of *O. smaragdina*, and these might be possible reason for low establishment (Sundararaju 2004).

The role of *Dolichoderus thoracicus* to control *Helopeltis* spp. has been extensively studied and well understood (Way and Khoo 1991, 1992; Khoo 1992; Khoo and Ho 1992). The predatory ant, *O. smaragdina*, was also found in high numbers for each observation, and no *H. antonii* was found on cashew plants occupied by this ant. In Northern Australia, *O. smaragdina* has been used to control *H. pernicialis* on cashews (Peng et al. 1995, 1997a, b, 1999a, b). Other predators frequently found in quite high numbers were arachnids and to a lesser extent mantids and coccinellids (Siswanto 2008a). Among the five species of ants, viz. *Camponotus* sp., *Anoplolepis longipes*, *Crematogaster* spp., *Paratrechina longicornis* and *O. smaragdina* observed under cashew trees, only *Oecophylla* could control TMB effectively, while other ants were found to feed on extrafloral nectaries and as scavengers (Sundararaju 2000a). *Crematogaster wroughtonii* Forel (Formicidae) has been recorded as a predator of nymphs of TMB (Ambika and Abraham 1979).

Forty-nine species of ants (Fam: Formicidae) belonging to five subfamilies were recorded having multiple roles like predators, pollinators, scavengers, extrafloral nectarine feeders, etc. Ants belonging to Myrmicinae subfamily were dominant (22 species) followed by Formicinae (13 species). Among the ant species, *Oecophylla smaragdina* (Fabricius) and *Anoplolepis gracilipes* Smith were most abundant, while *Camponotus compressus* and *C. sericeus* were found throughout the year. The activities of most ant species are predominant during flowering and fruiting period (November–April) and pre-monsoon period (May), while during heavy rain, i.e. southwest monsoon, activities of *Myrmicaria brunnea*, *C. sericeus*, *Prenolepis naoroji* and *C. angusticollis* were only seen. In a single tree, foraging activities of maximum of seven species were found at a time especially during flowering and initial fruiting season (Vanitha et al. 2015).

Ants foraging on (**a**) cashew shoot. (**b**) *Diacamma* sp. (**c**) Queen ant of *Oecophylla smaragdina* and its eggs. (**d**) *Myrmicaria brunnea*

(E) Reduviids (Hemiptera: Reduviidae: Harpactorinae) Reduviids are recorded as potential natural enemies of *Helopeltis* spp. (Stonedahl 1991; Sundararajau 1996). Five species of reduviids, viz. *Sycanus collaris* Fab., *Sphedanolestes signatus* Dist., *Endochus inornatus* Stal, *Irantha armipes* Stal and *Occamus typicus* Dist., were reported as predators of *Helopeltis antonii* Sign. on cashew in India (Sundararaju 1984). All these predate on tea mosquito nymphs. A total of 16 species of reduviids belonging to the subfamily Harpactorinae, viz. *Alcmena* sp., *Biasticus* sp., *Cydnocoris gilvus* Burmeister, *Endochus albomaculatus* Stal, *Endochus* sp., *Epidaus bicolor* Distant, *Evagoras plagiatus* Burmeister, *Irantha armipes* Stal, *Lanca* sp., *Panthous bimaculatus* Distant, *Rhynocoris fuscipes* Fabricius, *Rihirbus trochantericus* Stal var. *sanguineous*, *Rihirbus trochantericus* Stal var. *luteous*, *Sphedanolestes signatus* Distant and *Sycanus galbanus* Distant were recorded from cashew ecosystem. The damage to cashew trees by tea mosquito bug can be reduced by the introduction of assassin bugs (Sundararaju 1984; Bhat et al. 2013c).

Cydnocoris gilvus	*Panthous bimaculatus* Distant	*Epidaus bicolor*

15.5 Pollinators

In Brazil, honeybee (*A. mellifera* L.) and solitary bee are the efficient pollinators of cashew (Freitas 1997). The fruit set is mainly influenced by activity of the pollinators (Reddy 1993). Devasahayam (1986) had reported that halictid bee [*Pseudaspis oxybelloides* (Smith)], another species of bee [*Pithitis smaragdula* (Fabr.)], honeybee (*Apis cerana indica* Fabr.) and wasp (*Odynerus* sp., Fam: Eumenidae) are pollinators of cashew and the pollen grains were detected on their legs and bodies (Sundararaju 2000b). These pollinators remove nectar or sticky pollen grains of cashew by resting directly or hovering on newly opened cashew flowers. Application of insecticides during flowering had not affected the fruit set.

Honeybee visiting cashew flowers

15.6 Future Challenges

The perennial nature of cashew is ideal for assemblage of a vast number of arthropods throughout the year. The gap in the area of bioecology of certain pests and biological control of key pests needs to be filled up. It is essential to have concerted efforts to popularise plant protection measures in cashew. Basically, cashew farmers have to be trained about the nature of the initial damage symptoms for correct identification of the pests. Need-based and timely application of pesticides is effective which needs to be based on surveillance of pests of endemic nature, since the indiscriminate sprays may result in elimination of natural enemies, mainly arthropods harbouring in cashew.

References

Abraham V (1958) Pests of cashew (*Anacardium occidentale*) in South India. Indian J Agric Sci 28:531–544

Abraham CC, Nair GM (1981) Effective management of the tea mosquito bugs for breaking the yield barriers in cashew. Cashew Causerie 3(1):6–7

Ambika B, Abraham CC (1979) Bioecology of *Helopeltis antonii* Sign. (Miridae: Hemiptera) infesting cashew trees. Entomon 4:335–342

Ananthakrishnan TN (1984) Host relationships and damage potential of thrips infesting cashew. In: Rao EVVB, Khan HH (eds) Cashew research and development. Indian Society for Plantation Crops, CPCRI, Kasargod, pp 132–134

Anikwe JC, Okelana FA, Otuonye HA, Hammed LA, Aliyu OM (2007) The integrated management of an emerging insect pest of cashew: a case study of the cashew root and stem borer, *Plocaederus ferrugineus* In Ibadan, Nigeria. J Agric For Soc Sci 5(1):1597–0906

Arjuna Rao, Dharmaraju E, Ayyanna T (1976) Biology of *Metanas triahyrtaca* Cram., a serious pest of cashew. Andhra Agric J 23:61–68

Asogwa EU, Hammed LA, Ndubuaku TCN (2008) Integrated production and protection practices of cashew (*Anacardium occidentale*) in Nigeria. Afr J Biotechnol 7(25):4868–4873

Ayyanna T, Tejkumar S, Ramadevi M (1985) Distribution and status of pests of cashew in coastal districts of Andhra Pradesh. Cashew Causerie 7:4–5

Ayyar TVR (1942) Insect enemies of cashew plant (*Anacardium occidentale* L.) in South India. Madras Agric J 30:223–226

Basu Choudhuri JC (1962) Preliminary investigation on insect pests of cashew plantations in Kerala. Indian For 88:516–522

Basu Choudhuri JC, Misra MP (1973) *Hyalopsila leuconeurella* Ragnot (Lepidoptera: Pyralidae) and *Anarsia epotias* Meyrick (Lepidoptera: Gelechidae) two pests of cashew apple and nut in south India with a note on their control. J Plant Crops 1(S):168–170

Bhat PS, Raviprasad TN (1996) Pathogenicity of entomopathogenic fungi against cashew stem and root borer *Plocaederus ferruginues* L. (Coleoptera: Cerambycidae). J Plant Crops 24:265–271

Bhat PS, Raviprasad TN (2008) Sex pheromone of tea mosquito bug *Helopeltis antonii* Signoret (Miridae:Heteroptera). J Plant Crops 36(3):451–453

Bhat PS, Srikumar KK (2012) Record of *Erythmelus helopeltidis* Gahan (Hymenoptera: Mymaridae), an egg parasitoid from *Pachypeltis maesarum* (Heteroptera: Miridae) infesting cashew. Pest Manag Hortic Ecosyst 18(1):103–104

Bhat PS, Srikumar KK (2013) Occurrence of man- faced stink bug, *Catacanthus incarnatus* Drury on cashew in Puttur region of Karnataka. Insect Environ 19(1):32–34

Bhat PS, Sundararaju D, Ravi Prasad TN (2002) In: Singh HP, Balasubramanian PP, Hubbali VN (eds) Integrated management of insects, pests and diseases in cashew, Indian Cashew Industry. DCCD, Cochin

Bhat PS, Srikumar KK, Raviprasad TN (2013a) Seasonal diversity and status of spiders (Arachnida: Araneae) in cashew ecosystem. World Appl Sci J 22(6):763–770

Bhat PS, Srikumar KK, Raviprasad TN, Vanitha K, Rebijith KB, Asokan R (2013b) Biology, behaviour, functional response and molecular characterization of *Rihirbus trochantericus* Stal var. *luteous* (Hemiptera: Reduviidae: Harpactorinae) a potential predator of tea mosquito bug (*Helopeltis* spp.). Int J Trop Insect Sci 123(4):264–271

Bleicher E, Rodrigues SMM, Melo QMS, Pinho JH (2010) Minimal effective dose of phosphine to control the cashew root borer, *Marshallius bondari* Rosado-Neto (Coleoptera: Curculionidae). Rev Ciênc Agron 41(2):315–318, Fortaleza, Apr/June. http://dx.doi.org/10.1590/S1806-66902010000200021

Butani DK (1979) Insects and fruits. International Book Distributors, Dehra Dun

Chatterjee ML (1997) A study on leaf miner infestation and their control by insecticides. Cashew Bull 4:9–11

Chin D, Brown H (2008) Red-banded thrips on fruit Trees. Agnote. http://www.nt.gov.au/dpifm/Content/File/p/Plant_Pest/719.pdf. 19 Aug 2008

Desai MA, Kulkarni SV, Rodrigues A (1977) Report on the marketing survey on cashew in Goa. Government of Goa, Deman, pp 11–17

Devasahayam S (1986) Some observation on insects visiting cashew inflorescence. Cashew Bull 23(5):1–5

Devasahayam S (1989) *Erythmelus helopeltidis* Gahan(Hymenopera:Mymaridae) a new parasite of *Helopeltis antonii* Sign. on cashew. J Bombay Nat Hist Soc 86:113

Devasahayam S, Nair CPR (1986) The tea mosquito bug *Helopeltis antonii* Signoret on cashew in India. J Plant Crops 14:1–10

Dharmaraju E, Rao PA, Ayyanna T (1974) A new record of *Nephopteryx* sp. as an apple and nut borer on cashew in Andhra Pradesh. J Res Andhra Pradesh Agric Univ 1(4 &5):198

Dharmaraju E, Ayyanna T, Sreeramulu C (1976) *Nephopteryx* sp. as an apple and nut borer on cashew. Indian Cashew J 10:9–11

Freitas BM (1997) Number and distribution of cashew (*Anacardium occidentale*) pollen grains on the bodies of its pollinators, *Apis mellifera* and *Centris tarsata*. J Apic Res 36:15–22

George MV, Singh V, Peter T (1984) Aerial spraying against tea mosquito bug in cashew. In: Bhaskar Rao EVV, Khan HH (eds) Cashew research and development. Indian Society for Plantation Crops, CPCRI, Kasaragod, pp 120–125

Giesberger G (1983) Biological control of the *Helopeltis* pest of cocoa in Java. Cocoa Research in Indonesia 1900–1950 American Cocoa Research Institute, International Office of Cocoa and Chocolate, pp 91–181

Godse SK, Dumbre RB, Karat SB (1990) Bionomics of cashew stem and root borers, *Plocaederus ferrugineus* L. Cashew Bull 27:1–17

Jacob TK, Belvadi VV (1990) The cashew leaf miner, its pest status and larval size relationship with leaf area damage in Andamans. Cashew 4:10

Jena BC (1990) Cashew stem-root and wood borers and their management. Cashew 4:5–7

Jena BC, Satpathy CR (1989) Reaction of cashew varieties to the leaf miner (*Acrocercops syngramma* M.) incidence. Cashew 3:15–17

Jena BC, Patnaik NC, Satapathy CR (1985a) Insect pests of cashew. Cashew Causerie 7(3):10–11

Jena BC, Patnaik NC, Satapathy CR (1985b) Insect fauna of cashew nut in Orissa. Sci Cult 51(11):385–386

Karmawat IE, Savitri TH, Wahyono TE, Laba IW (1999) Dinamika populasi *Helopeltis antonii* Sign. pada tanaman jambu mente. J Penelitian Tanaman Ind 4(6):163–167

Khoo KC (1992) Manipulating predators for biological control with special reference to tropics. In: Ooi PAC, Lim GS, Teng PS (eds) Biological control issues in the tropics. Malaysian Plant Protection Society, Kuala Lumpur, pp 15–22

Khoo KC, Ho CT (1992) The influence of *Dolichoderus thoracicus* (Hymenoptera: Formicidae) on losses due to *Helopeltis theivora* (Heteroptera: Miridae), black pod disease, and mammalian pests in cocoa in Malaysia. Bull Entomol Res 82:485–491

Misra MP, Basu Choudhuri JC (1985) Control of *Plocaederus ferrugineus* L. (Coleoptera: Cerambycidae) through field hygiene. Indian J Agric Sci 55:290–293

Mohapatra LN, Mohapatra RN (2004) Distribution, intensity and damage of cashew stem and root borer, *Plocaederus ferrugineus* L. in Orissa. Indian J Entomol 66(1):4–7

Mohapatra LN, Behara AK, Satapthy CR (1998) Influence of the environmental factors on the cashew nut shoot tip caterpillar, Hypotima haligramma Meyr. Cashew Bull 35:17–18

Mohapatra RN, Jena BC, Lenka PC (2007) Effect of some IPM components in management of cashew stem and root borer. J Plant Prot Environ 4(2):21–25

Muhamad R, Chung GT (1993) The relationship between population fluctuation of *Helopeltis theivora* Waterhouse, availability of cocoa pods and rainfall pattern. Pertanika J Trop Agric Sci 16(2):81–86

Ohler JG (1979) Cashew. Department of Agricultural Research. Royal Tropical Institute, Amsterdam, p 260

Patnaik HP, Das MS, Sahoo KC, Senapati B (1986) Preliminary studies on nut drying disease in cashew and its control. Cashew Causerie 8:16–17

Patnaik HP, Satapathy CR, Sontakke BK, Senapathy AHD (1987) Flower thrips of cashew (Anacardium occidentale), their seasonal incidence and assessment of damage in coastal Orissa. Cashew 1:11–13

Peng RK, Christian K, Gibb K (1995) The effect of the green ant, *Oecophylla smaragdina* (Hymenoptera: Formicidae), on insect pests of cashew trees in Australia. B Entomol Res 85:279–284

Peng RK, Christian K, Gibb K (1997a) Control threshold analysis for the tea mosquito bug, *Helopeltis pernicialis* (Hemiptera: Miridae) and preliminary results concerning the efficiency of control by the green ant, *Oecophylla smaragdina* (Hymenoptera: Formicidae) in Northern Australia. Int J Pest Manag 43(3):233–237

Peng RK, Christian K, Gibb K (1997b) Distribution of the green ant, *Oecophylla smaragdina* (F.) (Hymenoptera: Formicidae), in relation to native vegetation and the insect pests in cashew plantations in Australia. Int J Pest Manag 43:203–211

Peng RK, Christian K, Gibb K (1998) Management of *Oecophylla smaragdina* colonies in cashew plantations to control the main cashew insect pests. Sixth Australasian Applied Entomological Research conference, Brisbane, Australia 2, p 332

Peng RK, Christian K, Gibb K (1999a) The effect of colony isolation of the predacious ant, *Oecophylla smaragdina* (F.) (Hymenoptera: Formicidae), on protection of cashew plantations from insect pests. Int J Pest Manag 45:189–194

Peng RK, Christian K, Gibb K (1999b) Utilisation of green ants, *Oecophylla smaragdina*, to control cashewe insect pests. Rural Industries Research and Development Corporation, Canberra, p 88

Pillai GB, Dubey OP, Singh V (1976) Pests of cashew and their control in India. A review of current status. J Plant Crops 4:37–50

Pillai GB, Singh V, Dubey OP, Abraham VA (1979) Seasonal abundance of tea mosquito *Helopeltis antonii* on cashew tree in relation to meteorological factors. Proceeding of The International Cashew Symposium, Cochin, pp 103–110

Rai PS (1984) Handbook on cashew pests. Research Co Publications, New Delhi

Ramadevi M, Krishnamurthy PR (1983) Schedule of pest occurrence on cashew (*Anacardium occidentale* L.). Cashew Causerie 5(1):14–16

Ramaseshaiah G, Bali R (1987) On the identity of a lymantrid defoliator of cashew and cocoa in South India. Curr Sci 56:1191–1192

Rao GSLHYP, Raju GTT, Puttaswamy, Krishnakumar KN, Tony X (2002) Forecasting tea mosquito bug (*Helopeltis* antonii) of cashew (*Anacardium occindentale* L.). J Agrometeorol 4:45–52

Raviprasad TN, Bhat PS (1998) Laboratory rearing techniques for cashew stem and root borer, *Plocaederus ferrugineus* Linn. (Coleoptera: Cerambycidae). Abst. PLACROSYM XIII Coimbatore, 16–18 Dec 1998, p 47

Raviprasad TN, Bhat PS, Sundararaju D (2009) Integrated pest management approaches to minimize incidence of cashew stem and root borers (*Plocaederus* spp.). J Plant Crops 37:185–189

Rebijith KB, Asokan R, Krishna Kumar NK, Srikumar KK, Ramamurthy VV, Bhat PS (2012a) DNA barcoding and development of species-specific markers for the identification of Tea Mosquito Bugs (Miridae: Heteroptera) in India. Environ Entomol 41(5):1239–1245

Rebijith KB, Asokan R, Krishna V, Krishna Kumar NK, Ramamurthy VV (2012b) Development of species-specific markers and molecular differences in mitochondrial and nuclear DNA sequences of *Aphis Gossypii* and *Myzus Persicae* (Hemiptera: Aphididae). Fla Entomol 95(3):674–682

Reddy EUB (1993) Pollination studies of cashew in India: an overview. In Veeresh GK, Uma Shankar R, Ganeshaiah KN (eds). Proc Int Symp Polln Trop, pp 321–324

Remamony KS (1965) Biology of *Anarsia epotias* Meyr. (Lepidoptera: Gelichidae). Agric Res J Kerala 3:46–47

Riechert SE, Bishop L (1990) Prey control by an assemblage of generalist predators: spiders in garden test systems. Ecology 71:1441–1450

Samiyyan K, Palaniswamy KP, Ahmed Sah, Manivannan K (1991) Effect of prophylactic measures against cashew stem and root borer *Plocaederus ferrugineus* L. Cashew 5(1):16–17

Sathiamma B (1978) Occurrence of insect pests on cashew. Cashew Bull 15(4):9–10

Siswanto RM, Omar D, Karmawati E (2008a) Population fluctuation of *Helopeltis* antonii Signoret on Cashew *Anacarcium occidentalle* L., in Java, Indonesia. Pertanika J Trop Agric Sci 31(2):191–196

Siswanto RM, Omar D, Karmawati E (2008b) Life tables and population parameters of *Helopeltis* antonii (Hemiptera: Miridae) reared on cashew (Anacardium occidentale L.). J Biosci 19(1):91–101

Srikumar KK, Bhat PS (2013a) Biology and feeding behaviour of *Helopeltis antonii* (Hemiptera: Miridae) on Singapore cherry (*Muntingia calabura*)-a refuge host. J Ent Res 37(1):11–16

Srikumar KK, Bhat PS (2013b) Demographic parameters of *Helopeltis antonii* Signoret (Heteroptera: Miridae) on neem, cocoa and henna. Afr J Agric Res 8(35):4466–4473
Srikumar KK, Bhat PS (2013c) Biology of the tea mosquito bug, *Helopeltis theivora* Waterhouse on *Chromolaena odorata* (L.) and notes on its egg parasitoids. Chil J Agric Res 73(3):309–314
Stonedahl GM (1991) The oriental species of *Helopeltis* (Heteroptera: Miridae): a review of economic literature and guide to identification. Bull Entomol Res 81:465–490
Subba Rao V, Rajasekhar P, Rama Subba Rao V, Shrinivasa Rao V (2006) Seasonal incidence and control of shoot tip and inflorescence caterpillar on cashew. Ann Pl Prot Sci 14(1):245–246
Sundararajau D (1996) Studies on *Helopeltis* spp. with special reference to *H.antonii* S. in Tamil Nadu. Ph.D. thesis, T.N.A.U. Coimbatore, p 202
Sundararaju D (1984) Studies on cashew pests and their natural enemies in Goa. J Plant Crops 12:38–46
Sundararaju D (1993a) Studies on the parasitoids of the mosquito bug, *Helopeltis* antonii Sign. (Hymenoptera: mymaridae) on cashew with special reference to *Telenomous* sp. (Hymenoptera: Scelionidae). J Biol Control 7:6–8
Sundararaju D (1993b) Compilation of recently recorded and some new pests of cashew in India. Cashew 7:15–19
Sundararaju D (2000a) Insects associated with extra floral nectaries of cashew leaves. J Plant Crops 28(2):175–178
Sundararaju D (2000b) Foraging behaviour of pollinators on cashew. The Cashew 14(4):17–20
Sundararaju (2004) Influence of spiders and insect predators on the incidence of tea mosquito bug in cashew. The Cashew 18(1):9–13
Sundararaju D (2005) Seasonal abundance and extent of damage of tea mosquito bug on cashew. J Plant Crops 33(1):53–58
Sundararaju D (2007) Pest and disease management in Cashew. Presented in 6th national seminar: Indian cashew in the next decade – challenges and opportunities. 18–19 May, Raipur, pp 53–63
Sundararaju D, Bakthavatsalam N (1994) Pests of cashew. In: Chadha KL, Rethinam P (eds) Advances in horticulture, vol 10. Malhotra Publishing House, New Delhi, pp 759–778
Sundararaju D, Sundara Babu RC (1998) Life table studies of *Helopeltis antonii* sign (Heteroptera: Miridae) on neem, guava and cashew. J Entomol Res 22(3):241–244
Sundararaju D, Sundara Babu RC (1999a) Species composition of *Helopeltis* (Heteroptera: Miridae) in South India and their host range. Insect Environ 4(4):13–14
Sundararaju D, Sundara Babu RC (1999b) GC-MS analysis of virgin females of *Helopeltis* antonii Sign. (Heteroptera: Miridae). Insect Environ 4(4):3–4
Sundararaju D, Yadukumar N, Bhat PS, Raviprasad TN, Venkatakumar R, Sreenath Dixit (2006) Yield performance of 'Bhaskara' cashew variety in coastal Karnataka. J Plant Crops 34:216–219
Thirumalaraju GT, Gowda MC, Krishnappa KS, Narayana Reddy MA (1990) Seasonal incidence of flower thrips under eastern dry zone of Karnataka. Cashew 4:3–4
Vanitha K, Srikumar KK, Bhat PS (2014) Record of weed flora of cashew plantations as hosts of tea mosquito bug. Ecoscan 8(3&4):221–224
Vanitha K, Bhat PS, Raviprasad TN, Srikumar KK (2015) Species composition of ants in cashew plantations and their inter-relationship with cashew. Proc Natl Acad Sci India Sect B Biol Sci. doi:10.1007/s40011-0150600-3
Venkata R (2009) Record of *Helopeltis antonii* (Homoptera: Miridae) on the fruits of *Annona* spp. Pest Manag Hortic Ecosyst 15:74–76
Way MJ, Khoo KC (1991) Colony dispersion and nesting habits of the ants, *Dolichoderus thoracicus* and *Oecophylla smaragdina* (Hymenoptera: Formicidae) in relation to their success as biological control agents on cocoa. Bull Entomol Res 81:341–350
Way MJ, Khoo KC (1992) Role of ants in pest management. Annu Rev Entomol 37:479–503

16 The Coconut Mite: Current Global Scenario

N.S. Aratchige, A.D.N.T. Kumara, and N.I. Suwandharathne

Keywords
Coconut mite • Invasive pest • Colonization • Population dynamics

16.1 Introduction

Coconut (*Cocos nucifera* L.), which is one of the major oil crops in the world, is cultivated in more than 120 million ha in the world (Anon 2013). It is predominantly produced in tropical regions in Asia (9.7 million ha), Africa (1.2 million ha) and America (0.6 million ha) (FAOSTAT 2013). The major coconut producers in the world are Indonesia, the Philippines, India, Brazil, Sri Lanka, Thailand, Mexico, Vietnam, Papua New Guinea and Malaysia (FAOSTAT 2013). Coconut is used in day-to-day lives of millions of people as a source of food, drink and cooking oil. It is also used for medicinal purposes, fibre, mats and fuel and as raw material for many industries such as charcoal, desiccated coconut, coir products, timber and handicrafts. Because of its diverse uses, coconut is called the 'tree of heaven' and 'tree of life'.

Coconut production is affected by many biotic and abiotic factors. Among the biotic factors, a wide range of pests and diseases play a major role in limiting the coconut production. Major pests such as red palm weevil (*Rhynchophorus ferrugineus* Olivier), rhinoceros beetle (*Oryctes rhinoceros* L.), coconut black-headed caterpillar (*Opisina arenosella* Walker) and coconut scale insect (*Aspidiotus destructor* Signoret) and several major diseases caused by fungi, bacteria, phytoplasma, nematodes and viroids are known to occur worldwide. During the last few decades, new

N.S. Aratchige (✉) • A.D.N.T. Kumara • N.I. Suwandharathne
Crop Protection Division, Coconut Research Institute of Sri Lanka,
Lunuwila 61150, Sri Lanka
e-mail: nsaratchige@gmail.com

A.K. Chakravarthy, S. Sridhara (eds.), *Economic and Ecological Significance of Arthropods in Diversified Ecosystems*, DOI 10.1007/978-981-10-1524-3_16

invasive pests and diseases associated with coconut are becoming a great threat to the world coconut industry. They include the lethal yellowing disease caused by phytoplasma that caused extensive damage in the Caribbean and African regions (CABI 2015a); the coconut leaf beetle, *Brontispa longissima* Gestro, that wreaked havoc in more than 25 countries in Asia (including the Pacific Islands) and Oceania (CABI 2015b); the red palm mite, *Raoiella indica* Hirst, that caused severe damage in the Caribbean region and North and South America (CABI 2015c); and the coconut mite, *Aceria guerreronis* Keifer, that damages coconut in America, Africa and Asia (Navia et al. 2013).

16.2 World Distribution of Coconut Mite

Coconut mite was first described by Keifer (1965) from the specimens collected in the state of Guerrero, Mexico. Later, it was reported in several other regions of the American continent (Mariau and Julia 1970; Ortega et al. 1967; Hall and Espinosa Becerril 1981; Zuluaga and Sánchez 1971). Recently it has been found in Colombia (Estrada and Gonzalez 1975), Dominica (Moore and Alexander 1985), Saint Lucia (Moore et al. 1989), Costa Rica (Schliesske 1988), Jamaica (Howard et al. 2001) and Puerto Rico (Howard et al. 1990). In Africa, it has been observed in many coconut-growing countries such as São Tomé e Principe, Benin, Cameroon, Nigeria and Togo (Cabral and Carmona 1969; Mariau 1969), Ivory Coast (Mariau 1977) and Tanzania (Seguni 2002). In Asia, the coconut mite was first reported in Sri Lanka in 1997 (Fernando et al. 2002) and India in 1998 (Sathiamma et al. 1998). After its invasion of India, it has spread to all coconut-growing states of India (Muthiah 2007). In Sri Lanka, it was first reported in 1997 in the Kalpitiya Peninsula (North Western Province, dry zone) (Fernando et al. 2002). Within 2 years, it has spread to almost all coconut-growing areas in the dry and intermediate zones of the country and few coconut-growing areas in the wet zone. At present, the coconut mite has invaded all districts except Nuwara Eliya, which is mainly a hilly area where coconut is not extensively grown (Fig. 16.1). However, the incidence of coconut mite varies from district to district with higher incidences in the dry and intermediate zones than in the wet zone. It is also reported in Oman (Al-Shanfari et al. 2010), Bangladesh (Alam and Islam 2014), Pakistan (Solangi 2014) and the Maldives (Ahmed 2014).

16.3 Origin of the Pest

Out of the 29 populations collected worldwide, higher diversity of mitochondrial DNA 16S ribosomal sequences (six of a total of seven haplotypes) and the nuclear ribosomal internal transcribed spacer (ITS) in populations collected in Brazil suggested the American origin of the coconut mite (Navia et al. 2006). All samples collected from Africa and Asia were identical or very similar suggesting that the coconut mite invaded Africa and Asia later (Navia et al. 2006).

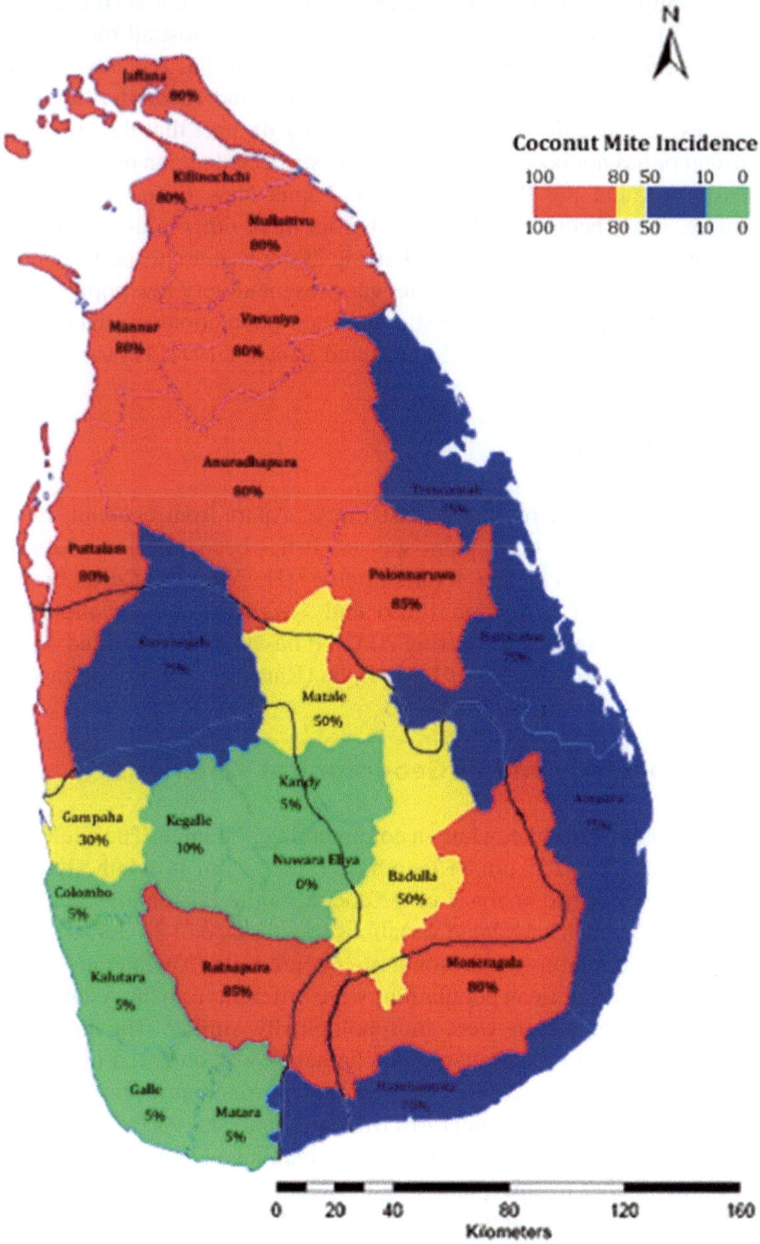

Fig. 16.1 Incidence of coconut mite in different districts in Sri Lanka (Source: Aratchige 2014)

The coconut mite always causes serious damages to coconut fruits, and it is by far the most important coconut fruit-infesting pest in almost all the coconut mite-invaded countries. It is believed that the coconut mite has recently shifted to the coconut palm from an unknown host plant (Moore and Howard 1996; de Moraes and Zacarias 2002). This is further supported by the fact that the coconut mite is either absent or has not been reported from the suspected region of origin of coconut (Southeast Asia and the Pacific Island of Papua New Guinea) (Persley 1992). Furthermore, it has been reported at least from three other non-coconut palm species (Flechtmann 1989; Ansaloni and Perring 2002; Ramaraju and Rabindra 2002). It is also possible that the coconut mite was present at very low populations before the 1960s and became a serious pest only when populations are increased due to unknown reasons (Doreste 1968; Zuluaga and Sánchez 1971; Howard et al. 1990).

16.4 Host Range

The coconut mite has a restricted host range. Apart from coconut, it has been reported only from three other palm species. It has been reported to attack palms and seedlings of *Lytocaryum weddellianum* (H. Wendl.) *Cocos weddelliana* H. Wendl in Brazil (Flechtmann 1989) and *Syagrus romanzoffiana* (Cham.) in California, USA (Ansaloni and Perring 2002). It has also been reported on the Asian palmyra palm, *Borassus flabellifer* L., in India (Ramaraju and Rabindra 2002).

16.5 Morphology and Its Geographical Variations

Coconut mites are worm like, white in colour and very minute. The adult females of coconut mite are 205–255 μm in length and 36–52 μm in width (Keifer 1965). Geometric morphometric analysis of 27 coconut mite populations collected from America, Asia and Africa revealed variations in the shape of prodorsal shield, coxigenital and ventral regions of idiosoma of coconut mite (Navia et al. 2006, 2009). These characters in American populations were different from the Asian and African populations, both of which were morphologically similar. The highest within-geographical variation of morphometric characters was observed in the American populations (Navia et al. 2006). Multivariate morphometric analyses (principal component analysis and canonical discriminant analysis) on the body length and width, length of the scapular setae and the coxigenital and opisthosomal setae, distance and the number of microtubercles between ventral setae and the number of dorsal and ventral rings of the same populations have also shown clear differences and greater diversity in the American populations compared to the Asian and African population (Navia et al. 2009).

Table 16.1 Duration and sizes (μm) of different stages of life cycle of *A. guerreronis*

Stage of life cycle	Mean no. of days ±SE	Mean size ±SE
Egg	2.9 ± 0.17	34.37 ± 2.42
Larva	1.9 ± 0.26	82.69 ± 3.33
First inactive stage (nymphochrysalis)	1.4 ± 0.1	115.62 ± 4.54
Nymph	2.5 ± 0.17	157.14 ± 4.61
Second inactive stage (imagochrysalis)	1 ± 0.12	168.75 ± 11.96
Adult	5.3 ± 0.59	195 ± 9.35

Source: Wickramananda et al. (2004)

16.6 Life Cycle

The life stages of the coconut mite (Table 16.1) include eggs, larvae, nymphs and adults (Manson and Oldfield 1996). All post-embryonic, mobile life stages look similar except for the size and the presence of genital openings in adults. Immature development of the coconut mite is completed in 8–10.5 days (Mariau 1977; Wickramananda et al. 2004; Shobha and Haq 2011). On tender coconut leaf tissues, six stages (with three mobile and three immobile stages) in the life cycle of coconut mite were reported (Wickramananda et al. 2004). Eggs hatched to larval stage in 2–3 days, and the larval stage lasted for 2 days which was followed by an inactive nymphochrysalis stage of 1 day (Fig. 16.2). Then the second active stage (nymphal stage) which lasted for about 2–3 days was started, and after passing through another resting stage of 1 day, it attains the adult stage of 5 days (Wickramananda et al. 2004). Mean length of the life cycle from egg to adult was 9.7 days at 27 °C (Table 16.1) (Wickramananda et al. 2004). Durations of egg to adult development on pieces of meristematic tissue of young *S. romanzoffiana* leaves at different temperatures were 30.5, 16.0, 11.5, 8.1 and 6.8 days at 15, 20, 25, 30 and 35 °C, respectively (Ansaloni and Perring 2004). Minimum, optimum and maximum temperatures for the development from egg to adult were 9.3, 33.6 and 40 °C, respectively (Ansaloni and Perring 2004). It has also been able to withstand short periods of frost and periods of temperature closer to 0 °C (Howard et al. 1990). Fertilized females lay up to 51 eggs and from eggs of unfertilized females, only males are produced (Arrhenotokous) (Ansaloni and Perring 2004).

16.7 Damage Symptoms and Economic Importance

Feeding of coconut mite on the meristematic zone of the coconut fruit causes physical damage which is first visible as a triangular white patch on the fruit surface next to the margin of the perianth (Fig. 16.3a). As the infested fruit grows, the surface becomes necrotic and suberised (Fig. 16.3b) with deep, longitudinal fissures and gummy exudates (Fig. 16.3c). At severe damage stages, fruit is distorted, stunted and unevenly grown if the coconut mite infestation is concentrated on one side of

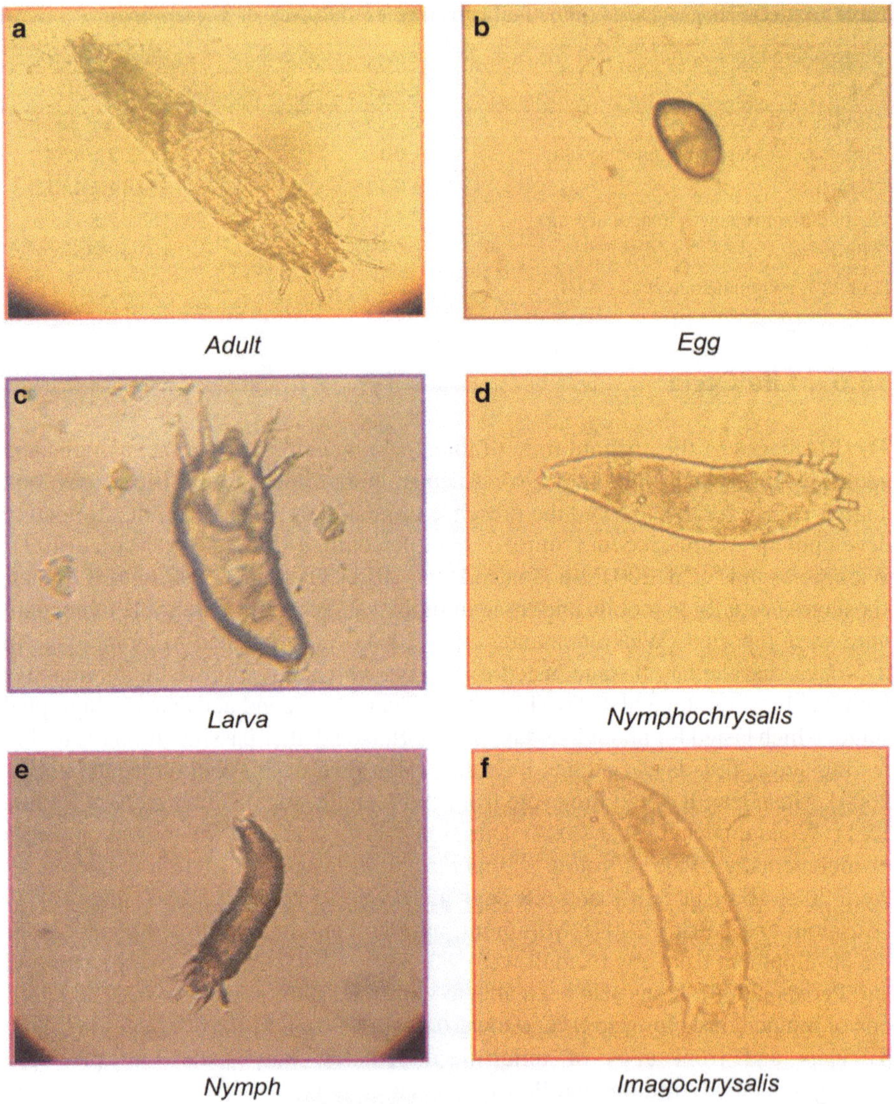

Fig. 16.2 Different stages of life cycle of *A. guerreronis* (Source Wickramananda et al. 2004)

the fruit surface (Fig. 16.3d). On the queen palm *S. romanzoffiana*, coconut mite damage has caused necrosis of meristematic tissues and mortality of young palms in nurseries (Ansaloni and Perring 2004).

Coconut mite damage can lead to reductions in copra yield (Hernández 1977; Howard et al. 2001; Moore and Howard 1996; Ramaraju et al. 2000; Muralidharan et al. 2001; Alam and Islam 2014), premature fruit drop (Doreste 1968; Nair 2002; Wickramananda et al. 2007), reduction in coconut fibre length and tensile strength

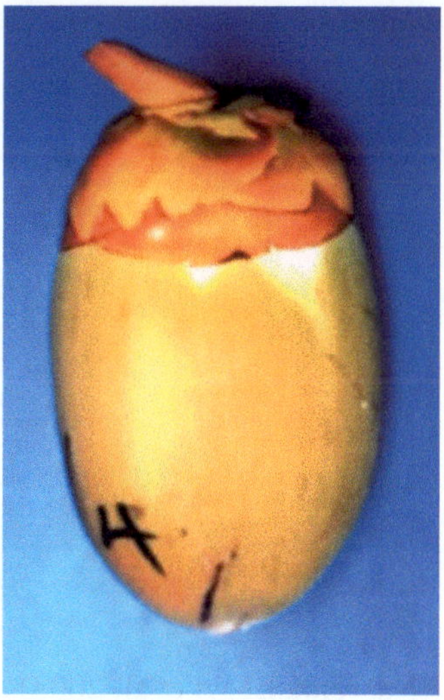

Fig. 16.3a *White* colour *triangular* patch of initial infestation of coconut mite on a young fruit

Fig. 16.3b Coconut bunch showing fruits with *brown* colour, necrotic patches of coconut mite damage

Fig. 16.3c Coconut fruit with deep, longitudinal fissures and gummy exudates due to coconut mite damage

Fig. 16.3d Undamaged fruits and severely stunted coconut fruits due to coconut mite damage

(Naseema Beevi et al. 2003) and reduction in husk availability for the coir industry (Wickramananda et al. 2007). It can also cause small and deformed fruits (Alam and Islam 2014) and reduction in yield of brown and white fibre from fruits (Kumar and Ramaraju 2010). But the level of coconut mite infestation in seed nuts has not significantly affected the performance of young coconut seedlings (Thomas et al. 2004).

16.8 Habitat and the Colonization

The coconut mite is inefficient in finding its host, but this is compensated for by their high reproductive rate and their rapid development (Moore and Howard 1996). Though not scientifically proven, probably the population on a coconut fruit starts from one or few inseminated females migrating from an infested fruit either of the same palm or of a nearby infested palm. Due to its minute size and wormlike body, the coconut mite can creep through the teplas and reach the meristematic zone covered by the perianth of the young coconut fruit. The perianth of young fruits prior to and soon after the fertilization is tightly adpressed to the coconut fruit surface giving little or no room for the coconut mite to enter the area under the perianth (Moore and Howard 1996; Howard and Abreu Rodriguez 1991). As the fruit develops, the microscopic gap between the perianth and the fruit surface increases providing access to coconut mites to enter the meristematic area where they feed and reproduce (Howard and Abreu Rodriguez 1991; Mariau 1977; Mariau and Julia 1970; Moore and Alexander 1987; Moore and Howard 1996; Aratchige 2007). Within this protected habitat, which is well protected from the external biotic and abiotic threats, the coconut mite can build up its population rapidly.

On a fruit, the coconut mite is more predominantly found in the area under the perianth than on the opened fruit surface (Lawson-Balagbo et al. 2008). Within the area under the perianth, more coconut mites have been observed on the fruit surface than on the surface of bracts (Thirumalai Thevan et al. 2004) but Varadarajan and David (2002) did not report any significant difference between the number of mobile coconut mites on the fruit surface and on the bract surface. Among the bracts, a higher number of coconut mites were observed on the inner bracts (Varadarajan and David 2002; Lawson-Balagbo et al. 2007).

Melo et al. (2014) have shown that the coconut mites are not attracted to specific parts (leaflets, spikelets or fruits) on the coconut palm by volatile chemicals. Nevertheless, a high number of coconut mites were observed at the ends of the arms of a cross-shaped arena, only when they were in contact with the epidermis discs of the meristematic area on the fruits (Melo et al. 2014). They have concluded that the coconut mites are arrested once they contact the substrate of specific sites.

The coconut mite can attack fruits of coconut almost throughout the whole development, but the population densities vary among bunches of different age. However, the coconut mite has not been observed on inflorescences (Mariau and Julia 1970; Moore and Alexander 1987; Lawson-Balagbo et al. 2008), but thereafter, it has been observed even on fruits up to 13 months after fertilization (Moore and Alexander 1987). In general, higher densities of coconut mites were observed on fruits of 3–7-month-old bunches (i.e. on fruits of 3–7 months after fertilization) (Moore and Alexander 1987; Fernando et al. 2003; Al-Shanfari et al. 2013). Peak densities were observed on fruits, 5 months after fertilization in Sri Lanka (Fernando et al. 2003), 3–5 months old in India (Varadarajan and David 2002; Mallik et al. 2003; Thirumalai Thevan et al. 2004), 4 months old in Brazil (Galvão et al. 2011) and 3–4 months old in Tanzania and Benin (Negloh et al. 2011). In Oman, higher densities of coconut mite have been observed on fruits of 4–5 months old compared to the fruits of 2–3 months old during summer (Al-Shanfari et al. 2013). Population decline on bunches of more than 4 months could be due to fruits becoming less favourable for coconut mite colonization and damage, possibly due to increased lignin content in tissues and region under the perianth becoming more exposed to the predators, which are in general larger than the coconut mite. Within bunch variation of coconut mite was lower on 6-month-old bunches (Fernando et al. 2003; Galvão et al. 2011).

16.9 Population Dynamics

Contradictory observations have been made in relation to the population dynamics of coconut mite worldwide. No significant correlation between the level of occurrence of the coconut mite and the abiotic factors has been observed in Brazil (Reis et al. 2008). Howard et al. (1990) reported the absence of a correlation between rainfall and coconut mite densities. In Sri Lanka, both amount and frequency of rainfall have not significantly correlated with the coconut mite densities (Aratchige et al. 2012a). The absence of a significant correlation between the coconut mite densities and levels of the abiotic factors may be due to the fact that the given abiotic factor alone is not significantly sufficient to regulate the coconut mite populations (Reis et al. 2008; Al-Shanfari et al. 2013). Concealed habitat underneath the perianth may also protect the coconut mites from the direct external abiotic stresses such as rainfall (Navia et al. 2013; Al-Shanfari et al. 2013). Correlations may also be masked by the effect of the abiotic factors that were not considered in these studies (Doreste 1968; Mariau 1969; Moore and Howard 1996).

However, in other studies, significant correlations were observed between the coconut mite densities and the abiotic factors. Higher coconut mite densities have been observed during dry seasons of the year in Ivory Coast (Julia and Mariau 1979; Mariau 1977), Trinidad and Tobago (Griffith 1984), India (Mallik et al. 2003; Nair 2002; Varadarajan and David 2002) and Brazil (Lawson-Balagbo et al. 2008). In Brazil, coconut mite population densities showed a positive correlation with high temperatures, low relative humidity and low accumulated precipitation (Souza et al. 2010). In Benin, coconut mite populations were higher in May to October (months with higher relative humidity) than in December to February (months with lower relative humidity) (Julia and Mariau 1979).

During the rainy season, reduction of coconut mite populations has been observed in Saint Lucia (Moore et al. 1989), Sri Lanka (Aratchige et al. 2012a) and India (Mathew et al. 2000; Nair 2002; Nampoothiri et al. 2002; Mallik et al. 2004). In Oman, the highest populations of coconut mites were observed during February (end of winter, when temperature and relative humidity are low) and May (mid of summer, when temperature and relative humidity are high), while the lowest populations were observed in August and November (beginning of winter) (Al-Shanfari et al. 2013). Coconut mite populations have been positively associated with temperature and negatively associated with relative humidity and rainfall in India (Pushpa and Nandihalli 2008). In Sri Lanka, drought length (number of days without rainfall of >5 mm) has positively influenced the coconut mite populations (Aratchige et al. 2012a).

High coconut mite populations during the months of low rainfall have been related to the slower growth rate of fruits, allowing the fruits to stay susceptible to the coconut mite attack for a longer time (Mariau 1986). It could also be due to their restricted movements and spending more time underneath the perianth to avoid desiccation during dry periods, resulting in higher numbers in the sampled fruit (Aratchige et al. 2012a). High rainfall may have a direct wash-off effect on migratory coconut mites causing low population levels. During dry seasons, sugar and amino acids, particularly proline concentration in the plant sap of coconut, are increased in order to maintain a negative osmotic potential in the palm (Ranasinghe and Jayasekara 1989). It can be expected that the nutrient composition of the sap of the surface cells of coconut fruits may also be different during the dry seasons compared to wet seasons. Cell sap may become more nutritious with high concentrations of sugars during dry seasons, increasing the reproductive rate of coconut mites (Aratchige et al. 2012a). More coconut mite infestations have been observed in dry regions than in wet regions in Colombia (Zuluaga and Sánchez 1971) and in Sri Lanka (Fernando and Aratchige 2010; Aratchige 2014).

16.10 Dispersal

Understanding the dispersal behaviour of the coconut mite is important in designing control measures, yet it was one of the poorly studied areas. Phoresy and aerial dispersal have been suggested as mechanisms of dispersal over long distances (Moore and Howard 1996; Galvão et al. 2012). On coconut palms, they can move into bunches or within bunches or within fruits by walking, and being negatively geotactic, they tend to move into younger bunches from older bunches (Moore and Alexander 1987). The coconut mite can walk at a rate of 20–125 μm per second (Moore and Howard 1996). The average distance travelled by the coconut mite within 30 min of observation period was 22.5 cm (Galvão et al. 2012). In a separate study, Melo et al. (2014) found that the average distance travelled by the coconut mite in 10 min was 135.2 mm.

The dispersal ability was dependent on the age and the state of the coconut fruit from which the mites were collected; mites collected outside the perianth of old fruits travelled over longer distances than mites collected under the perianth of young fruits, and they travelled over longer distances in the presence of food-related cues than in their absence (Melo et al. 2014). It was also observed that the number of coconut mites walking from the infested to non infested fruits was inversely proportional to the distance between the fruits (Galvão et al. 2012). Dispersal mostly occurs at night (Moore and Alexander 1987) mainly by the inseminated females (Moore and Howard 1996).

In a wind tunnel, the number of coconut mites trapped in a sticky trap was directly proportional to the wind speed (Galvão et al. 2012). Mariau and Julia (1970) and Mariau (1977) also related the increased proportion of coconut mite-infested fruits with the action of wind. Griffith (1984) has also shown that coconut mite is dispersed by wind.

Though the massive structure of the coconut palm provides a good target for the aerially dispersing coconut mite, the mortality associated with this is high (Moore and Alexander 1987; Moore and Howard 1996). But this may be compensated by their high reproductive rate and rapid development for their survival (Moore and Howard 1996).

Phoresy has also been identified as a possible mechanism of dispersal of coconut mite (Moore and Howard 1996). The coconut mite has been observed on bees visiting flowers (Griffith 1984), *Parisoschoenus obesulus* Casey, but not on *Apis mellifera* L. (Galvão et al. 2012). To obtain honey, bees are most likely to visit the unfertilized female flowers in which coconut mites are not colonized and it may at least partly explain the inefficiency of dispersal of the coconut mite through bees visiting coconut palms. Wind seems to be more influential than phoresy in dispersal of the coconut mite (Mariau and Julia 1970; Mariau 1977; Griffith 1984; Galvão et al. 2012).

Limits to the ambulatory displacement of coconut mite have been studied by Melo et al. (2014). The survival rate of the coconut mites outside the host decreased with increasing temperature levels from 18 to 33 °C, but it was not significantly different at 18, 21 and 24 °C. Survival time also showed a significant decline with increasing temperature (Melo et al. 2014). The survival rate of the coconut mites was increased with relative humidity from 10 % to 95 %, but it was not significantly different between 10 % and 25 %, and the survival time also increased with the increasing relative humidity (Melo et al. 2014).

16.11 Management

Its secluded habitat and high reproductive rate make the coconut mite one of the most intractable pests in the world (Mariau and Julia 1970; Lawson-Balagbo et al. 2007; Navia et al. 2013). Furthermore, many of the commercially grown coconut cultivars that are tall in stature do not permit feasible application of control measures for this pest. However, both chemical and non-chemical methods have been tested in many countries to manage the coconut mite.

16.11.1 Chemical Control

More than 50 chemicals and mixtures have been tested worldwide to control this pest, but only a handful of chemicals have been reported to be at least partially effective. Chinomethionate (Mariau and Julia 1970; Mariau and Tchibozo 1973; Hernández Roque 1977; Cabrera 1991), monocrotophos (Mariau and Tchibozo 1973; Hernández Roque 1977; Julia and Mariau 1979; Cabrera 1991; Fernando et al. 2002; Nair 2002; Sujatha et al. 2003), dicrotophos (Hernández Roque 1977; Cabrera 1991), fenpyroximate (Sujatha et al. 2003), triazophos (Ramaraju et al. 2002), methyl demeton 25 EC (Ramaraju et al. 2002), endosulfan and carbosulfan (Muthiah et al. 2001; Rethinam et al. 2003; Sujatha et al. 2003), dicofol and triazophos (Muthiah et al. 2001) and abamectin (Melo et al. 2012; Roseleen and Ramaraju 2012) were found to be effective in controlling the coconut mite damage, but frequent repeated applications were necessary. Systemic insecticides were more persistent, but the residues have been observed in fruits. Significant differences in the abamectin toxicity in two populations of the coconut mite have been observed in Brazil (Monteiro et al. 2012). Lima et al. (2012) found that fenpyroximate and chlorfenapyr are promising agents for managing the coconut mite in combination with *Neoseiulus baraki* Athias-Henriot, a predaceous mite, because both are selective and do not affect predators' instantaneous rate of increase.

Neem-based insecticides such as azadirachtin, neem seed kernel extract, neem oil, 2 % neem oil and garlic mixture and NeemAzal T/S (1 % azadirachtin) have also been tested and recommended (Sujatha et al. 2003; Pushpa and Nandihalli 2010; Nair et al. 2002; Ahmed 2014). However, azadirachtin has shown to impair the overall activity of *N. baraki*, male reproductive behaviour (e.g. failure to mount while attempting to mate), and reduce the daily fecundity which has a greater impact in the rate of population growth than the total fecundity (Lima et al. 2015).

The application of 30 % used engine oil in water, soap powder and wheat flour on the immature fruit surface was effective in controlling the pest and decreasing the damage incidence in treated bunches as well as in newly developed bunches in Sri Lanka (Chandrasiri and Fernando 2004). However, this treatment reduced *N. baraki* numbers in treated fruits. Though these chemicals were effective in controlling the pest under experimental conditions, growers' acceptance was low due to difficulty in application of the chemical.

An emulsion of 20 % vegetable oil and 0.5 % sulphur WP was found to be effective in controlling the coconut mite in Sri Lanka (Fernando and Chandrasiri 2010) and in the Maldives (Ahmed 2014). The application of this emulsion was less effective on *N. baraki* (Fernando and Chandrasiri 2010).

Chemical control is not always effective against the coconut mites that are excellently protected under the perianth. However, it has been suggested that the coconut mites are killed when they leave the perianth for migration and are in contact with the acaricidal residues on the fruit surface (Melo et al. 2012; Monteiro et al. 2012).

16.11.2 Biological Control

Natural enemies such as predacious mites and acaropathogenic fungi have been observed in association with the coconut mite. However, their effectiveness has been poorly evaluated (Moore and Howard 1996; de Moraes and Zacarias 2002) until the detection of coconut mites in Sri Lanka and India. Later countries such as Brazil, Benin, Tanzania and Oman also have intensified their research in this direction. Simulation study (ex ante) of the economic benefits of the biological control of coconut mites in Benin using a standard economic surplus model has shown that biological control of the coconut mite is a viable technology (Oleke et al. 2013).

The Use of Predacious Mites Extensive reviews of the predacious mites associated with the coconut mite have been published by de Moraes and Zacarius (2002) and Navia et al. (2013). Among the predaceous mites, *N. baraki*, *N. paspalivorus*, *Proctolaelaps bickleyi* Bram, *P. bulbosus* deMoraes, Reis and Gondim Jr., *Amblyseius largoensis* Muma, *N. mumai* Denmark, *Lasioseius* sp., *Proctolaelaps* sp. and *Typhlodromips sabali* De Leon were found to be associated more commonly with the coconut mite (de Moraes and Zacarias 2002; Navia et al. 2013 and the references therein). Of the commonly associated predacious mites, *N. baraki*, *N. paspalivorus* and *P. bickleyi* have been studied as biological control agents of the coconut mite.

In Sri Lanka, out of five reported predacious mites in association with the coconut mite (de Moraes et al. 2004), only *N. baraki* has been extensively evaluated as a prospective biological control agent. Its flat and elongated idiosoma with short distal setae and short legs (de Moraes and Zacarias 2002; de Moraes et al. 2004;

Aratchige 2007) which enable it to creep under the perianth, close association with the coconut mite (Fernando et al. 2003; Aratchige 2007; Aratchige et al. 2012a) and ability to feed and develop on coconut mites (Annual Report of the Coconut Research Institute of Sri Lanka 2003) were the key factors for selecting it to evaluate as a potential predaceous mite against the coconut mites. It is mass produced on *Tyrophagus putrescentiae* using tray-type (Aratchige et al. 2010) or sachet-type (Kumara et al. 2014) rearing methods for field releases (Fernando et al. 2010; Aratchige et al. 2012b). Single release of *N. baraki* resulted in significant increase of *N. baraki* with a mean number of 8.99 mites per fruit in the released palms compared to the unreleased palms (6.19 mites per fruit) and a reduction of coconut mite in released palms (1,264.77 per fruit) compared to unreleased palms (1,815.0 per fruit) (Fernando et al. 2010).

In another study in Sri Lanka, multiple releases of approximately 5,000 *N. baraki* mites per palm at 2- or 4-month intervals onto 25 % of palms of the plantation resulted in a higher percentage of normal-sized fruits in the harvest (85.6 and 88.4 % in two released blocks compared to 79.1 and 80.1 % in unreleased blocks) and a lower percentage of small-sized fruits (13.3 and 10.1 % in two released blocks compared to 20.0 and 17.2 % in unreleased blocks). The release of *N. baraki* in this manner for 2 years resulted in benefit-cost ratio of more than 1, confirming that the releases are cost effective (Aratchige et al. 2012b). Results of experiments on single and multiple releases of *N. baraki* have led to a recommendation of releasing 5,000 *N. baraki* at 3–4-month intervals to a quarter of the coconut plantation at least for 2 years. This is the first ever and so far the only recommendation of using predaceous mites for the control of coconut mite in the world (Aratchige 2014).

The Use of Acaropathogenic Fungi Until the late 1990s, the use of acaropathogenic fungi for the control of the coconut mite has been evaluated mainly in the American region (Cabrera 2002; Hall et al. 1980). After the invasion of coconut mite into the Asian region in the late 1990s, research on acaropathogenic fungi, particularly *Hirsutella thompsonii and H. nodulosa*, has been intensified, mainly in India. Laboratory and field evaluation of the commercial formulation of *H. thompsonii* 'MYCOHIT' has proven to be effective in controlling the coconut mite in India (Sreerama Kumar and Singh 2000; Gopal and Gupta 2001; Rabindra and Sreerama Kumar 2003; Sreerama Kumar 2002, 2010). Two liquid variants of the same product (MYCOHIT-LG20 and MYCOHIT-OS) were also found to be equally effective (Sreerama Kumar 2010). A mycelial application of an Indian isolate, MF(Ag) 66 of *H. thompsonii* with glycerol as the adjuvant, was found to be effective in reducing the coconut mite population by 85–97 % (Sreerama Kumar and Singh 2008).

A survey carried out in coconut mite-infested areas in Sri Lanka has shown that the natural incidence of *H. thompsonii* on coconut mite is low (Edgington et al. 2008). Four isolates (IMI 390486, IMI 391722, IMI 391942 and IMI 390486) outperformed the other isolates collected in this survey (Edgington et al. 2008) and used in environmental studies (temperature profiling and UV tolerance) (Edgington et al. 2008) and field evaluation (Fernando et al. 2007). Isolate IMI 391722 showed the highest efficacy in reducing coconut mite populations, but it did not persist on treated fruits sufficiently to cause significant epizootics after 4 weeks, suggesting frequent applications for long-term management of the coconut mite (Fernando et al. 2007).

16.11.3 Host Plant Resistance

Most of the commonly grown coconut varieties are damaged by the coconut mite. However, certain varieties are less damaged by the coconut mite (Julia and Mariau 1979; Schliesske 1988; Muthiah and Natarajan 2004; Nair 2002; Ramaraju et al. 2002; Thirumalai Thevan et al. 2004; Varadarajan and David 2003; Moore and Alexander 1990; Perera et al. 2013; Mohan et al. 2014). The shape and colour of the fruits and tightness of the perianth have been suggested as possible mechanisms of the varietal resistance against the coconut mite, but this aspect has been poorly understood. Usually, spherical-shaped fruits were less damaged by the coconut mite than elongated fruits (Moore 1986; Moore and Alexander 1990; Varadarajan and David 2003). Dark green fruits of Jamaica Tall cultivar have been less attacked by the coconut mite than lighter fruits (Moore and Alexander 1990), and in India, orange-coloured fruits have been reported to be less damaged by the coconut mite than green- and yellow-coloured fruits (Muthiah and Bhaskaran 2000; Varadarajan and David 2003). In Sri Lanka, Gon thambili, Yellow dwarf and Yellow dwarf x Tall hybrid (DYT) have been identified as the varieties having the highest putative tolerance to coconut mite (Perera et al. 2013). Less coconut mite damage has been observed on fruits with small perianth and when the angle between inner bracts was greater than 136° (Varadarajan and David 2003).

In India, 'Kalpa Haritha', a high yielding tall selection, has shown lesser coconut mite incidence, and Chowghat Orange Dwarf, Malayan Green Dwarf, Laccadive ordinary, Cochin China, Andaman ordinary, Gangabondum, Spicata and Kenthali have shown the maximum tolerance to coconut mite damage (Mohan et al. 2014). West Coast Tall, Laccadive Tall, East Coast Tall, Tiptur Tall and Chowghat Green Dwarf have recorded the maximum coconut mite incidence in the field (Mohan et al. 2014).

16.11.4 Cultural Methods

Phytosanitary measures (cleaning the crown of the palm, keeping the plantation clean and burning of all immature fruits fallen due to mite infestation) and agronomic practices that improve the plant health, *viz.* soil moisture conservation, application of recommended doses of fertilizers, irrigation, recycling of biomass by vermicompost and raising and incorporation of green manure crops have been recommended to control the coconut mite in India (http://www.coconutboard.nic.in/protect1.htm#erio: Accessed on 18-11-2015; Lokesh and Nandihalli 2009; Mohan et al. 2014) and in Bangladesh (Alam and Islam 2014). Coconut mite damage has been observed to be less in well-managed irrigated coconut plantations in India (Sujatha and Rao 2004). Increased nitrogen levels have increased the coconut mite damage (Moore et al. 1991), while increased potassium levels seem to decrease the coconut mite damage (Moore et al. 1991; Muthiah et al. 2001). Bunch pruning has been recommended for small-scale farmers in Brazil (Navia et al. 2013 and references therein). Multiple cropping systems (coconut with other crops) have also caused reduction in coconut mite damage (Moore et al. 1989; Muthiah et al. 2001; Varadarajan and David 2003; Rajan et al. 2012). The application of borax, calcium and organic manure with adequate supply of fertilizer has also shown to reduce the coconut mite damage (Muthiah and Natarajan 2004, 2005). Coconut mite damage has been low in palms treated with neem cake + bone meal + mill ash (Muthiah and Bhaskaran 2000).

Cultural practices have not always been able to control the coconut mite infestation. Melo et al. (2012) showed that the removal of bunches or the distal portion of spikelets is not an effective practice for the control of coconut mite in areas with high levels of infestation. After removal of even all fruits of bunches, damage severity has been restored within 2 months on the newly produced bunches.

References

Ahmed J (2014) Mite management of coconut in Maldives. In: Akter N, Azad AK, Alam MN (eds) Proceedings of the mite management workshop of coconut in SAARC member countries. SAARC Agriculture Centre (SAC), Dhaka, pp 59–66

Alam NS, Islam MN (2014) Mite management of coconut in Bangladesh. In: Akter N, Azad AK, Alam MN (eds) Proceedings of the mite management workshop of coconut in SAARC member countries. SAARC Agriculture Centre (SAC), Dhaka, pp 25–38

Al-Shanfari A, Gowen SR, Moore D, Hountondji FCC, Moraes GJ de (2010) The coconut mite *Aceria guerreronis* Keifer (Acari: Eriophyidae) in the sub-arid coastal region of Dhofar, Oman. In: de Moraes GJ, Castillo RC, Flechtmann CH (eds) Proceedings of XIII International Congress of Acarology. Abstracts, Recife, p 11

Al-Shanfari A, Hountondji FCC, Al-Zawamri H, Rawas H, Al-Mashiki Y, de Moraes GJ, Moore D, Gowen SR (2013) Occurrence and seasonal prevalence of the coconut mite, *Aceria guerreronis* (Eriophyidae), and associated arthropods in Oman. Exp Appl Acarol 60(2):139–151

Anon (2013) Coconut statistical year book of the Asia pacific Coconut Community (APCC), Jakarta

Ansaloni T, Perring TM (2002) Biology of *Aceria guerreronis* Keifer on queen palms. In: Morales-Malacara JB, Rivas G (eds) Proceedings of XI International Congress of Acarology. Abstract. Universidad Nacional Autónoma de México, Distrito Federal México, p 145

Ansaloni T, Perring TM (2004) Biology of *Aceria guerreronis* (Acari: Eriophyidae) on queen palm, *Syagrus romanzoffiana* (Arecaceae). Int J Acarol 30:63–70. doi:10.1080/01647950408684370

Aratchige NS (2007) Predators and the accessibility of herbivore refuges in plants. PhD thesis, University of Amsterdam

Aratchige NS (2014) Mite management of coconut in Sri Lanka. In: Akter N, Azad AK, Alam MN (eds) Proceedings of the mite management workshop of coconut in SAARC member countries. SAARC Agriculture Centre (SAC), Dhaka, pp 87–104

Aratchige NS, Fernando LCP, de Silva PHPR, Perera KFG, Hettiarachchi CS, Waidyarathne KP, Jayawaredena SMV (2010) A new tray-type arena to mass rear *Neoseiulus baraki*, a predatory mite of coconut mite, *Aceria guerreronis* in the laboratory. Crop Prot 29:556–560

Aratchige NS, Fernando LCP, Waidyarathne KP, Chandrasiri KAS (2012a) Population dynamics of *Aceria guerreronis* (Acari: Eriophyidae) and its predatory mite, *Neoseiulus baraki* (Acari: Phytoseiidae) in two coconut growing areas in Sri Lanka. Exp Appl Acarol 56:319–325. doi:10.1007/s10493-012-9519-z, ISSN 0168-8162

Aratchige NS, Fernando LCP, Kumara ADNT, Jayalath KVNN, Perera KFG, Suwandarathne NI (2012b) Biological control of coconut mite: effect of inundative release of the predatory mite, *Neoseiulus baraki* (Acari: Phytoseiidae) In: Hettiarachchi LSK, Abeysinghe ISB (eds) Proceedings of the 4th Symposium on Plantation Crop Research – Technological Innovations for Sustainable Plantation Economy. Tea Research Institute of Sri Lanka, St. Coombs, pp 125–133

CABI (2015a) http://www.cabi.org/isc/mobile/datasheet/38647. Accessed 8 Dec 2015

CABI (2015b) Invasive species compendium. http://www.cabi.org/isc/datasheet/10059. Accessed 27 Nov 2015

CABI (2015c) Invasive species compendium. http://www.cabi.org/isc/datasheet/46792. Accessed 27 Nov 2015

Cabral RVG, Carmona MM (1969) *Aceria guerreronis* Keifer (Acarina: Eriophyidae), uma espécie nova para S. Tomé e Príncipe. Port Acta Biol 10:353–358

Cabrera RI (1991) El ácaro del cocotero *Eriophyes guerreronis* su importancia económica y métodos de lucha. Reseña. Estación Nacional de Sanidad de los Cítricos y Otros Frutales, Habana

Cabrera RI (2002) Biological control of the coconut mite *Aceria guerreronis* (Acari: Eriophyidae) with the fungus *Hirsutella thompsonii* and its possible integration with other control methods. In: Fernando LCP, de Moraes GJ, Wickramanada IR (eds) Proceeding of international workshop on coconut mite (*Aceria guerreronis*). Coconut Research Institute, Lunuwila, pp 89–103

Chandrasiri KAS, Fernando LCP (2004) Effectiveness of used engine oil in the management of coconut mite (*Aceria guerreronis*) (Acari: Eriophyidae). COCOS 16:43–55

de Moraes GJ, Zacarias MS (2002) Use of predatory mites for the control of eriophyid mites. In: Fernando LCP, de Moraes GJ, Wickramanada IR (eds) Proceedings of international workshop on coconut mite (*Aceria guerreronis*). Coconut Research Institute, Lunuwila, pp 78–88

de Moraes GJ, Lopes PC, Fernando LCP (2004) Phytoseiid mites (Acari: Phytoseiidae) of coconut growing areas in Sri Lanka, with descriptions of three new species. J Acarol Soc Jpn 13:141–160

Doreste SE (1968) El ácaro de la flor del cocotero (*Aceria guerreronis* Keifer) en Venezuela. Agron Trop 18:370–386

Edgington S, Fernando LCP, Jones K (2008) Natural incidence and environmental profiling of the mite-pathogenic fungus *Hirsutella thompsonii* Fisher for control of coconut mite in Sri Lanka. Int J Pest Manag 54:123–127

Estrada J, Gonzalez M (1975) Daños causados en coco por *Aceria guerreronis* (Acarina: Eriophyidae) en Cuba. Rev Agric 8:30–34

FAOSTAT (2013) http://faostat3.fao.org/download/Q/QC/E. Accessed 27 Nov 2015

Fernando LCP, Aratchige NS (2010) Status of coconut mite *Aceria guerreronis* and biological control research in Sri Lanka. In: Sabelis MW, Bruin J (eds) Trends in acarology. Springer, Amsterdam, pp 379–384

Fernando LCP, Chandrasiri KAS (2010) Effectiveness of spraying vegetable oil-sulphur emulsion to manage coconut mite, *Aceria guerreronis*. In: Dharmakeerthi RS, Senevirathna AMWK (eds) Proceedings of the third symposium on plantation crop research – stakeholder empowerment through technological advances. Rubber Research Institute of Sri Lanka, Dartonfield, pp 241–250

Fernando LCP, Wickramanada IR, Aratchige NS (2002) Status of coconut mite, *Aceria guerreronis* in Sri Lanka. In: Fernando LCP, de Moraes GJ, Wickramanada IR (eds) Proceedings of the international workshop on coconut mite (*Aceria guerreronis*). Coconut Research Institute, Lunuwila, pp 1–8

Fernando LCP, Aratchige NS, Peiris TSG (2003) Distribution pattern of coconut mite *Aceria guerreronis* and its predator *Neoseiulus* aff. *paspalivorus* on coconut palms. Exp Appl Acarol 31:71–78

Fernando LCP, Manoj P, Hapuarachchi DCL, Edgington S (2007) Evaluation of four isolates of *Hirsutella thompsonii* against coconut mite (*Aceria guerreronis*) in Sri Lanka. Crop Prot 26:1062–1066

Fernando LCP, Waidyarathne KP, Perera KFG, de Silva PHPR (2010) Evidence for suppressing coconut mite *Aceria guerreronis* by inundative release of the predatory mite, *Neoseiulus baraki*. Biol Control 53:108–111

Flechtmann CHW (1989) *Cocos weddelliana* H. Wendl (Palmae: Arecaceae), a new host plant for *Eriophyes guerreronis* (Keifer 1965) (Acari: Eriophyidae) in Brazil. Int J Acarol 15:241

Galvão AS, Gondim MGC Jr, de Moraes GJ, Melo JWS (2011) Distribution of *Aceria guerreronis* and *Neoseiulus baraki* among and within coconut bunches in northeast Brazil. Exp Appl Acarol 54:373–384

Galvão AS, Melo JWS, Monteiro VB, Lima DB, de Moraes GJ, Gondim MGC Jr (2012) Dispersal strategies of *Aceria guerreronis* (Acari: Eriophyidae), a coconut pest. Exp Appl Acarol 57:1–13

Gopal M, Gupta A (2001) Has *Hirsutella thompsonii* the wherewithal to counter coconut eriophyid mite scourge? Curr Sci 80:831–836

Griffith R (1984) The problem of the coconut mite, *Eriophyes guerreronis* (Keifer), in the coconut groves of Trinidad and Tobago. In: Webb R, Knausenberger W, Yntema L (eds) Proceedings of 20th annual meeting of the Caribbean food crops society. Coll. Virgin Islands & Caribbean Food Crops Society, Virgin Islands, pp 128–132

Hall RA, Espinosa Becerril A (1981) The coconut mite *Eriophyes guerreronis* with special reference to the problem in Mexico. British Crop Protection Conference – Pests and Diseases. Proceedings British Crop Protection Council, Farnham, pp 113–120

Hall RA, Hussey NW, Mariau D (1980) Results of a survey of biological control agents of the coconut mite *Eriophyes guerreronis*. Oléagineux 35:395–398

Hernández Roque F (1977) Combate químico del eriófido del cocotero *Aceria* (*Eriophyes*) *guerreronis* (K.) en la Costa de Guerrero. Agric Tec Méx 4:23–38

Howard FW, Abreu Rodríguez E (1991) Tightness of the perianth of coconuts in relation to infestation by coconut mites. Fla Entomol 74:358–361

Howard FW, Abreu Rodríguez E, Denmark HA (1990) Geographical and seasonal distribution of the coconut mite, *Aceria guerreronis* (Acari: Eriophyidae), in Puerto Rico and Florida, USA. J Agric Univ Puerto Rico 74:237–251

Howard FW, Moore D, Giblin-Davis RM, Abad RG (2001) Insects on palms. CABI Publishing, Wallingford

Julia JF, Mariau D (1979) Nouvelles recherché en Côte d'Ivoire sur *Eriophyes guerreronis* K., acarien ravageur des noix du cocotier. Oléagineux 34:181–189

Keifer HH (1965) Eriophyid studies B-14. California Department of Agriculture, Bureau of Entomology, Sacramento (Special publication)

Kumar PP, Ramaraju K (2010) Impact of nut-infesting eriophyid mite, *Aceria guerreronis* Keifer, on the out-turn and quality of coconut fiber. In: de Moraes GJ, Castillo RC, Flechtmann CH (eds) Proceedings of the XIII International Congress of Acarology. Abstracts. Recife, p 214

Kumara ADNT, Fernando LCP, Suwandharathne NI, Aratchige NS (2014) Use of polypropylene bags for mass rearing *Neoseiulus baraki* (Acari: Phytoseiidae), a predatory mite of *Aceria guerreronis* Keifer (Acari: Eriophyidae). Biocontrol Sci Tech. doi:10.1080/0958307.2014.918245

Lawson-Balagbo LM, Gondim MGC Jr, de Moraes GJ, Hanna R, Schausberger P (2007) Refuge use by the coconut mite *Aceria guerreronis*: fine scale distribution and association with other mites under the perianth. Biol Control 43:102–110

Lawson-Balagbo LM, Gondim MGC Jr, de Moraes GJ, Hanna R, Schausberger P (2008) Exploration of the acarine fauna on coconut palm in Brazil with emphasis on *Aceria guerreronis* (Acari: Eriophyidae) and its natural enemies. Bull Entomol Res 98:83–96

Lima DB, Melo JW, Guedes RN, Siqueira HA, Pallini A, Gondim MGC Jr (2012) Survival and behavioural responses to acaricides of the coconut mite predator *Neoseiulus baraki*. Exp Appl Acarol 60(3):381–393

Lima DB, Melo JW, Guedes NMP, Gontijo LM, Guedes RNC, Gondim MGC Jr (2015) Bioinsecticide-predator interactions: azadirachtin behavioral and reproductive impairment of the coconut mite predator *Neoseiulus baraki*. PLoS ONE 10(2), e0118343. doi:10.1371/journal.pone.0118343

Lokesh BR, Nandihalli BS (2009) Management of coconut perianth mite, *Aceria guerreronis* (Keifer) through nutrition. Karnataka J Agric Sci 22(3-Spl.Issue):603–605

Mallik B, Gowda CC, Jayappa J, Guruprasad H, Onkarappa S (2003) Coconut eriophyid mite in India. Issues and strategies, In: Singh HP, Rethinam P (eds) Coconut eriophyid mite – issues and strategies. Proceedings of the International Workshop on Coconut Mite. Bangalore, pp 27–34

Mallik B, Gowda CC, Kalmath BS (2004) Can coconut mite (*Aceria guerreronis* K.) be managed in future? In: Peiris TSG, Ranasinghe CS (eds) Proceedings of the international conference of the Coconut Research Institute of Sri Lanka. Coconut Research Institute, Lunuwila, pp 200–206

Manson DCM, Oldfield GN (1996) Life forms, deuterogyny, diapause and seasonal development. In: Lindquist EE, Sabelis MW, Bruin J (eds) Eriophyoid mites – their biology, natural enemies and control. Elsevier, Amsterdam, pp 173–183

Mariau D (1969) *Aceria guerreronis* Keifer: récent ravageur de la cocoteraie Dahoméenne. Oléagineux 24:269–272

Mariau D (1977) *Aceria* (*Eriophyes*) *guerreronis*: un important ravageur des cocoteraies africaines et americaines. Oléagineux 32:101–108

Mariau D (1986) Comportement de *Eriophyes guerreronis* Keifer à l''egard de différentes variétés de cocotiers. Oléagineux 41:499–505

Mariau D, Julia JF (1970) L'acariose a *Aceria guerreronis* (Keifer), ravageur du cocotier. Oléagineux 25:459–464

Mariau D, Tchibozo HM (1973) Essais de lutte chimique contre *Aceria guerreronis* (Keifer). Oléagineux 28.133–13

Mathew TB, Paul A, Beevi SN, Saradamma K (2000) Population dynamics of coconut mite, *Aceria guerreronis* Keifer on developing nuts. Paper presented at Entomocongress 2000, Perspective plan for the new millennium, 5–8 November, Thiruvananthapuram, p 195

Melo JW, Domingos CA, Pallini A, Oliveira JEM, Gondim MGC Jr (2012) Removal of bunches or spikelets is not effective for the control of *Aceria guerreronis*. Hortscience 47(5):626–630

Melo JW, de Lima DB, Sabelis MW, Pallini A, Gondim MGC Jr (2014) Limits to ambulatory displacement of coconut mites in absence and presence of food-related cues. Exp Appl Acarol 62(4):449–461. doi:10.1007/s10493-013-9753-z

Mohan C, Thomas GV, Josephrajkumar A (2014) Mite management of coconut in India. In: Akter N, Azad AK, Alam MN (eds) Proceedings of the mite management workshop of coconut in SAARC member countries. SAARC Agriculture Centre (SAC), Dhaka, pp 43–58

Monteiro VB, Lima DB, Gondim MGC Jr, Siqueira HAA (2012) Residual bioassay to assess the toxicity of acaricides against *Aceria guerreronis* (Acari: Eriophyidae) under laboratory conditions. J Econ Entomol 105(4):1419–1425

Moore D (1986) Bract arrangement in the coconut fruit in relation to attack by the coconut mite *Eriophyes guerreronis* Keifer. Trop Agric 63:285–288

Moore D, Alexander L (1985) Outbreaks and new records. FAO Plant Prot Bull, Dominica, p 119

Moore D, Alexander L (1987) Aspects of migration and colonization of the coconut palm by the coconut mite, *Eriophyes guerreronis* (Keifer) (Acari: Eriophyidae). Bull Entomol Res 77:641–650

Moore D, Alexander L (1990) Resistance of coconuts in St. Lucia to attack by the coconut mite, *Eriophyes guerreronis* Keifer. Trop Agric 67:33–36

Moore D, Howard FW (1996) Coconuts. In: Lindquist EE, Sabelis MW, Bruin J (eds) Eriophyoid mites: their biology natural enemies and control. Elsevier, Amsterdam, pp 561–570

Moore D, Alexander L, Hall RA (1989) The coconut mite, *Eriophyes guerreronis* Keifer in St. Lucia: yield losses and attempt to control it with acaricides, polybutene and *Hirsutella* fungus. Trop Pest Manag 35:83–89

Moore D, Ridout MS, Alexander L (1991) Nutrition of coconuts in St Lucia and relationship with attack by coconut mite *Aceria guerreronis* Keifer. Trop Agric 68:41–44

Muralidharan K, Mathew CJ, Thampan C, Amamath CH, Anithakumari P, Mohan C, Vijayakumar K, Sairam CV, Nair CPR, Arulraj S (2001) Pilot sample survey on the incidence and yield loss due to eriophyid mite on coconut in Alappuzha District. Indian Coconut J 21:28–32

Muthiah C (2007) Estimation of yield loss caused by eriophyid mite on coconut. Ann Plant Prot Sci 15:484–486

Muthiah C, Bhaskaran R (2000) Major outbreak of eriophyid mite of coconut in India. Planter 76:243–246

Muthiah C, Natarajan C (2004) Varietal reaction and nutrient management of coconut eriophyid mite. Planter 80:159–169

Muthiah C, Natarajan C (2005) Management of coconut eriophyid mite with nutrients and biofertilizers. Indian Coconut J 25:16–19

Muthiah C, Bhaskaran R, Kannaiyan S (2001) Bioecology and control of eriophyid mite of coconut – an Indian experience. Planter 77:255–263

Nair CPR (2002) Status of coconut eriophyid mite *Aceria guerreronis* Keifer in India. In: Fernando LCP, de Moraes GJ, Wickramanada IR (eds) Proceedings of the international workshop on coconut mite (*Aceria guerreronis*). Coconut Research Institute, Lunuwila, pp 9–12

Nair MG, Mathew TB, Saradamma K, Naseema Beevi S (2002) Infestation of coconut mite (*Aceria guerreronis* K.) in the Southern Indian State, Kerala. In: Fernando LCP, de Moraes GJ, Wickramanada IR (eds) Proceedings of the international workshop on coconut mite (*Aceria guerreronis*). Coconut Research Institute, Lunuwila, pp 32–40

Nampoothiri KUK, Nair CPR, Kannaiyan S, Sabitha D, Saradamma K, Naseema Beevi S and Sreerama Kumar P (2002) Coconut eriophyid mite (*Aceria guerreronis* Keifer) – an update. In: Sreedharan K, Vinod Kumar PK, Jayarama and Basavaraj MC (eds) PLACROSYM XV Central Coffee Research Institute, Mysore, pp 487–496

Naseema Beevi S, Mathew TB, Bai H, Saradamma K (2003) Status of eriophyid mite in Kerala-Resume of work done. In: Singh HP, Rethinam P (eds) Proceedings of the International Workshop on Coconut Mite – issues and strategies. Bangalore, pp 64–75

Navia D, de Moraes GJ, Querino RB (2006) Geographic variation in the coconut mite, *Aceria guerreronis* Keifer (Acari: Eriophyidae): a geometric morphometric analysis. Int J Acarol 32:301–314

Navia D, de Moraes GJ, Querino RB (2009) Geographic pattern of morphological variation of the coconut mite, *Aceria guerreronis* Keifer (Acari: Eriophyidae), using multivariate morphometry. Braz J Biol 69:773–783

Navia D, Gondim MGC Jr, Aratchige NS, de Moraes GJ (2013) A review of the status of the coconut mite *Aceria guerreronis* (Acari: Eriophyidae), a major tropical mite pest. Exp Appl Acarol 59(1–2):67–94

Negloh K, Hanna R, Schausberger P (2011) The coconut mite, *Aceria guerreronis*, in Benin and Tanzania: occurrence, damage and associated acarine fauna. Exp Appl Acarol 55:361–374. doi:10.1007/s10493-011-9474-0

Oleke JM, Manyong V, Mignouna D, Isinika A, Mutabazi K, Hanna R, Sabelis MW (2013) Ex-ante economic analysis of biological control of coconut mite in Benin. AgBioforum 16(2):161–169

Ortega CA, Rodriguez VJ, Garibay CV (1967) Investigaciones preliminares sobre el eriófido del fruto del cocotero *Aceria guerreronis* Keifer, en la Costa Grande Guerrero (México). Oléagineux 6:371–372

Perera L, Sarathchandra SR, Wickramananda IR (2013) Screening coconut cultivars for tolerance to infestation by the coconut mite, *Aceria guerreronis* (Keifer) in Sri Lanka. CORD 29(1):46–51

Persley GJ (1992) Replanting the tree of life: towards an international agenda for coconut palm research. CAB International, Wallingford

Pushpa, V. and Nandihalli, B. S. 2008. Bioefficacy of botanicals in the management of coconut eriophyid mite,Aceria guerreronis Keifer. J. Plant Prot. Environ. 5(1): 90–94.

Pushpa V, Nandihalli BS (2010) Evaluation of pesticides and biopesticides against the coconut eriophyid mite, *Aceria guerreronis* Keifer under laboratory conditions. Karnataka J Agric Sci 23:178–179

Rabindra RJ, Sreerama Kumar P (2003) Prospects of biological control of coconut mite. In: Singh HP, Rethinam P (eds) Proceedings of the International Workshop on Coconut Mite – issues and strategies. Bangalore, pp 35–42

Rajan P, Mohan C, Chalapathy Rao NBV, Thomas GV (2012) Scenario of coconut eriophyid mite infestation in Andhra Pradesh. Indian Coconut J LV 2:25–31

Ramaraju K, Rabindra RJ (2002) Palmyra, *Borassus flabellifer* Linn. (Palmae): a host of the coconut eriophyid mite *Aceria guerreronis* Keifer. Pest Manag Hortic Ecosyst 7:149–151

Ramaraju K, Palaniswamy S, Annakodi P, Varadarajan MK, Muthukumar M, Bhaskaran V (2000) Impact of coconut eriophyid mite, *Aceria guerreronis* K. (Acari: Eriophyidae) on the yield parameters of coconut. Indian Coconut J 25:12–15

Ramaraju K, Natarajan K, Sundara Babu PC, Palaniswamy S, Rabindra RJ (2002) Studies on coconut eriophyid mite, *Aceria guerreronis* Keifer in Tamil Nadu, India. In: Fernando LCP, de Moraes GJ, Wickramanada IR (eds) Proceedings of the international workshop on coconut mite (*Aceria guerreronis*). Coconut Research Institute, Lunuwila, pp 13–31

Ranasinghe CS, Jayasekara C (1989) Physiological and biochemical characters of different varieties and cultivars of coconut. Abstract in Proc Sri Lanka Association for the Advancement of Science, December 1989, Colombo

Reis AC, Gondim MGC Jr, de Moraes GJ, Hanna R, Schausberger P, Lawson-Balagbo LE, Barros R (2008) Population dynamics of *Aceria guerreronis* Keifer (Acari: Eriophyidae) and associated predators on coconut fruits in Northeastern Brazil. Neotrop Entomol 37:457–462

Rethinam P, Singh HP, Vijayakumar H, Gopalakrishnan R (2003) Eriophyid mite in coconut. CoconutDevelopment Board, India

Roseleen SSJ, Ramaraju K (2012) Acaricidal effects and residues of profenofos and abamectin on the nut-infesting eriophyid mite, *Aceria guerreronis* Keifer (Acari: Prostigmata) on coconut. Int J Acarol 38:465–470

Sathiamma B, Radhakrishnan Nair CPR, Koshi PK (1998) Outbreak of a nut infesting eriophyid mite, *Eriophyes guerreronis* (K.) in coconut plantations in India. Ind Coconut J 29:1–3

Schliesske J (1988) On the gall mite fauna (Acari: Eriophyidae) of *Cocos nucifera* L. in Costa Rica. Nachrichtenblatt Duetschen Pflanzenschutzdienstes 40:124–127

Seguni Z (2002) Incidence, distribution and economic importance of the coconut eriophyid mite, *Aceria guerreronis* Keifer in Tanzania. In: Fernando LCP, de Moraes GJ, Wickramananda IR (eds) Proceedings of the international workshop on coconut mite (*Aceria guerreronis*). Coconut Research Institute, Lunuwilla, pp 54–57

Shobha TR, Haq MA (2011) Postembryonic development of the coconut mite, *Aceria guerreronis*, on coconut in Kerala, India In: de Moraes GJ, Proctor H (eds) Acarology XIII: Proceedings of the International Congress. Zoosymposia 6:68–71

Solangi AH (2014) Mite management of coconut in Pakistan. In: Akter N, Azad AK, Alam MN (eds) Proceedings of the workshop on mite management of coconut in SAARC member countries. SAARC Agriculture Centre (SAC), Dhaka, pp 73–86

Souza IV de, Ramos ALR, Oliveira AR, Gondim Jr. MGC, Bittencourt MAL, Ferraz MIF (2010) Pattern of occurrence of *Aceria guerreronis* Keifer (Acari: Eriophyidae) and other mites associated to coconut fruits in Una-BA, Brazil. In: de Moraes GJ, Castillo RC, Flechtmann CH (eds) Proceedings of XIII International Congress of Acarology. Abstracts. Recife, p 265

Sreerama Kumar P (2002) Development of a biopesticide for the coconut mite in India. Proceedings of the British crop protection conference – pests and diseases. British Crop Protection Council, Brighton, pp 335–340

Sreerama Kumar P (2010) *Hirsutella thompsonii* as a mycoacaricide for *Aceria guerreronis* on coconut in India: research, development and other aspects. In: Sabelis MW, Bruin J (eds) Trends in acarology. Springer, Amsterdam, pp 441–444

Sreerama Kumar P, Singh SP (2000) *Hirsutella thompsonii*: the best control option for the management of the coconut mite in India. Indian Coconut J 31:11–17

Sreerama Kumar P, Singh L (2008) Enabling mycelial application of *Hirsutella thompsonii* for managing the coconut mite. Exp Appl Acarol 46:169–182

Sujatha A, Rao NBVC (2004) Studies on coconut eriophyid mite in Andhra Pradesh. Indian Coconut J 34:8–10

Sujatha A, Kumar DA, Srinivasulu B (2003) Efficacy of certain pesticides against coconut eriophyid mite, *Aceria guerreronis* (K.) through root feeding. J Appl Zool Res 14:51–53

Thirumalai Thevan PS, Muthukrishman N, Manoharan T, Thangaraj T (2004) Influence of phenotypic and biochemical factors of coconut on eriophyid, *Aceria guerreronis* Keifer and predatory, *Amblyseius* spp. mites. J Entomol Res 28:291–299

Thomas RJ, Nair RV, Nair CPR, Nampoorthiri CK, Mathews C (2004) Growth and vigour of coconut seedlings raised from eriophyid mite, *Aceria guerreronis* Keifer infested seed nuts. J Plant Crop 32:25–29

Varadarajan MK, David PMM (2002) Population dynamics of the coconut mite *Aceria guerreronis* Keifer (Acari: Eriophyidae) and associated arthropods in Tamil Nadu, India. Insect Sci Appl 22:47–59

Varadarajan MK, David PMM (2003) Effect of ground vegetation and nut characteristics on the severity of infestation by *Aceria guerreronis* in coconut. Entomon 28:361–365

Wickramananda IR, de Silva PHPR, Sarathchandra SR, Kumara ADNT, Rajapaksha RHS (2004) A mehod for in vitro culture of coconut mite, *Aceria guerreronis* Keifer (Acari: Eriophyidae) in immature leaf tissues of coconut (*Cocos nucifera* L). In: Peiris TSG, Ranasinghe CS (eds) Proceedings of the International Conference of the Coconut Research Institute of Sri Lanka – part 2 (Contributed papers). Coconut Research Institute of Sri Lanka. Lunwila, pp 229–236

Wickramananda IR, Peiris TSG, Fernando MT, Fernando LCP, Edgington S (2007) Impact of the coconut mite (*Aceria guerreronis* Keifer) on the coconut industry in Sri Lanka. CORD 23:1–16

Zuluaga CI, Sánchez PA (1971) La roña o escoriación de los frutos del cocotero (*Cocos nucifera* L.) en Colombia. Acta Agron 21:133–139

Arthropod Communities in Coffee: A Habitat Mimicking Tropical Forests

17

N.E. Thyagaraj, G.V. Manjunatha Reddy, S. Onkara Naik, and B. Doddabasappa

Abstract

Traditional coffee plantations contain a high biodiversity of plants and animals including arthropods. The biodiversity is significantly reduced when the system is modernised. Generally, populations of harmful species such as phytophagous pests on coffee are well regulated by the natural enemy complex consisting of parasites, parasitoids and predators which include arthropods, frogs, birds and small mammals. Long-term studies in Puerto Rico, Costa Rica, South India and Latin American countries have revealed that abundance of arthropod and species richness is higher in shaded coffee ecosystems than non-shaded. It is observed that several dominant predator groups such as spiders; tiger beetles; coccinellids and pollinators such as honeybees, megachilids and *Xylocopa*; beetles and butterflies dwell in coffee plantations. Arthropods in association with microorganisms responsible for nutrient cycling, conditioning and aeration of soil also inhabit coffee-cultivated systems. Many millipedes and spiders are endemic to coffee plantations and contribute significantly to soil nutrients. Habitat loss, deterioration and fragmentation and chemical pollution are the leading factors

The original version of this chapter was revised. An erratum to this chapter can be found at DOI 10.1007/978-981-10-1524-3_21

N.E. Thyagaraj (✉)
Department of Entomology, College of Agriculture, Madenur, Hassan, Karnataka, India
e-mail: nethyagaraj@gmail.com

G.V.M. Reddy
Department of Entomology, Central Coffee Research Institute,
Coffee Research Station, Chikmagalur 577177, Karnataka, India

S.O. Naik
Division of Entomology and Nematology, Indian Institute of Horticultural Research (IIHR),
Hessaraghatta Lake Post, Bengaluru 560089, Karnataka, India

B. Doddabasappa
Department of Entomology, College of Horticulture, Kolar, Karnataka, India

A.K. Chakravarthy, S. Sridhara (eds.), *Economic and Ecological Significance of Arthropods in Diversified Ecosystems*, DOI 10.1007/978-981-10-1524-3_17

causing significant biodiversity decline in coffee plantations. It can be generalised that conservation efforts to preserve biological diversity in coffee-cultivated ecosystems should include traditional coffee plantations as conservation units.

Keywords
Arthropod communities • Biodiversity • Conservation efforts • Traditional Coffee

17.1 Introduction

Coffee has a history of over 400 years. In India, coffee was first introduced in Chikmagalur district by saint Baba Budan during 1600 A.D. However, it was not until late 1820s that commercial plantations came into existence in South India with British enterprise. Currently, coffee is chiefly grown in South India. In India the commercially chief species, viz. Arabica and Robusta, are cultivated in almost the same proportions. Coffee is cultivated under a carefully trained canopy of shade trees, which greatly influences the microclimate in the coffee ecosystem. Cultivated coffee farms work much like forests, providing habitat for beneficial insects, birds and nitrogen-fixing plants. They also provide the greater opportunities for diversification by way of cultivation of associated crops like pepper, orange, banana and spices for additional returns. The genus *Coffea* (Rubiaceae) comprises 103 species (Davis et al. 2006), but only two, viz. *C. arabica* and *C. robusta*, are commonly cultivated in large scale. Coffee is grown in more than 10 million hectares in over 80 developing countries (http://faostat.fao.org/). The Western Ghats of India is considered one among the 25 biodiversity hot spots of the globe which comes under tropical climate. The Western Ghats extends the West coast of South India. Coffee is grown amidst nature in these mountainous areas. Coffee plantations in India have always maintained a symbiotic relationship with the surrounding biotic community and imitate the elements of evergreen tropical forests.

India, Vietnam, Colombia, Indonesia, Brazil, Honduras, Peru, Ethiopia, Guatemala and Mexico are the world's top ten coffee-producing countries. Coffee cultivation in Ivory Coast is important for the economy of the country, as coffee is the second largest export commodity. Ivory Coast is the largest Robusta producers in the world. Coffee plants were introduced into Ivory Coast in the nineteenth century by French colonisers. Brazil is responsible for about a third of all coffee, making Brazil the world's largest producer. The first coffee bush in Brazil was planted in 1727. The Portuguese were looking for a coffee market, and by 1920s, Brazil was nearly a monopolist of the international coffee market. The coffee plant spread to Colombia by 1790. Coffee-producing regions of these countries have heterogeneous landscapes with major variations in topography, latitude, climate soil and ecological conditions. Earlier slash-and-burn method was adopted to cultivate coffee, but currently it has been abandoned due to ecological damage. Thus, in over 200 years, many kinds and species of arthropods have colonised coffee. Some have even become endemic, some endangered. Population of arthropods is highly diverse in tropical climate especially insects. Coffee, being a perennial plant, is subject to attack by migratory as well as sedentary pests. Of the two major commercially cultivated species of coffee, Arabica coffee is more prone to pest attack. Only a few of the insects, invertebrates and mammals that injure coffee are of economic importance.

17.2 Plant Feeders and Natural Enemies

Over 1000 species of insects alone are recorded in coffee plantations. Among them, only few are pests incurring economic losses. The white stem borer, *Xylotrechus quadripes* Chevrolat (Coleoptera: Cerambycidae); (major pest of C. arabica) coffee berry borer, *Hypothenemus hampei* Ferrari (Coleoptera: Scolytidae); (major pest of C. robusta) shot-hole borer (Fig. 17.1), *Xylosandrus compactus* Eichhoff

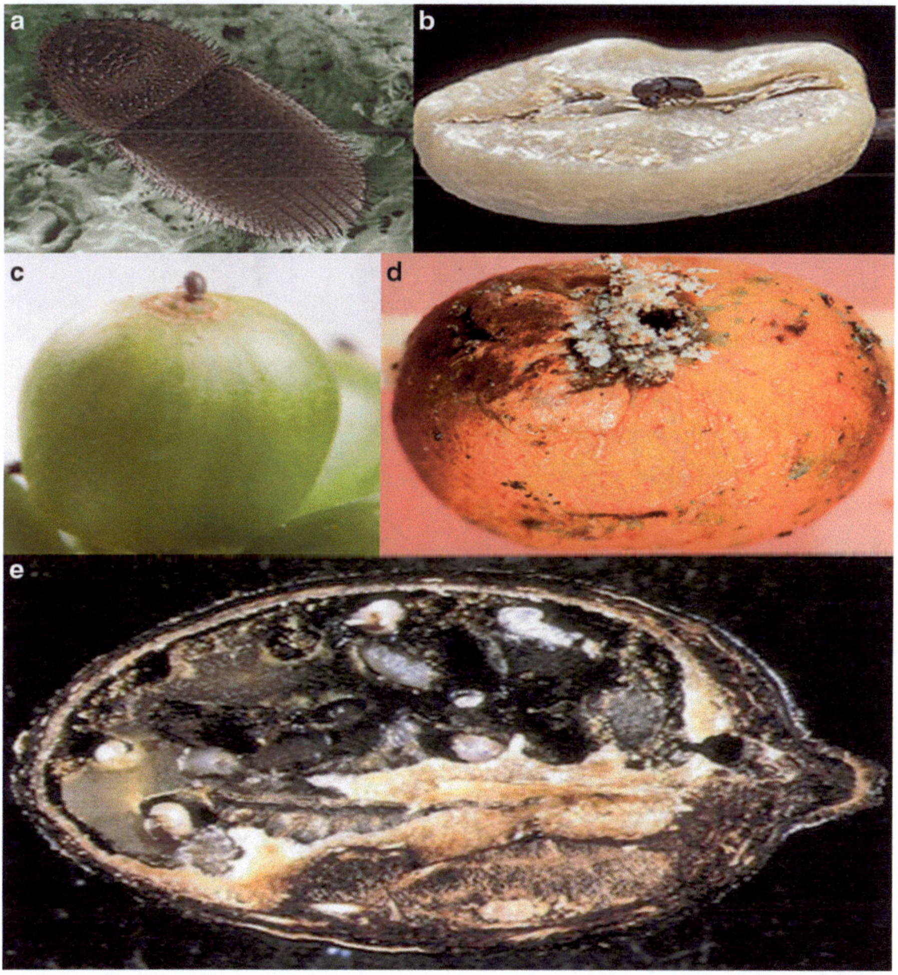

Fig. 17.1 Female coffee berry borer (**a**) walking over a coffee seed (**b**) to show the small size of the insect (ca. 2 mm long, 1 mm wide). Female coffee berry borer boring a hole in a coffee berry (**c**) with characteristic symptom of infestation revealing frass on the entrance hole (**d**). Damage caused by larval feeding inside the coffee berry (**e**). Credits: (**a**) E. Erbe, USDA, ARS; (**b**) P. Greb, USDA, ARS; (**c**) and (**e**) G. Hoyos, Cenicafé; (**d**) G. Mercadier, USDA, ARS (Vega et al. 2009)

(Coleoptera: Scolytidae) (major pest of C. robusta) (Sridharan et al. 1992); mealybugs, *Planococcus citri* Risso and *P. lilacinus Ckll.* (Homoptera: Pseudococcidae); green scale, *Coccus viridis* Green (Homoptera: Coccidae); root grub (*Holotrichia* spp.) (Fig. 17.2a); and the root-lesion nematode, *Pratylenchus coffeae* Zimmermann (Nematoda: Pratylenchidae) are the major pests on coffee in India (Sekhar 1964; Anonymous 1998). One of the major constraints to coffee production throughout the world is the damage caused by the coffee berry borer. Many natural enemies of the coffee berry borer have been reported (Figs. 17.3, 17.4 and 17.5).

Over 50 species of scales and mealybugs are reported to attack various parts of the coffee, viz. trees, roots, branches, leaves, flower clusters and berries where they suck the sap. *Planococcus citri* (Risso), *P. kenyae* (Le Pelley), *P. lilacinus* (Cockerell), *P. minor* (Maskell) and *Ferrisia virgata* (Cockerell) are mealybugs causing economic damage to coffee. *P. citri* and *P. lilacinus* are the most common.

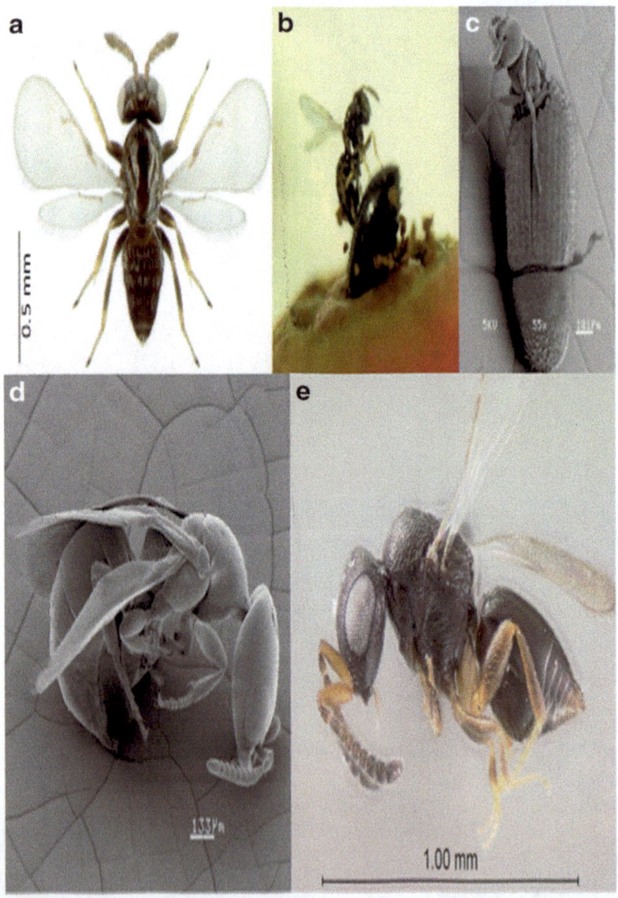

Fig. 17.2 Adult *Phymastichus coffea* (**a**) ovipositing in the coffee berry borer (**b**) with one adult parasitoid emerging from the insect (**c**). Adult *Cephalonomia stephanoderis* (**d**) and *Aphanogmus dictynna* (**e**), a hyperparasitoid of *Prorops nasuta*. Credits: (**a**) G. Goergen, IITA; (**b**) A. Castillo and F. Infante, Ecosur; (**c**) and (**d**), G. Nieto, ECOSUR; (**e**) M. Buffington and A. Simpkins, USDA, ARS (Vega et al. 2009)

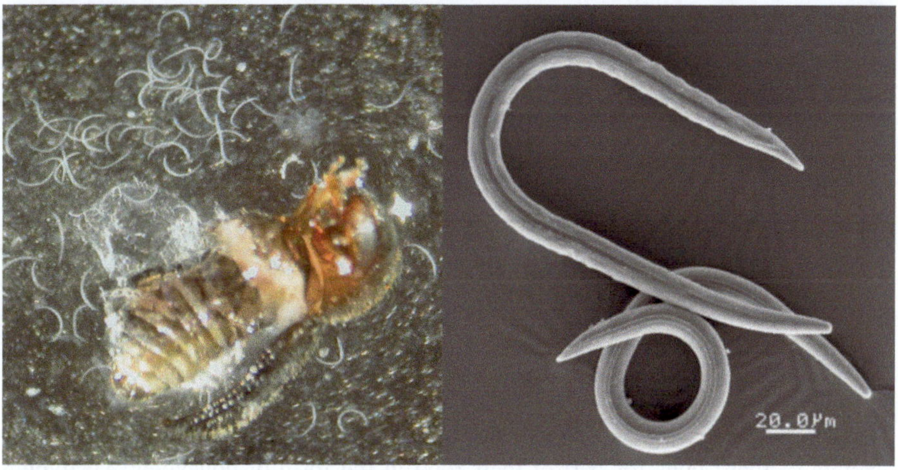

Fig. 17.3 Infective juveniles of *Metaparasitylenchus hypothenemi* emerging from an infected coffee berry borer (*left*) and detail of the infective juvenile (*right*). Credits: (**a**) A. Castillo, ECOSUR; (**b**) G. Nieto, ECOSUR (Vega et al. 2009)

Fig. 17.4 Spiders: common predators in coffee plantations – *Poecilotheria regalis*

These pests infest both Robusta and Arabica but prefer Arabica. *P. ficus*, *P. pacificus* and *P. minor* have been recorded on coffee as minor pests.

Coccids and pseudococcids are invariably associated with ants in coffee plantations all over the world. Three important subfamilies of the Formicidae, viz. Myrmicinae, Campotinae and Dolichoderinae, include many species of ants which

Fig. 17.5 Nest of *red* ants

attack aphids, coccids and some other bugs of homoptera, and they rely on the excreta of these bugs for part of their food in the form of honeydew. Mealybugs produce honeydew and ants of different species to it. Ants give mealybugs sanitation and protection from natural enemies like hymenopterans (Tables 17.1 and 17.2). The absence of ants enhances the numbers and mortality of mealybugs due to natural enemies and helps to trap the nymphs in the honeydew. Twenty seven ant species have been recorded all over the world on species of homoptera-attacking coffee. Thirteen species have so far been recorded from coffee-cultivated tracts of South India (Venkataramaiah and Rehman 1989). Of the species recorded, *Plagiolepis* sp. is widespread and seen in almost any estate in the coffee-growing region. *Acrophaga* sp. is documented from Coorg and has not been in Wyanad district. Unlike *Crematogaster* and *Oecophylla* (Fig. 17.6), this ant is harmless to human beings.

There are few other cosmopolitan pests on coffee, but as an illustration, natural enemies of only select insect pests have been listed above (Table 17.3).

17.3 Decomposers

Arthropods especially insects are responsible for much of the nutrient cycling, conditioning and aeration of the soil in association with microorganisms. Insects and other arthropods generally make up half the animal biomass in tropical forests (Fittkau and Klinge 1973). Some of the examples of decomposers are dung beetles, flies, carrion beetles, wood borers, millipedes, cockroaches, ants and termites.

This category included a diversity of litter fauna: detritivores such as ants (Formicate) and a whole range of fungal feeders. Dark-winged fungus gnats of family Sciaridae and species of Mycetophilidae and Drosophilidae (all Diptera) live in decaying vegetation and fungi. Coleopterans of family Nitidulidae and Scarabaeidae are also excellent decomposers. The larvae of Lauxaniidae live in decaying vegetation; in this study they were recorded only in organic coffee plantations. Although thrips are good pollinators of canopy trees, they were classified under decomposers since adults are fungal feeders too. Scavenger flies feed on carrion and inhabit moist terrestrial habitats (R).

Table 17.1 List of parasitoids recorded on mealybugs infesting coffee in India

Common name	Scientific name	Family	Pest	Reference
	Alamella flava Agarwal	Encyrtidae	Mealybugs	Reddy et al. (1990)
Redshank, *Ceratodon* moss, fire moss	*Aprostocetus purpureus* (Cameron)	Eulophidae	Cycad Aulacaspis scale	Reddy et al. (1990)
Nesoanagyrus	*Anagyrus agraensis* Saraswat	Encyrtidae	Scale insect	Reddy et al. (1990)
	Leptomastix nigrocoxalis Compere	Encyrtidae	Mealybug	Prakasan and Kumar (1985)
	Prochiloneurus sp.	Encyrtidae	Mealybug	–
	Tetracnemus indicus	Encyrtidae	Mealybug	–
	Tetracnemoidea indica (Ramakrishna Ayyar)	Encyrtidae	Mealybug	Pruthi and Mani (1940)
	Apanteles sp.nr. *sauros* Nixon	Braconidae	Leaf roller	Reddy et al. (1990)
Glassy-winged sharpshooter	*Gonatocerus* sp.	Mymaridae	*Homalodisca vitripennis*	Reddy et al. (1990)
	Belpyrus insularis (Cameron)	Encyrtidae		Reddy et al. (1990)
	Leptacis sp.	Platygastridae	*Hemileia vastatrix*	Reddy et al. (1990)
	Aenasius advena Compare	Encyrtidae	*Ferrisia virgata*	Balakrishnan et al. (1991)

Table 17.2 List of select predators on mealybugs infesting coffee in India

Scientific name	Family	Reference
Dicrodiplosis sp.	Cecidomyiidae	Reddy et al. (1990)
Pseudoscymnus pallidicollis (Mulsant)	Coccinellidae	Reddy et al. (1990)
Diadiplosis coccidivora	Cecidomyiidae	Reddy et al. (1990)
Domomyza perspicax (Knab)	Drosophilidae	Reddy et al. (1990)
Spalgis epius (Westwood)	Lycaenidae	Chacko and Bhat (1976)
Brumoides suturalis (Fabricius)	Coccinellidae	Le Pelley (1968)
Horniolus vietnamicus	Coccinellidae	Irulandi et al. (2001)
Allograpta javana (Weidemann)	Syrphidae	Balakrishnan et al. (1991)
Brumoides suturalis (Fabricius)	Coccinellidae	Le Pelley (1968)
Leucopis sp.	Chamaemyiidae	Balakrishnan et al. (1991)
Mallada sp.	Chrysopidae	Balakrishnan et al. (1991)
Scymnus sp.	Coccinellidae	Balakrishnan et al. (1991)
Diadiplosis coccidivora (Felt)	Cecidomyiidae	Balakrishnan et al. (1991)

Fig. 17.6 Coffee flowers in bloom

Table 17.3 Parasitoids of coffee stem borer *Xylotrechus quadripes* recorded in India

Scientific name	Family	Reference
Metapelma sp.	Eupelmidae	Subramaniam (1941)
Campylonerus sp.	Braconidae	Annual Report (1976)
Gasteruption sp.	Gasteruptiidae	Annual Report (1984)
Parallarhogas pallidiceps	Braconidae	Prakasan et al. (1986)
Doryctus coxalis	Braconidae	Shylesha et al. (1992)
Doryctus compactus	Braconidae	Shylesha et al. (1992)
Scleroderma vigilans	Braconidae	Shylesha et al. (1992)
Scleroderma sp.	Braconidae	Shylesha et al. (1992)
Eurytoma sp.	Eurytomidae	Shylesha et al. (1992)
Iphiaulax sp.	Braconidae	Venkatesha et al. (1997)
Apenesia sahyadrica	Bethylidae	Shylesha et al. (1992)
Avetianella sp.	Encyrtidae	Shylesha et al. (1992)
Apensia sp.	Bethylidae	Annual Report (2002)

Millipedes act as buffer stock for significant increase of calcium and magnesium and augment mineral recycling. *Arthrosphaera* is known for a narrow range of distribution and exhibits single-site endemism in the Peninsular India. *Arthrosphaera* prefers mixed litter than monolitter, and the mixed litter ingestion and faecal pellet production are higher than other tropical millipedes. Increased nitrogen, phosphorus and potassium, narrow C/N ratio and shift of pH towards neutral in mixed litter compost produced by *Arthrosphaera* reveal its value as alternative to vermicompost. Pill millipedes are hosts for a variety of microbes, and from food to faecal matter, the bacterial load increases and fungal load decreases. The presence of ergo sterol in the faecal pellets of *Glomeris* indicates intensive digestion of fungi in the gut passage. *Glomeris* consume up to 1.7–10 % of annual litter production, which amounts to ten times the body mass. Clear evidence has emerged on the invasion of

Arthrosphaera from evergreen and semievergreen forests to plantations in Western Ghats in India.

17.4 Butterflies: Indicator Species

Abundance and diversity of butterflies were documented in 12 coffee estates in Western Ghats, a tract of endemism and of genetic variability (Dolia et al. 2008). Distance from the Bhadra Wildlife Sanctuary was the most important factor for the abundance and richness of butterfly in coffee plantations. The closer the coffee plantations to the sanctuary, the higher the species diversity. The element of butterfly community (Fig. 17.7) in coffee estates also became less similar to that of native forest as distance from the sanctuary increased. The proportion of Australian *Grevillea robusta*, a fast-growing shade tree planted in place of native species in coffee estates, did not seem to affect butterfly diversity. Three or four species of shade trees dominated the coffee-cultivated areas, and none were attractive to butterflies. Coffee has traditionally been grown under native shade, but there has been an increasing use of fewer, often exotic, species and less shade in recent years. Nectar-feeding butterflies of larger species, which are strong fliers, may have disproportionally represented at estates away from the sanctuary. Butterfly larvae essentially feed on shrubs and plants, which may be not formed in coffee estates. Application of insecticides also has adverse effects on butterflies.

17.5 Spiders

Spiders are the other dominant predator group found in coffee plantations. They are the most predominant natural enemies in cultivated ecosystems, and in coffee at Western Ghats of India, they are often present in large numbers. They are generalist predators and play an important role in reducing insect pests attacking agriculture.

Fig. 17.7 (**a**) Coffee root grub. (**b**) Coffee twig showing infestation of root grub

Spiders are tiny, cryptic creatures, have extra-oral digestion and suck the fluids and contain amorphous gut contents. Coffee bush has complicated physical architecture and structure, harbouring rich spider diversity.

Senthil Kumar and Regupathy (2009) carried out gut content analysis of spiders in coffee ecosystem. Gut content analysis of freshly collected spiders (Fig. 17.8) from the coffee plantations of Horticultural Research Station, Yercaud, Tamil Nadu, was conducted through electrophoresis that is based on the detection of prey enzymes in homogenates of the predator after PAGE and staining for esterase activity to know preferential feeding habit under field conditions. Two different preys (*C. viridis* and an acridid, *Aularches* sp.) and eight species of spiders (*Leucauge decorate* (Blackwall), *Oxyopes* sp., *Dieta virens* (Thorell), *Olios milleti* (Pocock), *Telamonia dimidiata* (Simon), *Clubiona* sp., *Hippasa* sp. and *Plexippus* sp.) were deployed for the investigation.

The regal parachute spider *Poecilotheria regalis* (Pocock 1899) is one of the 14 described large-bodied parachute spiders. The raphosidae is confined to India and Sri Lanka, with seven described species each from the two nations. The large-sized group of parachute spiders is very scarcely documented in the wilderness, but is in demand by traders in America and Europe. *Poecilotheria regalis* was recorded from Arakkonam (Tamil Nadu) and later located in the Western Ghats, in Matheran, Dahanu, Coorg, Nilgiris, Anamalais, Mysore and Bangalore (Pocock 1899, 1900). *P. regalis* was first described from Arakkonam, Tamil Nadu. It was found in the timbers brought from Eastern Ghats (Pocock 1900). During the study *P. regalis* was the most commonly sighted parachute spider (Fig. 17.8), currently recorded from 13 localities in South India. So far the spiders have been recorded from 22 localities in India.

Fig. 17.8 Shot-hole borer damage on coffee

17.6 Pollination in Coffee

Pollination is a key service provided by arthropods to the coffee plant that cascades into the beneficial effects. About 80 % of the flowering plants on Earth (Fig. 17.9) are pollinated by insects such as honeybees, bumblebees, solitary bees, beetles, butterflies, flies and ants; Klein et al. (2003) recorded observations on bees visiting flowers of *Robusta coffee* (*Coffea canephora*) in Indonesia. A coffee estate with 20 bee species leads to a 95 % higher fruit set compared to an agroforestry system with fewer bee species. The abundance and diversity of social bees declined with an increase in distance from the forest. Observations suggested that a rich bee community plays a key part in the fruit set in coffee estates. Coffee estates located in the vicinity of forests or forest fragments may GET help from increased bee pollination and diversity. The following bee species visiting coffee were documented, viz. *Apis dorsata*, *Apis cerana indica*, *Tetragonula iridipennis*, *Apis florea*, *Braunsapis picitarsus*, *Ceratina hieroglyphica*, *Ceratina smaragdula*, *Amegilla* spp., *Thyreus* spp., *Xylocopa aestuans*, *Xylocopa latipes*, *Nomia iridescens*, *Megachile rotundata*, *Megachile bicolor* and *Lasioglossum* spp.

Jha and Vandermeer (2010) documented tropical bee communities within a deforested shade coffee-growing region in Chiapas, Mexico. The study indicated that the determinants for bee diversity were the density of trees, the number of tree species in bloom and the canopy cover of the local agroforestry landscape. Solitary bees were the most abundant in habitats with high and dense canopy cover. The social bees were abundant in areas where tree species richness, cavity-nesting and wood-nesting bee abundance was present in the estate. Ground-nesting bees were also abundant in niches with a large number of tree species in bloom. Results of studies indicated that among bee sociality groups, nesting guilds and tribes, the most critical factor influencing bee communities was the kind of vegetation in the estate. These results indicated the key role that agroforestry managers play for biodiversity conservation and the potential contribution they make by creating resource-rich agricultural matrices. Specifically, findings highlight the importance of diverse

Fig. 17.9 A nymphalid butterfly – indicator species

Fig. 17.10 A wasp nest in coffee estate

overstory tree management in supporting native bee communities within agroforestry situations. The great value of bees as pollinators of coffee plants has been known for many years, but unfortunately, this knowledge is not widely applied in increasing production. The scientific literature supporting the benefits of bee pollination for coffee is convincing. Bees may increase yield of Robusta coffee up to 83 % according to Central Bee Research and Training Institute, Pune. Majority of coffee species are diploid and self-infertile and therefore have to be cross-pollinated by wind and insects for better yield. *C. arabica* is tetraploid, self-fertile and at times cleistogamous and so relies less on cross-pollination. Cross-pollination also enhances biotic diversity including that of arthropods (Fig. 17.10).

17.7 Coffee Biodiversity

Coffee plantations with varying shade cover are a home for arboreal mammals of high conservation value. Since coffee plantations imitate tropical forests, many wildlife species frequent and take shelter in coffee-cultivated areas (Defler et al. 2003; IUCN 2010). The decrease in the population of night monkeys occurring naturally in Colombia and Ecuador, has been attributed to the conversion of high-elevation forests to agricultural lands. Radio telemetry observations revealed that monkeys spent the majority of time on natural forests and coffee plantations with 80 % shade cover. They rarely entered coffee plantations with 60 % shade cover. The findings suggest that night monkeys prefer to inhabit natural forest areas spending significant time foraging in coffee plantations with dense shade cover. Fruit trees in forests and coffee plantations are a critical part of the night monkey habitat (Hughell and Newsom 2013). Thus, fruit-bearing shade trees that yield alternative income to growers could be encouraged to be planted. For instance, *Prunus integrifolia* served as an important food item for monkeys and in coffee plantations. Researchers have concluded that shaded coffee plantations can be a source, good buffers for designated protected areas by providing habitat for a diverse species of mammals and arthropods (Hughell and Newsom 2013).

The coffee-producing region of Columbia is a heterogeneous landscape. In this heterogeneous rural landscape, the potentials and challenges for vertebrate and invertebrate conservation are multifaceted and complex. Shaded coffee plantation of Colombia provides habitat for migratory birds like cerulean warblers (*Setophaga cerulea*). Some coffee plantations in Columbia provide habitat for rare, endangered species of invertebrates and vertebrates. For wildlife, coffee estates seem to improve landscape connectivity. The heterogeneity of coffee landscape provides habitat that supports a rich and diverse flora and a fauna. Mammalian diversity in coffee landscapes will ensure conservation of arthropod and other biodiversity elements.

17.8 Biodiversity Decline

The coffee plantations, for instance, in South India is home to several wildlife species, national parks, tiger reserves and biodiversity plantations. Bhadra Wildlife Sanctuary, the Bandipur National Park flanked by Nagarhole National Park, Mudumalai Wildlife Sanctuary and Wayanad Wildlife Sanctuary together make up the protected Nilgiri Biosphere Reserve which is India's first biosphere reserve. This reserve is an important breeding landscape for tigers and elephants in the three South Indian states. Indian coffee is an important part of this biosphere reserve. The Indian coffee farmer has been an asset to the nation as well as to the global community by being a proactive a natural, nature conservationist.

Worldwide, habitat loss is the major factor for wildlife depletion. More prevalent than outright destruction, however, is habitat fragmentation, exhibited especially in tropical regions. Effects of these habitats, and the subsequent effect on local invertebrate populations, were less well studied so far. A small-scale study was conducted in western Jamaica in January 2009 by Alex Enrique assessing moth and species abundance between forest edges and crop interiors on a shade coffee plantation. Scientists utilise shade coffee as an example of sustainable agriculture that conserves migratory birds, and shade coffee plantations have been shown to act as a buffer area ("halo") between fragmented habitats for insects. The vegetation patches help moths to disperse between areas without suitable vegetation for host plants. To sample the local moth fauna, UV black light traps were used at the edge and interior of each site. There was no significant difference in morphospecies richness or abundance; there was a difference in diversity between sites, with very less duplication. The lack of significance suggests coffee habitat can serve as a buffer between less disturbed forest and disturbed human-altered areas. However, some species can be localised to certain habitats. Many moth species are host specific and can hold as useful indicators of ecosystem and vegetative health. Research on arthropods is key because of several ecosystem services they perform. Further research is necessary on arthropods in coffee in tropicals.

Perfecto et al. (1996) reported that under the canopy of *Erythrina poeppigiana*, the workers documented 30 species of ants, 103 species of other hymenopterans and 126 species of beetles. A second tree yielded 27 species of ants, 61 species of hymenopterans and 110 species of beetles. These data indicated that shaded coffee estates may hold local species diversity within the same order of magnitude as undisturbed forest. No doubt local multiple species of shade trees generally yield higher productivity and biomass turnover.

Mone et al. (2014) worked on ground insects and fruit-eating butterflies in 29 different plantations in Kodagu, Karnataka, South India. These included organic and conventional coffee and cardamom plantations using different levels of chemical fertilisers and pesticides. A total of 457 ground-dwelling insect species were collected using pitfall traps which included 92 species of ants and 123 species of beetles; 25 species of butterflies were collected using bait traps.

The Western Ghats in South India where coffee is cultivated is a habitat to thousands of animal species including at least 325 globally threatened species. Several are endemic species, especially in the amphibian, reptilian and Piscean classes. Thirty-two threatened species of mammals dwell in the Western Ghats region. Of the 16 endemic mammals, 13 are threatened (Mewa singh and Werner Kaumanns 2005; Malviya et al. 2011). There are at least 139 mammal species. Among critically endangered mammals in Western Ghats region is the nocturnal Malabar large-spotted civet (*Viverra civettina*), the arboreal lion-tailed macaque (*Macaca silenus*) and the purple frog (*Nasikabatrachus sahyadrensis*).

A majority of Karnataka's 500 species of birds are from the Western Ghats region. Bhadra Wildlife Sanctuary is located at the north of the Malabar ranges and the southern tip in the Sahyadri ranges, and bird species from both ranges can be seen here. There are at least 16 species of birds endemic to the Western Ghats including the endangered rufous-breasted laughing-thrush (*Glaucis hirsutus*), the vulnerable Nilgiri wood pigeon (*Columba palumbus*), white-bellied shortwing (Ardea insignis) and broad-tailed grassbird (*Schoenicola platyurus*), the near-threatened grey-breasted laughing-thrush (*Trochalopteron fairbanki*), black-and-rufous flycatcher (*Ficedula nigrorufa*), Nilgiri flycatcher (*Eumyias albicaudatus*) and Nilgiri pipit (*Anthus nilghiriensis*) and to a lesser extent the Malabar blue-winged parakeet (*Psittacula columboides*), Malabar grey hornbill (*Ocyceros griseus*), white-bellied treepie (*Dendrocitta leucogastra*), grey-headed bulbul (*Pycnonotus priocephalus*), rufous babbler (*Turdoides subrufa*), Wynaad laughing-thrush (*Garrulax delesserti*), white-bellied blue flycatcher (*Cyornis tickelliae*) and the crimson-backed sunbird (*Leptocoma minima*).

The coffee plantations in Costa Rica were earlier characterized by a high vegetational and formal diversity. Currently the coffee estate has undergone a major change to intensive, monocultural plantations, where all shade trees are removed. In an investigation, the patterns of arthropod biodiversity decline with this change were considered. Canopy arthropods were counted in three coffee estates in a traditional cultivated area with several shade trees, a moderately shaded estate with only *Erythrina poeppigiana* and a coffee-dominated plantation. An insecticidal fogging technique was utilised to consider coffee canopy and arthropods. Research indicated that the transformation of coffee ecosystem results in a significant decline of biological diversity of canopy arthropods as well as arthropods in coffee plants (Table 17.4). Percentage of species overlap was low (Table 17.5). Arthropod abundance on a coffee bush was not in the same order as threat reported for trees in tropical forests (Perfecto et al. 1997).

Destruction of habitats, fragmentation and climate change are the major factors transforming natural habitats into complex mosaic of natural, seminatural and modified habitats. As natural habitats diminish biodiversity in the future will depend on conservation potential of countryside habitats like cultivated patches, gardens, ponds and lakes. A major challenge to conservation is to develop the capacity of rural niches to support biodiversity and, conversely, the capacity of different taxa to explore such areas (Goehring et al. 2002). In a mixed agricultural landscape in South Cost Rica, the richness and composition of arthropod community in different

Table 17.4 Number of species (and individuals) of beetles, ants and non-formicid hymenopterans in the canopy of shade trees and coffee plants in coffee farms

Species (no.)	Type of farm	Beetles	Ants	Hymenoptera[a]
Erythrina poeppigiana (1)	Traditional	126(401)	30(333)	103
Erythrina fusca (2)	Traditional	110(393)	27(1105)	61
Annona cherimola (3)	Traditional	–	10(179)	63
Erythrina poeppigiana (4)	Moderately shaded	48(107)	5(64)	46
Coffea arabica (10 plants)	Traditional	39(76)	14(135)	34
Coffea arabica (10 plants)	Moderately shaded	29(82)	9(128)	31
Coffea arabica (10 plants)	Unshaded	29(92)	8(47)	30

[a]Not including ants (Perfecto et al. 1997)

Table 17.5 Percentage of species overlap for beetles and ants among sampled shade trees and coffee plants

(A) Ants' shade trees					Coffee bushes			
Tree number	1	2	3	4	Coffee management	Trad.[a]	Mod.[b]	Monoc.[c]
	100	18.7	11.1	6.1	Tand.	100	21.0	29.4
		100	8.8	3.2	Mod		100	30.8
			100	7.1	Monoc.			100
(B) Beetles' shade trees					Coffee bushes			
Tree number	1	2		4	Coffee management	Trad.[a]	Mod.[b]	Monoc.[c]
1	100	14.0		3.6	Tand.	100	9.6	18.1
2		100		9.7	Mod		100	11.5
4				100	Monoc.			100

Perfecto et al. (1997)
[a]Traditional coffee plantation
[b]Moderately shaded coffee
[c]Coffee monoculture

types of habitats, in disturbed habitats, were recorded. In a large forest fragment (227 ha) in nine sampling weeks, 1577 arthropods were collected. In small forest fragment (5.3 ha), 1933 arthropods and in a coffee plantation (3 ha), 2466 arthropods were recorded. Relative evenness in species diversity was considerably higher in forest and fragment forest sites than in coffee estates. A decline in the richness of beetles with increasing anthropogenic disturbances was recorded.

Coffee is a crop that is cultivated in and around forest tracts and is similar in characteristics to the natural forest ecosystems. Coffee ecosystems support rich biodiversity and there is a potential for conservation of biotic elements including arthropods. Also technology is available to protect coffee bush from maladies of pests and diseases like cultivation of disease and pest-resistant coffee and integrated pest and disease management. More research and participation of growers will provide tools for conservation in this massive, productive landscape. Participatory research and extension programmes are an effective way to interest communities in conservation. Many coffee plantations in the tropics and subtropics provide habitat for rare endemic endangered species. Coffee landscape is a home for many migratory birds and arthropods. The future of biodiversity on planet Earth will depend to a large extent on the conservation value of urban areas and seminatural areas. In this context coffee landscapes will play a key role in conservation of biodiversity which will require combined efforts by coffee growers, consumers, researchers, foresters and biologists.

References

Anonymous (1998) A compendium on pests and diseases of coffee and their management in India. Coffee Board Research Department, India, pp 67

Balakrishnan MM, Sreedharan K, Venkatesha MG, Krishnamurthy P, Bhat PK (1991) Observations on *Ferrisia virgata* (Ckll) (Homoptera: Pseudococcidae) and its natural enemies on coffee, with the records of predators and host plants. J Coffee Res 21:11–19

Chako MJ, Bhat PK (1976) Record of *Ferrisia virgata* and its natural enemy *Spalgis epius* on coffee in India. J Coffee Res 6:56–57

Davis AP, Govaerts R, Bridson DM, Stoffelen P (2006) An annotated taxonomic conspectus of the genus Coffee (Rubiaceae). Bot J Linn Soc 152:465–512

Defler TR, Rodríguez-Mahecha JV, Hernández-Camacho JI (2003) Conservation priorities for Colombian primates. Primate Conserv 19:10–18

Dolia J, Devy MS, Aravind NA, Kumar A (2008) Adult butterfly communities in coffee plantations around a protected area in the Western Ghats, India. Anim Conserv 11:26–34

Fittkau EJ, Klinge H (1973) On biomass and trophic structure of the central Amazonian rain forest ecosystem. Biotropica 5(1):2–14

Goehring MD, Daily CG, Cagan SH (2002) Distribution of ground-dwelling arthropods in tropical countryside habitats. J Insect Conserv 6:83–91

Hughell David, Newsom Deanna (2013) Impacts of rainforest alliance certification on coffee farms in Colombia. Executive summary. www.rainforest-alliance.org

Irulandi S, Kumar PKV, Seetharama HG, Sreedharn K (2001) Bioefficacy of neem formulations alone and in combinations with synthetic insecticide against *planococcus citri* on coffee. J Coffee Res 29:248–254

IUCN (2010) Red list of threatened species. IUCN, Gland, www.iucnredlist.org

Jha S, Vandermeer JH (2010) Impacts of coffee agroforestry management on tropical bee communities. Biol Conserv 143(2010):1423–1431
Klein AM, Steffan-Dewenter I, Tscharntke T (2003) Pollination of *Coffea canephora* in relation to local and regional agroforestry management. J Appl Ecol 40:837–845
Le Pelley RH (1968) Pests of coffee. Longmans, Green and Co. Ltd, London, p 590
Malviya M, Srivastav A, Nigam P, Tyagi PC (2011) Indian national study book of Nilgiri Langur (*Trachypithecus johnii*). Wildlife Institute of India, Dehradun and Central Zoo Authority, New Delhi
Mewa singh, Werner kaumanns (2005) Behavioural studies: a necessity for wildlife management. Curr Sci 89(7):1233
Perfecto I, Rice RA, Greenberg R, Van der Voort ME (1996) Shade coffee: a disappearing refuge for biodiversity. Bioscience 46(83):598–608
Perfecto I, Vandermeer J, Hanson P, Cartin V (1997) Arthropod biodiversity loss and the transformation of a tropical agro-ecosystem. Biodivers Conserv 6:935–945
Pocock RI (1899) The genus *Poecilotheria*: its habits, history and species. Ann Mag Nat Hist Ser 3(S):82–96
Pocock RI (1900) The fauna of British India, including Ceylon and Burma. Arachnida. Taylor & Francis, London, p 190
Prakasan CB, Kumar MG (1985) New record of natural enemies on Coffee mealy bug and green scale. J Coffee Res 15:53–54
Prakasan CB, Sreedharan K, Krishnamoorthy P (1986) New record of a parasite of coffee white stem borer *Xylotrechus quadripes* Chev. from India. J Coffee Res 16(1&2):38–40
Purthi HS, Mani MS (1940) Biological note on Indian parasitic Chalcidoidea. Miscellaneous Bulletins. Indian Council of Agriculture Research 30, p 44
Reddy KB, Sreedharan K, Prakasan CB, Bhat PK (1990) New records of natural enemies of *Planococcus* spp., *Coccus viridis* (Green) and *Ferrisia virgata* (Cockerell) on coffee in India. J Coffee Res 20:153–156
Sekhar PS (1964) Pests of coffee In: Entomology in India. Entomological Society of India, New Delhi, p 529
Senthil Kumar CM, Regupathy A (2009) Gut content analysis of spiders in coffee ecosystem. Curr Sci 96(7):10–893
Shamika Mone KM, Kusha, Devcharan J, Musthak A, Anurag G (2014) Comparison of insect biodiversity between organic and conventional plantations in Kodagu, Karnataka. India J Threat Taxa 6(9):6186–6194
Shylesha AN, Veeresh GK, Siddegowda DK (1992) New record of egg parasite and survey of natural enemies complex of white stem borer of coffee *Xylotrechus quadripes* Chevr. (Cerambycidae: Coleoptera). J Plant Crop 20(Supplement):138–139
Sridharan K, Balakrishanan MM, Bhat PK (1992) *Callimerus* sp. (Coleoptera: Cleridae) a predator of shot hole borer, *Xylosandrus compactus* (Eichh). J Coffee Res 22:139–142
Subramaniam TV (1941) The coffee stem borer, bull. Department of Agriculture, Mysore. Ent Ser Bangalore 11:18
Vega FE, Infante F, Castillo A, Jaramillo J (2009) The coffee berry borer, *Hypothenemus hampei* (Ferrari) (Coleoptera: Curculionidae): a short review, with recent findings and future research directions. Terr Arthropod Rev 2:129–147
Venkataramaiah GH, Rehman A (1989) Ants associated with the Mealy bugs of coffee. Indian Coffee 43:13–14
Venkatesha MG, Seetharama HG, Sreedharan K (1997) *Iphiaulax* sp.- A new braconid parasitoid of coffee white stem borer, *Xylotrechus quadripes* (Chevr.) (Coleoptera : Cerambycidae). Pest Manag Hortic Ecosyst 3:112–113

Arthropod Pests and Natural Enemy Communities in Tea Ecosystems of India

18

Narayanannair Muraleedharan and Somnath Roy

Abstract

Tea is grown as a perennial monoculture over large contiguous areas in Northeast India. Arthropods occupy a variety of functional niches and microhabitats and play a major role in the sustainable and healthy functioning of tea ecosystem. Conventional tea cultivation has often accomplished high yields and stable crop production, but has been heavily dependent on continuous and often excessive chemical pesticides, which lead to pest resistance, resurgence and destruction of natural enemies. This communication collates diversity of foliage arthropods on tea plants and tea ecosystems in India. A total of 553 arthropod species comprising 439 species of insects in 89 families and 11 orders and 114 species of arachnids in 30 families and 2 orders are associated with the tea ecosystem. In the tea ecosystem, abundance of phytophagous pests was dominated by Lepidoptera, Hemiptera and Coleoptera. Based on feeding habits, 62.57 % of the arthropods recorded were natural enemies, dominated by spiders, coccinellids and hymenopteran parasitoids. This article is expected to provide useful foundation for exploring integrated pest management strategies appropriate for organic, conventionally grown tea.

Keywords

Foliage Arthropods • Diversity • Natural enemies • Spider • Tea ecosysytem

N. Muraleedharan • S. Roy (✉)
Department of Entomology, Tocklai Tea Research Institute, Tea Research Association, Jorhat 785008, Assam, India
e-mail: somnathento@gmail.com

A.K. Chakravarthy, S. Sridhara (eds.), *Economic and Ecological Significance of Arthropods in Diversified Ecosystems*, DOI 10.1007/978-981-10-1524-3_18

18.1 Introduction

The agroecosystem of tea comprises of tea plants, shade trees and auxiliary crops with biotic and abiotic components. The tea crop has unique characters, which influence the arthropod ecology in a special way (Calnaido 1973). Tea plantations are evergreen and perennial (over 100 years) (Banerjee 1983), comprising genetically diverse cultivars, interplanted with shade trees, particularly in Southeast Asia (Deka et al. 2006). Tea plantations roughly resemble a 'single-species forest' (Cranham 1966a, b), and arthropod species are thought to coexist by intratree distribution or well-defined stratification on ecological niche formation (Banerjee 1979, 1983).

Tea in each geographic region has its own distinctive pest fauna. The number of insects associated with the tea plants depends on the duration of its cultivation in that region. The area under tea becomes important only after allowing for the 'age effect', and latitude has no influence on species richness. In large tea areas, saturation in species richness reached over a period of 100–150 years. The North Asian countries with the history of longest period of tea growing have the largest number of insect and mite species, while the South American countries where tea was rather recently introduced have few insect pests. The accumulation of arthropod species on tea is influenced by the age of plants; older tea plantations harbour maximum number of insect species (Banarjee 1983). In the present communication, the arthropod assemblage on tea plants in different parts of India is reviewed for providing a general description of the arthropod pests and natural enemies and their utilization in sustainable tea cultivation.

A rich insect fauna comprising 432 species belonging to 87 families and 11 orders; 114 species of arachnids under 30 families constituting a total of 546 arthropod species were recorded from the tea fields in India (Figs. 18.1 and 18.2).

From the Indian tea ecosystem, a total of 207 species belonging to 52 families (4 acarines and 48 insects) and 11 orders of phytophagous arthropods were recorded (Table 18.1). The phytophagous guild was dominated by Lepidoptera followed by Hemiptera and Coleoptera. The remaining phytophagous arthropods comprised Acari, Diptera, Isoptera, Orthoptera, Thysanoptera, Hymenoptera and Neuroptera (Table 18.1). The natural enemy guild was dominated by predators. A total of 200 predatory arthropod species were recorded from the tea ecosystem (Table 18.2). Spiders were the most abundant predatory group with 70 species followed by Coleoptera, Acari (35 spp.), Neuroptera and Hemiptera (Table 18.2). The parasitoid guild comprising 146 species of insects was dominated by hymenopterans (133 spp.), mainly Eulophidae, Braconidae and Ichneumonidae. The dipteran parasitoid species discovered from the tea field belonged to Tachinidae (Table 18.3). The overall species composition reflects the richness of arthropod natural enemies, i.e. predators and parasitoids of tea pests, where the natural enemy to pest ratio is 1.7:1 (Figs. 18.3, 18.4a, 18.4b, and 18.4c). It is estimated that more or less 1000 species of arthropods (Hazarika et al. 2009) occur on tea all over the world (Chen and Chen 1989) as pests, casual visitors, predators and parasitoids of pests (Muraleedharan

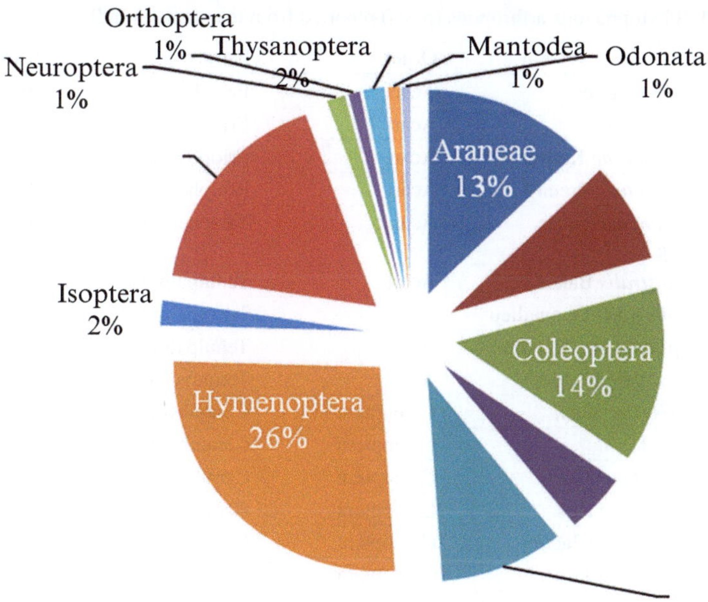

Fig. 18.1 Relative composition of families of arthropods in tea plantations, India

Fig. 18.2 Relative composition of pests and natural enemies in tea ecosystem, India

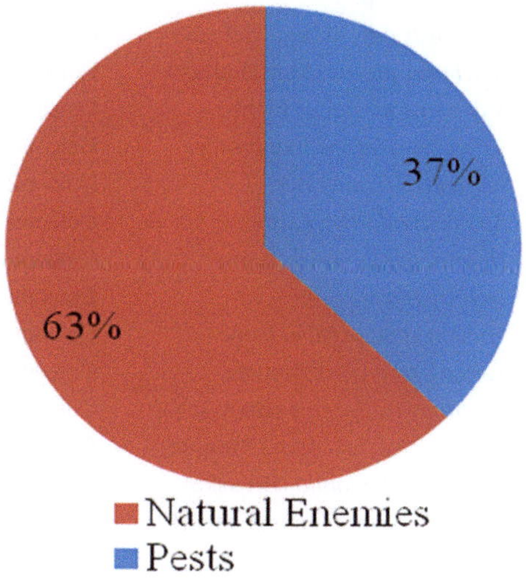

and Chen 1997); only about 300 species of insects and mites are reported from India (Muraleedharan 2010).

The dynamic adaptations of arthropods have enabled them to occupy every part of the tea plant, and the maximum numbers of arthropods occur on the foliage throughout the world (Chen and Chen 1989). The leaves, stems, shoots, flowers and

Table 18.1 Phytophagous arthropods (pests) reported from tea plants in India

Species	Order	Family	Distribution
Acaphylla indiae Keifer	Acari	Eriophyidae	NEI
Acaphylla theae Watt[a]	Acari	Eriophyidae	NEI, SI
Acaphyllisa parindiae Keifer	Acari	Eriophyidae	SI
Calacarus carinatus (Green)[a]	Acari	Eriophyidae	NEI, SI, HP
Polyphagotarsonemus latus (Banks) Ewing	Acari	Tarsonemidae	NEI, SI
Brevipalpus australis Baker	Acari	Tenuipalpidae	SI
Brevipalpus obovatus Donnadieu	Acari	Tenuipalpidae	NEI, HP
Brevipalpus phoenicis (Geijskes)	Acari	Tenuipalpidae	NEI, HP
Oligonychus coffeae Nietner[b]	Acari	Tetranychidae	NEI, SI, HP
Haplothrix griseatus Gah.	Coleoptera	Cerambycidae	NEI
Batocera rubus L.	Coleoptera	Cerambycidae	NEI
Cyrtognathus indicus Hope	Coleoptera	Cerambycidae	NEI
Melanauster verteegi Rits.	Coleoptera	Cerambycidae	NEI
Chrysolampra flavipes Jacoby	Coleoptera	Chrysomelidae	NEI
Chrysolampra indica Jacoby	Coleoptera	Chrysomelidae	NEI
Diapromorpha melanopus Lecord	Coleoptera	Chrysomelidae	NEI
Astycus chrysochlorus Wield	Coleoptera	Curculionidae	NEI
Asticus lateralis Fabr.	Coleoptera	Curculionidae	NEI
Myllocerus sp.	Coleoptera	Curculionidae	SI
Xyleborus approximate Schedl	Coleoptera	Curculionidae	NEI
Xyleborus piceus (Motschulsky)	Coleoptera	Curculionidae	NEI
Xyleborus torquatus Eichh.	Coleoptera	Curculionidae	NEI
Xyleborus elegans Sampson	Coleoptera	Curculionidae	NEI
Adoretus versutus Her.	Coleoptera	Scarabaeidae	NEI
Anomala bilobata Arrow	Coleoptera	Scarabaeidae	NEI
Holotrichia impressa Burm.	Coleoptera	Scarabaeidae	NEI
Holotrichia sp.	Coleoptera	Scarabaeidae	SI
Mimela xanthorrhina Hope	Coleoptera	Scarabaeidae	SI
Serica assamensis Brenske	Coleoptera	Scarabaeidae	NEI
Sophrops iridipennis (Brenske)	Coleoptera	Scarabaeidae	NEI
Sophrops sp. nr. *cotesi* (Brenske)	Coleoptera	Scarabaeidae	NEI
Sophrops plagiatula (Brenske)	Coleoptera	Scarabaeidae	NEI
Euwallacea fornicatus Eichhoff[a]	Coleoptera	Scolytidae	SI
Xyleborus fornicates Eichh.	Coleoptera	Scolytidae	NEI
Agromyza theae (Bigot) Meij	Diptera	Agromyzidae	NEI
Tropicomyia theae (Cotes)	Diptera	Agromyzidae	SI
Toxoptera aurantii (Boyer de Fonscolombe)[a]	Hemiptera	Aphididae	NEI, SI, HP
Empoasca flavescens Fabricius[a]	Hemiptera	Cicadellidae	NEI, SI, HP
Huechys sanguinea De Geer.	Hemiptera	Cicadidae	NEI

(continued)

Table 18.1 (continued)

Species	Order	Family	Distribution
Ceroplastes cerifera (Anderson)	Hemiptera	Coccidae	NEI
Ceroplastes floridensis (Comstock)	Hemiptera	Coccidae	NEI
Ceroplastes rubens (Maskell)	Hemiptera	Coccidae	NEI
Ceroplastodes cajani (Maskell)	Hemiptera	Coccidae	NEI
Ceroplastodes chiton (Green)	Hemiptera	Coccidae	NEI
Chloropulvinaria floccifera (Westwood)	Hemiptera	Coccidae	NEI
Coccus discrepens (Green)	Hemiptera	Coccidae	NEI
Coccus hesperidum Linnaeus	Hemiptera	Coccidae	NEI
Coccus viridis (Green)	Hemiptera	Coccidae	NEI, SI
Eriochiton theae (Green)	Hemiptera	Coccidae	NEI, SI
Eucalymnatus tessellates (Signoret)	Hemiptera	Coccidae	NEI
Saissetia coffeae (Walk.)	Hemiptera	Coccidae	NEI, SI
Saissetia formicarri (Green)	Hemiptera	Coccidae	NEI, SI
Saissetia nigra (Nietner)	Hemiptera	Coccidae	NEI
Saissetia oleae (Barnard)	Hemiptera	Coccidae	NEI
Saissetia watti (Green)	Hemiptera	Coccidae	NEI
Vinsonia stellifera (Westwood)	Hemiptera	Coccidae	NEI
Zeuzera coffeae Nietner	Hemiptera	Coccidae	NEI, SI
Abgrallaspis cyanophylli (Signoret)	Hemiptera	Diaspididae	NEI
Abgrallaspis sp. *pictor* (Williams)	Hemiptera	Diaspididae	NEI
Andaspis dasi Williams	Hemiptera	Diaspididae	NEI
Aonidiella aurantii (Maskell)	Hemiptera	Diaspididae	NEI
Aspidiotus destructor Signoret	Hemiptera	Diaspididae	NEI
Aspidiotus spinosus (Signoret)	Hemiptera	Diaspididae	NEI
Chrysomphalus pinnulifer (Maskell)	Hemiptera	Diaspididae	NEI
Chrysomphalus aonidium (*-ficus*) Ashmead	Hemiptera	Diaspididae	NEI
Chrysomphalus dictyospermi Morgan	Hemiptera	Diaspididae	NEI
Clavaspis sp.	Hemiptera	Diaspididae	NEI
Fiorinia theae Green	Hemiptera	Diaspididae	NEI
Hemiberlesia lataniae (Signoret)	Hemiptera	Diaspididae	NEI
Hemiberlesia rapax (Comstock)	Hemiptera	Diaspididae	NEI
Lepidosaphes sp.	Hemiptera	Diaspididae	NEI
Lindingaspis ferrisi (McKenzie)	Hemiptera	Diaspididae	NEI
Morganella longispina (Morgan)	Hemiptera	Diaspididae	NEI
Parlatoria proteus (Curtis)	Hemiptera	Diaspididae	NEI
Phenacaspis sp.	Hemiptera	Diaspididae	NEI
Phenacaspis manni (Green)	Hemiptera	Diaspididae	NEI

(continued)

Table 18.1 (continued)

Species	Order	Family	Distribution
Pinnaspis theae (Maskell)	Hemiptera	Diaspididae	NEI
Pseudaonidia duplex (Cockerell)	Hemiptera	Diaspididae	NEI
Pseudaonidia trilobitiformis (Green)	Hemiptera	Diaspididae	NEI
Velataspis serrulata Ganguli	Hemiptera	Diaspididae	NEI
Lawana conspersa Wlk.	Hemiptera	Flatidae	NEI
Helopeltis theivora Waterhouse[b]	Hemiptera	Miridae	NEI, SI
Lygus sp.	Hemiptera	Miridae	NEI, SI
Poecilocoris latus Dall.	Hemiptera	Pentatomidae	NEI
Nipaecoccus vastator (Maskell)	Hemiptera	Pseudococcidae	NEI
Nipaecoccus viridis (Newstead)	Hemiptera	Pseudococcidae	NEI, SI
Pseudococcus theaecola (Green)	Hemiptera	Pseudococcidae	NEI, SI
Crisicoccus sp.	Hemiptera	Pseudococcidae	NEI, SI
Rhizoecus sp.	Hemiptera	Pseudococcidae	NEI, SI
Orasema assectator Kerrich	Hymenoptera	Eucharitidae	NEI
Orasema initiator Kerrich	Hymenoptera	Eucharitidae	NEI
Orasema sp.	Hymenoptera	Eucharitidae	SI
Odontotermes sp.	Isoptera	Termitidae	NEI, SI
Coptotermes heimi (Wasm)	Isoptera	Termitidae	NEI
Microcerotermes heini (Wasmann)	Isoptera	Termitidae	NEI
Microcerotermes pakistanicus Ahmed	Isoptera	Termitidae	NEI
Microcerotermes sp.[a]	Isoptera	Termitidae	NEI
Microtermes sp.	Isoptera	Termitidae	NEI, SI
Neotermes buxensis Roonwal and Sensharma	Isoptera	Termitidae	NEI
Odontotermes assamensis Holmgren	Isoptera	Termitidae	NEI
Odontotermes feae (Wasm)	Isoptera	Termitidae	NEI
Odontotermes parvidus Holmgren	Isoptera	Termitidae	NEI
Odontotermes redemanni (Wasmann)	Isoptera	Termitidae	NEI
Amsacta lactinea Cram	Lepidoptera	Arctiidae	NEI
Amsacta lineola Fab. (= *emittens* Wlk.)	Lepidoptera	Arctiidae	NEI
Pericallia ricini Fab.	Lepidoptera	Arctiidae	NEI
Andraca bipunctata Walker[a]	Lepidoptera	Bombycidae	NEI
Ascotis sp.	Lepidoptera	Geometridae	NEI

(continued)

Table 18.1 (continued)

Species	Order	Family	Distribution
Biston (=*Buzura*) *suppressaria* Guen[a]	Lepidoptera	Geometridae	NEI, SI
Cleora sp.	Lepidoptera	Geometridae	NEI
Ectropis bhurmitra (Walker)	Lepidoptera	Geometridae	SI
Ectropis sp.	Lepidoptera	Geometridae	NEI
Hyposidra infixaria (Walker)[a]	Lepidoptera	Geometridae	NEI
Hyposidra talaca (Walker)[a]	Lepidoptera	Geometridae	NEI
Gracillaria theivora Walsm.	Lepidoptera	Gracillariidae	NEI
Caloptilia theivora (Walsingham)	Lepidoptera	Gracillariidae	SI
Sahyadrassus malabaricus (Moore)	Lepidoptera	Hepialidae	**SI**
Simplicia caeneusalis (Walker)	Lepidoptera	Herminiidae	**NEI**
Hypsa alciphron Cram.	Lepidoptera	Hypsidae	NEI
Indarbela quadrinotata Walk.	Lepidoptera	Indarbelidae	NEI
Indarbela theivora Hamps.	Lepidoptera	Indarbelidae	NEI
Estigena pardalis Walk.	Lepidoptera	Lasiocampidae	NEI
Gastropacha sp.	Lepidoptera	Lasiocampidae	NEI
Taragama sp.	Lepidoptera	Lasiocampidae	NEI
Trabala vishnou Lef.	Lepidoptera	Lasiocampidae	NEI
Belippa lalaena Moore	Lepidoptera	Limacodidae	SI
Cania bilinea Walk.	Lepidoptera	Limacodidae	NEI
Cheromettia apicata Moore	Lepidoptera	Limacodidae	NEI
Narosa conspersa Walk.	Lepidoptera	Limacodidae	NEI
Parasa pastoralis Butler	Lepidoptera	Limacodidae	NEI
Phocederma velutinum Koll	Lepidoptera	Limacodidae	NEI
Praesetora divergens Moore	Lepidoptera	Limacodidae	NEI
Susica pallida Walk.	Lepidoptera	Limacodidae	NEI
Thosea cana Walk.	Lepidoptera	Limacodidae	NEI
Thosea cervina Moore	Lepidoptera	Limacodidae	NEI, SI
Thosea cervina Moore	Lepidoptera	Limacodidae	
Thosea cotesi Swinh.	Lepidoptera	Limacodidae	NEI
Thosea cruda Walk.	Lepidoptera	Limacodidae	NEI
Thosea sinensis Walk.	Lepidoptera	Limacodidae	NEI
Thosea sp. *nr. bisura* Moore	Lepidoptera	Limacodidae	NEI
Trichogyia nigrimargo Her.	Lepidoptera	Limacodidae	NEI
Latoia lepida (Cramer)	Lepidoptera	Limacodidae	SI
Darna nararia (Moore)	Lepidoptera	Limacodidae	SI
Thosea recta Hampson	Lepidoptera	Limacodidae	SI
Belippa lalaena Moore	Lepidoptera	Limacodidae	SI
Eumeta crameri (Westwood)	Lepidoptera	Limacodidae	SI
Arctornis submerginata (Walker)	Lepidoptera	Lymantridae	NEI

(continued)

Table 18.1 (continued)

Species	Order	Family	Distribution
Arnabi punctata Hampson	Lepidoptera	Lymantridae	NEI
Dasychira horsfieldi Saund	Lepidoptera	Lymantridae	NEI
Dasychira mendosa Hbn.	Lepidoptera	Lymantridae	NEI
Dasychira securis Hbn.	Lepidoptera	Lymantridae	NEI
Dasychira thwaitesi Moore	Lepidoptera	Lymantridae	NEI
Euproctis subnotata Walk.	Lepidoptera	Lymantridae	NEI
Euproctis divisa Walk.	Lepidoptera	Lymantridae	NEI
Euproctis latifascia Walk.	Lepidoptera	Lymantridae	NEI
Lymantria albulunata Mre.	Lepidoptera	Lymantridae	NEI
Orgyia postica Walk.	Lepidoptera	Lymantridae	NEI
Orgyia sp.	Lepidoptera	Lymantridae	NEI
Redoa submarginatta Walk.	Lepidoptera	Lymantridae	NEI
Agrotis ipsilon Hufn.	Lepidoptera	Noctuidae	
Prodentia litura Fab.	Lepidoptera	Noctuidae	NEI
Spodoptera litura (Fabricius)	Lepidoptera	Noctuidae	SI
Neostauropus alternus (Walker)	Lepidoptera	Notodontidae	SI
Stauropus alternus Walk.	Lepidoptera	Notodontidae	NEI
Casmara patrona Meyr.	Lepidoptera	Oecophoridae	NEI
Brachycyttarus subtalbata Hmps.	Lepidoptera	Psychidae	NEI
Cathopsyche reidi Watt	Lepidoptera	Psychidae	NEI
Chalioides ferevitrea Joan	Lepidoptera	Psychidae	NEI
Chaloides vitrea Hmps.	Lepidoptera	Psychidae	NE
Clania antrami Hmpsn	Lepidoptera	Psychidae	NEI
Clania cramerii (Westwood)	Lepidoptera	Psychidae	NEI
Clania destructor (Dudgeon)	Lepidoptera	Psychidae	NEI
Clania mahanti Das	Lepidoptera	Psychidae	NEI
Clania Sikkima (Moore)	Lepidoptera	Psychidae	NEI
Clania vaulogeri Heyl.	Lepidoptera	Psychidae	NEI
Dappula tertius Templ.	Lepidoptera	Psychidae	NEI
Mahasena theivora Dudg.	Lepidoptera	Psychidae	NEI
Manatha assamica Watt	Lepidoptera	Psychidae	NEI
Metisa plana Walk.	Lepidoptera	Psychidae	NEI
Oiketioides bipars Walk.	Lepidoptera	Psychidae	NEI
Orophora triangularis Das	Lepidoptera	Psychidae	NEI
Pteroma plagiophleps Hmps.	Lepidoptera	Psychidae	NEI
Sylepta balteata Felder	Lepidoptera	Pyralidae	NEI
Ereboenis saturata Meyrick	Lepidoptera	Pyralidae	SI
Attacus atlas (Linnaeus)	Lepidoptera	Saturniidae	NEI
Striglina glareola Felder	Lepidoptera	Thyrididae	NEI
Ptochoryctis simolenta Meyr.	Lepidoptera	Tineidae	NEI

(continued)

Table 18.1 (continued)

Species	Order	Family	Distribution
Agriophora rhombata Meyrick.	Lepidoptera	Tineidae	NEI
Cerace tetraonis Butler	Lepidoptera	Tortricidae	NEI
Cydia leucostoma Meyrick	Lepidoptera	Tortricidae	SI
Homona coffearia (Nietner)	Lepidoptera	Tortricidae	NEI, SI
Odites sp.	Lepidoptera	Xyloryctidae	NEI
Comocritis pieria Meyrick	Lepidoptera	Yponomeutidae	NEI
Eterusia aedea virescens (Butler)[a]	Lepidoptera	Zygaenidae	SI
Eterusia magnifica Butl.[a]	Lepidoptera	Zygaenidae	NEI
Eterusia aedae L.s. sp. *edocla* Doubl.	Lepidoptera	Zygaenidae	NEI
Trypanophora semihyalina Kollar	Lepidoptera	Zygaenidae	NEI
Ascalaphus sp.	Neuroptera	Ascalaphidae	NEI
Schistocerca gregaria Forsk.	Orthoptera	Acrididae	NEI
Xenocatantops humilis (Serville)	Orthoptera	Acrididae	NEI
Orthacris incongruens Carl	Orthoptera	Pyrgomorphidae	SI
Orthacris robusta Kevan	Orthoptera	Pyrgomorphidae	SI
Brachytrupes portentosus Licht.	Orthoptera	Gryllidae	NEI
Gryllotalpa africana P. de B.	Orthoptera	Gryllotalpidae	NEI
Scirtothrips bispinosus (Bagnall)	Thysanoptera	Thripidae	SI
Scirtothrips dorsalis Hood[b]	Thysanoptera	Thripidae	NEI
Taeniothrips lefroyi (Bagnall)	Thysanoptera	Thripidae	NEI
Taeniothrips setiventris Bagnall	Thysanoptera	Thripidae	NEI, HP

NEI Northeast India, *SI* South India, *HP* Himachal Pradesh
[a]Pest of regional significance
[b]Pest of national significance

fruits offer habitats for several species, though the flowers and fruits are not of much economic significance in tea. However, tea seeds, especially biclonal seeds, are being used as planting material.

The branches of tea plants are frequented by many species of stem-boring and bark-eating caterpillars, scolytid beetles, etc. Similarly, there are several arthropods associated with the shade trees grown in tea plantations. In almost all the tea-growing regions, seasonal pests such as mites, thrips, mirids and cicadellids are responsible for considerable crop losses.

The tetranychid, *Oligonychus coffeae*; the tenuipalpid, *Brevipalpus phoenicis*; and the eriophyid, *Calacarus carinatus*, are the important mite pests. Several species of *Helopeltis* are reported from India, but *H. theivora* is one of the most destructive pests of tea. Similarly, *Scirtothrips* spp. inflict severe damage to tea plantations in India. Lepidopterous pests, including open foliage feeders; leaf folders and stem borers; the looper caterpillar complex; viz. *Buzura suppressaria*, *Hyposidra talaca*

Table 18.2 Predatory arthropods reported from tea plantations in India

Order	Family	Predators	Reference
Araneae	Araneidae	*Araneus mitifica* (Simon)	Hazarika and Chakraborti (1998)
Araneae	Araneidae	*Argiope pulchella* Thorell	Hazarika and Chakraborti (1998) and Roychaudhuri (2011)
Araneae	Araneidae	*Cyclosa bifida* (Doleschall)	Roychaudhuri (2011)
Araneae	Araneidae	*Cyclosa confraga* Thorell	Hazarika and Chakraborti (1998) and Roychaudhuri (2011)
Araneae	Araneidae	*Cyclosa fissicauda* Simon	Hazarika and Chakraborti (1998)
Araneae	Araneidae	*Cyclosa hexatuberculata* Tikader	Das et al. (2010)
Araneae	Araneidae	*Cyclosa insulana* Costa	Hazarika and Chakraborti (1998)
Araneae	Araneidae	*Cyclosa mulmeinensis* (Thorell)	Hazarika and Chakraborti (1998) and Roychaudhuri (2011)
Araneae	Araneidae	*Cyclosa quinqueguttata* (Thorell)	Roychaudhuri (2011)
Araneae	Araneidae	*Cyclosa simony* Tikader	Hazarika and Chakraborti (1998)
Araneae	Araneidae	*Cyclosa* sp.	Hazarika and Chakraborti (1998)
Araneae	Araneidae	*Cyclosa spirifera* Simon	Hazarika and Chakraborti (1998)
Araneae	Araneidae	*Cyrtophora cicatrosa* (Stoliczka)	Hazarika and Chakraborti (1998) and Roychaudhuri (2011)
Araneae	Araneidae	*Cyrtophora feae* (Thorell)	Roychaudhuri (2011)
Araneae	Araneidae	*Cyrtophora moluccensis* (Doleschall)	Roychaudhuri (2011)
Araneae	Araneidae	*Gasteracantha diadesmia* Thorell	Roychaudhuri (2011)
Araneae	Araneidae	*Gasteracantha kuhli* Koch	Das et al. (2010) and Roychaudhuri (2011)

(continued)

Table 18.2 (continued)

Order	Family	Predators	Reference
Araneae	Araneidae	*Hippasa* sp.	Hazarika and Chakraborti (1998)
Araneae	Araneidae	*Neoscona mukerjei* Tikader	Das et al. (2010)
Araneae	Araneidae	*Neoscona* sp.	Das et al. (2010) and Roychaudhuri (2011)
Araneae	Araneidae	*Parawixia dehaani* (Doleschall)	Roychaudhuri (2011)
Araneae	Araneidae	*Zygiella* sp.	Das et al. (2010)
Araneae	Gnaphosidae	*Gnaphosa* sp.	Das et al. (2010)
Araneae	Homalonychidae	*Homalonychus* sp.	Das et al. (2010)
Araneae	Lycosidae	*Lycosa* sp.	Das et al. (2010)
Araneae	Lycosidae	*Pardosa birmanica* Simon	Das et al. (2010)
Araneae	Lycosidae	*Pardosa minutus* Tikader and Malhotra	Das et al. (2010)
Araneae	Miturgidae	*Cheiracanthium sadanai* Tikader	Das et al. (2010)
Araneae	Miturgidae	*Cheiracanthium* sp.	Das et al. (2010)
Araneae	Oxyopidae	*Oxyopes birmanicus* Thorell	Das et al. (2010)
Araneae	Oxyopidae	*Oxyopes pandae* Tikader	Hazarika and Chakraborti (1998)
Araneae	Oxyopidae	*Oxyopes ratnae* Tikader	Das et al. (2010)
Araneae	Oxyopidae	*Oxyopes shweta* Tikader	Hazarika and Chakraborti (1998), Das et al. (2010), and Roychaudhury (2011)
Araneae	Oxyopidae	*Oxyopes* sp.	Anon (1989) and Das et al. (2010)
Araneae	Philodromidae	*Philodromus bhagirathi* Tikader	Das et al. (2010)
Araneae	Pisauridae	*Pisaura* sp.	Das et al. (2010)
Araneae	Salticidae	*Euophrys chiriatapuensis* Tikader	Das et al. (2010)
Araneae	Salticidae	*Euophrys* sp.	Hazarika and Chakraborti (1998)
Araneae	Salticidae	*Hyllus bengalensis* (Tikader)	Das et al. (2010)
Araneae	Salticidae	*Marpissa bengalensis* Tikader	Das et al. (2010)
Araneae	Salticidae	*Marpissa* sp.	Das et al. (2010)

(continued)

Table 18.2 (continued)

Order	Family	Predators	Reference
Araneae	Salticidae	*Marpissa tigrina* Tikader	Das et al. (2010)
Araneae	Salticidae	*Phidippus pateli* Tikader	Roychaudhuri (2011)
Araneae	Salticidae	*Phidippus* sp.	Das et al. (2010)
Araneae	Salticidae	*Plexippus paykulli* Audouin	Das et al. (2010)
Araneae	Salticidae	*Plexippus* sp.	Das et al. (2010) and Roychaudhuri (2011)
Araneae	Salticidae	*Rhene* sp.	Hazarika and Chakraborti (1998)
Araneae	Salticidae	*Salticus* sp.	Hazarika and Chakraborti (1998)
Araneae	Salticidae	*Telamonia dimidiata* (Simon)	Hazarika and Chakraborti (1998)
Araneae	Salticidae	*Lyssomanes* sp	Das et al. (2010)
Araneae	Sparassidae	*Heteropoda* sp.	Das et al. (2010)
Araneae	Sparassidae	*Heteropoda venatoria* L.	Das et al. (2010)
Araneae	Sparassidae	*Sparassus* sp.	Das et al. (2010)
Araneae	Struikzakspinnen (Clubionidae)	*Clubiona drassodes* Cambridge	Das et al. (2010)
Araneae	Tetragnathidae	*Leucauge bistriata* Gravely	Hazarika and Chakraborti (1998)
Araneae	Tetragnathidae	*Leucauge celebesiana* (Walckenaer)	Roychaudhuri (2011)
Araneae	Tetragnathidae	*Leucauge decorata* (Blackwall)	Roychaudhuri (2011)
Araneae	Tetragnathidae	*Leucauge decorata* Blackwall	Das et al. (2010)
Araneae	Tetragnathidae	*Leucauge* sp.	Hazarika and Chakraborti (1998) and Das et al. (2010)
Araneae	Tetragnathidae	*Leucauge tessellate* (Thorell)	Roychaudhuri (2011)
Araneae	Tetragnathidae	*Tetragnatha mandibulata* Walckenaer	Hazarika and Chakraborti (1998)
Araneae	Tetragnathidae	*Tetragnatha* sp.	Das et al. (2010) and Roychaudhuri (2011)
Araneae	Theridiidae	*Chrysso* sp.	Roychaudhuri (2011)
Araneae	Theridiidae	*Theridula angula* Tikader	Roychaudhuri (2011)

(continued)

Table 18.2 (continued)

Order	Family	Predators	Reference
Araneae	Thomisidae	*Philodromus* sp.	Das et al. (2010)
Araneae	Thomisidae	*Runcinia affinis* Simon	Das et al. (2010)
Araneae	Thomisidae	*Mesumena* sp.	Hazarika and Chakraborti (1998)
Araneae	Thomisidae	*Diaea* sp.	Das et al. (2010)
Araneae	Uloboridae	*Uloborus khasiensis* Tikader	Roychaudhuri (2011)
Araneae	Uloboridae	*Uloborus krishnae* Tikader	Roychaudhuri (2011)
Acari	Acaridae	*Acarus* sp.	Muraleedharan and Chandrsekaran (1981)
Acari	Anystidae	*Anystis* sp.	Gupta (1989)
Acari	Ascidae	*Lesioseius* sp.	Gupta (1989)
Acari	Bdellidae	*Bdella* sp.	Muraleedhran (1989)
Acari	Bdellidae	*Cyta* sp.	Gupta (1989)
Acari	Cunaxidae	*Cunaxa* sp.	Borthakur and Das (1988)
Acari	Cunaxidae	*Neocunaxoides* sp.	Gupta (1989)
Acari	Eupalopsellidae	*Exothorhis caudate* Summers	Borthakur and Das (1988)
Acari	Iolinidae	*Tydeus* sp.	Muraleedharan and Chandrsekaran (1981)
Acari	Iolinidae	*Parapronematus* sp.	Gupta (1989)
Acari	Ascidae	*Lasioseius* sp	Gupta (1989)
Acari	Phytoseiidae	*Amblyseius arecae* Gupta	Gupta (1989)
Acari	Phytoseiidae	*Amblyseius coccosoclus* Ghai and Menon	Chakraborty et al. (2005)
Acari	Phytoseiidae	*Amblyseius herbicolus* Chant	Muraleedharan and Chandrsekaran (1981)
Acari	Phytoseiidae	*Amblyseius largoensis* Muma	Somchoudhury et al. (1995)
Acari	Phytoseiidae	*Amblyseius longispinosus* Evans	Rahman et al. (2012)
Acari	Phytoseiidae	*Amblyseius maai* Tseng	http://www.myetymology.com/encyclopedia/Oligonychus_coffeae.html
Acari	Phytoseiidae	*Amblyseius multidentatus* Chant	Gupta (1978)

(continued)

Table 18.2 (continued)

Order	Family	Predators	Reference
Acari	Phytoseiidae	*Amblyseius multidentatus* Swirski and Shechter	Gupta (1978)
Acari	Phytoseiidae	*Amblyseius rhabdus* Denmark	Muraleedharan and Chandrsekaran (1981)
Acari	Phytoseiidae	*Amblyseius* sp.	Muraleedharan and Chandrsekaran (1981) and Gupta (1989)
Acari	Phytoseiidae	*Amblyseius* sp. *nr. sapienticola* Gupta	Gupta (1978)
Acari	Phytoseiidae	*Amblyseius taiwanicus* Ehara	http://www.myetymology.com/encyclopedia/Oligonychus_coffeae.html
Acari	Phytoseiidae	*Euseius ovalis* Evans	Somchoudhury et al. (1995)
Acari	Phytoseiidae	*Neoseiulus longispinosus* Evans	Rahman et al. (2012)
Acari	Phytoseiidae	*Phytoseiulus persimilis* Evans	Anantakrishnan (1960)
Acari	Phytoseiidae	*Typhlodromus darjeelingensis* Gupta	Gupta (1989)
Acari	Phytoseiidae	*Typhlodromus neotransvaalensis* Gupta	Gupta (1978)
Acari	Phytoseiidae	*Typhlodromus rhododendroni* Gupta	Gupta (1978)
Acari	Phytoseiidae	*Typhlodromus* sp.	Anantakrishnan (1960)
Acari	Stigmaeidae	*Agistemus* sp.	Borthakur and Das (1988)
Acari	Stigmaeidae	*Agistemus fleschneri* Summers	Gupta (1989)
Acari	Stigmaeidae	*Agistemus* sp. *nr. fleschneri*	Gupta (1989)
Acari	Stigmaeidae	*Ledermulleria* sp.	Gupta (1989)
Acari	Tydeidae	*Pronematus* sp.	Borthakur (1981)
Coleoptera	Carabidae	*Ophionea indica* Thumb	Muraleedharan (1982) and Das et al. (2010)
Coleoptera	Carabidae	*Calleida* sp.	Muraleedharan (1982) and Das et al. (2010)
Coleoptera	Carabidae	*Cicindela sexgutta* Fab.	Das et al. (2010)

(continued)

Table 18.2 (continued)

Order	Family	Predators	Reference
Coleoptera	Carabidae	*Cicindela collicia* Acciavatti and Pearson	Das et al. (2010)
Coleoptera	Coccinellidae	*Afidentula mandertiernae* Muls	Das et al. (2010)
Coleoptera	Coccinellidae	*Aspidimerus circumflexa* Muls	Das et al. (2010)
Coleoptera	Coccinellidae	*Caelophora* sp.	Rao et al. (1970) and Das et al. (2010)
Coleoptera	Coccinellidae	*Callineda decemnotata* Fab.	Das et al. (2010)
Coleoptera	Coccinellidae	*Chilocorus circumdatus* (Gyllenhal)	Das (1979) and Das et al. (2010)
Coleoptera	Coccinellidae	*Chilocorus nigritus* (F.)	Das (1979)
Coleoptera	Coccinellidae	*Coccinella repanda* Thumb	Das (1974) and Das et al. (2010)
Coleoptera	Coccinellidae	*Coccinella septempunctata* L. *var. divaricata* O.	Das (1974)
Coleoptera	Coccinellidae	*Coccinella transversalis* Fab.	Muraleedharan et al. (1988) and Das et al. (2010)
Coleoptera	Coccinellidae	*Coclophora sexareata* Muls.	Das et al. (2010)
Coleoptera	Coccinellidae	*Coclophora unicolor* Muls.	Das et al. (2010)
Coleoptera	Coccinellidae	*Coleophora biplagiata* (Swartz)	Das (1974)
Coleoptera	Coccinellidae	*Crytogonus bimaculatus 4-guttatus* Weise	Rao et al. (1970)
Coleoptera	Coccinellidae	*Crytogonus bimaculatus* Kapur	Rao et al. (1970) and Das et al. (2010)
Coleoptera	Coccinellidae	*Crytogonus orbiculus* (Gyllenhal)	Muraleedharan (1987)
Coleoptera	Coccinellidae	*Crytogonus quardriguttatus* Weise	Das et al. (2010)
Coleoptera	Coccinellidae	*Harmonia* sp.	Das et al. (2010)
Coleoptera	Coccinellidae	*Henosepilachna septima* Dieke	Das et al. (2010)
Coleoptera	Coccinellidae	*Jauravia opace* Weise	Rao et al. (1970) and Das et al. (2010)

(continued)

Table 18.2 (continued)

Order	Family	Predators	Reference
Coleoptera	Coccinellidae	*Jauravia pubescens* (F.)	Muraleedharan (1987)
Coleoptera	Coccinellidae	*Jauravia quadrinotata* Kapur	Das (1959) and Das et al. (2010)
Coleoptera	Coccinellidae	*Jauravia soror* (Weise)	Rao et al. (1970) and Das et al. (2010)
Coleoptera	Coccinellidae	*Jauravia* sp.	Das (1959)
Coleoptera	Coccinellidae	*Leis dimidiate* F.	Das (1974) and Das et al. (2010)
Coleoptera	Coccinellidae	*Lemnia bissellata* (Mulsant)	Das (1974) and Radhakrishnan et al. (1988)
Coleoptera	Coccinellidae	*Menochilus sexmaculatus* Fab.	Andrews (1928) and Das et al. (2010)
Coleoptera	Coccinellidae	*Micraspis discolour* (Fab.)	Das et al. (2010)
Coleoptera	Coccinellidae	*Oenopia kirbyi* Muls.	Das et al. (2010)
Coleoptera	Coccinellidae	*Oenopia luteopustulata* Muls.	Das et al. (2010)
Coleoptera	Coccinellidae	*Oenopia sexareata* Muls.	Das et al. (2010)
Coleoptera	Coccinellidae	*Ola* sp.	Das et al. (2010)
Coleoptera	Coccinellidae	*Pharoscymnus horni* (Weise)	Das and Gope (1981)
Coleoptera	Coccinellidae	*Pseudaspidimerus circumflexus* (Motschulsky)	Das et al. (2010)
Coleoptera	Coccinellidae	*Scymnus nubilus* Mulsant	Rao et al. (1970)
Coleoptera	Coccinellidae	*Scymnus pyrocheilus* Mulsant	Das (1974)
Coleoptera	Coccinellidae	*Scymnus* sp.	Andrews (1928) and Das et al. (2010)
Coleoptera	Coccinellidae	*Stethorus aptus* (Kapur)	Barua et al. (2013)
Coleoptera	Coccinellidae	*Stethorus gilviforn* Muls.	Rao et al. (1970) and Das et al. (2010)
Coleoptera	Coccinellidae	*Stethorus* sp.	Das (1959)
Coleoptera	Coccinellidae	*Stictobura* sp.	Rao et al. (1970)
Coleoptera	Coccinellidae	*Vernia vincta* Gorh	Das (1959) and Das et al. (2010)
Coleoptera	Dermestidae	*Aspectus indicus* Arrow	Rao et al. (1970)
Coleoptera	Dermestidae	*Orphinus fucundus* Arrow	Rao et al. (1970)

(continued)

Table 18.2 (continued)

Order	Family	Predators	Reference
Coleoptera	Nitidulidae	*Cybocephalus* sp.	Rao et al. (1970)
Coleoptera	Staphylinidae	*Oligota* sp.	Rao et al. (1970)
Coleoptera	Staphylinidae	*Oligota pygmaea* (Solier)	Perumalsamy et al. (2009)
Coleoptera	Staphylinidae	*Oligota flaviceps* Sharp	Babu et al. (2008)
Coleoptera	Staphylinidae	*Paederus fuscipes* Curtis	Das et al. (2010)
Thysanoptera	Aeolothripidae	*Aeolothrips intermedius* Bagnall	Muraleedharan et al. (1988)
Thysanoptera	Aeolothripidae	*Mymarothrips garuda* (Ramak. and Marg.)	Muraleedharan et al. (1988)
Thysanoptera	Thripidae	*Scolothrips asura* Ramakrishna and Margabandhu	Babu et al. (2010)
Thysanoptera	Thripidae	*Scolothrips rhagebianus* Priesner	Radhakrishnan et al. (1992) and Babu et al. (2010)
Thysanoptera	Thripidae	*Scolothrips* sp.	Rao et al. (1970)
Neuroptera	Chrysophidae	*Chrysopa madestes* Banks	Rao et al. (1970)
Neuroptera	Chrysopidae	*Chrysoperla carnea* (Stephens)	Sarkar et al. (2007)
Neuroptera	Chrysopidae	*Chrysoperla* sp.	Muraleedharan (1990)
Neuroptera	Chrysophidae	*Mallada basalis* (Walker)	Babu et al. (2011a)
Neuroptera	Chrysophidae	*Mallada boninensis* Okamoto	Vasanthakumar et al. (2012)
Neuroptera	Chrysophidae	*Mallada desjardinsi* Navas	Vasanthakumar and Babu (2013)
Neuroptera	Hemerobiidae	*Micromus timidus* Hegan	Rao et al. (1970) and Das et al. (2010)
Hemiptera	Anthocoridae	*Orius* sp.	Rao et al. (1970)
Hemiptera	Anthocoridae	*Anthocoris* sp.	Muraleedharan (1988b)
Hemiptera	Lygaeidae	*Geocoris ochropterus* (Fieber)	Das et al. (2010) and Sannigrahi and Mukhopadhyay (1992)
Hemiptera	Pentatomidae	*Cantheconidea furcellata* (Wolf)	Muraleedharan (1983) and Muraleedharan et al. (1988)
Hemiptera	Pentatomidae	*Eocanthecona furcellata* (Wolff)	Das et al. (2010)

(continued)

Table 18.2 (continued)

Order	Family	Predators	Reference
Hemiptera	Rhyparochromidae	*Pseudopachybrachius guttus* (Dallas)	Muraleedharan (1986)
Hemiptera	Reduviidae	*Opistoplatys* sp.	Muraleedharan et al. (1988)
Hemiptera	Reduviidae	*Acanthaspis quinquespinosa* Fab.	Das et al. (2010)
Hemiptera	Reduviidae	*Epidaurus* sp.	Das et al. (2010)
Hemiptera	Reduviidae	*Rhynocoris marginatus* F.	Das et al. (2010)
Hemiptera	Reduviidae	*Sycanus* sp.	Das (1965)
Diptera	Cecidomyiidae	*Triommata* sp.	Nagarkatti et al. (1979)
Diptera	Syrphidae	*Allobaccha nubilipennis* (Austen)	Muraleedharan and Radhakrishnan (1986)
Diptera	Syrphidae	*Betasyrphus adligatus* (Wiedemann)	Anon. (1936)
Diptera	Syrphidae	*Betasyrphus serarius* (Wiedemann)	Muraleedharan and Radhakrishnan (1986)
Diptera	Syrphidae	*Dideopsis aegrota* (Fab.)	Muraleedharan and Radhakrishnan (1986)
Diptera	Syrphidae	*Episyrphus balteatus* (De Geer)	Muraleedharan and Radhakrishnan (1986)
Diptera	Syrphidae	*Ischiodon scutellaris* (F.)	Das et al. (2010)
Diptera	Syrphidae	*Paragus atratus* De Mejere	Anon. (1936)
Diptera	Syrphidae	*Paragus indicus* Brun.	Das (1974)
Diptera	Syrphidae	*Paragus scratus* (Fab.)	Muraleedharan et al. (2001)
Diptera	Syrphidae	*Paragus yerburiensis* Stuckenberj	Muraleedharan et al. (2001)
Mantodea	Mantidae	*Amantis* sp.	Das et al. (2010)
Mantodea	Mantidae	*Elmantis* sp.	Das et al. (2010)
Mantodea	Mantidae	*Hierodula* sp.	Das et al. (2010)
Mantodea	Liturgusidae	*Humbertiella indica* Saus	Das et al. (2010)
Mantodea	Amorphoscelidae	*Amorphoscelis* sp.	Das et al. (2010)
Odonata	Aeshnidae	*Anax* sp.	Das et al. (2010)
Odonata	Coenagrionidae	*Ceriagrion* sp.	Das et al. (2010)
Odonata	Coenagrionidae	*Pseudagrion* sp.	Das et al. (2010)
Odonata	Gomphidae	*Ictinogomphus* sp.	Das et al. (2010)

Table 18.3 Parasitoids reported from tea plantations in India

Order	Family	Parasitoids	Reference
Diptera	Tachinidae	*Atherigona orientalis* (Schin.)	Das (1974)
Diptera	Tachinidae	*Austrophorocera grandis* Macquart	Rao (1978)
Diptera	Tachinidae	*Blepharella* sp.	Muraleedharan (1983)
Diptera	Tachinidae	*Carcelia* sp.	Muraleedharan (1983)
Diptera	Tachinidae	*Compsilura concinnata* (Meigen)	Rao (1974)
Diptera	Tachinidae	*Cylindromyia* sp.	Banerjee (1979)
Diptera	Tachinidae	*Microdon bellus* Brun.	Das (1974)
Diptera	Tachinidae	*Nealsomyia rufella* (Bezzi)	Rao (1978)
Diptera	Tachinidae	*Palexorista solennis* (Walker)	Muraleedharan (1984)
Diptera	Tachinidae	*Paribaea* sp.	Rao (1974)
Diptera	Tachinidae	*Peribaea orbata* (Wiedemann)	Muraleedharan et al. (1988)
Diptera	Tachinidae	*Syrphus balteatus* De Geer	Das (1965)
Diptera	Tachinidae	*Winthemia* sp.	Muraleedharan (1983)
Hymenoptera	Aphelinidae	*Aphelinus* sp.	Banerjee (1967)
Hymenoptera	Aphelinidae	*Aphterencyrtus microphagus* (Mayr.)	Das (1979)
Hymenoptera	Aphelinidae	*Aphytis fioriniae* Rosen and Rose	Rosen and Rose (1989)
Hymenoptera	Aphelinidae	*Aphytis* sp.	Muraleedharan et al. (1988)
Hymenoptera	Aphelinidae	*Aphytis* sp. *chrysomphali* (Mercet)	Das (1979)
Hymenoptera	Aphelinidae	*Aphytis sp. nr. mazalae* De Bach and Rosen	Nagarkatti et al. (1979)
Hymenoptera	Aphelinidae	*Aphytis theae* (Cameron)	Nagarkatti et al. (1979)
Hymenoptera	Aphelinidae	*Aspidiotiphagus citrinus* Craw	Das (1979)
Hymenoptera	Aphelinidae	*Aspidiotiphagus* sp.	Nagarkatti et al. (1979) and Das (1979)
Hymenoptera	Aphelinidae	*Coccophagus acanthosceles* Waterstone	Das (1974)
Hymenoptera	Aphelinidae	*Coccophagus bivittatus* Comp.	Nagarkatti et al. (1979)
Hymenoptera	Aphelinidae	*Coccophagus tachiachii* Madhihassan	Das (1974)
Hymenoptera	Aphelinidae	*Coccophagus cowperi* Gir.	Muraleedharan et al. (1988)
Hymenoptera	Aphelinidae	*Prospaltella* sp.	Nagarkatti et al. (1979)

(continued)

Table 18.3 (continued)

Order	Family	Parasitoids	Reference
Hymenoptera	Aphididae	*Aphidius colemani* Viereck	Muraleedharan (1986)
Hymenoptera	Aphididae	*Lipolexis scutellaris* Mackauer	Muraleedharan (1986)
Hymenoptera	Aphididae	*Trioxys* sp.	Anon. (1936)
Hymenoptera	Aphididae	*Trioxys indicus* Subba Rao and Sharma	Muraleedharan et al. (1988)
Hymenoptera	Aphidiidae	*Aphidius colemani* Viereck	Muraleedharan (1986)
Hymenoptera	Aphidiidae	*Lipolexis scutellaris* Mackauer	Muraleedharan (1986)
Hymenoptera	Aphidiidae	*Trioxys* sp.	Anon. (1936)
Hymenoptera	Aphidiidae	*Trioxys indicus* Subba Rao and Sharma	Muraleedharan et al. (1988)
Hymenoptera	Platygastridae	*Fidiobia* sp.	Das et al. (2010)
Hymenoptera	Braconidae	*Agathidinae* sp.	Das et al. (2010)
Hymenoptera	Braconidae	*Apanteles aristaeus* Nixon	Muraleedharan and Chandrasekaran (1981), Muraleedharan and Selvasundaram (1989)
Hymenoptera	Braconidae	*Apanteles coedicius* Nixon	Muraleedharan et al. (1988)
Hymenoptera	Braconidae	*Apanteles nr. malevolus* Wilkinson	Muraleedharan et al. (1988)
Hymenoptera	Braconidae	*Apanteles rufulus* Wilk.	Das (1974)
Hymenoptera	Braconidae	*Apanteles* sp. (ater group)	Rao et al. (1970), Das (1979), and Muraleedharan (1986)
Hymenoptera	Braconidae	*Apanteles sp.* (Octonarius group)	Subbiah (1988)
Hymenoptera	Braconidae	*Apanteles* sp. (ultor group)	Rao (1978)
Hymenoptera	Braconidae	*Apanteles taprobanae* Cameron	Das (1979)
Hymenoptera	Braconidae	*Asogaster* sp.	Muraleedharan et al. (1988)
Hymenoptera	Braconidae	*Batotheca nigriceps* (Cameron)	Das (1974)
Hymenoptera	Braconidae	*Bracon* sp.	Rao (1970)
Hymenoptera	Braconidae	*Bracon sp. nr. greeni* Ashmead	Muraleedharan (1984)
Hymenoptera	Braconidae	*Bracon sp. nr. hebetor* Say	Muraleedharan et al. (1988)
Hymenoptera	Braconidae	*Chelonus indicus* Cameron	Das et al. (2010)

(continued)

Table 18.3 (continued)

Order	Family	Parasitoids	Reference
Hymenoptera	Braconidae	*Cotesia ruficrus* Haliday	Das et al. (2010)
Hymenoptera	Braconidae	*Cotesia* sp.	Das et al. (2010)
Hymenoptera	Braconidae	*Dolichogenidea* sp.	Das et al. (2010)
Hymenoptera	Braconidae	*Fornicia* sp.	Muraleedharan and Selvasundaram (1986)
Hymenoptera	Braconidae	*Meteoridinae* sp.	Das et al. (2010)
Hymenoptera	Braconidae	*Meteorus* sp. (group of versicolor)	Das (1974)
Hymenoptera	Braconidae	*Opius* sp.	Muraleedharan et al. (1988)
Hymenoptera	Braconidae	*Pambolus* sp.	Das et al. (2010)
Hymenoptera	Braconidae	*Spathius critolaus* Nixon	Das et al. (2010)
Hymenoptera	Braconidae	*Spinaria spinator* Guer	Das (1974)
Hymenoptera	Ceraphronidae	*Aphanogmus manila* (Ashmead)	Das (1979)
Hymenoptera	Ceraphronidae	*Aphanogmus* sp.	Das et al. (2010)
Hymenoptera	Chalcididae	*Brachymeria lasus* (Walker)	Muraleedharan (1985) and Das et al. (2010)
Hymenoptera	Diapriidae	*Trichopria* sp.	Das et al. (2010)
Hymenoptera	Encyrtidae	*Anagyrus dactylopii* (Howard)	Das (1974)
Hymenoptera	Encyrtidae	*Aprostocetus purpureus* (Cameron)	Muraleedharan et al. (1988)
Hymenoptera	Encyrtidae	*Blastothrix* sp.	Das (1974)
Hymenoptera	Encyrtidae	*Comperiella bifasciata* Howard	Das (1979)
Hymenoptera	Encyrtidae	*Encyrtus infelix* (Embleton)	Muraleedharan et al. (1988)
Hymenoptera	Encyrtidae	*Leptomastix nigrocoxalis* Compere	Muraleedharan (1992b)
Hymenoptera	Encyrtidae	*Microterys* sp.	Das (1979)
Hymenoptera	Encyrtidae	*Ooencyrtus ferriere* Shafi Alam and Agarwal	Das et al. (2010)
Hymenoptera	Encyrtidae	*Prospaltella* sp.	Das (1979)
Hymenoptera	Encyrtidae	*Blastothrix* sp.	Das (1974)
Hymenoptera	Encyrtidae	*Pseudorhopus* sp.	Das (1974)
Hymenoptera	Eulophidae	*Aprostocetus nowsherensis* Kurian	Das et al. (2010)
Hymenoptera	Eulophidae	*Asympiesiella india* Gir.	Rao et al. (1970) and Das (1974)
Hymenoptera	Eulophidae	*Asympiesiella* sp.	Sengupta (1967)

(continued)

Table 18.3 (continued)

Order	Family	Parasitoids	Reference
Hymenoptera	Eulophidae	*Brachymeria excrinata* Gahan	Muraleedharan (1985)
Hymenoptera	Eulophidae	*Closterocerus insignis* Waterston	Banerjee (1988)
Hymenoptera	Eulophidae	*Elachertus* sp.	Sengupta (1967)
Hymenoptera	Eulophidae	*Elasmus anamalaianus* Mani & Saraswat	Das et al. (2010)
Hymenoptera	Eulophidae	*Elasmus homonae* Ferr.	Rao (1970)
Hymenoptera	Eulophidae	*Elasmus homonae* Ferr.	Rao (1970)
Hymenoptera	Eulophidae	*Elasmus* sp.	Das et al. (2010)
Hymenoptera	Eulophidae	*Eriborus* sp.	Muraleedharan and Selvasundaram (1986)
Hymenoptera	Eulophidae	*Euplectromorpha* sp. *nr. salomonis* Ferr.	Das (1974)
Hymenoptera	Eulophidae	*Euplectromorpha* sp. *nr. uridiceps* Ferr.	Das (1965)
Hymenoptera	Eulophidae	*Euplectromorpha* sp. *nr. viridiceps* Ferr.	Das (1974)
Hymenoptera	Eulophidae	*Eurytoma* sp.	Muraleedharan et al. (1988) and Subbiah (1988)
Hymenoptera	Eulophidae	*Goniozus* sp.	Muraleedharan et al. (1988)
Hymenoptera	Eulophidae	*Gryon* sp.	Das et al. (2010)
Hymenoptera	Eulophidae	*Mestocharella javensis* Gahan	Rao (1978)
Hymenoptera	Eulophidae	*Metaplectrus thoseae* Ferr.	Das (1974)
Hymenoptera	Eulophidae	*Metaplectrus thoseae* (Ferr.)	Das (1974)
Hymenoptera	Eulophidae	*Microphanurus* sp.	Das (1974)
Hymenoptera	Eulophidae	*Nesolynx* sp.	Das et al. (2010)
Hymenoptera	Eulophidae	*Pediobius elasmi* (Ashmead)	Das et al. (2010)
Hymenoptera	Eulophidae	*Pediobius foveolatus* (Crawford)	Das et al. (2010)
Hymenoptera	Eulophidae	*Pediobius* sp.	Subbiah (1988)
Hymenoptera	Eulophidae	*Sparasion* sp.	Das et al. (2010)
Hymenoptera	Eulophidae	*Sympiesis dolichogaster* Ashmead	Muraleedharan and Selvasundaram (1986)
Hymenoptera	Eulophidae	*Sympiesis india* (Girault)	Rao (1978)
Hymenoptera	Eulophidae	*Sympiesis india* (Girault)	Rao (1978)
Hymenoptera	Eulophidae	*Telenomus* sp.	Rao (1978) and Das et al. (2010)

(continued)

Table 18.3 (continued)

Order	Family	Parasitoids	Reference
Hymenoptera	Eulophidae	*Tetrastichus epilachnae* (Giard)	Das et al. (2010)
Hymenoptera	Eulophidae	*Tetrastichus* sp. (miser-group)	Subbiah (1988)
Hymenoptera	Eulophidae	*Trissolcus* sp.	Das et al. (2010)
Hymenoptera	Eulophidae	*Euplectrus* sp. *nr. singularis*	Das (1974)
Hymenoptera	Ichneumonidae	*Apophua* sp.	Das et al. (2010) and Muraleedharan (1983)
Hymenoptera	Ichneumonidae	*Aptesis* sp.	Das et al. (2010)
Hymenoptera	Ichneumonidae	*Astomaspis* sp.	Rao et al. (1970)
Hymenoptera	Ichneumonidae	*Charops obtusus* Morl.	Das (1974)
Hymenoptera	Ichneumonidae	*Charops* sp.	Das et al. (2010)
Hymenoptera	Ichneumonidae	*Diadegma* sp.	Das et al. (2010)
Hymenoptera	Ichneumonidae	*Diadegma* sp.	Muraleedharan et al. (1988)
Hymenoptera	Ichneumonidae	*Meloboris* sp.	Rao et al. (1970)
Hymenoptera	Ichneumonidae	*Phytodietus* sp.	Muraleedharan et al. (1988)
Hymenoptera	Ichneumonidae	*Phytodietus spinipes* (Cameron)	Muraleedharan (1984), Muraleedharan and Selvasundaram (1991)
Hymenoptera	Ichneumonidae	*Plectochorus nr. iwatensis* Uchida	Rao et al. (1970)
Hymenoptera	Ichneumonidae	*Pristomerus* sp.	Rao (1970)
Hymenoptera	Ichneumonidae	*Trathala* sp.	Muraleedharan (1990)
Hymenoptera	Ichneumonidae	*Triclistus* sp.	Muraleedharan (1986)
Hymenoptera	Mymaridae	*Alaptus* sp.	Das et al. (2010)
Hymenoptera	Mymaridae	*Anagurus* sp.	Das et al. (2010)
Hymenoptera	Mymaridae	*Polynema* sp.	Das et al. (2010)
Hymenoptera	Pteromalidae	*Oxyharma* sp.	Muraleedharan (1989)
Hymenoptera	Pteromalidae	*Trigonogastra* sp	Banerjee (1988)
Hymenoptera	Platygastridae	*Synopeas* sp.	Das et al. (2010)
Hymenoptera	Platygastridae	*Leptacis indicus* Mukerjee	Das et al. (2010)
Hymenoptera	Pteromalidae	*Dipara* sp.	Das et al. (2010)
Hymenoptera	Pteromalidae	*Anysis* sp.	Das (1974)
Hymenoptera	Trichogrammatidae	*Trichogramma chilonis* Ishii	Hazarika et al. (1995)
Hymenoptera	Eulophidae	*Tetrastichus* sp	Das (1974)
Hymenoptera	Eulophidae	*Tetrastichus* sp	Das (1974)

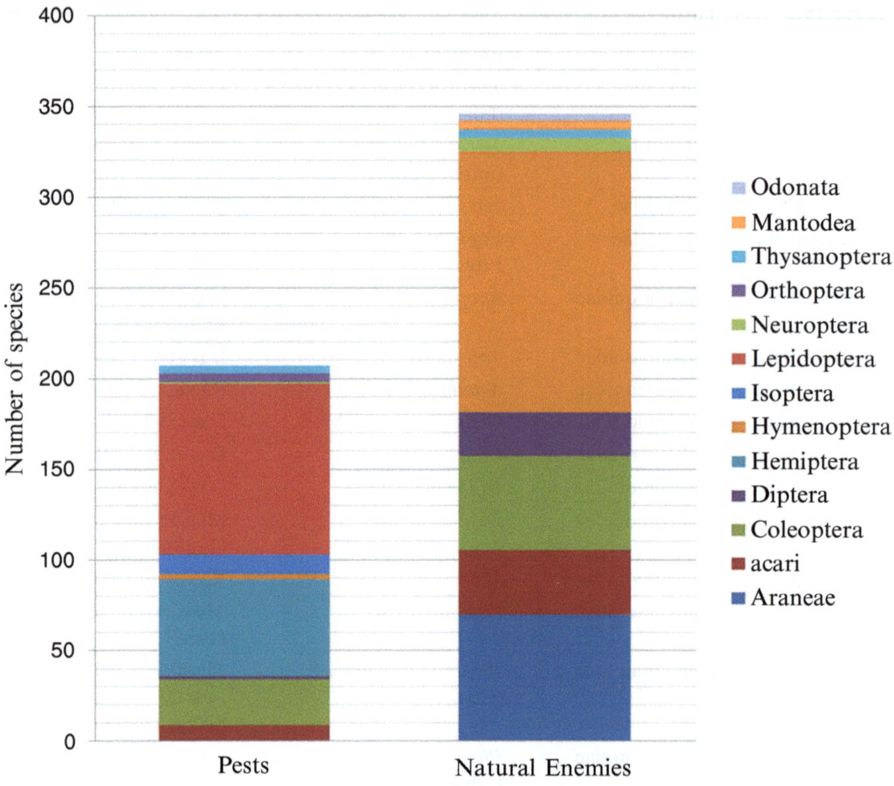

Fig. 18.3 Number of arthropod species (families) under different feeding guilds in tea ecosystem, India

and *Hyposidrainfixaria*; and the bunch caterpillar, *Andraca bipunctata*, cause considerable defoliation. *Homona coffearia*, *H. magnanima*, *Adoxophyes* sp. and *Cydia leucostoma* are the main tortricid pests of tea. The leaf folder, *Caloptilia theivora*; the stem borer, *Euwallacea fornicates* (*Xyleborus fornicates*); and the termite, *Odontotermes, Microtermes* and *Microcerotermes* species are of considerable importance in India.

The presence of insects feeding on tea plants and shade trees invariably has led to the build-up of several species of insect parasitoids and predators, and the foliage below the plucking surface is important as a refuge for natural enemies (Muraleedharan et al. 2001). In fact, the minor status of several pests is due to these natural enemies. Two such cases of excellent natural regulation are of the scale insect *Fiorinia theae* and the aphid *Toxoptera aurantii* (Nagarkatti et al. 1979; Radhakrishnan et al. 1988, 1992). Rao et al. (1970) catalogued predators and parasitoids of the flush worm *Cydia leucostoma* and the phytophagous mites affecting tea in South and Northeast India. Das (1974, 1979) prepared two directories of predators and parasitoids of pests of tea, shade trees and ancillary crops. An account

Fig. 18.4a Common arthropods in tea ecosystem of India

Fig. 18.4b Common arthropods in Tea ecosystem of India

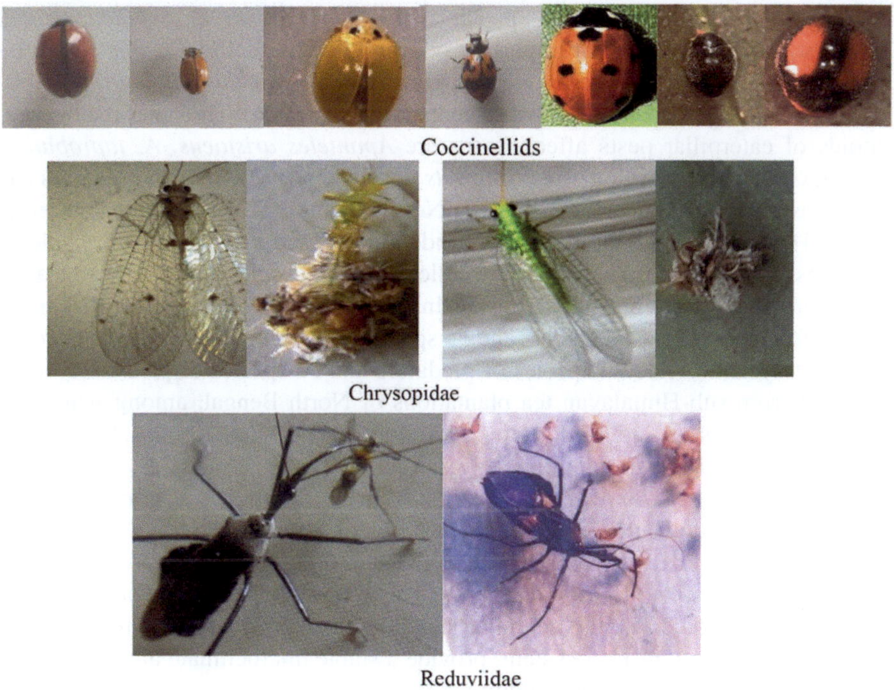

Fig. 18.4c Common arthropods in Tea ecosystem of India

of the predators and parasitoids occurring in the tea fields of South India with notes on their bioecology had also been published (Muraleedharan et al. 1988). Subsequently, the data on the biological control agents were compiled by Borthakur et al. (1992) and Hazarika et al. (1994). Somchoudhury et al. (1995) identified 38 species of predatory mites feeding on tea red spider mite in Northeast India. The tea plantations harbour several species of coccinellids such as *Cryptogonus bimaculatus*, *Juravia quandrinotata*, *J. opaca*, *Menochilus sexmaculatus* and *Stethorus gilvifrons*, which feed on eriophyid mites, spider mites, aphids and scale insects. Syrphid flies which are active in tea fields include *Episyrhus balteatus*, *Paragus tibialis*, *Betasyrphus serarius*, *Ischiodon scutellaris* and *Allobaccha nubilipennis*, with their larvae primarily feeding on the aphids. The hemerobiid, *Micromus timidus*, has a wide distribution in tea fields compared to chrysopids. The parasitoids, *Apanteles coedicius*, *Trioxys indicus* and *Aphidus colemani*, have several hosts like *Toxoptera aurantii* feeding on tea and *Aphis gossipii* and *Aphis craccivora* on weeds in tea fields.

In the tea plantations in Northeast India, coccinellids and spiders are the most dominant predators, and braconids, ichneumonids and tachinids are the most common parasitoids (Roy et al. 2005). The tea mosquito bug *H. theivora* is preyed upon by *Chrysoperla carnea*, *Oxyopes* sp., *Plexippus* sp., *Phidippus* sp., *Scymnus* sp. and mantids. *Adrynis* wasp is an important parasitoid of cicadellid *Empoasca flavescens*. Eggs of *H. theivora* were found parasitized by *Erythmelus helopeltidis* Gahan in South India (Sudhakaran and Muraleedharan 1998, 2006). The other major parasitoids of caterpillar pests affecting tea are *Apanteles aristaeus*, *A. taprobanea*, *Sympiesis dolichogaster* and *Mestocharells javensis*. The activities of predators and the parasitoids have been found high in Northeast India (Gurusubramaniam et al. 2008). Hazarika and Chakraborti (1998) identified 28 species of predatory spiders on mites, cicadellids and aphids. They collected these species not only from tea but also from shade trees and other plants. In the areas of North Bengal, Roy et al. (2005) found 35 species of spiders and 25 species of coccinellids as natural enemies in tea ecosystem, while 94 species of predators and 33 species of parasitoids were reported from sub-Himalayan tea plantations of North Bengal, among which the predators, spiders and ladybird beetles and among the parasitoid groups, Braconidae and Ichneumonidae, were dominant (Das et al. 2010). The braconids, *Cotesia* sp., and the tachinid, *Argyrophylax* sp., were active in the foothills of Darjeeling, causing 40–45 % parasitization of the geometrid, *Buzura suppressaria*.

This compilation revealed high species richness of arthropod is present in Indian tea ecosystem in spite of pesticides stress. The evergreen, perennial tea plantations with genetically diverse cultivars, interplanted with an array of shade trees species with the adjoining forest ecosystem, provide a stable microclimate and continuous supply of food for the arthropod communities.

An important principle of integrated pest management is the maximization of natural control, and, therefore, the temporal changes in arthropod abundance, diversity, species richness and community structures are important considerations in designing pest management strategies. After tea planting, arthropod species colonize the plantations and over the time progressively increase in diversity. Their communities vary with the environment, varieties, cropping patterns and cultivation practices. In order to achieve the objective of production of export-quality tea, *in situ* conservation and maintenance of natural enemies in the tea ecosystems are desirable with reduction in the use of insecticides: a biorational method of tea production. Large-scale and indiscriminate applications of broad-spectrum organosynthetic insecticides for control of pests eliminate natural enemies, as is evident from comparative studies on diversity of natural enemies between organic (with high diversity index) and pesticide-treated conventional tea gardens (Das et al. 2005; Hazarika et al. 2001). Protection, maintenance and enhancing efficacy of the existing population of natural enemies by practising eco-friendly operations and minimizing pesticide use constitute the main objectives of conservation biological control (CBC) (Jonsson et al. 2008). Plant diversification programmes help inhabitant manipulation by intercropping with shade trees and cover cropping of vacant land in tea plantations, which may contribute to the process of CBC, by providing shelter, nectar, pollen (Wackers et al. 2007) and alternative host/prey to the natural enemies (Zehnder et al. 2007).

Acknowledgements Authors are indebted to Mr. Dwiban Pujari, Tocklai Tea Research Institute and Jorhat for his secretarial help.

References

Ananthakrishnan NR (1960) Observations on red spider of tea in south India. Bull UPASI Tea Sci Dept 19:40–48

Andrews EA (1928) Red spider. Q J Indian Tea Assoc 4:206–219

Anonymus (1936) Report of the entomologist. Annual Report, UPASI Tea Science Department, pp 14–24

Anonymus (1989) Plant protection Department. Annual Scientific Report, Tea Research Association, Jorhat pp 66–74

Babu A, Kumar AR, Perumalsamy K, James SP (2008) New record of a predator of RSM. Newsl UPASI Tea Res Found 18(1):4

Babu A, Sundaravadiveln C, Kumar P, Vasanthakumar D (2010) Acaricidal activity of plant extract on red spider mite. Newsl UPASI Tea Res Found 20(1):3

Babu A, Vasantha Kumar D, Rahman VKJ, Roobak Kumar A, Sundaravadivelan C (2011) Potential of *Mallada boninensis* Okamoto (Neuroptera: Chrysopidae) as a biocontrol agent of *Oligonychus coffeae* Nietner (Acari: Tetranychidae) infesting tea. J Plant Crop 39(1):193–195

Banerjee B (1967) Annual Scientific Report. Entomology Department, Tea Research Association (Jorhat) 68–76

Banerjee B (1979) Intra tree variation in the distribution of the tea red spider mite *Oligonychus coffeae* (Nietner). Acarologia 21:216–220

Banerjee B (1983) Arthropod accumulation on tea in young and old habitats. Ecol Entomol 18(2):339–342

Banerjee B (1988) Why leaf miner *Agromyza theae* (Bigot) Meij (Agromyzidae: Diptera) populations are low on tea? Trop Ecol 29(2):73–78

Barua A, Babu A, Rajkhowa R (2013) Study on life cycle parameters of Stethorus aptus (Kapur) (Coleoptera: Coccinellidae): A new predator of tea Red spider mite Oligonychus coffeae (Nietner) (Tetranychidae). Int J Sci Res 2(12):566–567

Borthakur M (1981) Biological notes on a mite predator of the scarlet mite, *Brevipalpus phoenicis* (Geijskes). Two Bud 78(1):18–19

Borthakur M, Das SC (1988) Studies on acarine predators of phytophagous mites on tea in north-east India. Two Bud 34:21–24

Borthakur M, Borthakur MC, Kakoty NN and Das SC (1992) Prospects on biologicxal control of tea pests. In Proceedings of 31 st Tocklai Conference, Tea Research Association, Tocklai Experimental Station, Jorhat, pp 118–130

Calnaido D (1973) New outlook on the pest management of tea. J Natl Sci Counc Sri Lanka 1:97–109

Chakraborty S, Dey PK, Bhattacharya DK, Chakraborty G, Sarkar PK, Somchoudhury AK (2005) Laboratory evaluation of a predatory mite, *Amblyseius coccosocius* Ghai and Menon against red spider mite, *Oligonychus coffeae* (Nietner) in tea. J Entomol Res 29(4):309–312

Chen ZM, Chen XF (1989) An analysis of world tea pest fauna. J Tea Sci 9(1):73–88

Cranham JE (1966a) Tea pests and their control. Annu Rev Entomol 11:491–514

Cranham JE (1966b) Insect and mite pest of tea in Ceylon and their control. Monographson Tea Prod Ceylon Tea Res Inst Ceylon 6:122

Das GM (1959) Bionomics of the red spider, *Oligonychus coffeae* (Nietner). Bull Entomol Res 50(2):265–274

Das GM (1965) Pests of tea in north east India and their control. Tocklai Experimental Station, Cinnamara, Assam, pp 115

Das SC (1974) Parasites and predators of pests of tea, shade trees and ancillary crops in Jorhat circle. Two Bud 21(1):17–21

Das SC (1979) Parasites and predators of tea pests. Two Bud 26(2):72–73
Das SC, Gope B (1981) More about a less known predator *pharoscymnus horni* (Weise), a predator of *Chrysomphalus ficus* Ashm. Two Bud 31(1):46–47
Das SC, Sarker M, Mukhopadhyay A (2005) Changing diversity of hymenopteran parasitoids from organically and conventionally managed tea ecosystem of North Bengal. Indian J Environ Biol 26(3):505–509
Das SC, Roy S, Mukhopadhyay A (2010) Diversity of arthropod natural enemies in the tea plantations of North Bengal with emphasis on their association with tea pests. Cur Sci 99(10):1457–1463
Deka A, Deka PC, Mondal TK (2006) Tea. In: Parthasarathy VA, Chattopadhyay PK, Bose TK (eds) Plantation crops. Naya Udyog, Kolkata, pp 1–147
Gupta SK (1978) Studies on Indian Phytoseiidae (Acarina: Mesostigmata). Some Typholodromus mites from South India with description of new species. Bull Zool Surv India 1:45–54
Gupta SK (1989) Mites occurring on tea plants in India with a key for their identification. In: Progress in Acarology, Oxford & IBH, New Delhi, pp 173–176
Gurusubramanian G, Rahman A, Sarmah M, Roy S, Bora S (2008) Pesticide usage pattern in tea ecosystem, their retrospects and alternative measures. J Environ Biol 29(6):813–826
Hazarika LK, Chakraborti SK (1998) Spider complex of tea ecosystem of Assam. Research Bulletin No. AAU/DA/EI. Assam Agricultural University, Jorhat
Hazarika LK, Borthakur M, Singh K and Sannigrahi S (1994) Present status and future prospects of biological control of tea pests of North East India, In Proceedings of 32 nd Tocklai Conference, Tea Research Association, Tocklai Experimental Station, Jorhat, pp 167–177
Hazarika LK, Saikia MK, Sharma M, Borthakur M (1995) Parasitization by *Trichogramma chilonis* on tea Lasiocamped. Two Bud 42(1):32–34
Hazarika LK, Puzari KC, Wahab S (2001) Biological control of tea pests. In: Upadhyay RK, Mukerji KG, Chamola BP (eds) Biocontrol potential and its exploitation in sustainable agriculture, vol 2, Insect Pests. Kluwer Academic, New York, p 421
Hazarika LK, Bhuyan M, Hazarika BN (2009) Insect pests of tea and their management. Annu Rev Entomol 54:267–284
Jonsson M, Wratten SD, Landisn DA, Gurr GM (2008) Recent advances in conservation biological control of arthropods by arthropods. Biol Control 45:172–175
Muraleedharan N (1982) Entomology, Annual Report, UPASI Tea Science Department, pp. 86–106
Muraleedharan N (1983) Entomology, Annual Report, UPASI Tea Science Department, pp. 72–91
Muraleedharan N (1984) Entomology, Annual Report, UPASI Tea Science Department, pp. 64–82
Muraleedharan N (1985) Entomology, Annual Report, UPASI Tea Science Department, pp. 65–90
Muraleedharan N (1986) Entomology, Annual Report, UPASI Tea Science Department, pp. 89–118
Muraleedharan N (1987) Entomology, Annual Report, *UPASI* Tea Science Department, pp. 79–103
Muraleedharan N (1988b) Final technical report of the CSIR project "Studies on the distribution pattern and biology of the important insect and mite predators and parasites of caterpillars, aphids, thrips and mites infesting tea in South India", UPASI Tea Science Department, Tea Research Institute, Valpari, Mimeographed, p. 80
Muraleedharan N (1989) Entomology, Annual Report, UPASI Tea Science Department, pp. 21–29
Muraleedharan N (1990) Entomology, Annual Report, UPASI Tea Science Department, pp. 33–42
Muraleedharan N (1992) Entomology, Annual Report, UPASI Tea Science Department, pp. 69–92
Muraleedharan N (2010) An IPM package for tea in India. In Glimpses of tea research, published by United Planters' Association of Southern India, Coonoor, pp. 147–169
Muraleedharan N, Chandrasekaran R (1981) Observations on the seasonal variations of *Acaphylla theae* (Watt) and *Calacarus carinatus* (Green) in tea field at the Anamallais, South India). Pestology 5(6):11–15
Muraleedharan N, Chen ZM (1997) Pests and diseases of tea and their management. J Plant Crop 25:15–43
Muraleedharan N, Radhakrishnan B (1986) Syrphid predators of the tea aphid *Toxoptera aurantii* (Boyer de Fonscolombe) in the Anamallis. Indian J Agric Sci 56(4):307

Muraleedharan N, Selvasundaram R (1986) New record of parasites of flushworm and leaf roller of tea. Entomon 11(4):262–263

Muraleedharan N, Selvasundaram R (1989) Life history and seasonal abundance of *Apanteles aristaeus*, a larval parasitoid of *Cydia leucostoma*, the flushworm of tea. Entomon 14(1/2):139–142

Muraleedharan N, Selvasundaram R (1991) Bioecology of *Phytodietus spinipes* (Cameron) a parasitoid of *Homona coffearia* Nietner, the tea tortrix. J Plant Crop 19(1):26–32

Muraleedharan N, Selvasundaram R, Radhakrishnan B (1988) Natural enemies of certain tea pests occurring in Southern India. Int J Trop Insect Sci 9:647–654

Muraleedharan N, Selvasundaram R, Radhakrishnan B (2001) Parasitoids and predators of tea pests in India. J Plant Crop 29(2):1–10

Nagarkatti S, Sankaran T, Manjunath TM (1979) Natural control of the scale *Fiorinia theae* Green (Coccoidea: Diaspididae) in Assam. Prot Ecol 1:179–184

Perumalsamy K, Selvasundaram R, Roobakkumar A, Rahman VJ, Babu A, Muraleedharan N (2009) Life table and predation of *Oligota pygmaea* (Coleoptera: Staphylinidae) a major predator of the red spider mite, *Oligonychus coffeae* (Acari: Tetranychidae) infesting tea. Biol Control 51:96–101

Radhakrishnan B, Loganathan S, Murraleedharan N (1988) Coccinellid predators of *Toxoptera aurantii* (Boyer de Fonscolombe) infesting tea in Southern India. Newsl Aphidological Soc India 71:7–8

Radhakrishnan B, Loganathan S, Muraleedharan N (1992) Bioecology of the pink mite, *Acaphylla theae* (Watt) (Eriophyidae: Acarina) in southern India. J Plant Crop 20(Suppl):97–102

Rahman VJ, Babua A, Roobakkumara A, Perumalsamya K, Vasanthakumara D, Subramaniama MSR (2012) Efficacy, prey stage preference and optimum predator–prey ratio of the predatory mite, *Neoseiulus longispinosus* Evans (Acari: Phytoseiidae) to control the red spider mite, *Oligonychus coffeae* Nietner (Acari: Tetranychidae) infesting tea. Arch Phytopathol Plant Protect 45(6):699–706

Rao GN (1970) Tea pests in Southern India and their control. PANS Pest Articles News Summ 16:667–672

Rao GN (1974) Entomology, Annual Report, UPASI Tea Science Department, pp. 28–40

Rao GN (1978) Entomology, Annual Report, UPASI Tea Science Department, pp. 82–108

Rao VP, Dutta B, Ramaseshiah G (1970) Natural enemy complex of flush worm and phytophagous mites on tea in India. Tea Board, Calcutta, p 53

Rosen D, Rose M (1989) *Aphytis fioriniae*, sp. nov. (Hymenoptera: Aphelinidae), a parasite of tea scale, *Fiorinia theae* Green, form India. Orient Insects 23:269–273

Roy S, Talukdar T, Saha AS, Banerjee DK, Sannigrahi S, Gurusubramanian (2005) Species richness and seasonal abundance of spider and lady bird fauna in tea eco-system of North Bengal. In Proceedings of 34 th Tocklai Conference, Tea Research Association, Tocklai Experimental Station, Jorhat, pp. 347–351

Roychaudhuri D (2011) Assessment of spiders as second order of bio-control agents in tea ecosystem with special reference to Assam and Dooars of West Bengal. *Annual Scientific Report*, National Tea Research Foundation, Tea Board, Kolkata, India pp 38–47

Sannigrahi S, Mukhopadhyay A (1992) Laboratory evaluation of predatory efficiency of *Geocoris ochropterus* Fieber (Hemiptera: Lygaeidae) on some common tea pests. Sri Lanka J Tea Sci 61:39–44

Sarkar PK, Somchoudhury AK, Sekh K (2007) Role of predators in management of *Oligonychus coffeae* Nietner in tea in India. In The 5th International Symposium on Biocontrol and Biotechnology. November 1–3, 2007 at Khon Kaen University, Nong Khai Campus, Nong Khai, Thailand, pp. 15

Sengupta N (1967) Natural enemies of flushworm *Laspeyresia leucostoma* (Meyr.). Two Bud 14(4):170–182

Somchoudhury AK, Saha K, Sarkar PK, Choudharya, Bhattacharya A (1995) Approaches to integrated control of red spider mite, *Oligonychus coffeae* (Nietner) on tea. In Proce.1995 Int. tea Quality – Human Health Symposium, Shanghi, China.

Subbiah K (1988) New records of insect parasitoids of leaf roller and flushworm of tea. Curr Sci 57(12):686

Sudhakaran R, Muraleedharan N (1998) An egg parasitoid of Helopeltis theivora infesting tea. Newsl UPASI Tea Res Inst 8(2):4

Sudhakaran R, Muraleedharan N (2006) Biology of Helopeltis theivora (Hemiptera: Miridae) infesting tea. Entomon 31(3):165–180

Vasanthakumar D, Babu A (2013) Life table and efficacy of *Mallada desjardinsi* (Chrysopidae: Neuroptera), an important predator of tea red spider mite, *Oligonychus coffeae* (Acari: Tetranychidae). Exp Appl Acarol. doi:10.1007/s10493-013-9664-z

Vasanthakumar D, Roobakkumar A, Rahman VJ, Kumar P, Sundaravadivelan C, Babu A (2012) Enhancement of the reproductive potential of *Mallada boninensis* Okamoto (Neuroptera: Chrysopidae), a predator of red spider mite infesting tea: an evaluation of artificial diets. Arch Biol Sci 64(1):281–285

Wackers FL, Romeis J, van Rijn P (2007) Nectar and pollen-feeding by insect herbivores and implications for multitrophic interactions. Annu Rev Entomol 52:301–323

Zehnder G, Gurr GM, ¨uhne SK, Wade MR, Wratten SD, Wyss E (2007) Arthropod pest management in organic crops. Annu Rev Entomol 52:57–80

Forest Arthropod Communities in India: Their Role and Conservation

19

G. Mathew, K.P. Kumar, and M. Chandrashekaraiah

Abstract

Tropical forest arthropod fauna exhibits incredible diversity. Arthropods in forests are poorly documented in the tropics and subtropics. Because of this, forest arthropod's systematics and phylogeny are poorly documented. Since arthropods like spiders and millipedes are sensitive to changes and ecological disturbances, they are good indicators of environmental changes. Of arthropods, insects are dominant, and among insects, beetles are one of the most diverse and abundant groups on the planet. For every species of plants and birds in the forest, one can find 20, 83 and 312 species of arthropods, respectively. Arthropods carry out a vital role in ecosystem services and need to be conserved. But they are often ignored in conservation and management plans. Reducing habitat loss, human intervention, impact of global warming and invasive species would go a long way in arthropod conservation. Legislation, public participation, social acceptance and regional, national and international networking would substantially contribute to forest management in conserving biological diversity including arthropods.

G. Mathew (✉)
Forest Health Division, Kerala Forest Research Institute, Peechi 680653, Kerala, India
e-mail: mathew@kfri.org

K.P. Kumar
Department of Agricultural Entomology, University of Agricultural Sciences (UAS), Gandhi Krishi Vignana Kendra (GKVK), Bengaluru 560065, Karnataka, India
e-mail: kumarkp4830@gmail.com

M. Chandrashekaraiah
Zonal Office, Central Silk Technological Research Institute, Central Silk Board, 2nd Floor, Satyam Commercial Complex, Link Road, Bilaspur 495001, Chhattisgarh, India

A.K. Chakravarthy, S. Sridhara (eds.), *Economic and Ecological Significance of Arthropods in Diversified Ecosystems*, DOI 10.1007/978-981-10-1524-3_19

Keywords
Arthropods • Conservation • Diversity • Tropical forests

19.1 Introduction

India is unique in being the meeting place of the three major biogeographical realms, viz. Indo-Malayan, Eurasian and Afrotropical. As a result, the faunal and floral diversity of India is very high. India is one of the mega-diversity countries in the world on account of its unique biogeographic location, diversified climatic conditions and enormous eco-diversity. Altogether, about 67.701 million ha roughly constituting about 22.8 % of the total land area are under forests, out of which approximately 3.226 million ha represent 4.8 % forest plantations (FAO 2003). Sal, teak, mahogany, *Terminalia*, *Dalbergia*, *Grewia*, *Albizia* and pines are the major timber species in the natural forests. Plantations of a variety of tree species such as *Acacia* spp., *Eucalyptus* spp., *Albizia* spp., *Azadirachta indica*, *Casuarina equisetifolia*, *Dalbergia sissoo*, *Gmelina arborea*, *Populus* spp., *Prosopis* spp., *Shorea robusta*, *Terminalia* spp., *Cedrus deodara*, *Pinus roxburghii*, *P. patula*, *P. caribaea* and *Tectona grandis* are raised for meeting the timber requirements.

Arthropods except insects are not well documented in forest ecosystems in India. Even under insects, only Lepidoptera, Coleoptera and Hemiptera are well documented. Insects are grouped under 29 orders. The total number of insects is stated to be between 3 and 30 million. As per Gaston (1991), the estimate is around five million. This wide range in our estimates in insect numbers is mainly due to the uncertainty about their number in tropical forests since very few studies have been made in this region. Every year, about 10,000 new species of insects are being discovered worldwide.

No accurate estimates on the arthropod or insect fauna of India are available. According to Varshney (1998), 59,353 species forming nearly 6.83 % of the world insect fauna are already recorded from India (Tables 19.1 and 19.2; Fig. 19.2).

Table 19.1 Arthropod faunal diversity in India vis-à-vis world fauna

Taxonomic group	No. of species (world)	No. of species in India	Percentage in India
Onychophora	100	1	1.00
Crustacea	60,000	3,549	5.91
Insecta	1,020,007	63,423	6.22
Arachnida	73,451	5,850	7.96
Pycnogonida	600	17	2.83
Chilopoda	8,000	101	1.26
Diplopoda	7500	162	2.16
Symphyla	120	4	3.33
Merostomata	4	2	50.00
Arthropoda	1,181,398	74,175	6.28

Source: India's fifth national report to the Convention on Biological Diversity (CBD) (2014)

Table 19.2 Insect faunal diversity in India vis-à-vis world fauna

Insect order	No. of species in world	No. of species in India	Percentage of species in India
Thysanura	1,250	31	2.48
Diplura	355	16	4.50
Protura	260	20	7.69
Collembola	5,500	210	3.81
Ephemeroptera	2,200	106	4.81
Odonata	6,000	499	8.31
Plecoptera	2,100	113	5.38
Orthoptera	17,250	1,750	10.14
Phasmida	2,262	146	6.45
Embioptera	200	33	16.50
Dermaptera	2000	320	16.00
Blattaria	5,000	186	3.72
Mantodea	2,310	162	7.04
Isoptera	2,000	253	12.65
Psocoptera	2,500	90	3.60
Phthiraptera	3,000	400	13.33
Hemiptera	80,000	6,500	8.12
Thysanoptera	6,000	693	11.55
Neuroptera	5,000	335	6.70
Coleoptera	350,000	15,500	4.42
Strepsiptera	554	18	3.25
Mecoptera	350	15	4.28
Siphonaptera	2000	52	2.60
Diptera	100,000+	6,093	6.09
Lepidoptera	142,500	1,50,00	10.52
Trichoptera	7000	812	11.60
Hymenoptera	1,20,000	10,000	8.33
Total	867,391	59,353	6.83

According to Narendran (2001), about 60,000 insect species have been described from India, but many times more than this number are yet to be discovered.

19.2 Diversity of Forest Arthropods

Because of the availability of diversified ecosystems, the arthropod fauna of forests is highly diverse. Of the forest types, the tropical evergreen forests are seen in the Eastern Himalayas; wet evergreen and semievergreen forests are seen in the south along the Western Ghats, the Nicobar and Andaman Islands and all along the north-eastern region. Moist deciduous forests which are found throughout India harbour the maximum number of species. It is possible that majority of insects recorded from other habitats except for some specialized groups such as Grylloblattodea and Mantophasmatodea are present in the forests.

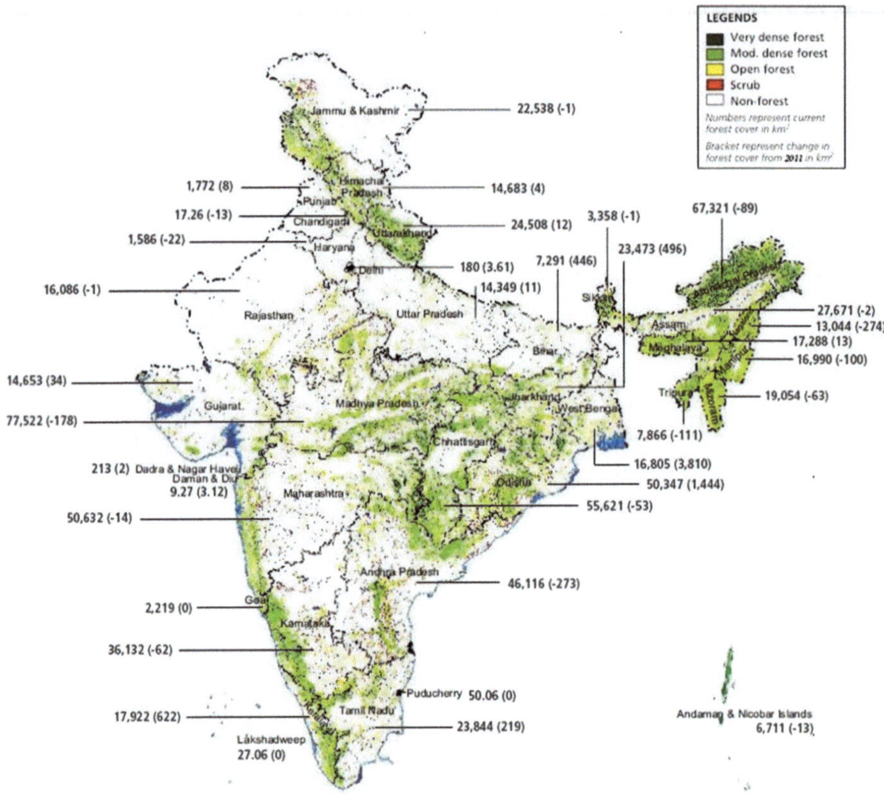

Fig. 19.1 Indian forest cover (Source: India State of Forest Report 2013)

19.2.1 Forest Insects

An excellent summary of work carried out on Indian forest insects is given by Nair (2007). The earliest publications on the forest insects of India seem to be on certain economically important insects such as termites (Koenig 1779) and lac insects (Kerr 1782). Several amateur entomologists who worked as army officers and doctors of the East India Company have made excellent contributions on the forest insects of India. This includes studies on the Indian moths by Hampson (1893–1896) and the work on Indian forest insects by Stebbing (1899) and Lefroy (1909). Stebbing, who was appointed as the first Forest Zoologist of the Imperial Forest Research Institute at Dehra Dun in 1906, published a monumental work on forest beetles of India (Stebbing 1914). Bhasin and Roonwal (1954) made a comprehensive list of 4,300 species of forest insects and their associated forest plants in India and its adjacent countries (Fig. 19.1). Another comprehensive reference book was published in 1968 (Browne 1968). Thakur (2000) prepared a book entitled *Ecology and Management of Forest Insects*. Nair (2007) consolidated all available information on tropical forest insect pests covering ecology and management options of the major pests associated with commercially important tree species (Fig. 19.2).

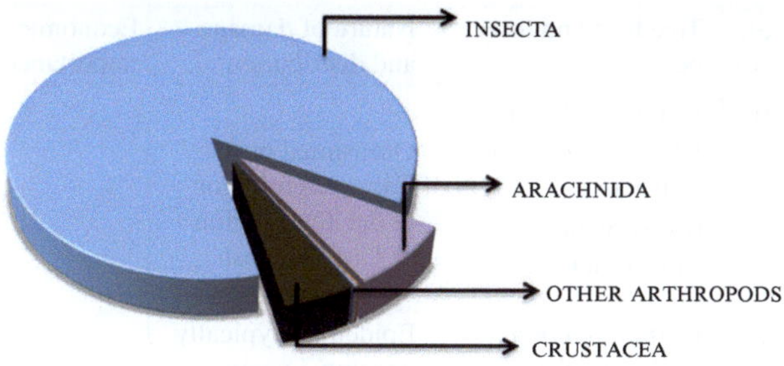

Fig. 19.2 Arthropod distribution in India

19.2.2 Insect Pests of Select Tree Species

Forest insects have been studied as pests of forest tree species. An account of major forest tree species and the insect pests associated with these species are presented below.

Sl. No.	Tree host and insect pest	Nature of damage and distribution	Economic importance	Alternative host
I. Teak (*Tectona grandis*)				
1	*Hyblaea puera* (Cramer) (teak defoliator) (Lepidoptera: Hyblaeidae)	Defoliation of teak over extensive areas annually causing	About 44 % losses in the potential volume increment (Nair 2001)	*Alstonia scholaris*, *Avicennia* spp., *Callicarpa* spp., *Pterocarpus macrocarpus*, *Rhizophora* spp., *Tectona grandis*, *Vitex* spp.
2	*Eutectona machaeralis* Walker (teak skeletonizer) (Lepidoptera: Pyralidae)	Major pest of teak. Larvae feed on leaf tissues leaving only the veins. Complete defoliation by the pests results in more or less leaflessness during most of the growing period	Causing loss of approximately 0.051 million ha annually (FAO 2003)	
II. Chir pine (*Pinus roxburghii*)				
1	*Cryptothecia crameri* Westwood (Chir pine defoliator) (Lepidoptera: Psychidae)	Serious pest of Chir pine. During 1989–1990, an outbreak of this pest was reported in Jammu and Kashmir causing tree mortality	Approximately 5 % tree mortality in about 2,000 ha area resulting in a net loss of 22.5 million rupees was reported (FAO 2003)	*Anthocephalus cadamba*, *Mangifera indica*, *Quercus incana*, *Q. serrata*, *Shorea robusta*, *Syzygium cumini*, *Terminalia arjuna*, *T. myriocarpa*

(continued)

Sl. No.	Tree host and insect pest	Nature of damage and distribution	Economic importance	Alternative host
III. Poplar (*Populus* spp.)				
1	*Chrysomela populi* Linnaeus (Coleoptera: Chrysomelidae)	Occasional pest. Causes defoliation in the Terai region of Uttar Pradesh and in Punjab		*Populus* spp., *Salix* spp.
2	*Clostera cupreata* Butler (poplar defoliator) (Lepidoptera: Notodontidae)	Epidemics typically develop 3 years after plantation establishment		
3	*Clostera fulgurita* (Walker) (poplar defoliator) (Lepidoptera: Notodontidae)	Defoliation. Epidemics typically develop 3 years after plantation establishment. Important pest of poplar plantations in Uttar Pradesh and Punjab		*Populus* spp.
4	*Apriona cinerea* Chevrolat (poplar stem borer) (Coleoptera: Cerambycidae)	Major pest. Generally infests young plants in the Northwest Himalayas		
IV. Deodar (*Cedrus deodara*)				
1	*Ectropis deodarae* Prout (deodar defoliator) (Lepidoptera: Geometridae)	A major pest causes defoliation of large areas of deodar forests in the Northwest Himalayas		
V. Spruce (*Picea* spp.)				
1	*Eucosma hypsidryas* Meyrick (spruce budworm) (Lepidoptera: Tortricidae)	Trees of all ages are attacked. Causes mortality of spruce trees in the Himalayas		
VI. Sal (*Shorea robusta*)				
1	*Hoplocerambyx spinicornis* (sal heartwood borer) (Coleoptera: Cerambycidae)	Serious pest of sal. Bores in the trunk causing mortality of the tree. Distributed in the central and northern regions		

(continued)

Sl. No.	Tree host and insect pest	Nature of damage and distribution	Economic importance	Alternative host
VII. Gamhar (*Gmelina arborea*)				
1	*Calopepla leayana* (gamhar defoliator) (Coleoptera: Chrysomelidae)	Throughout India. Young larvae feed mainly on the undersurface of gamhar (*Gmelina arborea*) leaves, leaving only the midribs and main veins intact. The adult beetle feeds on the leaf, cutting large circular holes, and also eats young buds and shoots	Heavy infestation leads to the drying up of shoots of young trees. The trees remain leafless for about 4 months of the growing season leading to ultimate death	
VIII. Mahogany (*Swietenia macrophylla*)				
1	*Hypsipyla robusta* Moore (mahogany shoot borer) (Lepidoptera: Phycitidae)	Major pest of mahogany. Caterpillars bore into the tips and shoots causing 100 % mortality of seedlings and young plantations (FAO 2003). The caterpillars destroy the apical shoot causing the tree to form many side branches and frequently a deformed trunk leading to a decreased value of the timber (FAO 2003)	Australia, Bangladesh, Nigeria, Pakistan, Sri Lanka and the West Indies. The caterpillars destroy the apical shoot causing the tree to form many side branches and frequently a deformed trunk leading to a decreased value of the timber (FAO 2003)	*Khaya* spp., *Cedrela* spp., *Cedrela toona*, *Toona ciliata*, *Tectona grandis*, *Swietenia macrophylla*
2	*Lymantria mathura* Moore (pink gypsy moth) (Lepidoptera: Lymantriidae)	No significant tree mortality occurs after defoliation of the sal tree, *Shorea robusta*, but tree vigour may be reduced and susceptibility to attack from other insect species may increase. In India, outbreaks are infrequent but extensive when they do occur		*Anthocephalus cadamba*, *Mangifera indica*, *Quercus incana*, *Q. serrata*, *Shorea robusta*, *Syzygium cumini*, *Terminalia arjuna*, *T. myriocarpa*

(continued)

Sl. No.	Tree host and insect pest	Nature of damage and distribution	Economic importance	Alternative host
3	*Lymantria obfuscata* Walker (Indian gypsy moth) (Lepidoptera: Lymantriidae)	*Lymantria obfuscata* is a defoliator of willows. Defoliation causes loss of increment. Trees may be killed if they are severely defoliated for more than 1 year (FAO 2003)		*Salix* spp.
Introduced insects				
1	*Heteropsylla cubana* Crawford (Homoptera: Psyllidae)	*Leucaena* psyllid, *Heteropsylla cubana*, appeared in Chengalpattu (Tamil Nadu) and caused severe defoliation and extensive death of young trees. By 1990, it had attacked all the *Leucaena* plantations in the country (FAO 2005b)	*Leucaena leucocephala* is a tree grown extensively in community forestry and agroforestry ecosystems for fodder and fuel throughout the tropics including India. The tree was almost pest-free in India until 1988	*Eucalyptus camaldulensis*, *E. tereticornis*, *E. grandis*, *E. deanei*, *E. globulus*, *E. nitens*, *E. botryoides*, *E. saligna*, *E. gunnii*, *E. robusta*, *E. bridgesiana*, *E. viminalis*
2	*Icerya purchasi* Maskell (Homoptera: Coccidae)	It was accidentally introduced into India in 1921 (FAO 2005b). The scale has done serious damage to plants in the Nilgiri hills and has since become well established throughout the country (FAO 2005b). *Rodolia cardinalis* (Coleoptera: Coccinellidae) was introduced for the control of this scale	It damages *Acacia decurrens* and *A. dealbata* besides several other plant species	*Acacia decurrens* and *A. dealbata*

(continued)

Sl. No.	Tree host and insect pest	Nature of damage and distribution	Economic importance	Alternative host
3	*Leptocybe invasa* Fisher and La Salle (the blue gum chalcid) (Hymenoptera: Eulophidae)	It is a gall-inducing wasp. It lays eggs in the bark of shoots or in the midribs of leaves. The eggs develop into minute, white, legless larvae within the host plant producing galls on the midribs, petioles and twigs. Severely attacked trees show stunted growth and dieback	A serious pest of eucalypts. Native of Australia. It got introduced to planted eucalypt forests in various parts of the world including Kenya, Morocco, New Zealand, Tanzania, Uganda and India (Jacob et al. 2007)	*Eucalyptus camaldulensis*, *E. tereticornis*, *E. grandis*, *E. deanei*, *E. globulus*, *E. nitens*, *E. botryoides*, *E. saligna*, *E. gunnii*, *E. robusta*, *E. bridgesiana*, *E. viminalis*
4	*Pineus pini* (Gmelin) (pine woolly aphid) (Hemiptera: Adelgidae)	The pine woolly aphid feeds on the shoots of *Pinus* spp., causing apical shoot dieback	First introduced to India in the 1970s. Severe damage to *Pinus patula* plantations in the Nilgiri hills has been reported (FAO 2005b)	*Pinus* spp. *Pinus patula*
5	*Quadraspidiotus perniciosus* (Comstock) (the San Jose scale) (Homoptera: Coccidae)	Generally attacks wood, but, in severe infestations, leaves and fruits may also be penetrated. The bark often cracks and exudes gum, resulting in a surrounding dark-brown gelatinous area	Native of China. *Quadraspidiotus perniciosus* reached India in 1911 and by 1933 had attained pest status in fruit orchards and plantations of poplars and willows (FAO 2005b)	*Populus* spp., *Salix* spp., *Aesculus* spp., *Alnus* spp., *Betula* spp., *Celtis* spp., *Fagus* spp., *Fraxinus* spp., *Morus* spp.

19.3 Arachnida

19.3.1 Spiders

Mangalavanam, an eco-sensitive mangrove forest in Kerala, is home for 51 species of spiders where *Pisauragitae* Tikader is the dominant species and the genus *Tapponia* is reported for the first time from India (Sebastian et al. 2006). The great Indian spider, *Ischnocolus* (*Poecilotheria*) *regalis* Pocock, is present in Vazhachal

Fig. 19.3 Giant wood spider (forest spider) *Nephila maculate* Fab (Source: Dinesh Valke)

forest in Kerala (Cheeran and Nagaraj 1997). A preliminary study on the spider fauna in Mannavan shola forest, Kerala, by Sudhikumar et al. (2005) revealed 72 species of spiders belonging to families Araneidae, Tetragnathidae, Salticidae and Thomisidae. Shegokar's (2012) investigation in Katepurna Sanctuary in Maharashtra revealed 16 families, 37 genera and 74 species of spiders. Prominent among them were *Phonognatha graeffei* Keyserling and *Cyclosa insulana* Costa. In Toranmal Sanctuary, Maharashtra, 17 spider species from 20 families and 55 genera were observed. *Neoscona* was the most abundant genus recorded (Meshram 2011).

De (2001) listed 19 species of spiders from the Dudhwa Tiger Reserve. The giant wood spider (forest spider) *Nephila maculata* (Fabricius) (Araneae: Araneidae) is the largest orb-weaving spider in India (Fig. 19.3). Surveys carried out from August 1997 to August 1999 in the forested tracts along the River Godavari in Kawal Wildlife Sanctuary and Eturnagaram Wildlife Sanctuary revealed an abundant number of giant wood spiders (Srinivasulu 2000). In Nanda Devi Biosphere Reserve (NDBR), Uttarakhand, in Western Himalaya, 244 species belonging to 108 genera and 33 families were collected by Uniyal et al. (2011).

19.3.2 Mites and Ticks

In India, Acarina is the soil and litter microarthropod. They are minute free-living arthropods and are the most abundant and dominant group in the soil-litter subsystem and play an important role in sustaining the forest ecosystem. They decompose and mineralize the leaf litter and thus maintain the edaphic factors in balance.

Oribatid mites *Tectocepheus velatus* Michael, *Lamellobates palustris* Hammer and the species of genus *Scheloribates* were dominant in Bodaganj forest in West Bengal (Moitra 2013). *Dermacentor auratus* Supino tick species, a carrier of Kyasanur forest disease (KFD), has been recorded from hosts like man, cattle, deer, buffalo, wild boar and small mammals in India (Fig. 19.4; Pattnaik 2006).

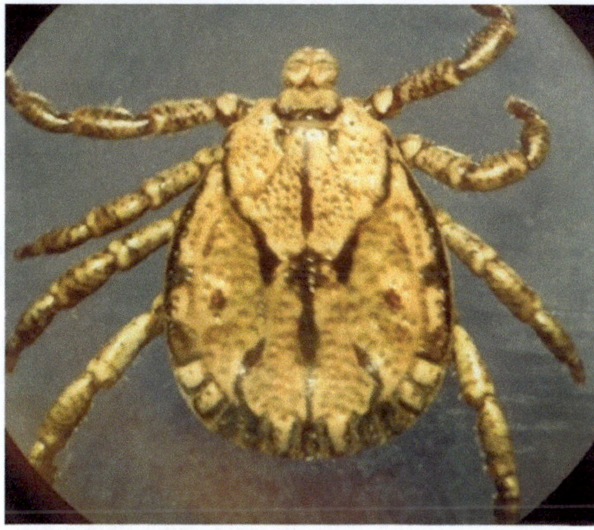

Fig. 19.4 *Dermacentor auratus* Supino, 1897 (Source: K G Ajithkumar)

Fig. 19.5 *Scorpiops leptochirus* (L), *Euscorpiops bhutanensis* (R) (Source: Aamod Zambre)

19.3.3 Scorpions

The Indian red scorpion is considered as one of the five most lethal scorpions in the world. A systematic account of five species of scorpions (Arachnida) was collected during a survey in Eaglenest Wildlife Sanctuary (EWS) in Arunachal Pradesh. *Euscorpiops bhutanensis* Tikader is a new record in India. Some other important species of this sanctuary are *Euscorpiops thenurus* Pocock, *Chaerilus pictus* Pocock and *Scorpiops leptochirus* (Fig. 19.5) (Zambre 2008). The buthid scorpion belonging to the genus *Lychas* is described from the degraded scrub of Sanjay Gandhi National Park, Mumbai (Mirza and Sanap 2010).

19.4 Diplopoda

In forests, millipedes are known to ingest 20–100% plant detritus and return 60–90% organic matter in the form of faecal pellets. They are conservative and sensitive to water deficiency and fail to overcome the limitation of even a single edaphic factor particularly soil texture and litter thickness.

Diversity of millipedes was recorded in Alagar Hills Reserve Forest in Tamil Nadu where the millipede species *Harpaphe haydeniana* Wood, *Xenobolus carnifex* (Fabricius), *Arthrosphaera magna*, *Aulacobolus newtoni* and *Spinotarsus colosseus* (Attems) were common (Alagesan and Ramanathan 2013). Four species were recorded in this forest (Figs. 19.6, 19.7 and 19.8). Diversity of the arthropods was reported in an evergreen forest of Rajgurunagar ecosystem of the Northern Western Ghats, viz. *Harpaphe haydeniana* Wood, *Narceus americanus* (Palisot de Beauvois), *Oxidus gracilis* (C. L. Koch) and *Trigoniulus corallinus* (Gervais) (Choudhari et al. 2014).

Fig. 19.6 *Harpaphe haydeniana* (Source: Wikipedia)

Fig. 19.7 *Arthrosphaera magna* (Source: Saandip Nanadagudi)

Fig. 19.8 *Aulacobolus newtoni* (Source: Barathan. N)

Fig. 19.9 *Uca* (*Austruca*) *lacteal annulipes* (Source: Chinmayisk)

19.5 Crustacea

The Mahi and Dhadhar estuarine region, Gujarat, is covered by mangrove forests. A total of 14 species of brachyuran crabs were recorded belonging to 11 genera and 9 families. Crab species, viz. *Uca* (*Austruca*) *lacteal annulipes* (Fig. 19.9), *Uca* (*Tubuca*) *dussumieri*, *Ashtoret lunaris*, *Scylla serrata*, *Parasesarma plicatum*, *Macropthalmus* (*Venitus*) *dentipes*, *Macrophthalmus* (*Mareotis*) *depressus* and *Cardisoma carnifex*, were reported from all sites and were common in the area (Figs. 19.9 and 19.10) (Shukla et al. 2013). The freshwater crab fauna in the forest of Kerala and Tamil Nadu are *Baratha pushta*, *Baratha peena*, *Vanni ashini*, *Vanni deepta*, *Vanni giri*, *Pilarta anuka*, *Snaha aruna*, *Travancorian akuleera*, *Travancorian acharu and Vela virupa.*

Fig. 19.10 *Scylla serrata* (Source: Wikipedia)

19.6 Onychophora

Onychophorans generally inhabit dark and moist microhabitats like forest litter soil and rotten logs by feeding on small invertebrates which are captured with the help of an adhesive substance. *Typhloperipatus williamsoni* Kemp is one of the important species from India (Monge-Najera 1995). Of the forest insects associated with different forest trees, nearly 20 species of insects are known to cause potential damage (Mohandas et al. 1990; FAO 2003, 2005a; Singh et al. 2005). As far as pest management practices are concerned, strategies involving chemical, biological and silvicultural measures are attempted to protect the forest plantations of commercially important tree species such as teak, mahogany and pines although no pest management strategies are adopted in natural forests. But arthropods have a vital role in ecosystem services inside the forests that include plant litter decomposition and humus formation (Giller 1996) but have unfortunately been ignored in conservational studies. They are also sensitive to disruption of their environments as well as specific to their altitudinal gradients.

19.7 Arthropod Management and Conservation

Tree felling destroys the habitats of spiders that inhabit tree trunks. Spiders can be conserved by making students, researchers and people aware about keeping the atmospheric carbon dioxide (CO_2) concentration below 350 ppm and maintaining the ecosystems of spiders lively and healthy. The primary cause of the decline of spider diversity is by habitat destruction. There is a loss of habitat of *Thrigmopoeus truculentus* Pocock in the Western Ghats. This spider is categorized as 'Near Threatened' as it does not meet the 'restricted distribution criteria'. At present, most of the spider species in India are classified as Data Deficient (Vankhede 2001). Preservation of spider diversity and better land management strategy require an understanding of the patterns of spider ecology at an appropriate regional scale (Skerl and Gillespie 1999).

Millipedes are conservative and sensitive to water deficit and fail to overcome the limitation of even a single edaphic factor particularly soil texture and litter thickness. The climate desiccation and human interference are the major threats leading to millipede extinction. Limited mobility of millipedes results in high degree of speciation and endemism. Many diplopod species are microendemics. So, they are vulnerable to the destruction of their natural habitat (Choudhari et al. 2014) and need to be conserved.

A large number of freshwater shrimps and crabs in the Western Ghats are in danger of becoming extinct due to increasing human activity in their habitat and irregularities in their classification and documentation. More than 90 % of gecarcinucid crabs found in the Western Ghats are endemic to the area (www.indiasendangered.com). The Western Ghats hotspot is confirmed as a globally significant centre of diversity and endemism for freshwater species. In the Western Ghats, about 16 % of the 1,146 freshwater taxa assessed are threatened with extinction, with a further 1.9 % assessed as near threatened. Although many protected areas are located within or near areas of the richest freshwater diversity, the region experiences the highest level of threat to freshwater species. So inventories and monitoring of freshwater fauna, impressed enforcement of pollution laws, environmental assessments and awareness programmes are urgently required in the Western Ghats region.

References

Alagesan P, Ramanathan B (2013) Diversity of millipedes in Alagar hills reserve forest in Tamil Nadu, India. Int J Biodiv 2013:5

Bhasin GD, Roonwal ML (1954) A list of insect pests of forest plants in India and the adjacent countries, part 1 and 2. Indian For Bull (New Series) 171(1):1–93

Browne FG (1968) Pests and diseases of forest plantation trees: an annotated list of the principal species occurring in the British commonwealth. Clarendon, Oxford

Cheeran JV, Nagaraj BN (1997) Occurrence of the great Indian spider, *Ischnocolus* (= *Poecilotheria*) *regalis* in Vazhachal forest, Kerala. Zoos Print 12(1).28–29

Choudhari CR, Dumbare YK, Theurkar SV (2014) Diversity of millipedes along the Northern Western Ghats, Rajgurunagar (MS), India (Arthropod: Diplopod). J Entomol Zool Studies 2(4):254–257

Convention on Biological Diversity (2014) Biodiversity profile of India. India's fifth national report to the convention on biological diversity, part-1. Available at: https://www.cbd.int/doc/world/in/in-nr-05-en.pdf

FAO (2003) Global forest resources assessment updates 2005 – FRA 2005 – Pilot study for country reporting: India. Forest Resources Assessment working paper 78. Available at: http://www.fao.org/docrep/007/ae354e/AE354E00.htm

FAO (2005a) Global forest resources assessment 2005 – India – country report. Forestry department, forest resources assessment 2005, Country Report 001, p 128

FAO (2005b) The unwelcome guests. Proceedings of the Asia-Pacific forest invasive species conference. In: McKenzie P, Brown C, Jianghua S, Jian W (eds) RAP Publication 2005/18. Bangkok, Thailand, FAO Regional Office for Asia and the Pacific, Available at: http://www.fao.org/docrep/008/ae944e/ae944e00.htm pp 172

Gaston JK (1991) The magnitude of global insect species richness. Conserv Biol 5:283–296

Giller SP (1996) The diversity of soil communities, the 'poor man's tropical rainforest'. Biodivers Conserv 5(2):135–168

Hampson FG (1983–96) Fauna of British India, Moths, vols 1–4. Taylor & Francis Ltd, London
Jacob JP, Devaraj R, Natarajan R (2007) Outbreak of the invasive gall-inducing wasp *Leptocybe invasa* on eucalypts in India. Invasives 8:4, Newsletter of the Asia- Pacific Forest Invasive Species Network (APFISN)
Kerr J (1782) Natural history of the insect which produces the gumlacca. Philos Trans R Soc Lond 71:374–382
Koenig JC (1779) Beschaftigungen der Berlinischen Gesellschaft Naturforchenderderunde. English translation: Fletcher TB (1921) Proceedings of the 4th Entomology Meet. Pusa, India, pp 312–333
Lefroy HM (1909) Indian insect life: a manual of the insects of the plants (Tropical India). Thacker, Spink & Co, Calcutta
Meshram A (2011) Spiders (Arachnida: Araneae) from Toranmal Sanctuary, Maharashtra, India. E-Int Sci Res J 3(4):2094–1749
Mirza ZA, Sanap RV (2010) Description of a new species of scorpion of the genus *Lychas* C.L. Koch, 1845 (Scorpiones: Buthidae) from Maharashtra, India. J Threatened Taxa 2(4):789–796
Mohandas K, Mathew G, Nair KSS, Menon ARR (1990) Pest incidence in natural forests – a study in moist deciduous and evergreen forests of India. In: Pests and diseases of forest plantations in the Asia-Pacific Region. Proceedings of the IUFRO Workshop. FAO, Regional Office for Asia and the Pacific (RAPA), Publication 1990/9, pp 129–134
Moitra MN (2013) On variation of diversity of soil oribatids (Acari, Oribatida) in three differently used soil habitats – a waste disposal site, a natural forest and a tea garden in the northern plains of Bengal, India. Int J Sci Res 3(11):1–12 (ISSN: 2250-3153)
Monge-Najera J (1995) Phylogeny, biogeography and reproductive trends in the Onychophora. In: Walker MH, Norman DB (eds) Onychophora: past and present. Zool J Linnean Soc 114:21–60
Nair KSS (2001) Pest outbreaks in tropical forest plantations. Is there a great risk for exotic tree species? Centre for International Forestry Research (CIFOR), Jakarta
Nair KSS (2007) Tropical Forest insect pests: ecology, impact and management. Cambridge University Press, New York, p 404
Narendran TC (2001) Taxonomic entomology: research and education in India. Curr Sci 81:445–447
Pattnaik P (2006) Kyasanur forest disease: an epidemiological view in India. Rev Med Virol 16:151–165
Sebastian PA, Murugesan S, Mathew MJ, Sudhikumar AV, Sunish E (2006) Spiders in Mangalavanam, an ecosensitive mangrove forest in Cochin, Kerala, India (Araneae). Acta Zool Bulg 1:315–318
Shegokar MM (2012) Spider diversity of Katepurna Sanctuary District Akola, Maharashtra. India 1(2):103–106
Shukla ML, Patel BK, Trivedi JN, Vachhrajani KD (2013) Brachyuran crab diversity of Mahi and Dhadhar estuaries, Gujarat, India. Res J Mar Sci 1(2):8–11
Singh AP, Bhandari RS, Verma TD (2005) Important insect pests of poplars in agroforestry and strategies for their management in northwestern India. Agroforest Syst 63(1):15–26
Skerl KL, Gillespie RG (1999) Spiders in conservation – tools, targets and other topics. J Insect Conserv 3:249–250
Srinivasulu C (2000) Giant wood spider *Nephila maculate* fabr. in forested tracts along river Godavari in Andhra Pradesh. Zoos' Print J 15(3):229–230
Stebbing EP (1899) Injurious insects of Indian forests. Government of India, New Delhi
Stebbing EP (1914) Indian forest insects of economic importance: coleoptera. Eyre and Spottiswoode Ltd, London
Sudhikumar AV, Mathew MJ, Sunish E, Murugesan S, Sebastian PA (2005) Preliminary studies on the diversity of spider fauna in Mannavan shola forest, Kerala, India. Acta Zool Bulg Suppl (1):319–327

Thakur ML (2000) Forest entomology: ecology and management. Sai Publishers, Dehra Dun
Uniyal VP, Sivakumar K, Quasin S (2011) Diversity of spiders in Nanda Devi biosphere reserve. Wildlife Institute of India, Dehradun (DST Project Completion Report)
Vankhede GN (2001) Conservation of Spiders in India. ENVIS Bulletin: arthropods and their conservation in India (Insects & Spiders). 14(1):54–59
Varshney RK (1998) Faunal diversity in India. In: Alfred JRB, Das AK, Sanyal AK (eds) Zoological Survey of India, Calcutta www.indiasendangered.com by Ramya Naresh, pp 145–228
Zambre A (2008) http://indianscorpions.blogspot.in/ Preliminary Report of Scorpion Fauna of Eagle nest Wildlife Sanctuary

Awareness on Pesticide Residues in Food Crops: A Challenge

20

G.V. Ranga Rao, B. Ratna Kumari, K.L. Sahrawat, and S.P. Wani

Abstract

A review of field research conducted at the ICRISAT in collaboration with national partners to monitor the insecticide residues on food crops (rice, chickpea, maize, pigeon pea) and vegetables and the recent studies by other researchers in this field revealed the presence of residues of selected pesticides on crops. These include monocrotophos, chlorpyrifos, endosulfan, and cypermethrin. Only 3 % of the rice had beta endosulfan residues, while 35 % of tomato and 56 % brinjal had residues of these insecticides; however, only 4 % of the samples had residues above the maximum residue limits (MRLs). The crop samples analyzed (56) for pesticide residues in 15 contact (nonchemical pesticide group of 41 farmers) and 5 noncontact (15 samples) villages revealed the presence of pesticide residues in 21 samples above 0.001 ppm, except for two *Dolichos* bean and tomato samples which had residues of monocrotophos and chlorpyrifos above MRL. Though the residues in pulses were observed at harvest, they were below detectable levels after processing, i.e., thrashing and splitting the seed into dhal, indicating their safety in food chain; however, their haulms had insecticide residues. Fields under integrated pest management (IPM) showed substantial reduction in pesticide use across crops, which reflected in the occurrence of low residues. Twenty percent of brinjal and tomato samples had residues compared to 47 % in non-IPM fields. Though the contamination levels in crops in IPM and non-IPM fields indicated substantial differences, the residue concentrations were below MRLs indicating safety to beneficial arthropods.

G.V.R. Rao (✉)
International Crops Research Institute for the Semi-Arid Tropics, Hyderabad, India
e-mail: g.rangarao@cgiar.org

B.R. Kumari • K.L. Sahrawat • S.P. Wani
Acharya NG Ranga Agricultural University, Hyderabad, India

A.K. Chakravarthy, S. Sridhara (eds.), *Economic and Ecological Significance of Arthropods in Diversified Ecosystems*, DOI 10.1007/978-981-10-1524-3_20

Keywords
Food crops • IPM • Pesticide residues • Plant protection

20.1 Introduction

Insects and pathogens are the integral part of all agricultural systems and are normally present at relatively low densities, causing little damage and having negligible impact on crop growth and vigor under natural competition. However as the pressure to increase productivity to meet the growing demand increases, some species outgrow rapidly, resulting in outbreaks. Such large populations may have adverse effects on crops and affect the livelihoods of farming communities.

Increasing human population and food demand are placing unprecedented pressure on agriculture and natural resources. But there is no point in producing food that is unsafe or will not last long enough to utilize. Chemicals should be used only if their benefits (in food production and storage) outweigh the risks to the people and livestock from their residues. Agricultural chemicals help countries to economically and efficiently feed their people and livestock. It is estimated that globally pests account for more than 40 % of the preharvest crop losses despite the use of chemicals. Of these losses, 15 % are attributable to insects and 13 % each to weeds and pathogens. An additional 10 % is lost during postharvest processing of the crop (Anon 2007: Pimentel 2009).

Chemical Pesticides Approximately 2.5 million tons of pesticides are used in agriculture annually throughout the world (Meena et al. 2008). Pesticides play an important role in agriculture, protecting the crops, improving the quality, and reducing the labor. In India, so far more than 240 insecticides/pesticides have been registered under Insecticide Act of 1968. Though pesticides over the years helped in increased food production, their injudicious use resulted in residues in food and feed, environmental pollution, pest resistance, resurgence, and reduction in nontargets. Of these ill effects, the presence of pesticide residues is of major concern especially to beneficial arthropods like pollinators, parasites, and predators.

The advantages of pesticides were realized early and only subsequently that the ill effects were felt. So the use of pesticides suddenly increased from 2.2 g ha^{-1} active ingredient (a.i.) in 1950 to 381 g ha^{-1} by 2007, i.e., about 270-fold (Anon 2009) of which more than 50 % is used on cotton and vegetables which are grown in less than 5 % area of cultivated crops. For effectiveness, agricultural chemicals must persist for a long period in the substrate. Persistence of toxic pesticide residues in fruits and vegetables when consumed fresh might create health hazards to the consumers due to their toxic residues. For example, contact with clothes made from plant fibers containing DDT (dichlorodiphenyltrichloroethane) residues can cause skin problems. Insecticides with higher doses in breast milk were affected by life habits of people in Greece. Women who consumed seven (or more) portions of fresh vegetables per week had gamma-BHC concentrations in breast milk that exceeded 0.15 mu g/l (odds ratio = 1.23 [95 % confidence interval = 1.05, 1.44]; $p = 0.006$). The loads of DDT derivatives were associated with the portions of fish, chicken, fruits, milk, and potatoes consumed each week (Schinas et al. 2000). The studies were from Anupgarh in Rajasthan, where intensive agriculture was practiced and growers applied heavy doses of pesticides to increase crop productivity; the exposure of humans to the toxic chemicals directly in the fields and indirectly through contaminated diet resulted in the deposits of organochlorine residues in blood (3.3–6.3 mg l^{-1}) and milk (3.2–4.6 mg l^{-1}) of the lactating women (Kumari et al. 2005). These chemical residues get concentrated in food chain and result in adverse affects on arthropods. Heavy doses of chemical residues (15–605 times) were recorded in blood samples of cotton farmers from four villages in Punjab, North India (Anon 2005).

There have been several investigations determining the ill effects of pesticide exposure (McCauley et al. 2006). The World Health Organization and the United Nations Environment Programme (UNEP) estimate that each year, three million farm workers in the developing world experience severe pesticide poisoning of which about 18,000 were fatal (Miller 2004). The study with 23 school children shifted to organic food from routine food dramatically reduced the levels of organophosphorus pesticide levels (Lu et al. 2006).

As per Pesticide Action Network Asia Pacific (1999), about 51 % of food material is contaminated with residues in the developing world countries, 21 % worldwide, of which 20 % were above MRL prescribed by the Food and Agriculture Organization (FAO) standards (Anon 1999). This means, 20 % of food is unfit for human consumption and still being consumed in the developing countries. An indiscriminate use of pesticides has increased mortality and morbidity of humans in the developing countries (Wilson and Tisdell 2001). The World Health Organization and the United Nations Environment Programme (UNEP) estimated that every year, three million agricultural workers in the developing world experienced severe pesticide poisoning in which about 18,000 were fatal (Miller 2004). An excessive dependence on the chemical pesticides also leads to the resistance development in insect pests to insecticides (Kranthi et al. 2002).

Information from India revealed that about 51 % food material is polluted with pesticide residues compared to 21 % worldwide, of which 20 % were above MRL (Anon 1999). This toxic food was not discarded in developing countries, but was consumed due to ignorance and innocence. Lack of awareness on the consequences of pesticide-contaminated food could be one of the reasons for increased incidences of cancers in developing countries. Besides the damage to human health, indiscriminate use of toxic chemicals adversely affects the natural biodiversity that resulted in the reduction of natural enemies and other beneficials (Ranga Rao et al. 2005).

Food for Thought In order to utilize the full potential of pesticides in agriculture and public health without adverse effect on the environment, it is essential to study the facts about pesticide behavior and their persistence under different environmental conditions. In general, about 50 % of the chemical pesticides that are applied to the crops directly go into the soil and other non target species. The chemical residues from the soil find their way to the aquatic systems or accumulate in plant products or aquatic arthropods and other fauna. Chlorinated compounds are more persistent in nature than organophosphorus, carbamate, and pyrethroid compounds. The basic problem in plant protection is that the safety measures are over looked and the residues are not monitored (Ranga Rao et al. 2002). Pesticide residues in food commodities and their entry into food chain have become a major concern all over the world. With a view to develop protocols for safe use of pesticides, good agricultural practices (GAPs), fixing maximum residue levels and the preharvest interval/safe waiting period through the analysis of residues on various crops and environment came into existance. Due to increased public awareness and legality aspects to meet the exports, monitoring pesticide loads in food materials is mandatory to address the safety of food feed and environment (Sharma 2013).

Analysis of 26,932 samples of vegetables and 10,419 samples of fruits during 2013 in India indicated the presence of pesticides above MRLs in 2.7 % of the samples in vegetables and 0.8 % in fruits. Consolidated data of 76,000 samples of food commodities brought out the presence of residues in 1.6 % samples above MRLs. This type of monitoring to regulate pesticide residues in food commodities and also to eliminate the trade barriers following pre-and postharvest measures plays an important role to address the existing ill effects.

The main objectives of residue monitoring are to:

- Monitor contaminants in products
- Detect any unauthorized treatments
- Establish reliable measures for detection
- Ensure healthy products for exports
- Establishment of state-of-the-art laboratories with appropriate human capacities

The MRLs have been established for pesticide residues in food for individual compounds; however, it is important to note that consumers often ingest several residues by taking various foods. Residues of several active substances interacting with each other are also found on fruits and vegetables. The cumulative risk assessment including defining MRLs needs to be well established; otherwise, the ongoing individual MRLs may not have much value in protecting the human health.

Studies in Asia Studies organized to monitor the incidence of residues in plant products from Asia indicated their presence in fruits and vegetables from Southeast Asia. A total of 721 samples of 63 different commodities in Southeast Asia were collected in 2011. The products were imported to Denmark, Finland, Norway, and Sweden from ten countries (about 80 % were imported from Thailand). In 60 % of the samples, residues were not found, while 28 % had residues below or at the

MRLs. Results also revealed that the MRLs were found in 12 % of the samples (Table 20.1). In comparison 6 % of surveillance samples from third countries and 1.5 % of surveillance samples from the EU and EFTA countries exceeded the MRL in the EU monitoring program in 2011 (Skretteberg et al. 2015).

People across the globe are getting increasingly alerted on food safety problems and health risks accompanying pesticide residues, a fact which is justified by the higher increase in public demands for organic products in Norway and several countries and the popularity of safe vegetable shops in Vietnam. Whereas agricultural imports cover around 50 % of domestic agricultural consumption in Norway, Vietnam is a major exporter of paddy, coffee, nuts, and fruits and has a great export potential in the European market. However, the product must comply with the EurepGAP regulations to be accepted by Europeans (Marit et al. 2006). By adopting good agricultural practices, the risks in agricultural production – e.g., pesticide concerns – will become less. In Norway, regulating pesticide residues in domestic and imported fruit and vegetables brought out 250 samples contaminated with pesticides out of 1,600 samples that were analyzed; however, only 2.2 % had residues at a level above the maximum permitted MRL (Mattilsynet 2009). It was found that the imported samples contained higher pesticides (51 %) than the domestic produce (21 %). Imported vegetable samples from Asia (Thailand, Vietnam, India, Pakistan, and Sri Lanka) were of particular concern, as 24.5 % of the samples had residues above MRL. Eight out of 14 samples from Vietnam had pesticide residues, and three samples (21 %) had levels above MRL (cypermethrin, chlorpyrifos, and endosulfan). Still to safeguard the exports, a national monitoring program for residues in crops is yet to be functional in Vietnam.

Myanmar, in early 1990s, had trade problems concerning pesticide residues in food. The residues detected were mainly organochlorine. The information until 2,000 revealed that in the early 1990s the violations of MRLs (National and Codex) were due to organochlorine pesticide residues. With the enactment of Pesticide Law in recent years, the use and import of several organochlorine pesticides has been banned or restricted in this country. Currently, the pesticides applied for crops are mainly rapidly deteriorating organophosphate (OPs) and synthetic pyrethroids; as a result the loads in food crop cultivated in Myanmar are well below the MRLs established by the joint WHO/FAO Codex Alimentarius Commission (Mya Thwin and Thet Thet Mar 2002).

Studies in India Studies on the pesticide residues in vegetable (brinjal, cucumber, okra, ridge gourd, and tomato) and water samples collected from Kothapally, Adarsha watershed in Ranga Reddy district, Telangana, India, during 2007, detected monocrotophos (range 0.001–0.044 mg kg^{-1}), chlorpyrifos (0.001–5.154 mg kg^{-1}), cypermethrin (0.001–0.352 mg kg^{-1}), and endosulfan (0.001–0.784 mg kg^{-1}). The residues of monocrotophos and endosulfan were below MRL in all the 59 vegetable samples, while the loads of chlorpyrifos were above MRL in four samples and cypermethrin in two samples. A comparison of the residue levels in test samples of the above five vegetable crops indicated higher residues of chlorpyrifos and

Table 20.1 Data on pesticide residue levels from crop samples in contact and noncontact villages under DM 2005 project (ICRISAT)

SNo	Crop	Farmer name	Pesticide residue levels (ppm)			
			Chlorpyrifos	Cypermethrin	Endosulfan	Monocrotophos
Contact villages						
1	Brinjal	G Krishna	<0.001	<0.001	<0.001	**0.011**
2	Ridge gourd	G Gopal	<0.001	<0.001	<0.001	<0.001
3	Tomato	G Ramaiah	<0.001	<0.001	<0.001	<0.001
4	Carrot	G Gopal	<0.001	<0.001	<0.001	<0.001
5	Chillies	Ch Narsimha Reddy	<0.001	<0.001	<0.001	<0.001
6	Brinjal (white)	Ch Narsimha Reddy	<0.001	<0.001	**0.033**	<0.001
7	Ridge gourd	D Chandraiah	<0.001	**0.060**	<0.001	<0.001
8	Brinjal	D Ailaiah	<0.001	<0.001	<0.001	<0.001
9	Tomato	D Chandraiah	<0.001	<0.001	<0.001	<0.001
10	Brinjal (white)	D Ailaiah	<0.001	<0.001	**0.041**	<0.001
11	Tomato	M Galaiah	<0.001	<0.001	<0.001	<0.001
12	Brinjal (white)	M Galaiah	<0.001	<0.001	**0.031**	<0.001
13	*Dolichos* (beans)	M Galaiah	<0.001	<0.001	<0.001	<0.001
14	Ridge gourd	M Chandraiah	<0.001	<0.001	<0.001	<0.001
15	Carrot	M Chandraiah	<0.001	<0.001	<0.001	<0.001
16	Onion	M Chandraiah	<0.001	<0.001	<0.001	<0.001
17	Tomato	Sailu	<0.001	<0.001	<0.001	**0.105**
18	Brinjal	Sailu	<0.001	<0.001	**0.017**	**0.090**
19	*Dolichos* (beans)	Sailu	<0.001	<0.001	**0.028**	**0.697**
20	Chillies	Sailu	<0.001	<0.001	<0.001	**0.012**
21	Ridge gourd	G Muthyam Reddy	<0.001	**0.026**	<0.001	<0.001
22	Tomato	G Muthyam Reddy	**0.126**	**0.132**	<0.001	<0.001
23	Brinjal (white)	S Raghava Reddy	<0.001	<0.001	**0.020**	<0.001
24	Cucumber	S Anji Reddy	<0.001	**0.033**	<0.001	<0.001
25	Tomato	B Siddimallaiah	<0.001	<0.001	<0.001	<0.001
26	Ladies finger	B Siddimallaiah	<0.001	<0.001	<0.001	<0.001
27	Ridge gourd	B Siddimallaiah	<0.001	<0.001	<0.001	<0.001

(continued)

Table 20.1 (continued)

SNo	Crop	Farmer name	Pesticide residue levels (ppm)			
			Chlorpyrifos	Cypermethrin	Endosulfan	Monocrotophos
28	Ridge gourd	R Raji Reddy	<0.001	<0.001	<0.001	<0.001
29	Cucumber	R Raji Reddy	<0.001	<0.001	<0.001	<0.001
30	*Dolichos*	B Bhaskar Reddy	<0.001	<0.001	<0.001	<0.001
31	Tomato	P Ravinder Reddy	<0.001	<0.001	<0.001	<0.001
32	Ridge gourd	M Narsimha Reddy	<0.001	<0.001	<0.001	<0.001
33	Cucumber	M Narsimha Reddy	<0.001	<0.001	<0.001	<0.001
34	Ladies finger	M Narsimha Reddy	<0.001	<0.001	<0.001	<0.001
35	Chillies	K Pomya	<0.001	<0.001	<0.001	<0.001
36	Tomato	K Narya	<0.001	<0.001	<0.001	<0.001
37	Tomato	D Anjaiah	<0.001	<0.001	<0.001	<0.001
38	Chillies	M Jangaiah	<0.001	<0.001	<0.001	<0.001
39	Tomato	B Sugunamma	<0.001	<0.001	<0.001	<0.001
40	Brinjal	P Jangamma	<0.001	<0.001	<0.001	<0.001
41	Tomato	Shiva Ramulu	<0.001	<0.001	<0.001	<0.001
Noncontact villages						
1	Tomato	–	<0.001	<0.001	<0.001	<0.001
2	Ladies finger	–	<0.001	<0.001	<0.001	**0.009**
3	Brinjal	–	<0.001	<0.001	**0.023**	<0.001
4	Brinjal (white)	–	<0.001	<0.001	**0.022**	<0.001
5	Chillies	–	<0.001	<0.001	<0.001	<0.001
6	Brinjal	A Mallesham	<0.001	<0.001	**0.022**	<0.001
7	Chillies	A Mallesham	<0.001	<0.001	**0.011**	<0.001
8	Ladies finger	A Mallesham	<0.001	<0.001	<0.001	<0.001
9	Ridge gourd	A Mallesham	<0.001	<0.001	<0.001	<0.001
10	Tomato	R Narsaiah	<0.001	<0.001	<0.001	<0.001
11	Brinjal	S Danji	<0.001	<0.001	<0.001	<0.001
12	Brinjal (white)	M Prem	<0.001	<0.001	<0.001	<0.001
13	Tomato	M Prem	<0.001	<0.001	<0.001	<0.001
14	Brinjal	Narayana Reddy	<0.001	<0.001	<0.001	<0.001
15	Chillies	Narayana Reddy	<0.001	<0.001	<0.001	<0.001

Table 20.2 Pesticide residues in vegetable samples collected from farmers' fields, Kothapally village, Ranga Reddy district during 2007

Crop (no. of samples)	Range of pesticide residue level (mg kg^{-1})			
	Monocrotophos	Chlorpyrifos	Endosulfan	Cypermethrin
Brinjal (10)	0.003	0.008	0.019	0.052
	(<0.001–0.007)	(<0.001–0.040)	(<0.001–0.089)	(<0.001–0.283)
Cucumber (10)	0.004	0.066	0.019	0.010
	(0.001–0.011)	(0.001–0.330)	(0.002–0.030)	(0.001–0.034)
Okra (10)	0.013	0.605	0.130	0.025
	(<0.001–0.044)	(0.001–5.154)	(0.001–0.784)	(<0.001–0.112)
Ridge gourd (6)	0.015	0.050	0.021	0.086
	(<0.001–0.041)	(0.001–0.223)	(0.002–0.061)	(0.001–0.352)
Tomato (23)	0.005	0.035	0.032	0.024
	(<0.001–0.025)	(<0.001–0.151)	(<0.001–0.466)	(<0.001–0.141)

< 0.001 = below detectable levels, (values in the parenthesis denote the range)

endosulfan in okra and monocrotophos and that of cypermethrin in ridge gourd (Table 20.2) (Ranga Rao et al. 2009).

The main reason for low levels of monocrotophos and endosulfan could be due to low utilization of these conventional chemicals in the recent years. Research results from other locations and countries also revealed similar inferences on the occurrence of residues. For example, in vegetables in the Shaanxi area of China, five organophosphorus pesticides ranging from 0.004 to 0.257 mg kg^{-1} and 18 of 200 samples had residue levels in excess of MRLs (Yahong Bai et al. 2006). In another case, Jagadishwar Reddy et al. (1997) found endosulfan residues in vegetables collected from Srikakulam, and cypermethrin residues in vegetables from farmers' fields around Hyderabad and Guntur, but they were below MRLs. The findings of Mukherjee (2003) also brought out contamination of vegetables in Delhi with various pesticides, of which 31 % samples had residues above MRL.

20.1.1 Occurrence of Residues at Different Intervals

Results presented in Table 20.3 reveal the presence of residues of the four pesticides at different intervals. Monocrotophos residues ranged from 0.006 to 0.009 mg kg^{-1} during May to July. The residues of chlorpyrifos and endosulfan increased from 0.023 to 0.015 in May to 1.329 and 0.215 mg kg^{-1} in July, respectively. While the residues of cypermethrin decreased from 0.039 in May to 0.006 mg kg^{-1} in July. In general, the insecticide residue concentration increased over the season, which could be due to the cumulative effect of pesticide application during the vegetable-growing season. Similar findings were also reported by Neela Bakore et al. (2002), who detected maximum organochlorine insecticide residues in the tomato crop (17.1 mg kg^{-1}) at the end of the season in Rajasthan, India.

Table 20.3 Residues of four pesticides in samples of tomato and cucumber sampled at different intervals in Kothapally village, Ranga Reddy district, 2007

	Residue levels (mg kg^{-1})			
Month (no. of samples)	Monocrotophos	Chlorpyrifos	Endosulfan	Cypermethrin
May (30)	0.006	0.023	0.015	0.039
June (25)	0.009	0.084	0.051	0.025
July (4)	0.005	1.329	0.215	0.006

Table 20.4 Pesticide residues in two vegetable samples collected from IPM and farmers' practice plots in Kothapally village, Ranga Reddy district, combined Andhra Pradesh, 2007

	Treatment (no.	Residue levels (mg kg^{-1})			
Crop	of samples)	Monocrotophos	Chlorpyrifos	Endosulfan	Cypermethrin
Tomato	IPM (18)	0.005	0.034	0.012	0.023
Tomato	Non-IPM (5)	0.005	0.041	0.101	0.028
Cucumber	IPM (5)	0.004	0.027	0.011	0.009
Cucumber	Non-IPM (5)	0.005	0.106	0.026	0.012

20.1.2 Role of Integrated Pest Management in Reducing Residues

In the IPM system, in developing countries especially, pesticide use is the first choice, when other control measures are ineffective. This suggests that an IPM system does not prevent farmers from using pesticides. However, in certain situations where it is necessary, pesticides should be used judiciously, sticking to legal requirements and ensuring appropriate action.

Comparison of residues of these four pesticides in samples collected from IPM and non-IPM fields of tomato indicated 0–741 % higher residues in non-IPM samples (Table 20.4). Monocrotophos residues were similar in both the treatments, while endosulfan residues were 741 % higher in non-IPM samples compared to IPM samples of tomato. The residues of chlorpyrifos and cypermethrin were 21 % and 22 % higher in non-IPM samples of tomato compared to IPM samples. In cucumber, the non-IPM samples recorded 25–292 % higher residues compared to samples collected from IPM fields. In the case of cucumber samples, monocrotophos residues were 25 % higher in non-IPM samples compared to IPM. Chlorpyrifos residues were 292 % higher in non-IPM samples compared to IPM. Endosulfan residues were 136 % higher in non-IPM samples compared to IPM. The residues of cypermethrin were 33 % higher in non-IPM samples than IPM. The residues in IPM fields ranged from 0.004 to 0.027, while it was 0.005–0.106 in non-IPM fields. In IPM fields the presence of residues can be due to the leftover residues in the soil and water.

20.1.3 Think of the Devil:"Cancer Train"

Due to the regular use of pesticides in Punjab, North India, farmers are plagued with cancer; most are poor patients from Punjab needing to catch the 9:30 pm train toward the city of Bikaner for cancer treatment. The train's sole reserved compartment, with a capacity of 72, is occupied by 30 cancer patients. Earlier, this train was called as the "TB Train" (for tuberculosis patients); in the past few years, it has been named as "Cancer Express." "There has to be a comprehensive survey on the prevalence of cancer in Punjab." "People don't disclose someone in their houses has cancer fearing stigmatization." Critics say government is ignoring the problem, even today. Although pesticides such as endosulfan and DDT are banned globally, they are still available in the country. This problem is due to the misuse of pesticides. The consumption of pesticides is very high in Punjab, compared to other states in India.

Plant protectionists are facing a complex challenge to provide healthy food for growing populations with the available land and limited resources under adverse environment. Using modern technologies such as cultivating hybrids, enhancing natural enemies, efficient cultural management tools, and judicious use of pesticides and their integrated approach (IPM) are relatively simple, even in resource-limited situations. The ill effects of pesticides have become evident. Markets have become more quality conscious and are reluctant to buy produce with quality constraints. The major hurdles in the Asian exports are aflatoxin and pesticide residues. The nonavailability of MRLs for all the registered pesticides on all important/key crops is a major requirement. In the absence of international MRLs for various chemicals, at least regular monitoring of residues in various raw and processed food commodities helps to contain the pesticide residue by promoting good agricultural practices (GAPs) and by following appropriate safe waiting period. For maintaining the quality of environment and products, it is necessary to keep the produce safe from residues of pesticides and other agrochemicals. A zero level/below detectable level residue in the finished product is desired. It is also required for good health and environment. Hence, it is time to promote safer and alternative practices in agriculture to reduce the application of poisonous pesticides through the existing integrated pest management programs. Though it takes substantial time to rectify the spoiled situations, it is time to awake and work together in a participatory approach to realize this dream to ensure healthy food for all.

References

Anon (1999) Pesticide action network Asia and Pacific "Warning: pesticides are dangerous to your health"

Anon (2005) Very high levels of pesticides" found in blood samples in Punjab, India. The Hindu daily, 8 June 2005

Anon (2007) World Bank – ICRISAT – DM project: traditional technology with a modern twist. Project completion report June 2005 – May 2007

Anon (2009) Indian pesticides industry, vital for ensuring food security. https://www.bharatbook.com
Jagdishwar Reddy D, Narasimha Rao B, Narasimha Reddy K, Sultan MA, Ramesh Babu T (1997) Monitoring of pesticide residues in river, tank and canal water. Pesticide Res J 9(1):97–100
Kranthi KR, Jadhav DR, Kranthi S, Wanjari RR, Ali S, Russell D (2002) Insecticide resistance in five major insect pests of cotton in India. Crop Prot 21:449–460
Kumari B, Singh J, Singh S, Kathpal TS (2005) Monitoring of butter and ghee (clarified butter fat) for pesticidal contamination from cotton belt of Haryana, India. Environ Monit Assess 105:111–120
Lu C, Toepel K, Irish R, Fenske RA, Barr DB, Bravo R (2006) Organic diets significantly lower children's dietary exposure to organophosphorus pesticides. Environ Health Perspect 114(2):260–263
Marit Almvik, Nina O, Svendsen, Vuong Truoung Giang (2006) Food safety: pesticide residue analysis in Norway and Vietnam. http://www.bioforsk.no/ikbViewer/Content/31551/Pesticide%20residue%20analysis%20in%20Norway%20and%20Vietnam%20Abstract.pdf
Mattilsynet (2009) Rester av plantevernmidler i næringsmidler 2008. Report June 2009. In: Norwegian with summary in English. Available at mattilsynet.no. http://www.bioforsk.no/ikbViewer/Content/104495/TEMA_7-4_Vietnam-web.pdf
McCauley LA, Anger WK, Keifer M, Langley R, Robson MG, Rohlman D (2006) Studying health outcomes in farm worker populations exposed to pesticides. Environ Health Perspect 114(3):953–960
Meena RP, Meena SS, Meena SR (2008) Use of agrochemicals in agriculture, their consequences and possible solution. Green Farming 1(4):54–55
Miller GT (2004) Sustaining the Earth, 6th edn. Thompson Learning, Pacific Grove, pp 211–216. Chap. 9. http://www.worstpolluted.org/projects_reports/display/82
Mukherjee I (2003) Pesticide residues in vegetables in and around Delhi. Environ Monit Assess 86(3):265–271
Mya Thwin, Thet Mar (2002) Paper presented at FAO/WHO Global Forum of Food Safety Regulators, Marrakech, Morocco, 28–30 January 2002 on the current status of pesticides residue analysis of food in relation with food safety by Myanmar. http://www.fao.org/docrep/meeting/004/ab429e.htm
Neela Bakore, John PJ, Bhatnagar P, Bakore N, Bhatnagar P (2002) Evaluation of organochlorine insecticide residue levels in locally marketed vegetables of Jaipur city, Rajasthan, India. J Environ Biol 23(3):247–252
Pimental D (2009) Innovations development process, chapter 3. Springer, Dordrecht, pp 83–87
Ranga Rao GV, Pande S, Reddy YVR, Narayana Rao J Rameshwar Rao V (2002) Insect pests and diseases of agricultural importance and their management in India. In: The proceedings (volume I) of the national workshop on Resource Management in Plant Protection, 14–15 November 2002. Plant Protection Association of India, NPPTI Campus, Rajendranagar, Hyderabad, pp 85–103
Ranga Rao GV, Rameswar Rao V, Reddy YVR, Murthy KSRK (2005) Status of safety in pesticide use in Asia: Options for improvement. In: Proceedings of the National Seminar on Pesticide Residues and their Risk Assessment organized by Food and Drug Toxicology Research Centre, National Institute of Nutrition, Hyderabad, January 20–21 2005, pp 129–136
Ranga Rao GV, Sahrawat KL, Srinivasa Rao CH, Das B, Kirankumar Reddy K, Bharath BS, Rameshwar Rao V, Murthy KVS, Wani SP (2009) Insecticide residues in vegetable crops grown in Kothapally watershed, Andhra Pradesh, India: A case study. Indian J Dry Land Agric Res Dev 24(2):21–27
Schinas, Vassilios, Leotsinidis, Michalis, Alexopoulos, Athanasios, Tsapanos, Vassilios, Kondakis, Xenofon Gerasimos (2000) Organochlorine pesticide residues in human breast milk from Southwest Greece: Associations with Weekly Food Consumption Patterns of Mothers. http://connection.ebscohost.com/c/articles/3850412/organochlorine-pesticide-residues-

human-breast-milk-from-southwest-greece-associations-weekly-food-consumption-patterns-mothers

Sharma KK (2013) Status of pesticide residues in agricultural procedure-scenario in India. Proceedings of the International conference on emerging issues on health effects of pesticide residues in food and environment – ultimate challenges and research opportunities, 12–13 December 2013 organized by Food and Drug Toxicology Research Centre – National Institute of Nutrition, Indian Council of Medical Research, Tarnaka, Hyderabad

Skretteberg LJ, Lyrån B, Holen B, Jansson A, Fohgelberg P, Siivinen K, Andersen JH, Jensen BH (2015) Pesticide residues in food of plant origin from Southeast Asia – a Nordic project. Food Control 51:225–235

Wilson C, Tisdell C (2001) Why farmers continue to use pesticides despite environmental, health and sustainability costs. Ecol Econ 39:449–462

Yahong Bai, Ling Zhou, Jiang Wang (2006) Organophosphorus pesticide residues in market in Shaanxi area, China. Food Chem 98(2):240–242

ERRATUM

Arthropod Communities in Coffee: A Habitat Mimicking Tropical Forests 17

N.E. Thyagaraj, G.V. Manjunatha Reddy, S. Onkara Naik, and B. Doddabasappa

A.K. Chakravarthy, S. Sridhara (eds.), *Economic and Ecological Significance of Arthropods in Diversified Ecosystems*, DOI 10.1007/978-981-10-1524-3_17

DOI 10.1007/978-981-10-1524-3_21

The Figures captions in Chapter 17 were wrongly set which have now been rectified. The correct Figure captions for each Figure should reas as below:

Fig. 17.1 Female coffee berry borer (**a**) walking over a coffee seed (**b**) to show the small size of the insect (ca. 2 mm long, 1 mm wide). Female coffee berry borer boring a hole in a coffee berry (**c**) with characteristic symptom of infestation revealing frass on the entrance hole (**d**). Damage caused by larval feeding inside the coffee berry (**e**). Credits: (**a**) E. Erbe, USDA, ARS; (**b**) P. Greb, USDA, ARS; (**c**) and (**e**) G. Hoyos, Cenicafé; (**d**) G. Mercadier, USDA, ARS (Vega et al. 2009)

Fig. 17.2 Adult *Phymastichus coffea* (**a**) ovipositing in the coffee berry borer (**b**) with one adult parasitoid emerging from the insect (**c**). Adult *Cephalonomia stephanoderis* (**d**) and *Aphanogmus dictynna* (**e**), a hyperparasitoid of *Prorops nasuta*. Credits: (**a**) G. Goergen, IITA; (**b**) A. Castillo and F. Infante, Ecosur; (**c**) and (**d**), G. Nieto, ECOSUR; (**e**) M. Buffington and A. Simpkins, USDA, ARS (Vega et al. 2009)

The updated original online version for this chapter can be found at DOI 10.1007/978-981-10-1524-3_17

A.K. Chakravarthy, S. Sridhara (eds.), *Economic and Ecological Significance of Arthropods in Diversified Ecosystems*, DOI 10.1007/978-981-10-1524-3_21

Fig. 17.3 Infective juveniles of *Metaparasitylenchus hypothenemi* emerging from an infected coffee berry borer (*left*) and detail of the infective juvenile (*right*). Credits: (**a**) A. Castillo, ECOSUR; (**b**) G. Nieto, ECOSUR (Vega et al. 2009)

Fig. 17.4 Spiders: common predators in coffee plantations – *Poecilotheria regalis*

Fig. 17.5 Nest of *red* ants

Fig. 17.6 Coffee flowers in bloom

Fig. 17.7 (**a**) Coffee root grub. (**b**) Coffee twig showing infestation of root grub

Fig. 17.8 Shot-hole borer damage on coffee

Fig. 17.9 A nymphalid butterfly – indicator species

Fig. 17.10 A wasp nest in coffee estate

Figure captions errors in the original chapter are regretted.

MIX
Papier aus verantwortungsvollen Quellen
Paper from responsible sources
FSC® C105338

If you have any concerns about our products, you can contact us on
ProductSafety@springernature.com

In case Publisher is established outside the EU, the EU authorized representative is:
Springer Nature Customer Service Center GmbH
Europaplatz 3, 69115 Heidelberg, Germany

Printed by Libri Plureos GmbH
in Hamburg, Germany